CHALLENGES AND INNOVATIONS IN OCEAN IN SITU SENSORS

CHALLENGES AND INNOVATIONS IN OCEAN IN SITU SENSORS

Measuring Inner Ocean Processes and Health in the Digital Age

Edited by

ERIC DELORY
Oceanic Platform of the Canary Islands (PLOCAN), Telde, Spain

JAY PEARLMAN
FourBridges, Port Angeles, WA, United States
IEEE, Paris, France

ELSEVIER

Elsevier
Radarweg 29, PO Box 211, 1000 AE Amsterdam, Netherlands
The Boulevard, Langford Lane, Kidlington, Oxford OX5 1GB, United Kingdom
50 Hampshire Street, 5th Floor, Cambridge, MA 02139, United States

Notices
Knowledge and best practice in this field are constantly changing. As new research and experience broaden our understanding, changes in research methods, professional practices, or medical treatment may become necessary.

Practitioners and researchers must always rely on their own experience and knowledge in evaluating and using any information, methods, compounds, or experiments described herein. In using such information or methods they should be mindful of their own safety and the safety of others, including parties for whom they have a professional responsibility.

To the fullest extent of the law, neither the Publisher nor the authors, contributors, or editors, assume any liability for any injury and/or damage to persons or property as a matter of products liability, negligence or otherwise, or from any use or operation of any methods, products, instructions, or ideas contained in the material herein.

Library of Congress Cataloging-in-Publication Data
A catalog record for this book is available from the Library of Congress

British Library Cataloguing-in-Publication Data
A catalogue record for this book is available from the British Library

ISBN: 978-0-12-809886-8

For information on all Elsevier publications visit our website at
https://www.elsevier.com/books-and-journals

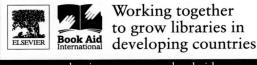
Working together
to grow libraries in
developing countries

www.elsevier.com • www.bookaid.org

Publisher: Candice Janco
Acquisition Editor: Louisa Hutchins
Editorial Project Manager: Hilary Carr
Production Project Manager: Bharatwaj Varatharajan
Cover Designer: Christian J. Bilbow

Typeset by TNQ Technologies

Contents

List of Contributors

Eric P. Achterberg   GEOMAR Helmholtz Centre for Ocean Research Kiel, Kiel, Germany

Simon Allen   Spatial Analytics, Hobart, TAS, Australia

José C. Alves   INESC TEC, FEUP–DEEC, Porto, Portugal

David Aragon   Center of Ocean Observing Leadership, Department of Marine and Coastal Sciences, School of Environmental and Biological Sciences, Rutgers University, New Brunswick, NJ, United States

Douglas Au   Monterey Bay Aquarium Research Institute (MBARI), Moss Landing, CA, United States

Christopher R. Barnes   School of Earth and Ocean Sciences, University of Victoria, Victoria, BC, Canada

Carole Barus   Laboratoire d'Etudes en Géophysique et Océanographie Spatiales, UMR 5566, Université de Toulouse, CNRS, CNES, IRD, UPS Toulouse Cedex 9, France

Alex Beaton   National Oceanography Center, United Kingdom

Ryan J. Bell   Owner Beaver Creek Analytical, LLC, Lafayette, CO, United States

Pierre Blouch   Météo-France, Brest, France

Patrice Brault   nke Instrumentation, Hennebont, France

Filipa Carvalho   Center of Ocean Observing Leadership, Department of Marine and Coastal Sciences, School of Environmental and Biological Sciences, Rutgers University, New Brunswick, NJ, Canada

Pablo Cervantes   Centro Tecnológico Naval y del Mar (CTN), Fluente Álamo (Murcia), Spain

D. Chen Legrand   Laboratoire d'Etudes en Géophysique et Océanographie Spatiales, UMR 5566, Université de Toulouse, CNRS, CNES, IRD, UPS Toulouse Cedex 9, France

Florent Colas   Ifremer, REM/RDT/LDCM, F-29280 Plouzané, France

Timothy Cowles   College of Earth, Ocean, and Atmospheric Sciences, Oregon State University, Corvallis, OR, United States

Nuno A. Cruz   INESC TEC, FEUP–DEEC, Porto, Portugal

Arnaud David   nke Instrumentation, Hennebont, France

Joaquin del Rio Fernandez   UPC, Universitat Politècnica de Catalunya, Vilanova i la Geltrú, Spain

Laurent Delauney   Detection, Sensors and Measurement Laboratory Manager, Research and Technological Development Unit, Ifremer, Plouzane, France

Eric Delory   Oceanic Platform of the Canary Islands (PLOCAN), Telde, Spain

Boris Dewitte   Laboratoire d'Etudes en Géophysique et Océanographie Spatiales, UMR 5566, Université de Toulouse, CNRS, CNES, IRD, UPS Toulouse Cedex 9, France; Center for Advanced Studies in Arid Zones (CEAZA), La Serena, Chile; Facultad de Ciencias del Mar, Departamento de Biología Marina, Universidad Católica del Norte, Coquimbo, Chile; Millennium Nucleus for Ecology and Sustainable Management of Oceanic Islands (ESMOI), Coquimbo, Chile

Bruno M. Ferreira   INESC TEC, FEUP–DEEC, Porto, Portugal

Albert Fischer   Intergovernmental Oceanographic Commission of UNESCO, Paris, France

Veronique Garçon   Laboratoire d'Etudes en Géophysique et Océanographie Spatiales, UMR 5566, Université de Toulouse, CNRS, CNES, IRD, UPS Toulouse Cedex 9, France

Scott Glenn   Center of Ocean Observing Leadership, Department of Marine and Coastal Sciences, School of Environmental and Biological Sciences, Rutgers University, New Brunswick, NJ, United States

Robert Harcourt   Sydney Institute of Marine Science, Mosman, NSW, Australia; Department of Biological Sciences, Macquarie University, North Ryde, NSW, Australia

Michelle R. Heupel   Australian Institute of Marine Science, Townsville, QLD, Australia; Centre for Sustainable Tropical Fisheries and Aquaculture and College of Science and Engineering, James Cook University, Townsville, QLD, Australia

Simon Jirka   52° North Initiative for Geospatial Open Source Software GmbH, Münster, Germany

Justyna Jońca   Laboratoire d'Etudes en Géophysique et Océanographie Spatiales, UMR 5566, Université de Toulouse, CNRS, CNES, IRD, UPS Toulouse Cedex 9, France

Clayton Jones   Teledyne Webb Research, Falmouth, MA, United States

Benoit Jugeau   nke Instrumentation, Hennebont, France

Gottfried P.G. Kibelka   CMS Field Products, OI Analytical, Pelham, AL, United States

S. Kim Juniper   Ocean Networks Canada, University of Victoria, Victoria, BC, Canada

Josh Kohut   Center of Ocean Observing Leadership, Department of Marine and Coastal Sciences, School of Environmental and Biological Sciences, Rutgers University, New Brunswick, NJ, United States

Chrysi Laspidou   Civil Engineering Department, University of Thessaly, Thessaly, Greece

Adam Leadbetter   Marine Institute, Oranmore, Ireland

Emilie Leblond   Ifremer - Centre de Bretagne, Plouzané, France

E.J.I. Lédée   Fish Ecology and Conservation Physiology Lab, Carleton University, Ottawa, ON, Canada

Hassan Mahfuz   Department of Ocean and Mechanical Engineering, Florida Atlantic University, Boca Raton, FL, United States

Carmem-Lara Manes   Microbia Environnement, Marine Station Banyuls sur Mer, France

Enoc Martinez   UPC, Universitat Politècnica de Catalunya, Vilanova i la Geltrú, Spain

Sergio Martinez   LEITAT Technological Center, Barcelona, Spain

Aníbal C. Matos   INESC TEC, FEUP–DEEC, Porto, Portugal

Janice McDonnell   Center of Ocean Observing Leadership, Department of Marine and Coastal Sciences, School of Environmental and Biological Sciences, Rutgers University, New Brunswick, NJ, Canada

Scott McLean   Ocean Networks Canada, University of Victoria, Victoria, BC, Canada

Simone Memè   Oceanic Platform of the Canary Islands (PLOCAN), Telde, Spain

Daniel Mihai Toma   UPC, Universitat Politècnica de Catalunya, Vilanova i la Geltrú, Spain

Travis Miles   Center of Ocean Observing Leadership, Department of Marine and Coastal Sciences, School of Environmental and Biological Sciences, Rutgers University, New Brunswick, NJ, Canada

Seyed Morteza Sabet   Department of Ocean and Mechanical Engineering, Florida Atlantic University, Boca Raton, FL, United States

Matthew Mowlem   National Oceanography Center, United Kingdom

P. Muñoz Parra   Facultad de Ciencias del Mar, Departamento de Biología Marina, Universidad Católica del Norte, Coquimbo, Chile

Tom O'Reilly   Monterey Bay Aquarium Research Institute (MBARI), Moss Landing, CA, United States

Klas Ove Möller   Institute of Coastal Research, Helmholtz-Zentrum Geesthacht, Centre for Materials and Coastal Research, Geesthacht, Germany

Jay Pearlman   FourBridges, Port Angeles, WA, United States; IEEE, Paris, France

Wilhelm Petersen   Helmholtz-Zentrum Geesthacht, Institute of Coastal Research, Geesthacht, Germany

Benoît Pirenne   Ocean Networks Canada, University of Victoria, Victoria, BC, Canada

Paul Poli   Centre de Météorologie Marine, Météo-France, Brest, France

Hervé Precheur   Sensorlab, Gran Canaria, Spain

Loïc Quemener   Ifremer - Centre de Bretagne, Plouzané, France

Emily Ralston   Florida Institute of Technology, Melbourne, FL, United States

Marcel Ramos   Center for Advanced Studies in Arid Zones (CEAZA), La Serena, Chile; Facultad de Ciencias del Mar, Departamento de Biología Marina, Universidad Católica del Norte, Coquimbo, Chile; Millennium Nucleus for Ecology and Sustainable Management of Oceanic Islands (ESMOI), Coquimbo, Chile

Anja Reitz   GEOMAR Helmholtz Centre for Ocean Research Kiel, Kiel, Germany

Matthes Rieke   52° North Initiative for Geospatial Open Source Software GmbH, Münster, Germany

Hugh Roarty   Center of Ocean Observing Leadership, Department of Marine and Coastal Sciences, School of Environmental and Biological Sciences, Rutgers University, New Brunswick, NJ, United States

Ivan Romanytsia   Laboratoire d'Etudes en Géophysique et Océanographie Spatiales, UMR 5566, Université de Toulouse, CNRS, CNES, IRD, UPS Toulouse Cedex 9, France

Adrian Round   Ocean Networks Canada, University of Victoria, Victoria, BC, Canada

Pablo Ruiz   Centro Tecnológico Naval y del Mar (CTN), Fluente Álamo (Murcia), Spain

Grace Saba Center of Ocean Observing Leadership, Department of Marine and Coastal Sciences, School of Environmental and Biological Sciences, Rutgers University, New Brunswick, NJ, United States

Allison Schaap National Oceanography Center, United Kingdom

Oscar Schofield Center of Ocean Observing Leadership, Department of Marine and Coastal Sciences, School of Environmental and Biological Sciences, Rutgers University, New Brunswick, NJ, United States

Greg Seroka Center of Ocean Observing Leadership, Department of Marine and Coastal Sciences, School of Environmental and Biological Sciences, Rutgers University, New Brunswick, NJ, Canada

Christoph Stasch 52° North Initiative for Geospatial Open Source Software GmbH, Münster, Germany

Nicolas Striebig Observatoire Midi-Pyrénées, Toulouse Cedex 9, France

R. Timothy Short Center for Security and Survivability, SRI International, St. Petersburg, FL, United States

Strawn K. Toler Center for Security and Survivability, SRI International, St. Petersburg, FL, United States

Vinay Udyawer Arafura Timor Research Facility, Australian Institute of Marine Science, Darwin, NT, Australia

Maria Valladares Center for Advanced Studies in Arid Zones (CEAZA), La Serena, Chile; Facultad de Ciencias del Mar, Departamento de Biología Marina, Universidad Católica del Norte, Coquimbo, Chile

Martin Visbeck GEOMAR Helmholtz Centre for Ocean Research Kiel, Kiel, Germany; Christian-Albrechts Universität zu Kiel, Kiel, Germany

Ian Walsh Sea-Bird Scientific, Philomath, OR, United States

Karen Wild-Allen CSIRO Oceans and Atmosphere, Hobart, TAS, Australia

Patrice Woerther Ifremer - Centre de Bretagne, Plouzané, France

Mathieu Woillez Ifremer - Centre de Bretagne, Plouzané, France

Jochen Wollschläger Institute for Chemistry and Biology of the Marine Environment, Carl von Ossietzky University of Oldenburg, Wilhelmshaven, Germany; Institute of Coastal Research, Helmholtz-Zentrum Geesthacht, Centre for Materials and Coastal Research, Geesthacht, Germany

Xu Yi Center of Ocean Observing Leadership, Department of Marine and Coastal Sciences, School of Environmental and Biological Sciences, Rutgers University, New Brunswick, NJ, United States; IMBER Regional Project Office, State Key Laboratory of Estuarine and Coastal Research, East China Normal University, Shanghai, China

Kelli Zargiel Hunsucker Florida Institute of Technology, Melbourne, FL, United States

Foreword

INTRODUCTION

The last two decades have witnessed some remarkable innovations for in situ sensors that measure ocean processes and ultimately provide for better understanding of the health of the oceans. These developments have not been without significant technological and commercial challenges, along with those for successful deployment and operation of observing systems and networks, especially those in the deep ocean. These have emerged at a most critical time in human history when our understanding of ocean dynamics and ecosystems is essential to the Earth's sustainability. There has been increasing acceptance of the severity and cost of climate and sea-level change facing society. This is a combination of environmental and social change as the number of major hurricanes has increased, as well as the increasing percentage of humans who inhabit coastal regions that are most vulnerable to hurricanes, with the concomitant consequence of increasing impact on both human life and coastal infrastructure.

The climate is evolving with continued records of warming. In 2016, the global CO_2 level exceeded 400 ppm, one of the key greenhouse gas benchmarks for global warming. The Arctic Ocean has experienced a marked reduction in the areal extent of sea ice, as have the ice caps and glaciers of Greenland and West Antarctica, which impact ocean-circulation patterns and polar communities. The impacts also include a migration trend for fish and marine mammals toward the poles and a change in the vitality of nonmobile sea life. Ocean circulation will continue to evolve and a more-detailed assessment of these dynamics is essential to forecast future environments.

Although increased resources are necessary to observe the ocean, the path forward is challenging. Most nations, both more and less developed, have increasing social needs that have limited expansion of ocean observations. One consequence is declining financial capacity to deal with the impending scale of environmentally induced catastrophes. For climate change, the intermediate effects are seen as issues of desertification in many parts of Africa and major droughts in the eastern Mediterranean, India, Indonesia, Cambodia, and California. Over longer timescales, sea-level rise is impacting low-lying areas of many Pacific and Indian Ocean islands, countries (Bangladesh, Vietnam, the Netherlands), and coastal regions/cities (Florida/Miami, New Jersey/New York, Venice). This will inevitably result in mass migration and resettlement of peoples on a far greater scale than recently has been seen.

The oceans are the dominant controlling factor in the Earth's climate. They contain much of the Earth's surface heat in contrast to the atmosphere and absorb a substantial part of the CO_2 being added to the atmosphere through anthropogenic processes. This, in turn, leads to significantly increasing ocean acidification, reef destruction, and expansion of oceanic dead zones (hypoxia; [1]). Repeated scientific measurement of ocean conditions in both space and time is paramount to properly understanding ocean processes and assessing the health of the world's oceans. In recent decades, this only partially has been achieved by ship-borne observations limited by weather (sea state) conditions and commonly by an inability to make repeated observations from specified sites or transects. Newer programs and technologies in both platforms and sensors have helped to address these deficiencies with, for example, open-ocean moorings and buoys (e.g., OceanSITES program [http://www.oceansites.org; December 2016]) and drifting floats (http://www.aoml.noaa.gov/phod/dac/index.php; March 2018). Profiling floats such as the Array for Real-Time Geostrophic Oceanography (ARGO) floats are extending observations with new strategies (http://www.argo.net; December 2016). ARGO floats routinely operate to 2000-m depth and have significantly extended the coverage of ocean observation. New extensions of ARGO for deep observations, and with new sensors for chemical observations, are emerging. Glider technology may become a cost-effective tool for global monitoring, creating opportunities for development and integration of innovative sensors (http://www.ego-network.org; April 2018). Some international programs such as the Global Ocean Ship-based Hydrographic Investigations Program (GO-SHIP) are providing sustained high-quality observations over decadal timescales to examine trends in the ocean (http://www.go-ship.org; April 2018). However, the scientific community to this stage has not captured many long time series of data from the deeper parts of the world's oceans (i.e., between about 2000 and 7000 m) to fully understand ocean-circulation changes and conveyor-belt systems.

Other changes have been occurring in the ocean that are significant and must be addressed as part of a long-term solution leading to a sustainable ecosystem. Overfishing is a recognized problem that has been addressed through national-level regulations. Such regulations can be effective, but only part of the waters are under such jurisdiction. To understand alternatives for fishery management, there is an essential need to understand the dynamics of the upper trophic levels, the fish that are usually consumed by humans. The need does not remain at the upper levels as the whole of the food chain starting from plankton to whales needs to be understood.

The monitoring of ocean biology is less mature than the monitoring of physical ocean characteristics. This offers opportunities for significant advancement of in situ instrumentation. We see the emphasis on biological monitoring increasing, though the implementation of standard monitoring configurations is a work in progress. The distribution of nutrients plays a role in ecosystem dynamics. New techniques in chemical measurements (e.g., nitrates, phosphates) are showing promise and maturing to the point that they are becoming feasible for sustained applications. Some of the first applications may be in the area of aquaculture, which would benefit from harmful algal bloom warnings and improved water-quality monitoring. With the increase in fish farming, water quality will need to be managed, and this management further drives the needs for advanced sensors with good reliability and extended life under operation. Issues such as biofouling also become critical for stable, sustained observations. Significant improvements are occurring through the use of electrolysis and ultraviolet illumination. Although these have been demonstrated in situ, their application to a broad range of sensors is only beginning.

Other factors that will be changing the ocean environment are mineral extraction from the sea floor and energy extraction from waves, tides, and thermal gradients. Each of these causes some change in the local (and perhaps larger) environments. Mining may take place in local hot spots where communities of benthic creatures survive due to such local conditions. Ecological baselines to enable mining while minimizing impacts require more information than is currently available about most of the sea floor. New survey techniques are becoming available and seabed mapping at a global scale is an important initiative of the ocean research and application community. Energy extraction from the sea also causes acoustic issues in the surrounding waters. As low-frequency acoustic waves can propagate across entire ocean basins, being able to assess and characterize the contribution of human activities to the ocean soundscape is becoming an environmental issue.

Thus, we see increasing human interaction with the ocean from a three-dimensional perspective. This, in turn, is motivating more measurements in areas that have not been generally accessible for routine observations. As mentioned previously, deep ARGO profilers, deep gliders, and more-effective autonomous sensor/platform operations are areas that are currently being pursued.

These are, however, only some of the challenges that need to be addressed. In response, oceanography has evolved over the last century with the investment of nations in ocean science and technologies. The need for continuing investment is driven, in part, by the significant changes that have occurred as the human population has expanded and pressure on ocean resources make imperative an increased understanding of the ocean as part of a linked ecosystem.

ABOUT THE BOOK

Improving today's and future capacities to observe the ocean greatly relies on our ability to integrate the scientific and technical expertise inherited from early and modern oceanographers, scientists, and engineers into autonomous systems. Although several solutions are now available to navigate or work autonomously in the ocean, our ability to observe from fixed or mobile platforms still requires innovations in sensor technology. This book exposes emergent needs, the derived challenges, innovative sensors for in situ monitoring, and applications including integration and examples of deployments on fixed and mobile platforms. Several chapters result from collaborations of industry and academia in identifying the key bottlenecks and providing new solutions. A new generation of sensors is presented that addresses innovative sensing techniques (e.g., for new or essential variables of high impact and low concentration), higher reliability (e.g., against biofouling or corrosion), better integration on platforms in terms of size and communication, and data flow across domains. State-of-the-art developments are presented, showcasing a broad diversity of measuring techniques and technologies.

AUDIENCE

The book is aimed at advanced students as well as professionals in academia, industry, and government who are addressing new research and data challenges that cannot be met with the current set of tools and the need to employ a new generation of ocean sensors, platforms, data services, and products. The book will give the reader an understanding of new technology advances and how these can be applied to their research and applications. The book is written at a level that both technologists and technology administrators should find understandable and valuable.

CONTENT

The book covers different sensor approaches and technologies for in situ measurements, including platform options, applications, and quality monitoring, subjects that underlie the ability to monitor and the requirements that can realistically be addressed. This also includes new technologies that could drive future capabilities. As in the Framework for Ocean Observing, there is a balance between what sensors/observations are available, what requirements need to be met, and what the societal impacts of the observations are. From this perspective, the book addresses what is available now (mature), what is emerging (maturing), and what concepts are under development or envisioned for the next generation of measurements.

Chapter 1 serves as introduction to the book and provides an overview of the ocean-observing framework. Section 1.1 examines the interfaces between in situ observing and other forms of monitoring the marine environment. The unique attributes of each capability and observation approach are explored in the context of coastal and ocean processes, physical, chemical, and biological, and the suitability of the observation method to deliver meaningful insight across a range of spatial and temporal scales. Section 1.2 reviews the needs for an integrated system of ocean observing at the requirements, design, data flow, and information production levels to improve information quality, quantity, and accessibility to more effectively contribute to societal economic well-being.

The book then takes the reader for a tour on current and emerging challenges, addressed with innovative sensor approaches and technologies. Focus was set on sensors for biogeochemical and biological sensors, addressed in Chapters 2 and 3, respectively. For biogeochemistry, covered aspects are optical sensing for high-accuracy measurement of acidity (Section 2.1), the implementation of underwater mass spectrometry in an autonomous unmanned vehicle (Section 2.2), the development of new electrochemical sensors for nutrients and microfluidics-based sensors (Section 2.3).

Chapter 3 includes plankton monitoring through spectral absorption (Section 3.1), and the use of surface plasmon resonance or biosensors for sensing low-concentration toxins and biological compounds (Sections 3.2 and 3.3).

Some crosscutting innovations are covered in Chapter 4, with acoustics (passive-acoustics [Section 4.1] and active-acoustics telemetry [Section 4.2]), as well as some important recent contributions to improve reliability with new materials and antifouling techniques (Sections 4.3 and 4.4).

Chapter 5 has eight sections (Sections 5.1–5.8) dedicated to the innovations that have emerged for observing platforms. Such innovations include the fixed or mobile systems that carry the sensors and provide power, as well as communication with the outer world, in which the main focus and common denominator is increasing cost-efficiency as an enabler of greater spatial and temporal resolution for observations on a large scale.

Measuring ocean processes in the digital age increasingly requires harmonizing data flow and safeguarding the information required for future exploitation of collected data, independently of sensor or platform manufacturers and operators. Chapter 6 walks the reader through related innovations and standards, from sensor to users (Sections 6.1 and 6.2), and looks into the future of standardization for sensors and data systems (Sections 6.3 and 6.4), including best practices.

Chapter 7 focuses on operating and building in situ ocean-observing systems. These require important efforts in the definition of mission requirements to optimize resources and costs based on existing capabilities (Section 7.1), and in the integration of sensors on platforms, paying special attention to the evaluation of sensor and platform candidates and the derivation of engineering needs and solutions given specific mission objectives and requirements (Section 7.2).

Chapter 8 completes the book with a chapter on use-case scenarios based on glider technology. The examples highlight different aspects including sensor integration, exploring extreme environments, and how glider technology is enabling efforts to entrain the next generation of ocean scientists.

FINAL COMMENTS

Although the book addresses a substantial number of challenges and innovations, some limitations (mostly due to relative volume or size) and priorities had to be set from the beginning. For example, biogeochemical and biological oceanography were prioritized over physical oceanography. For most scientists this may seem an obvious choice, yet for newcomers to the field of oceanography this may require a brief but necessary justification. The reason for this choice was mainly based on maturity and societal drivers. As physical oceanography sensors have been around for decades with the early development and integration of sensors that measure Conductivity, Temperature, and Depth (CTDs), sensors have reached a high level of maturity—or high Technology Readiness Level—and miniaturization. This also happens to be the case for several optical sensing techniques for chemical compounds, with the increasing use of integrated light-emitting diodes (LEDs) and optical spectrometry technologies. On the other hand, sensors for biogeochemistry and biological compounds, in most cases, are at a lower maturity level. They are, thus, an excellent focus for a book on in situ innovations for ocean observations.

Although we have tried to offer a comprehensive view of innovations for in situ monitoring, the book should not be considered a complete picture. For example, Laser-Induced Breakdown Spectroscopy (LIBS) is not included in the book, despite its potential use for marine applications. There are likely others.

The continuing expansion in the use of Universal Resource Locators (URLs) and Digital Object Identifiers (DOIs) leads to the expanded inclusion of web resources within the text and in bibliographies. Some URLs may have become obsolete some years from the time of publication, such as those corresponding to time-limited projects and initiatives. We apologize for this and ask for the reader's understanding. When possible and available, DOIs were provided in references.

Reference

[1] Breitburg D, Levin LA, Oschlies A, Grégoire M, Chavez FP, Conley DJ, Garçon V, Gilbert D, Gutiérrez D, Isensee K, Jacinto GS, Limburg KE, Montes I, Naqvi SWA, Pitcher GC, Rabalais NN, Roman MR, Rose KA, Seibel BA, Telszewski M, Yasuhara M, Zhang J. Declining oxygen in the global ocean and coastal waters. Science 2018;359.

Acknowledgments

The editors acknowledge the interest and contributions of the chapter authors, whose ideas were the seeds for many interesting discussions. Even while writing the book, the importance of oceans has increased and will likely continue to do so. We recognize the excitement permeating the ocean-research community as we move to address core global challenges. The innovative technologies play a major role in what we can do and when. The Oceans of Tomorrow projects, of which the NeXOS project was a part, have set standards for innovation and maturing technologies over a relatively short time. It was a privilege to be part of these efforts.

Eric Delory would like to acknowledge the support of the Oceanic Platform of the Canary Islands (PLOCAN) provided to him in coordinating the NeXOS project, which has been key for the book initiative. In particular, the management team for the coordination of the project, formed by Simone Memè, Ayoze Castro, Joaquín Hernández Brito, and Octavio Llínas. Jay Pearlman would like to thank René Garello and the Institute of Electrical and Electronics Engineers (IEEE) France for their support.

The editors would like to dedicate this book to their families. Eric dedicates the book to his family, his parents, Gercende and their daughter Maé for their love and patience, in particular during the numerous hours spent on this work on weekends and holidays. Jay acknowledges the support and contributions of his wife and partner, Françoise, who read the book in preparation and offered many good suggestions from the perspective of having recently completed editing her own book on the impact of environmental information on society (GEOValue).

CHAPTER

1

Introduction

CHAPTER

1.1

Ocean In Situ Sampling and Interfaces With Other Environmental Monitoring Capabilities

Simon Allen[1], Karen Wild-Allen[2]

[1]Spatial Analytics, Hobart, TAS, Australia;
[2]CSIRO Oceans and Atmosphere, Hobart, TAS, Australia

OUTLINE

Challenges and Innovations in Ocean In Situ Sensors
https://doi.org/10.1016/B978-0-12-809886-8.00001-6

1.1.1 WHY WE NEED TO UNDERSTAND OUR OCEAN

During 2017, we pass the population milestone of seven and one-half billion people. We have entered the Anthropocene [1], and our collective impact on our planet is being observed and measured. Seventy-one percent of the surface of this planet is covered by oceans and seas and the changes we are effecting in them are no less profound than those being observed on land and in the atmosphere, they are just more difficult to observe.

Life on this planet started in the oceans and the life in our oceans continues to be imperative to our survival, because the ocean's flora provide somewhere between 55 and 85% of the oxygen we breathe. The oceans have stored over 90% of the excess heat trapped by increased greenhouse gases and absorbed over 30% of the extra CO_2 created by humanity since the start of the Industrial Age [2].

The oceans moderate the extremes of climate in coastal regions and over half the global population lives in the 10% of land that is considered coastal. In 2000, 17 of the world's 24 megacities were coastal. Looking at this vertically, 10% of the global population lives in the 2% of land with an elevation of less than 10 m [3].

Humanity relies on the oceans for the oxygen we breathe; we seek out their climate-moderating properties and huddle around its edges, yet anthropogenic forces are changing them in ways we are only just beginning to understand and find difficult to observe. In the coastal zone, global climate change meets localized anthropogenic stressors. Our systems understanding needs to take into account these stressors with their very different spatial and temporal scales if we are to effectively and sustainably manage our environment (Fig. 1.1.1).

As the global population continues to grow, the role of the ocean as a source of farmed food will become increasingly important, from farming herbivorous fish, rather than carnivorous fish, to the growing of photosynthetic product for human consumption. Herbivorous fish convert feed to protein three times more efficiently than herbivorous land animals, but we need a healthy and living ocean if we wish to farm there.

1.1.2 MONITORING OR OBSERVING?

Major large-scale observing systems have been created: the Australian Integrated Marine Observing System, Global Ocean Observing System, American Integrated Ocean Observing System, European Multidisciplinary Seafloor and Water Column Observatory. Note that these, at an ocean level, are observing systems; monitoring, however, implies targeted observation with the possibility of effective intervention, something not considered possible currently over ocean scales. It is only within the coastal oceans, regional seas, or estuaries and lagoons that sufficient consensus of the current condition and likely future state is achieved. The smaller the area and the lower the number of interested parties, the greater the clarity with respect to the desired outcome of any intervention. The Pitt Water lagoon in Tasmania, Australia, for example, at 4000 ha, has less than 20 separate groups between which a shared optimum condition is managed. The Californian Water Directory, which is by no means complete, lists 10 Federal agencies, 33 State bodies, 36 environmental organizations, 15 other organizations, 11 legislative committees, and 39 other water associations and groups, all of which have a view on the optimum state of California's coasts and waterways. Ocean processes do not recognize administrative or jurisdictional boundaries, but it is within these boundaries that our returns on investment, for monitoring as part of a feedback loop for intervention, can be articulated. Within the Coastal Ocean Observing Systems, we see the move from agency-driven observation to coastal ocean monitoring for intervention and environmental management. It is within these coastal information systems that we see the need for real-time data and sense-making management overlays. Within these systems, predictive model outputs and modeled possible future scenario outputs are being presented to provide regional environmental managers with the tools to see both broad-scale influence and local stressors working together. Within the remaining oceanic waters, 93% of the CO_2 drawdown occurs that ultimately regulates global CO_2 levels; these are the waters around which consensus condition and needed actions seem hard to pin down. Although observations of dead fish are still used as an indicator of estuarine health, in many coastal systems we are moving toward the point at which we have environmental management levers that we can understand, monitor, and activate to avoid dead fish. Whether we can activate sufficient management levers to avoid a significantly altered planet remains to be seen.

1.1.3 WHY IN SITU SAMPLING?

The first step to understanding is quantitative observation and measurement. In situ sampling is the first link in that measurement chain. Measuring the components of a process on location, with the spatial and temporal resolution to fully describe that process increases the applicability of the observation and reduces the uncertainties that degrade the accuracy and reliability of the derived understanding. To borrow a phrase from our terrestrial colleagues, "nothing beats ground truth."

The ocean presents a significant challenge to satellite remote sensing as it is effectively opaque to many wavelengths of electromagnetic radiation. The average water depth of the 71% of the planet that is water is 3688 m; 99% of solar light in all visible wavelengths is absorbed within the top 250 m. Satellite remote sensing of the ocean at visible wavelengths can return information only from the depth of water that returns light of greater intensity than the reflectance of the surface. This is typically 40 m and can vary from about 100 m in very clear oceanic waters to less than 1 m in some turbid coastal waters. Thus, 98.8% of the ocean volume is unobservable by passive optical satellite-based remote sensing and the majority of that which is observed can only be observed for bulk optical properties. This optimistic figure disregards the impact of clouds on surface visibility, sun glint, and other surface effects.

The continental shelf makes up approximately 8% of the ocean surface area and has an average depth of 60 m; this generally drops away to the continental margin at about 140-m water depth. The continental shelf can be considered the area where the bottom matters in terms of nutrient resuspension and benthic primary productivity; yet for this domain, satellite-derived information must be based on bulk water properties with assumptions on bottom type and depth, or vice-versa. Many remotely sensed products share the same Achilles heel, they are inferred from a mix of remotely observed measurements and assumed or in situ measured ocean properties; thus, in situ sampling is the key to conversion from reflected solar radiation measured in space to phenomena of interest in the ocean.

Focused in situ sampling of clearly defined phenomena can be an end in its own right if the system being monitored is understood well enough to allow key sampling locations to be defined and the meaning of variance in the measured parameter linked to a key environmental condition. Often, however, in situ sampling is the first layer in a multilayered monitoring program, the layer which because of its position at the base must deliver observations that are of a defined quality with a clearly defined level of certainty and a full understanding of any possible confounding influences on the results. The aggregating and sense-making elements of the overall monitoring "system" depend completely on these in situ foundations, which provide the context to enable the joining of the dots on what is typically a very sparse map.

1.1.4 SAMPLING STRATEGIES FOR IN SITU MEASUREMENT

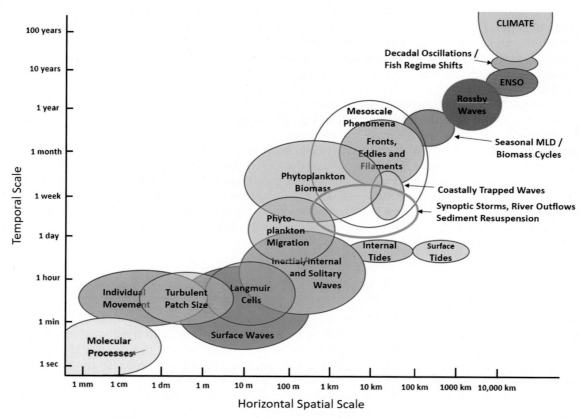

FIGURE 1.1.1   Tommy Dickey's Stommel diagram of ocean processes [4].

1.1.4.1 Broad-Scale Environmental Observing Systems

1.1.4.1.1 Satellite Sensors

The key trends in the placement of new Earth observation satellites in orbit have been the steady increase in the bandwidth and spectral resolution of the observation and decrease in ground-sample dimensions. A secondary trend is the move to regionally focused higher-altitude geostationary satellites that provide higher temporal differentiation within a specific region at the cost of spatial resolution [5].

Satellite observation-derived products for sea surface temperature and sea surface salinity deliver precisely that, sea surface values.

Satellite altimetry enables us to measure the height of the ocean for the entire planet over a period of days, and from the humps and bumps coupled with the atmospheric pressure, derive the drivers of large-scale circulations. Synthetic Aperture Radar gains us insight into waves and surface currents and, by inference, wind velocities. Recent gravity missions have created inferred maps of ocean bathymetry and determined features with a wavelength as short as 10 km, and as recently as 2014 were discovering thousands of seamounts that had previously been unmapped by ocean-based surveys or lower-resolution orbital sensors.

The satellite data can provide us with a comprehensive picture of the surface of the ocean and how it is moving. This, coupled with surface temperature, salinity, and optical properties, allows us to divide the surface up into water masses. Satellite altimetry coupled with improved topography allows water masses to be modeled in movement.

It is the understanding of the mechanics of why the water is moving and the evolution of these water masses physically, chemically, and biologically that prompts the need for sampling in situ.

1.1.4.1.2 Surface Radar for Waves and Currents

The United States has a network of 140 coastal radar stations providing regional insight into coastal currents by delivering both near real-time observations of surface currents and short-term modeled predictions of tidal currents. The evolution has started from delivery of the near real-time data to modeled data products that provide insight into the now and the near future. It is this fusion that makes the products salient to operational decision-making and moved coastal radar to the forefront of search and rescue and spill mitigation operational analysis [6].

1.1.4.1.3 Ocean Acoustics

Horizontal active acoustic remote sensing has been shown in field trials to deliver 100-km diameter snapshots of the pelagic environment and, over successive sensing periods, determine the movement of schools of fish to a resolution of 1 angular degree horizontally and tens of meters in range. The technique is currently qualitative enabling study of school dynamics. Although the method requires transducers placed in the water, the data collected are "remote." Acoustics muddies the waters between remote and in situ sampling, especially when used to collect basin-scale vertical transects of pelagic biomass [7].

Basin-scale transects rely on multifrequency active acoustics to quantify and differentiate object-size classes within the water column. With knowledge of the animal species in the area and their acoustic signatures, it is possible to estimate species distribution and biomass. This knowledge is gained through the collection of samples and imagery at discrete depths along with the cocollection of acoustic data and stereo imagery on profiling platforms towed behind the vessel collecting the full water column acoustic data. Here in one example, we see the interrelationships between scales and the need to constrain uncertainty as scales are changed [8].

Basin-scale acoustic tomography has for many years provided insight into ocean structure using in situ sources and sensors but inferring properties for water masses of many thousands of square kilometers.

1.1.4.1.4 Simple Models

Simple models are not just tools for understanding. When the problem and the goal are clearly defined and the drivers for either desirable or undesirable conditions are monolithic, then simple models can be used to predict, manage, and mitigate biological conditions (see Fig. 1.1.2). This has been demonstrated in Florida's Indian River Lagoon where the US Army Corps of Engineers adopted a simple modulated water-release schedule aimed at working with the tidal cycle to reduce water residence times within the lagoon system and reduce the volume of water vulnerable to harmful freshwater algal blooms. In developing this solution, the researchers involved emphasized the role of real-time in situ data alongside the models in allowing reactive data collection and creating intense interest when sensor values showed basins to have reached key states or passed trigger levels [9].

1.1.4.1.5 Complex Models

Numerical models of the physical, chemical, and biological ocean allow us to codify and test our understanding and, once understanding is sufficient, predict future states. These predictive models balance computational and morphological complexity against performance. In the learning phase, they require in situ samples for initialization, calibration, and evaluation. Once understanding has been developed and codified with a degree of rigor, the near real-time predictive models need in situ sampling to initialize and assimilate. These models, once calibrated and proven, provide the tools to test future scenarios and deliver insight into our collective future. The models go far beyond predicting the observed variables and short-term forecasting. A calibrated and validated coupled physical, chemical, and biogeochemical model codifies, encapsulates, and demonstrates the dynamic system hypothesis.

Within these numerical models, broad-scale and regional effects can be quantified, assessed, and monitoring moved toward management. The human scale of effective administration limits the management outcomes. By developing scenarios of potential future states, the Commonwealth Scientific and Industrial Research Organization (CSIRO) Coastal Environmental Modeling group determined the likely future impact of differing loads of fish-farm feed on the D'Entrecasteaux Channel in Southeast Tasmania. These scenarios directly influenced the management of that industry [10]. As scale increases, the difficulty is not with modeling, but with the agreed consensus of the optimum outcome. Recent work, again by CSIRO, has created a 1- and 4-km resolution model of the Australian Great Barrier Reef. These models are run routinely and, at a management level, provide insight into the impact of individual rivers, storms, chronic loads, and episodic events [11]. For ongoing validation and data assimilation, ocean

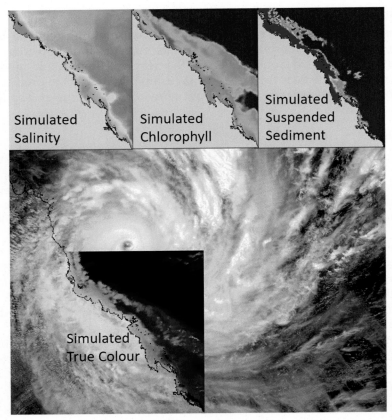

FIGURE 1.1.2   Using models to see through the storm. Simulated true color, surface salinity, chlorophyll, and suspended sediment in the central Great Barrier Reef, Australia, during Cyclone Yasi on the February 6, 2011 from the CSIRO 4-km model "eReefs" overlaid on Moderate-Resolution Imaging Spectroradiometer (MODIS) image. Note the contribution of coastal river plumes, suspended sediment, phytoplankton chlorophyll, and bottom reflectance to the simulation of true color.

color observations are compared directly with water-leaving radiance, which is simulated by a spectral optical model that integrates the absorption and scattering of optically active substances from the seabed and throughout the water column. This allows direct comparison of models and remotely sensed observations with the same units without employing statistical models or proxies with their confounding errors.

1.1.4.2 Array for Real-Time Geostrophic Oceanography

The Argo profiler array is perhaps the only singularly in situ oceanic observing system that was designed to answer a global process question and has successfully done so, to provide a

"quantitative description of the changing state of the upper ocean and the patterns of ocean climate variability from months to decades, including heat and freshwater storage and transport [12]".

The Argo data now support many broader ocean process inferences when coupled with additional, more parameter-diverse, in situ, and remotely sensed data sources (Fig. 1.1.3).

The difference between the data collected by an individual sensor/platform and the scales being addressed by the Argo network show how important temporal "oversampling" is to the understanding of larger-scale processes.

Once we have moved from *"observing to develop understanding"* toward *"monitoring of state or condition with capability to intervene,"* it is assumed that we are building a multi-tiered monitoring system that considers the temporal and spatial scales of the processes of interest.

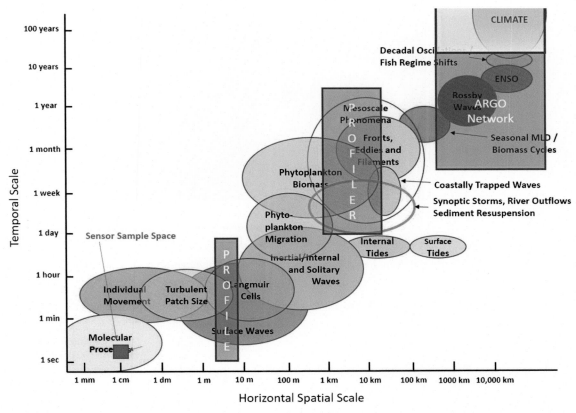

FIGURE 1.1.3 Tommy Dickey's Stommel diagram of ocean processes [4] overlaid with the footprints of the Argo network, a single Argo profiler, a single Argo profile, and a single sensor measurement.

The emerging broad-area observation networks show that real-time transmission of observations does not need to occur at the Nyquist frequency of the most rapid processes being observed. The observation strategy should collect information that will allow the processes and their drivers to be understood but transmission of those data may occur at rates related more to the broader-area processes. Vertical gradients that may be many orders of magnitude greater than horizontal gradients must also be considered.

Nyquist frequency–The Nyquist Theorem dictates that when digitizing an analog signal the sampling frequency must be at least twice that of the highest analog frequency component.

In Fig. 1.1.4, the vertical temperature gradient between the surface and 500 m in the Tropics of the Atlantic Ocean is greater than the horizontal surface temperature gradient across 50 degrees, or 5,000,000 m of latitude [13].

In a stratified estuary, surface measurements in freshwater tell us little of the processes that may be occurring below the pycnocline in the marine waters, but these waters may be deprived of oxygen, making them inhospitable to many forms of life. This highlights the need for in situ measurements at depth, not just in the surface layers. With the high cost of those in situ samples, how should we sample to better interface to the broader-scale observation and modeling methods?

If in situ sampling remains spatially undersampled, we can modify the temporal sampling strategy to allow better interfacing to the broader-scale environmental monitoring methods. Within the limited spatial constraints of the in situ sampling, we can gain knowledge of the spatial variability around our point sensors by sampling a greater density temporally with the water flow through the site, which provides access to a larger sample space (see Fig. 1.1.5).

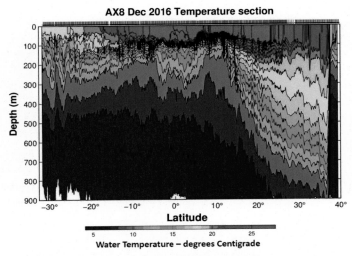

FIGURE 1.1.4   Single-ship high-density expendable bathythermograph transect through the Atlantic [13].

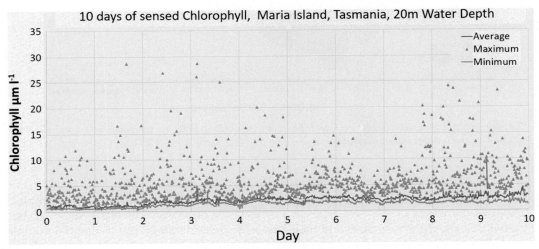

FIGURE 1.1.5   Chlorophyll derived from fluorescence sampled at 20-m water depth at the Australian National Reference Station at Maria Island, Tasmania, Australia. Values returned at 15 min intervals showing the mean, minimum, and maximum of a 1-min, 1-Hz observation window.

1.1.5   WHAT ARE WE SAMPLING?

For us to make the best use of our expensive in situ data we must be sure of exactly what it is we are sampling.

Very rarely does a sensor measure directly the phenomenon of interest; more often, a relationship is developed between a measurable and digitizable signal and that phenomenon. This may be a simple physics based model or a more complex empirical relationship based on observations. When the relationship is tight we can, almost, forget the relationship; however, when the relationship is loose, or dependent on a cascade of alternate measurements or presupposed conditions, uncertainty increases. For each variable that we deem of sufficient value that we are prepared to measure in situ, we must be clear what it is we are measuring. Only once we have that clarity can we fully understand the context and reliability of that measurement. Three examples are discussed below:

1.1.5.1   Temperature

Direct water-temperature measurement is an example of a very tightly coupled relationship. The measured resistance through a highly engineered platinum resistor (thermocouple) is the method used by most ocean research–grade in situ sensors. These sensors deliver results that are most commonly converted to temperature within the instrument and deliver absolute accuracies on the order of ±0.001°C and stability on the order of 0.001°C per month. These sensors are still measuring resistance through a resistor however, and, while moving through water with significant temperature gradients, the readings can exhibit hysteresis associated with the sensor's mass, or more importantly, the temperature reading of the thermistor may not be representative of the temperature of another sensor reliant on it for parameterization. Having ancillary information with respect to the rate of flow past the sensors allows partial corrections to be applied for the temperature hysteresis of the secondary sensor, thus the accuracy of the temperature of the secondary sensor can be more precisely determined.

1.1.5.2 Nitrate

Some in situ nitrate sensors use ultraviolet spectroscopy and the absorption characteristics of inorganic compounds in the 200–400 nm wavelengths to derive values for the nitrate content of water. The absorption characteristics vary with hydrochemistry, which often varies consistently with water mass; therefore, if temperature and salinity are collected alongside ultraviolet absorption, a value for nitrate may be derived. In the open ocean, this sensor has proved invaluable in providing continuous profiles of nitrate between the physical samples collected at specific depths through a rosette system. In coastal waters, the absorption characteristics of color dissolved organic material (CDOM) overlap with those of nitrate and this overlap confounds the nitrate quantification. CDOM does not have a single unique profile, it is dependent on regional variations; therefore, to derive nitrate from spectral absorption in coastal waters, complementary physical samples and analysis are needed to determine the impact of the confounders; this work is ongoing.

1.1.5.3 Salinity

The Thermodynamic Equation of Seawater 2010 [14] provides a very tight relationship that allows the derivation of salinity from conductivity, temperature, and pressure and onward to density in oceanic waters. Density gradients drive global oceanic circulations [15]. In oceanic waters, it can be assumed that the cocktail of salts is relatively consistent and therefore the salinity derived from conductivity is consistent. As we move into the coasts and up into estuaries, the focus of the conductivity measurement changes; it is not used primarily to derive density but as an indicator of the estuary or river health based on more localized definitions of "normal." Salinity calculation is heavily dependent on concurrent pressure and temperature, such that physical water-transport lags between sensors can (for example) create errors in gliders moving at speeds of 0.5 m/s of up to 0.3 practical salinity units (PSU) in which the environmental range is only observed to be 0.5 PSU [16].

1.1.6 WHERE ARE WE SAMPLING?

A temperature sensor samples the temperature of the water in contact with the probe. A fluorescence sensor measures the aggregate fluorescence returned from a small volume of water, typically 1 cm^3. An acoustic Doppler current profiler measures the Doppler shift due to water motion in narrow beams at 30–50 degree angles to each other through the water column and converts the readings to depth-binned water velocities; although the beams may have a footprint of hundreds of m, the returned results are typically presented for a vertical column above or below the sensor.

Moored (Eulerian) sensors placed at a specific depth may have a large horizontal representative footprint; in oceanic environments this may be on the order of hundreds of square km, but a vertical representativeness measured in tens of m.

In designing our network of monitoring capabilities, it is important that we consider the sensor volume, the platform footprint, and the network resolution, which as seen in the Argo example are very different, but complementary. That said, when we are monitoring with a view to intervention, there is an assumption that we know enough about a system to identify smaller areas in which change may be a precursor to broader-area change. Because discrete points may be directly influenced by natural variability and anthropogenic forcing, our network may therefore be built evenly across environmental gradients rather than evenly spatially distributed. In many cases, a long history of observations at a location provides a significant impetus to continue the observations. At these locations, the significance of change can be quantified against the historical context.

1.1.7 VARIABILITY IN SAMPLE SPACE

With in situ samples as the start of a cascade of sensing and sense making of our environment, it is not enough just to provide numerical ground truth of value and uncertainty, we must provide information about the natural variability that occurs over the time and space scales adjacent to that of the process of interest. This is especially true when the data feeds into modeling for validation and understanding development rather than as a well-structured assimilation. In this space, it is not so much that *"all models are wrong, but some are useful,"* more, that a modeled representation of a specified volume of water, normally many orders of magnitude larger than the sampled volume, may not have the process complexity to resolve all scales of natural variability shown by the sampling, or that a free-running (or loosely constrained) model may be offset in space or time due to model resolution, nonlinear responses, or imperfect initial conditions.

The National Oceanographic and Atmospheric Administration (NOAA) High-Frequency (HF) Radar network has this to say about the accuracy of its remote sensed surface current product: *"For the observed surface currents, although the precise amount of error is difficult to quantify, HF Radar data is generally expected to be accurate to within $10\,cm\,s^{-1}$ of current speed and 10 degrees of current direction. It is important to note, that the presented values are spatial and time averages - so that they may not be representative of the currents of a specific point within a grid cell (particularly near shore) or of an instant in time during the observed hourly period."*

Where in situ measurements have the capacity to provide insight into local variability, our sampling strategy should be developed to measure it; for example, the Australian National Reference Stations sample key variables at 1 Hz over a period of 1 min every 15 min, the mean, standard deviation, minimum, and maximum values are returned. Although not perfect, this is a start toward describing short-term variability and the range of values across which monitoring systems may intersect at different scales [17].

1.1.8 PLATFORMS FOR SENSORS

1.1.8.1 Eulerian or Lagrangian?

Marine researcher Tom Malone once likened looking at the results from a mooring, which is fixed in one place and therefore Eularian, to watching an opera that is being performed on a drifting barge. As the tide goes out, Figaro is happily contemplating marriage to Susannah, as the tide comes in you see a selection of actors you do not know and Figaro is claiming to have jumped out of a window. Finally, as the tide goes back out, Figaro is proclaiming his love for the Countess, not Susannah. You do not get to see the ending. The story you create to fill in the gaps will not match the real plot. On the other hand, a Lagrangian platform, a drifter, would move with the barge for the whole performance and see the whole story unfold, split into segments only by the sampling frequency.

Although it is easier to interpret water-mass evolution from a Lagrangian platform, it is easier to monitor change from fixed locations; interpreting the meaning of that change is another story.

In the past the options for fixed-station monitoring required moorings with a surface signature, a ship on site, a sensor on a jetty or under the ocean on a cabled observatory. However, the last decade has seen a wide variety of platforms that were operating in the realms of technology labs now being operated as part of routine observing and monitoring.

Moorings and undersea observatories still provide key long-term fixed platforms, a capability which is hard to replicate. However, semi-Eulerian deployments of gliders and profiling floats have been shown to offer alternative strategies that can provide insights into both the long-term evaluation of a specific location and the short-term evolution of water masses, with the added bonus of continuous vertical profiles. (See Chapter 5).

Surface autonomous craft have proven themselves in the last 5 years. The Liquid Robotics Waveglider Pacific crossing in 2012 [18] proved the endurance of wave-/solar-powered surface craft, and their survivability has been proven in many deliberate hurricane deployments. Now other wave-powered craft are entering the market without the size limitations of the Waveglider, which is constrained in size by the way it harvests wave energy for movement. The Autonaut, for example is available with hull forms up to 7 m in length, higher achievable hull speeds, and greater station-keeping ability [19]. The greater size allows greater solar energy collection and the potential for winch-based profiling. This evolution creates the options for hybrid moorings, where the subsea sections are deployed by ship and designed for exceptionally long-term deployments. The surface part of the mooring can self-deploy from port and replace itself as necessary. The surface part has a primary function of acting as a communications gateway from acoustics-in-water transmissions to electromagnetic satellite communications, but can collect transit data and surface data while at sea.

1.1.8.2 Established Platforms

Ships of opportunity have long been used for collecting marine environmental data. Increased automation of data collection systems is allowing a wider variety of parameters to be measured at higher densities from flow-through surface sensors and automated profiling tools. Organizations such as the World Ocean Council are building a base of willing platform operators, but past experience with a volatile shipping market has pushed the focus of current developments toward making better use of established ships of opportunity rather than an expansion of the network. With over 2500 surface drifters deployed globally, delivering close-to-surface water temperature and providing research-grade data with limited hysteresis, the need for basic ship-based surface temperature is not evident. However, the ship of opportunity has a clear niche in collecting either multiparameter surface data or profiles from expendable or autonomous profiling systems.

Instrumented moorings have been a staple of ocean observing for decades. Satellite communications and inductive and acoustic modems have enabled moorings to deliver in near real time from anywhere on the planet. In the past, the variability with depth was addressed by multiple sensors along the mooring line; this is changing as the breadth of observed parameters increases, making it more cost-effective to deploy a profiling multiple sensor platform from either the surface down or the bottom up. The latter has been enabled by the delivery of power and high-bandwidth communications to subsea locations as part of cabled observatories (see Chapter 5.5).

1.1.8.3 Underwater Gliders

The original concept for the underwater glider was powered by an engine utilizing the difference in thermal expansion properties between mediums (initially paraffin wax and salt water). The early designs never reached their theoretical potential, so the current operational glider fleet is driven by electrically powered pistons. These electric piston gliders can travel over 4000 km at speeds of half a knot. Recent reworking of the original concept of a thermal glider at Tianjin University has shown promising results with demonstrated mission duration of 27 days and 677-km range [20].

As with all things, size does matter. The bigger the glider, the faster it can move. The Liberdade glider has been on the edge of breakthrough for the past decade. This glider with its 6.1-m wingspan and speeds of up to 2 knots has a published range of 1500 km [21].

1.1.8.4 Animal Oceanographers

With the miniaturization of sensors and communications systems and the reduction in power needed for communications, an increasing number of species of marine animal are being instrumented. Fish tags record information internally and the data are harvested when the fish is caught. Other tags are simple transmitters that allow the fish to be identified when it passes within range of receivers. More recently, tags with the ability to network and pass information between tags have been developed to allow greater collection rates. Perhaps the most interesting tags as part of monitoring programs are those mounted on air-breathing marine creatures because, on surfacing, they have the relative broadband luxury of being able to transmit data in near realtime.

Surface transmission of data cannot be relied upon for all marine creatures and for those not prone to visit the surface, curtains of acoustic listening stations have been installed around the globe. For the majority of the tags deployed, simple presence/absence is detected by coded acoustic transmitters. These data can be used to detect movement and migration. In line with the ongoing miniaturization of everything, some of the newer tags can carry pressure, temperature, or accelerometer sensors and transmit this information for reception when within an acoustic receiver network.

Animal Oceanographers tend to congregate around the oceanic gradients, or biological hotspots found using their own "onboard sensors." Animal oceanographers present a very interesting but heavily biased view of the environment; the challenge is interpreting the records with the limited information that is digitized on the platform.

1.1.8.5 Project Loon

The Google X Project Loon is not a marine-sensing project, it is a low-cost digital internet solution for remote areas. Since the project launch in 2013, it has developed three things of significance to the marine monitoring community: (1) Its primary product, a low-power communications system using stratospheric balloons and existing cell phone technology; (2) an auto launcher capable of deploying a new balloon every 30 min; (3) an automated balloon fleet control system in which controlling the position of the balloons vertically within the jet stream controls their horizontal flight path. In the coastal space, the extended range of cell phone-based digital communications will be useful. The other components are interesting as they point to a more automated use of Lagrangian platforms in future marine monitoring. We are seeing projects out of places like the Queensland University of Technology that uses autonomous surface craft as robotic delivery mechanisms for small coastal profiling floats and simulations that are experimenting with profiling float-path planning using depth to control velocity [22].

1.1.9 PROVENANCE

In situ sampling represents the foundations of an observation and understanding framework and it is an absolute necessity that measurement, the measurement uncertainty, and possible confounding factors are understood, mitigated where possible and communicated to those that develop understanding based on the in situ results. In traditional oceanography, data pathways were vertical, usually with an individual or within a laboratory, and understanding of the measurement system could be assumed; in many cases the scientist who planned the data collection program was the individual using the data. Increasingly, in situ sampling data are accessible through web and machine-to-machine interfaces. For this use case, data need to be delivered as packages of complementary in situ measurements and measurement dependencies codified so that uncertainty can be assessed by the new user. In the Internet of Things (IoT) future, automated decision trees and machine learning will take data and deliver, among other things, multiparameter views and automated actions based on data that may not even be stored. As we move into this future, it is imperative we deliver realistic estimates of uncertainty of measure with all samples; this will allow repurposing far more effectively than quality flags.

1.1.10 THE SENSORS

Much will be written about sensor developments in later sections of this book. Suffice to say here that sensors are continuing to home in on the parameters of interest, becoming less bound by proxies, but in many cases now measure multiple parameters in situ to derive the parameter of interest; this colocation of sampling increases the certainty of the derived value of the parameter of interest.

1.1.10.1 Sensor Fouling

Sensor fouling is primarily an issue in the euphotic zone. The Argo Program, again, points to an operational solution; it effectively avoids the fouling issue by parking Argo profiling floats at 1000-m depth for 10 days between active profiles. This has resulted in the longevity of each platform's useful life being limited by battery life, not fouling. Some floats have managed lifetimes of over 6 years (225 profiles) still delivering conductivity, pressure, and temperature values with well-defined certainty.

At the surface where power can be harvested from the sun or at depth within the euphotic zone as part of cabled observatories, ultraviolet (UV) light is proving effective in keeping sensor surfaces free from biofouling. As this method is refined and the power required to prevent settling organisms from growing in situ is better understood, this option is becoming viable for battery-powered platforms [23]. Further biofouling mitigation strategies are discussed in Chapter 4.3.

Biofouling of the sensor platform is an ongoing issue. Slow-moving buoyancy-driven gliders fail to achieve the speeds needed for fouling release coatings to perform properly, although ablative or pesticide-based coatings and copper are still very effective, with limited life.

1.1.11 TECHNOLOGICAL TRAJECTORY AND TRANSACTION COST

The marine environment is undersampled spatially and parametrically. As we, the global population, seek to manage our oceans sustainably, we must recognize that we still do not understand our oceans well enough to monitor and manage them optimally. Therefore, internationally, observation networks for enhanced understanding will continue to be our focus for any foreseeable future. When we seek to monitor as part of management, we may not need a greater density of observations but we will need spatially focused and sustained monitoring. As we seek to manage more of our marine environment we need focused near real-time sampling that can be interpreted in the context of broader-scale remote sensing or modeling.

Reduction in transaction cost per focused observation, delivered in near real-time to sense-making analytical platforms, should be our single aim. This can be achieved by enhancing the certainty of the observation, deploying sensors in a coherent, systematic, and ongoing manner, and finally, ensuring interoperability between data warehouses from multiple agencies or data sources. Data strategies will be dealt with in detail in Chapter 6. The Australian Integrated Marine Observing System (IMOS) Draft Strategic Plan 2017–22 has this to say: *"Initially, the emphasis was on reducing the cost per observation. As the system has matured over time we have placed increasing emphasis on the relationship between cost and impact."*

Focused in situ observations deliver insight; by planning in situ sampling as part of broader environmental analytics, we can increase the impact of that insight, but still drive down the cost per observation by reducing human engagement at all stages of the sampling and analysis process. We are reducing the need for ships to deploy sensors by extending the capabilities of our robotic platforms, both surface and subsurface. We are increasing the longevity of deployments through both operational strategies and new sensor designs.

The IoT is delivering many ways to reduce data transmission volumes through edge computing, which enables signal-aware compression, event detection, and event-based variable density transmission. The generic IoT tools are providing robust and adaptable frameworks for data distribution in a big-data, machine-to-machine world with variable data provenance. The capacity of the marine community to adopt and take advantage of these new frameworks is limited by our custodianship of benchmark datasets, yet each more generalized standard we adopt decreases transaction costs and exposes our data to a broader set of analytical tools and new sets of eyes. Scalable cloud compute, coupled with new software tools allow the same programmatic analysis to be performed on one million, one billion, or one trillion data points with the same code used for single-point calculations, thus complex analyses can be moved seamlessly from points to three-dimensional cubes and four-dimensional data structures.

Edge Computing—The movement of compute resources away from the center toward the platforms or into the sensors to enable reduced communications bandwidth and perform localized analytics in real time.

In situ measurement in the ocean has its own unique set of challenges, but many are becoming shared with more mainstream ventures. The more general the solution, the more the benefits that can be obtained from the economies of scale in a global market. Low-cost, general purpose, multiple input/output processors (known as "maker" electronics) are now delivering low-power consumption solutions to allow deployment in marine sensors, thus oceanic in situ sampling will see more benefits over the next decade by embracing that which is common in mainstream sensing and analytics than in emphasizing the uniqueness of our problems.

References

[1] Steffen W, Persson Å, Deutsch L, et al. AMBIO 2011;40:739. https://doi.org/10.1007/s13280-011-0185-x.

[2] IPCC. In: Core Writing Team, Pachauri RK, Meyer LA, editors. Climate change 2014: synthesis report. contribution of working groups I, II and III to the Fifth assessment report of the intergovernmental panel on climate change. Geneva (Switzerland): IPCC; 2014. 151 p.

[3] Pelling M, Blackburn S. Megacities and the coast. Abingdon (Oxon): Routledge; 2014.

[4] Dickey TD, Bidigare RR. Interdisciplinary oceanographic observations: the wave of the future. Sci Mar 2005;69(S1):23–42.

[5] Belward AS, Skøien JO. Who launched what, when and why; trends in global land-cover observation capacity from civilian earth observation satellites. ISPRS J Photogramm Remote Sens 2015;103:115–28.

[6] https://www.ioos.noaa.gov/wp-content/uploads/2015/12/sarops_hfr_info2012.pdf.

[7] Makris NC, Jagannathan S, Ignisca A. Ocean acoustic waveguide remote sensing: visualizing life around seamounts. 2010.

[8] Kloser RJ, et al. Acoustic observations of micronekton fish on the scale of an ocean basin: potential and challenges. ICES J Mar Sci 2009. https://doi.org/10.1093/icesjms/fsp077.

[9] Walsh I. Seabird electronics. Pers. Coms.

[10] Wild-Allen K, et al. Applied coastal biogeochemical modelling to quantify the environmental impact of fish farm nutrients and inform managers. J Mar Syst 2010;81(1):134–47.

[11] Baird ME, et al. Remote-sensing reflectance and true colour produced by a coupled hydrodynamic, optical, sediment, biogeochemical model of the Great Barrier Reef, Australia: comparison with satellite data. Environ Model Softw 2016;78:79–96.

[12] Roemmich D, et al. ARGO: the global array of profiling floats. CLIVAR Exch 13 1999;4(3):4–5.

[13] http://www.aoml.noaa.gov/phod/hdenxbt/ax_home.php?ax=8.

[14] McDougall TJ, et al. The international thermodynamic equation of seawater 2010 (TEOS-10): calculation and use of thermodynamic properties Global Ship-based Repeat Hydrography Manual, IOCCP Report No 14. 2009.

[15] IOC, SCOR. IAPSO, 2010: The international thermodynamic equation of seawater–2010: calculation and use of thermodynamic properties. 196.

[16] Garau B, et al. Thermal lag correction on Slocum CTD glider data. J Atmos Ocean Technol 2011;28(9):1065–71.

[17] https://portal.aodn.org.au/.

[18] Villareal TA, Wilson C. A comparison of the pac-X trans-pacific wave glider data and satellite data (MODIS, aquarius, TRMM and VIIRS). PLoS One 2014;9(3):e92280.

[19] Willett L. AutoNaut remote USV aims to make waves. Jane's Int Def Rev 2014;47:36–7.

[20] Ma Z, et al. Ocean thermal energy harvesting with phase change material for underwater glider. Appl Energy 2016;178:557–66.

[21] D'Spain GL, et al. Underwater acoustic measurements with the Liberdade/X-Ray flying wing glider. J Acoust Soc Am 2005;117(4):2624.

[22] Katikala S. Google™ project Loon. InSight Rivier Acad J 2014;10(2).

[23] http://www.thejot.net/?page_id=837&show_article_preview=601&jot_download_article=601.

CHAPTER

1.2

Opportunities, Challenges and Requirements of Ocean Observing

Anja Reitz[1], Martin Visbeck[1,2], Albert Fischer[3]

[1]GEOMAR Helmholtz Centre for Ocean Research Kiel, Kiel, Germany; [2]Christian-Albrechts Universität zu Kiel, Kiel, Germany; [3]Intergovernmental Oceanographic Commission of UNESCO, Paris, France

1.2.1 INTRODUCTION

1.2.1.1 Why Do We Need Integrated Ocean Observing?

The ocean regulates global climate and provides resources such as food, materials, and energy. It facilitates 90% of global trade and enables recreational and cultural activities. Human development, population, and economic growth together with essentially free access to ocean resources cause increased pressures on marine systems. Examples of the most pressing issues include overfishing, unsustainable and polluting resource extraction, alteration of coastal zones, land-based pollution, and climate change, leading to ocean conditions of increasing temperatures, growing acidity, reduced oxygen, and rising sea levels. Although the prosperity of the global society depends on the ocean, its importance is not yet matched by our knowledge and understanding of the ocean. Hence, innovation in ocean observing and more international cooperation is needed to enable increased understanding of the ocean system and how it is changing. Such knowledge is also needed to derive ocean governance regimes based on the best available science to ensure more sustainable use of marine resources and protection of the marine environment to safeguard equitable prosperity for current and future generations globally.

Ocean in situ and satellite observations are indispensable to support ocean science, assessment, forecasting, services, and an increased range of societal benefit areas in the context of sustainable development. Currently, several largely independent ocean-observing initiatives and networks have been developed to serve the needs of specific sectors and their scientific communities and information users. However, most of the challenges the Earth is facing are by nature multidisciplinary and, in addition, the limited resources (funding, platforms, and technology) for ocean observing provide significant opportunity for more cooperation and would benefit from more sustainability. The OceanObs'09 conference articulated this opportunity and led to the development of the *Framework for Ocean Observing* (https://doi.org/10.5270/OceanObs09-FOO), which provides guidance toward an integrated system of ocean observing. It articulates the need to set requirements, establish essential ocean variables (EOVs), expand observing networks, improve data sharing, flows, and issue-driven information delivery. The *Framework* encourages ocean scientists and observation providers to work with the users of ocean information and to enhance the current system toward more and better-coordinated supply of quality data. It also articulates the need for system innovation by facilitating a pathway for the infusion of new technology including increased sensor capabilities and to promote best practice and metrological standards.

One part of the ocean-observing community that sustains global-scale ocean observing is strongly linked to and supports the information needs for climate information (Global Climate Observing System [GCOS] [1]) and research [2].

Climate researchers vitally depend on sustained ocean observations with adequate coverage in space and time to document changes in the ocean state of heat, salt, sea level, and its large-scale circulation. Such information enables long-term climate assessments by the Intergovernmental Panel on Climate Change (IPCC), which must include the physical role of the ocean in climate and climate impacts, vulnerabilities, and adaptation opportunities. Long-term assessments are also the basis for short-term predictions of rainfall, drought, and sea level changes in the context of climate services. The close interaction and collaboration of the ocean observing and climate research communities have their roots in the World Climate Research Programs (WCRP) core project Ocean & Climate: Variability, Predictability, and Changes (CLIVAR) [3]. Compared to the well-coordinated and organized atmospheric observations and space-based capabilities, in situ ocean observing is fragmented, despite the fact that the World Ocean Circulation Experiment (WOCE, 1990–2002) and the CLIVAR community have introduced some relevant sustained ocean-observing programs. These include (1) the global repeat hydrographic surveys, (2) the profiling float array, and (3) the tropical moored buoy array, which are all key contributions to the Global Ocean Observing System (GOOS) [2]. GOOS is cosponsored by four United Nations agencies, the Intergovernmental Oceanographic Commission (IOC) of the United Nations Educational, Scientific, and Cultural Organization (UNESCO), the World Meteorological Organization (WMO), the United Nations Environment Program (UNEP), and the International Council for Science (ICSU). The Intergovernmental Oceanographic Commission (IOC) of UNESCO is executing the GOOS program; however, its success relies on the coordinated contributions of several individuals, institutions, nations, regions, and organizations worldwide. GOOS is the key ocean-observing element of the "Oceans and Society: Blue Planet" Initiative of the Group on Earth Observations (GEO).

GOOS utilizes the Framework for Ocean Observing to guide its implementation of an integrated and sustained ocean-observing system. This systems approach, designed to be flexible and to adapt to evolving scientific, technological, and societal needs, helps deliver an ocean-observing system with maximized user base and societal impact.

Beyond the physical climate, a broad range of other needs for sustained ocean observing have been articulated. Those include the need to document the uptake of carbon by the ocean (International Ocean Carbon Coordination Project [IOCCP]), to assess the regional abundance of fish stock (International Council for the Exploration of the Sea [ICES], The North Pacific Marine Science Organization [PICES], to discover and document changes in the biodiversity of the ocean's ecosystem via the Group on Earth Observations Biodiversity Observation Network [GEO-BON]), or to provide the database that allows protection and more effective governance of the marine environment. For example, the Sustainable Development Goal 14, to "Conserve and sustainably use the oceans, seas and marine resources for sustainable development," (SDG14; e.g., Ref. [4]; find links to relevant websites at acronym list) under the United Nations Agenda 2030, lays out new requirements similar to those articulated regionally for Europe under the Marine Strategy Framework Directive (MSFD), 2008. Ocean mapping and the information needed to support safe and environmental friendly maritime operations are growing areas in need of coordinated ocean observations.

Thus the motivation and need to seriously engage with the proposed Framework for Ocean Observing to arrive at a more systematic and fit-for-purpose system of integrated observing is growing. A particular exciting opportunity exists in the area of innovation at the sensor, platform, data processing, and delivery levels, with new opportunities in the areas of miniaturization, energy efficiency, machine learning, and interoperable data systems.

1.2.1.2 History of Ocean Observing

The history of open ocean observation mainly started in the late 19th century with the cruise of HMS Challenger (1872–76; Figs. 1.2.1A), that lasted 4 years and covered more than 68,000 nautical miles (e.g., Refs. [3,5]). Systematic ocean observing tremendously profited from the improvement in navigation and the development of the chronometer to ascertain longitude at sea. These two improvements were the pathway to enable the estimation of surface currents, and proved important as well for ocean temperature and salinity measurements. To investigate the link between the physical properties of the ocean and fisheries in the North Atlantic, the International Council for the Exploitation of the Sea (ICES) was established in 1902. The most significant contribution of ICES to physical oceanography was the standardization of salinity detection in seawater by titration against silver nitrate solution and the resulting establishment of the Standard Seawater Service by ICES [3]. By the 1920s, many subsurface water sample designs, mainly water sample bottles clamped to a wire and sequentially triggered to close, were tested and used, and led to the establishment of the Nansen bottle as the generally used standard. The next step was the establishment of the determination of dissolved oxygen according to Winkler [6] used by Wüst during the Meteor Expedition (1925–27; Fig. 1.2.1B), enabling identification of the spreading of subsurface water masses [7]. Another important development in the first half of the 20th century was the mechanical bathythermograph invented in 1937 [8]. By this

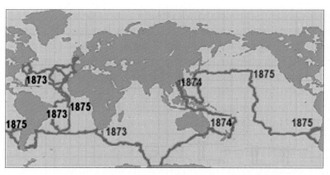

FIGURE 1.2.1A   The route of HMS *Challenger*. The expedition lasted 1000 days and covered more than 68,000 nautical miles. *Source:* http://oceanexplorer.noaa.gov/explorations/03mountains/background/challenger/challenger.html.

FIGURE 1.2.1B   Survey Vessel *Meteor* (1925–27). *Meteor* sailed over 67,500 nautical miles, made 67,000 depth soundings, established 310 observation points, and released 800 observation balloons. *Source: Into the Unknown: The story of Expeditions (The National Geographic Society. Washington (DC); 1987 and* http://www.ipy.org/index.php?/ipy/detail/a_victory_in_peace_the_german_atlantic_expedition_1925_27/.

instrument it was possible to decipher the thermal stratification of the upper 150 m of the ocean. New insights were possible with the development of new technology, for example neutrally buoyant floats [9] were the first battery-powered electronic instruments to measure deep currents. The use of these floats, hence, brought the first evidence of the occurrence and abundance of mesoscale variability in the ocean [10].

Improvements in navigation in the 1960s and 1970s providing accuracy better than 1 km added up to the fact that ocean scientists precisely knew where observation had been made, resulting in extremely improved surface-current measurements. During this time salinity titration was replaced by electronic conductivity measurements via salinometers [11], and the water bottle/thermometer combination to measure temperature was replaced by continuous temperature profile measurements via CTD instruments measuring conductivity (salinity), Temperature, and Depth [3]. Advances in sensor technology again enabled innovation in ocean observing and significant advances in capabilities that allowed new scientific questions to be addressed. The mechanical bathythermograph was replaced in the late 1960s by the eXpendable BathyThermograph (XBT), which estimated depth as a function of time via a fall-rate algorithm and measured temperature with a thermistor. The XBT is deployed by a ship via a two-conductor copper wire and relays its data back to the ship. Measurements by XBT probes greatly improved our knowledge about the temperature variability of the upper ocean before they were superseded for broad deployment by a new technology, profiling floats (e.g., Refs. [12,13]). XBTs remain useful in capturing strong thermal gradients in the oceans, such as across boundary currents.

Measuring sea surface temperature (SST) has been done since the beginning of commercial shipping and ocean research. However, ship-based measurements are sparse in time and space, and changes at the sea surface happen fast due to the interaction with the atmosphere, swift wind-driven currents or energetic ocean mesoscale eddies. The advent of satellite-based sensors, complemented by ground truthing with drifting surface buoys, enabled enormous progress [14]. The production of global climatology of SSTss was now possible by the combination of in situ and satellite temperature measurements [15]. This climatology and real-time data sharing and processing for the long-term mean are produced and constantly improved by the Global Ocean Data Assimilation Experiment (GODAE) High-Resolution Sea Surface Temperature consortium (GHRSST). They provide the backbone for climate assessments and the boundary conditions for ocean and climate forecast models.

Moored instruments equipped with microelectronic current meters that internally record the data significantly advanced the records of subsurface ocean currents expanding exploration and mapping of the mesoscale variability of the ocean [16,17] and today are used to document the mean and changes of ocean boundary currents. Global instrument tracking, particularly the path of surface drifters and subsequently the development of the Airborne Remote Geographic/Oceanographic System tracking system was provided by more accurate satellites in the 1970s [3,18,19]. This enhancement with some experiments culminated in an international deployment of 300 drifters in the Southern Ocean (1978–79), contributing to the first Global Atmospheric Research Project (GARP) and Global Experiment (FGGE), collecting surface temperature and atmospheric pressure [20].

All these enhanced technologies and methods led to a range of regional experiments in the 1980s, e.g., the Tropical Ocean Global Atmosphere (TOGA) project collecting data from the equatorial Pacific combined with numerical modeling intending to decipher and predict the evolution of the El Niño–Southern Ocean (ENSO) phenomenon. Besides enhancement of data collection and distribution, the project led to the deployment of the Tropical Atmosphere Ocean/Tropical Atmosphere Ocean–Triangle Trans-Ocean Buoy Network (TAO/TAO–TRITON) array mooring reporting upper-ocean and atmospheric real-time data [21].

The realization of oceanic variability and long-term changes by continuously evolving methods, the awareness of shortcomings in data interpretation of sparsely sampled in-situ data, and the perspective of satellites delivering comprehensive data sets led to the development of the World Ocean Circulation Experiment (WOCE). The WOCE is a global-scale program, intended to improve the ocean model and unravel the ocean's role in climate [22]. The comprehensiveness of temperature, salinity, and ocean chemistry data collected remotely and in situ in the framework of the WOCE Hydrographic Program, in collaboration with the Joint Global Ocean Flux Study, was pioneering in scope [23]. A significant development of the program was the neutrally buoyant float with positioning and data relay by satellite, making it independent of acoustic tracking appropriate for global-scale deployment [24]. Documentation of various transport estimates of deep and shallow boundary currents, flow-through passages, and transbasin areas were corrected by several WOCE arrays of moorings [3].

To enable observing of ocean velocity profiles, the moored and ship-mounted Acoustic Doppler Current Profiler (ADCP) was developed in the late 1980s and early 1990s. The ADCP records ocean currents at many different water depths simultaneously via acoustic beam that travels through the water column. The full potential of this new technology depended on the development of Global Positioning System (GPS). When GPS became increasingly available for scientific use it provided absolute positioning to meter accuracy and consequently enabled underway upper-ocean velocity observation (to 800 m). As a result, determination of velocity throughout the water column was possible due to ADCP incorporation into CTD/multisampler packages (e.g., Ref. [25]).

In the late 1990s, scientists and service providers increasingly realized the growing demand for and the potential of observational data to deliver reliable short-term ocean and weather forecasts and manage marine operations. Hence, the idea emerged to develop a framework for a sustained ocean-observing system. In 1995, CLIVAR, focusing on the coupled ocean–atmosphere system, was established as a component of the WCRP, pooling many of the physical observations established by then. The Global Ocean Observing System (GOOS) developed out of a dawning scientific understanding of the ocean's role in climate in the late 1980s. It built on the heritage of large-scale ocean-observing experiments, outlined earlier, and involved the Intergovernmental Oceanographic Commission (IOC) and the World Meteorological Organization (WMO). The IOC agreed in 1991 to lead development of sustained GOOS for physical ocean properties, which was broadened to include chemical properties and biological coastal monitoring. A GOOS program support office was established at the IOC with the WMO as a cosponsor. The UN Environment Programme (UNEP) and the International Council for Science (ICSU) joined as cosponsors in 1997.

This chapter reveals that innovation in ocean-observing sensors and platforms continues unabated. Over the last few years, a community of research organizations and commercial shipping companies, the Voluntary Observing Ship fleet, has agreed and arranged for a variety of instruments to be operated on commercial vessels, such as ferries, cruise ships, and container vessels. This fleet will enhance the data inventory of surface temperature and salinity observations, meteorological variables, and surface-layer chlorophyll, pCO_2, and upper ocean velocity data (e.g., Chapter 5.2 and 5.3).

The rapid development of autonomous ocean-observing platforms has been stimulated by the increasing demand for ocean observing and the corresponding high costs for operations with research vessels. Furthermore, extreme conditions in the Arctic and Southern Oceans, where key climate processes occur, have been technologically and logistically demanding but are now increasingly explored with modified and novel uses of existing observing platforms [3]. An example of these optimized technologies is the Argo floats with ice detection algorithms and rugged bodies [26] enabling profile extension to high-latitude oceans. Others include moored profiling systems developed to study ocean properties under ice shelves (e.g., Ref. [27]; Chapter 5), miniaturized CTD systems capable of being deployed on animals to provide observations in high-latitude oceans [28], self-propeled ocean gliders that can navigate the upper ocean (upper 1000 m) with battery endurance up to 1 year (e.g., Chapters 5 and 8). In addition, new technologies include autonomous underwater vehicles (AUVs) with shallow- and deep-diving systems, autonomous observation platforms with fast satellite communication and efficient power cabling onboard supporting multidisciplinary ocean observatories, and wave-glider systems that use surface-wave energy to move the vehicles.

Multidisciplinary oceanographic sensor systems can also be deployed on cabled sea-floor observatories facilitating real-time measurements (for further details see Chapter 5). Autonomous observing systems are often equipped with ocean sensors. These require sensors with demonstrated low power consumption, with low-calibration drift and compact physical configuration. Over the last three decades, a range of optical sensors measuring dissolved oxygen, particulate carbon, and dissolved organic matter, became available. Currently, a number of extremely small analytical systems for in situ wet chemistry and even mass spectrometry are maturing (for further details and specific applications see Chapter 3). Novel biological and chemical sensors taking automatic image analysis and conduct pattern detection to facilitate identification of specific plankton types are underway (e.g., Ref. [3]).

1.2.2　TOWARD A SUSTAINED OBSERVING SYSTEM FOR CLIMATE AND BEYOND

Several authors (e.g., Ref. [29]) emphasized the importance of sustained ocean observing on short- and long-time scales for regional and global ocean predictions and models. WOCE, furthermore, accentuated the need for strong international collaboration among research institutes and funding agencies toward a coordinated ocean-observing program including a system to collate quality control and distribute data. In 1999, CLIVAR, GOOS, and the Global Climate Observing System (GCOS) sponsored the first OceanObs'99 conference that was held to sustain and grow the success of ocean observation and to develop plans for networks to sustain ocean observing for the next decade, building on strong support from the ocean satellite agencies. The community aimed to provide a framework and set clear and feasible objectives for a sustained ocean-observing system [30].

Some major observational programs were evaluated during OceanObs'99: (1) the Argo program that was striving to implement a global-scale array of floats collecting freely available data profiles (in real time and quality controlled) up to 2000 m water depth, (2) a global XBT network focusing on high-resolution transects collecting temperature profiles in the upper ocean to a depth of 1000 m mainly by commercial vessels via the Ship of Opportunity Program, (3) a Surface Drifter Program providing quality controlled SST and surface velocity data, and (4) selected locations of moored observatory stations recording regularly repeated observing over long-time scales providing insights into physical and biogeochemical ocean variability of the upper and deeper ocean.

During OceanObs'99, the strong need for continued systematic, ship-based, global ocean surveys was highlighted by Gould et al. [31] and supported by CLIVAR, GOOS, and the IOCCP and resulted accordingly in a program of hydrographic sections at 5–10 year intervals [3]. Accordingly, in 2007 the Global Ocean Ship-based Hydrographic Investigations Program (GO-SHIP; [32]) was established to coordinate and maintain sustained ship-based hydrography and address the need for chemical measurements to monitor the oceans' carbon content. Additionally, OceanSITES—a network of full-depth and surface time series at key climate-relevant locations—was initiated at OceanObs'99 [33] and has since developed into a global mooring network incorporating arrays in all global oceans. Consequently, crucial progress has been made in global data interoperability by means of the standardization that OceanSITES has fostered through the implementation of the OceanSITES Network Common Data Format (NetCDF), essential for long-term data assimilation. Even though biogeochemistry plays an important role in the ocean ecosystem, the observation of biogeochemical parameters has not reached the density and global scale of physical measurements, even as the readiness for sustained observing continues to grow.

In 2009, the ocean-observing community reconvened at the second OceanObs'09 conference to review the progress of the past decade, and articulated the need for a more systematic, strategic, and interdisciplinary approach to global ocean observing [34]. As an outcome, experts developed a strategy under the name "Framework for Ocean Observing" (FOO; [35]), which became the strategic framework for GOOS in 2015. FOO has influenced the development of several regional ocean-observing strategies, including the tropical Pacific (Tropical Pacific Observing System [TPOS2020]), the Atlantic (Optimising and Enhancing the Integrated Atlantic Ocean Observing Systems [AtlantOS]), and Arctic (Integrated Arctic Observation System [INTAROS]).

The European Horizon 2020 project AtlantOS is an international contribution for assessing and defining a more effective and integrative Atlantic observing system. The AtlantOS team includes coordinated projects of the United States, Canada, the European Union, Brazil, Argentina, and South Africa to implement a basin-wide observing network for monitoring variability and changes in the Atlantic. AtlantOS, furthermore, cooperates with the North and South Atlantic Ocean research alliances that formed because of the signing of the Galway and Belem Statements. These statements on Transatlantic Ocean Cooperation enhance collaboration to better understand the Atlantic Ocean and sustainably use and protect its resources, and govern human activities. For the Southern Ocean, more specifically, system design and implementation is coordinated by the Southern Ocean Observing Systems (SOOS), with expansions through two newly deployed Ocean Observation Initiative nodes in the Argentine Basin and Southern Ocean, and the six-year Southern Ocean Carbon and Climate Observations and Modeling (SOCCOM) project, deploying 200 biogeochemical profiling floats. In the Pacific, the Tropical Pacific Observing System (TPOS2020) defines the strategy to implement a more sustainable and Integrated Ocean-Observing System (IOOS). For pan-Arctic observation, the Sustaining Arctic Observing Networks (SAON) aim to enhance Arctic-wide observing activities by facilitating partnerships and synergies among existing observing and data networks. The process is further supported by the European Union's Horizon 2020 project INTAROS. In the Indian Ocean, the Indian Ocean Global Ocean Observing System (IOGOOS) aims to implement an integrated basin-wide observing system. GOOS itself builds on global-scale observing networks coordinated under the Joint WMO-IOC Technical Commission for Oceanography and Marine Metrology (JCOMM) observations coordination group, and the efforts of 13 GOOS Regional Alliances.

1.2.2.1 The Framework for Ocean Observing

A key outcome of the OceanObs'09 conference was the realization that the demand for ocean observing was increasing both in quantity and quality. At the same time, more marine sectors were requesting similar information and the need for interpolated ocean information available to a wide range of applications was identified. However, the current design and management of the more sectorial or technology-driven networks seemed in need of reform. Accordingly, the conference set in motion a process to develop a systems-based approach to ocean observing and developed the Framework for Ocean Observing [36]. The objectives of the FOO were to: (1) ensure easy access to ocean information; (2) build on and enhance international, multidisciplinary, and integrated ocean-observing networks and activities; (3) respond to the needs of more-comprehensive ocean system models to enable improved climate and ocean service predictions; and (4) support national and international assessment and policy processes. Thus, the main intention of FOO was to provide strategic guidance to the ocean-observing community and to provide a systematic structure to foster harmonization of observing networks, initiatives, and communities based on existing structures where possible [35].

The FOO calls for the identification of requirements for ocean observing by specifying a finite set of "Essential Ocean Variables" (EOVs). Those will be observed by a finite web of networks taking advantage of existing and new in situ observing platforms, either ship-based or using more autonomous capabilities. The evolution of the networks takes into account the technology readiness level of its components [37]. Mature elements can be implemented at the full and global scale. Lower readiness levels will benefit from pilot implementation and improvements to ensure reliable and sound high-quality performance. Finally, all data need to be quality controlled, assembled, and globally shared and archived in a free, transparent, authoritative, and openly sustained manner.

Thus, the FOO follows a simple user interface in which the *input box* evaluates the requirements essential to address a particular societal issue or scientific challenge, the *process box* refers to the observing as such by global networks, and the *output box* contains data and information flows and their aggregation into products to respond to the requirements for information (Fig. 1.2.2; for details regarding the implementation of the methodology, see also Chapter 7).

Following this approach, integration can be stimulated across disciplines, platforms, and initiatives improving communication within the ocean-observing community, and to stakeholders such as users (e.g., service providers), funders, industry, and others. The FOO aims to avoid any unnecessary duplication and allows for intercalibration and elimination of systemic bias. It encourages adoption of common standards for data collection and flows. In

Framework for Ocean Observing Process Diagram

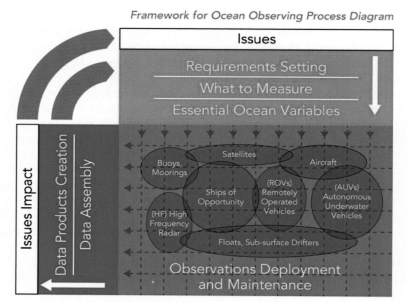

FIGURE 1.2.2   Simple systems (box model) approach of the Framework for Ocean Observation showing the corner parts of the framework (i) input—requirements, (ii) process—several observation initiatives, (iii) output—data and products including the inner and outer evaluation levels.

addition, the continuous involvement of key users (research community, service providers, decision makers, and others), assessing the performance and value of the observing and information system, leads to clear goals and priority requirements for system evolution. FOO process fosters coordination between networks and observing and data systems to optimize the implementation of sustained observation and to establish a compendium of best practices [38].

One first step within FOO is the identification of the EOVs that have to be measured to serve societal requirements to support climate and ocean research and services, real-time services, as well as sustained management of the ocean's state. To participate in a global, integrated, and fit-for-purpose system, the individual observing elements measuring a single EOV need to take responsibility to adopt standards and best practices in methods and data streams (see also Chapter 6). To evaluate whether the observing system is delivering on its requirements, the end-to-end observing processes need to be assessed on two levels: (1) an internal check to trace whether the requirements to observe EOVs are fulfilled by current observing elements and data management arrangements, and (2) an external check to evaluate whether the outputs have the expected scientific and societal impact, information impacts decision-making and policy, and the output is fit for purpose (Figs. 1.2.3 and 1.2.4).

1.2.2.2 The Ocean-Observing Value Chain

Another perspective on systematic ocean observing can be gained by reflecting on its value proposition. A recent report from the Organization for Economic Co-operation and Development (OECD) [39] explores the growth prospects for the ocean economy, its capacity for future employment creation and innovation, and its role in addressing global challenges. The report devotes special attention to the emerging ocean-based industries in light of their high growth and innovation potential and contribution to addressing challenges such as energy security, environment, climate change, and food security. Similar value statements have been given in the context of the GEO "Oceans and Society: Blue Planet" [40] Implementation Plan or the G7 science ministerial declaration on Ocean Observing in Italy [41]. A common thread is to understand ocean observing as a part of a value chain linking observation, research, and innovation to facilitate societal benefits. The most relevant societal benefit areas include: (1) advancement in ocean research, technology, analysis, data management systems, syntheses, and information products; (2) information support for commercial activities and operational services; and (3) advancement in environmental management and security at the national, regional, and global levels.

The base of the value chain (Fig. 1.2.5) is the process of identification of societal needs across many sectors. From those a set of EOVs can be developed taking into account feasibility, impact on information, and implementation costs. Subsequently, data collection, assembly, and quality control provide the grounds for subsequent data fusion,

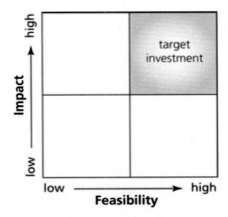

FIGURE 1.2.3   Essential Ocean Variables by feasibility and impact, after [35].

integration, and assimilation. Consequently, this will lead to a flow of a wide range of information products and applications.

Over the last two decades, common standards on data formats, real-time and delay-mode quality control, and data distribution have been developed (e.g., Ref. [42]), and infrastructures have been improved (see also Chapter 6). By these efforts and the focus on EOVs, data collection has become more consistent and comprehensive; however, efficiency of data collection can be enhanced by synthesis of multiplatform approaches, multivehicle operations, and more active integration. Continuous efforts in ocean observing are needed to expand the global geographic coverage based on GCOS and GOOS targets from about 65 to 100%.

The rhythm and frequency of in situ ocean observations compared to satellite observations vary based on the costs involved. Ship-based observations are less frequent than autonomous observations due to their capital-intensive nature. Global ship-based ocean-observing activities are estimated to be just a few more than 100 profiles per year. However, due to the great variety and high accuracy of the ship-based parameters they are justified [43].

Data management is a further essential part of the value chain as data are often distributed among several global repositories and Data Assembly Centers. Analysis and processing of these various data resources benefit from automated, processed, and combined efforts of multiple organizations. Several regional efforts facilitate a more coordinated approach. In the United States as part of the Ocean Archive System, the National Center for Environmental Information maintains the official archives for observational data collected by the IOOS. The Australian Ocean Data Network (AODN) is an interoperable online network of marine and climate data resources that connects the national IMOS program with the six Australian Commonwealth agencies. The European Marine Observation and Data Network (EMODnet) assembles fragmented marine resources (data, products, and metadata) from more than 160 organizations to make it available to users that depend on quality-assured, standardized, and harmonized data that are interoperable and open access. Further examples are Pan-European Infrastructure for Ocean and Marine Data Management (SeaDatanet) that provides data from 36 countries for service providers and the IOC International Oceanographic Data and Information Exchange (IODE) that integrates, archives, and assesses the quality of ocean observation data and best practices of more than 80 oceanographic data centers [44]. Thus, the value chain from ocean observing to societal benefit describes the added value from scientific analysis, assessment, information products, and public marine core services that can be turned into immediate societal value.

Scientific applications of ocean observing are mostly short and long-term forecasts about the ocean, tsunamis, ocean waves, sea–ice interaction, weather, and future climate changes; typical stakeholders are from academia, service providers, and governments. An example of operational applications are weather forecasts to ensure safer and improved outcomes of commercial and economic activities, including fisheries and aquacultures, maritime traffic, and energy industries. Providers for scientific and operational applications might interact with data management centers. The enhanced knowledge about the ocean and climate change supports preparation of society for risks such as rainfall anomalies, wet seasons, droughts, storms, and tsunamis (e.g., Ref. [45]). For the private sector, improved information and better predictions support decisions for commercial operations. In the ocean, several regions are establishing ocean information centers. Several countries around the world are beginning to establish ocean services. In the United States, there is the NOAA National Ocean Service, in Australia

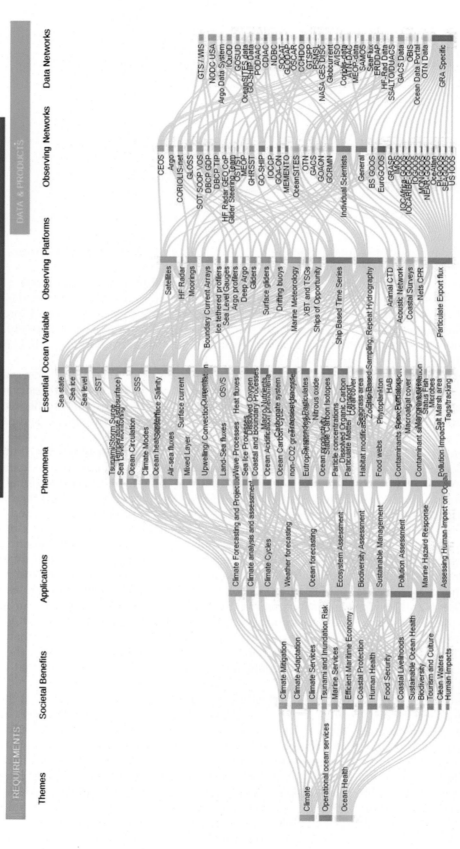

FIGURE 1.2.4   The GOOS strategic mapping tool to visualize the system and how it links to the Essential Ocean Variables identified by the GOOS Expert Panels (http://www.goosocean.org/index.php?option=com_content&view=article&id=120&Itemid=277).

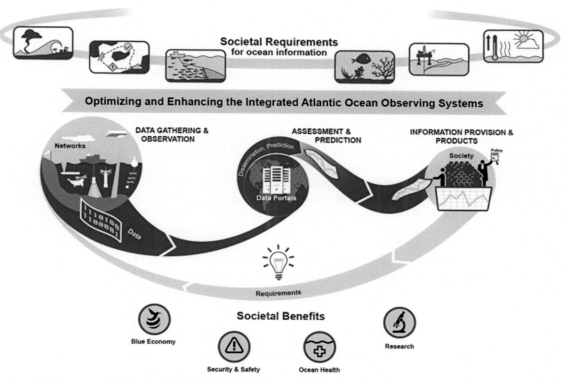

FIGURE 1.2.5 The value chain from the definition of societal requirements, via ocean observation, assessment, and prediction information, and product generation to societal, scientific, and economic benefit.

IMOS/BlueLink, and China the National Marine Environmental Forecasting Center (see GODAE Ocean View https://www.godae-oceanview.org/science/ocean-forecasting-systems for an extensive list). For Europe, the Copernicus Marine and Environmental Monitoring Service (CMEMS) has been designed to support the needs of the environmental, business, and scientific sectors. Using information from both satellite and in situ observations, it provides daily state-of-the-art analyses and forecasts, which offer an unprecedented capability to observe, understand, and anticipate marine environmental events. The CMEMS delivers a core information service to any user related to four areas of benefits, be they service providers or end-users from the commercial sector or from the R&D sector: (1) Maritime Safety, (2) Coastal and Marine Environment, (3) Marine Resources, and (4) Weather, Seasonal Forecasting, and Climate activities.

Currently, the value of ocean information is the reduced uncertainty of a situation that leads to expected benefits or reduced costs compared to the situation when this information was not available, and this added value is based on the enhanced information that is gained by improved forecasts, nowcasts, and real-time information [46]. Accordingly, the improved knowledge supports societal decisions as well as decisions related to commercial activities and investments [47]. Due to the growing and diverse parts involved in the value chain it is difficult to name the values of specific segments in the total benefit of ocean observing. The large initial investment for infrastructures comes mainly from the public sector but is in some cases cofunded by the private sector [50]. A common assumption is that the benefit from ocean-observation information exceeds the costs and might be much higher than expected but currently a robust socioeconomic justification delivering a net present-value figure is missing (e.g., Ref. [48]).

So far, the impacts of ocean information data and applications on decisions are not fully understood and assessed on a quantitative base to allow for a solid assessment of the socioeconomic benefits and impacts. However, recent studies give some hard numbers. A study of US Business Activity in Ocean Measurement, Observation & Forecasting (Ocean Enterprise; [49]) concluded that the overall revenue for businesses that provide Ocean Enterprise-related products and services as part of their activity is estimated at $ (US) 58 billion. About 25% of this overall revenue is estimated to be from Maritime-related activities. Total employment is estimated 250,000 employees in the United States. In addition, globally, the sustainable growth and management of blue/ocean economics are currently valued at $ (US) 1.5 trillion and expected to double by 2030 (GEO Strategic Plan 2016–25: Implementing Global Earth Observing System of Systems [GEOSS] reference document).

1.2.3 SUMMARY

Over the last decades, there have been substantial improvements in ocean observation capabilities for sensors, platforms, and information management. This is leading to more-comprehensive coverage in observations not only spatially but also across physical, chemical, and biological ocean and ecosystem characteristics. As the observations become more widespread with greater societal impact, there is increasing need for coordination within and across end-to-end systems. This is done, in part, through the Framework for Ocean Observing (FOO), which lays out a strategy for the development of systematic and sustained observation of the ocean including uniform approaches, recommended standards, and best practices. However, for a number of practical considerations, implementation of ocean observation is often best pursued at the basin scale. Good examples are the SOOS in the Southern Ocean, TPOS2020 for the Tropical Pacific, or AtlantOS for the Atlantic Ocean.

The global Framework for Ocean Observing will provide a global ambition and perspective and stimulate partnerships between research and operational communities and among actors of basin-scale networks to enhance and improve the readiness level of observing technology and methods as well as the data system and assessment approach for each EOV. Hence, among the various independent observing initiatives and elements, the FOO will foster an integrated, consistent, and comprehensive approach to assess the readiness, harmonize information and product sharing, and identify best practices. For a more integrated and fit-for-purpose ocean-observing system, one of the main goals is to enhance the efficiency and capability of all observing networks with a particular focus on strengthening the international partnerships within each of the networks to cover the global ocean.

References

[1] Global Climate Observing System GCOS. The global observing system for climate: implementation needs. 2016. https://doi.org/10.13140/RG.2.2.23178.26566. Report number: 200.
[2] Visbeck M, Lindstrom E, Fischer A. Introduction to CLIVAR exchanges issue for sustained ocean observing and information in support of ocean and climate research. CLIVAR Exch 2015;19(2):2–3. No. 67.
[3] Gould J, Sloyan B, Visbeck M, In-situ ocean observations: a brief history, present status and future directions. In: Siedler G, Griffies SM, Gould J, Church, JA, editors. Ocean circulation and climate: A 21st century perspective. International geophysics series, vol. 103. Oxford, GB: Academic Press; 2013 pp. 59–82. ISBN 978-0-12-391851-2.
[4] Visbeck M, Kronfeld-Goharani U, Neumann B, Rickels W, Schmidt J, van Doorn E, Matz-Lück N, Ott K, Quaas M. Securing blue wealth: the need for a special sustainable development goal for the ocean and coasts. Mar Policy 2014;48:184–91. https://doi.org/10.1016/j.marpol.2014.03.005.
[5] Wyville Thomson C, Murray J. The voyage of H.M.S. challenger 1873–1876. Narrative Vol. I. First part. Ch. III. Johnson Reprint Corporation; 1885. Available at: http://archimer.ifremer.fr/doc/00000/4751/.
[6] Winkler l. Die Bestimmung des in Wasser Gelösten Sauerstoffes. Ber Dtsch Chem Ges 1888;21(2):2843–55.
[7] Wüst G. Schichtung und Zirkulation des Atlantisches Ozeans. Das Bodenwasser und die Stratosphäre. Wissenshaftliche Ergebnisse der Deutschen Atlantik Expedition 'Meteor'. 1925–1927, 6: 288 p., Berlin. 1935.
[8] Spilhaus A. A bathythermograph. J Mar Res 1938;1:95–100.
[9] Swallow JC. A neutral-buoyancy float for measuring deep currents. Deep Sea Res 1955;3:74–81.
[10] Crease J. Velocity measurements in the deep water of the western north Atlantic, summary. J Geophys Res 1962;67:3173–6.
[11] Park K, Burt WV. Electroytic conductance of sea water and the salinometer (Part2). J Oceanogr Soc Jpn 1965;21(3):124–32.
[12] Koblinsky CJ, Bernstein RL, Schmitz Jr WJ, Niiler PP. Estimates of the geostrophic stream function in the western North Pacific from XBT surveys. J Geophys Res 1984;89(C6):10,451–60. https://doi.org/10.1029/JC089iC06p10451.
[13] Talley LD, White WB. Estimates of time and space scales at 300m in the mid-latitude North Pacific from the TRANSPAC XBT program. J Phys Oceanogr 1987;17:2168–88.
[14] Donlon CJ, Minnett PJ, Gentemann C, et al. Toward improved validation of satellite sea surface skin temperature measurements for climate research. J Clim 2002;15:353–69.
[15] Reynolds RW. A real-time global sea surface temperature analysis. J Clim 1988;1:75–86.
[16] Freeland, Gould. Objective analysis of mesoscale ocean circulation features. Deep Sea Res 1976;23:915–23.
[17] Kamenkovich VM. Synoptic eddies in the ocean. Kluwer; 1986. ISBN: 90-277-1925-X. 435 p.
[18] McNally G, Patzert W, Kirwan AD, Vastano A. The near-surface circulation of the North Pacific using satellite tracked drifting buoys. J Geophys Res 1983;88(C12). https://doi.org/10.1029/0JGREA000088000C12007507000001. 0148–0227.
[19] Richardson P. Eddy kinetic energey in the North Atlantic from surface drifters. J Geophys Res 1983;88:4355–67.
[20] Garrett JF. The availability of the FGGE drifting buoy system data set. Deep Sea Res 1980;27:1083–6.
[21] McPhaden MJ, et al. The Tropical Ocean-Global Atmosphere observing system: a decade of progress. J Geophys Res 1998;103:14,169–240.
[22] Thompson BJ, Crease J, Gould J. The origins, development and conduct of WOCE. In: Siedler G, Church J, Gould J, editors. Ocean circulation and climate. Academic Press; 2001.
[23] King BA, Firing E, Joyce TM. In: Siedler G, Church J, Gould J, editors. Shipboard observations during WOCE in ocean circulation and climate. 1st ed. Academic Press; 2001.
[24] Davis RE, Webb DC, Regier LA, Dufour J. The autonomous, lagrangiam circulation explorer (ALACE). J Atmos Ocean Technol 1992;9:264–85.
[25] Fischer J, Visbeck M. Deep velocity profiling with self-contained ADCPs. J Atmos Ocean Technol 1993;10:764–73.

[26] Wong APS, Riser SC. Profiling float observations of the upper ocean under sea ice off the Wilkes Land Coast of Antarctica. J Phys Oceanogr 2011;41:1102–15. https://doi.org/10.1175/2011JPO4516.1.

[27] Timmermans M–L, Krishfield R, Laney S, Toole J. Ice-tethered profiler 1111 measurements of dissolved oxygen under permanent ice cover in the Arctic ocean. J Atmos Ocean Technol 2010;27:1936–49.

[28] Boehme L, Meredith MP, Thorpe SE, Biuw M, Fedak M. Antarctic Circumpolar Current frontal system in the South Atlantic: monitoring using merged Argo and animal-borne sensor data. J Geophys Res Oceans 2008;(113):C09012. https://doi.org/10.1029/2007JC004647.

[29] Siedler G, Church J, Gould J. Ocean circulation & climate-observing and modelling the global ocean. In: International gepophysics series number 77. San Diego: Academic Press; 2001. 715 p.

[30] Koblinsky CJ, Smith NR. Observing the oceans in the 21st century – a strategy for global ocean observations. Australian Bureau of meteorology; 2001. ISBN: 0642 70618 2. 604 p.

[31] Gould WJ, Toole JM, co-authors. Investigating ocean climate variability: the need for systematic hydrographic observations within CLIVAR/GOOS. In: Koblinsky C, Smith N, editors. Observing the oceans in the 21st century. Melbourne: Bureau of Meteorology; 2001. p. 259–84.

[32] Hood EM, Sabine CL, Sloyan BM, editors. The GO-SHIP repeat hydrography manual: a collection of expert reports and guidelines. 2010. IOCCP Report Number 14, ICPO Publication Series Number 134. Available online at: http://www.go-ship.org/HydroMan.html.

[33] Send U, Weller R, Cunningham S, Eriksen C, Dickey T, Kawabe M, Lukas R, McCartney M, Østerhus S. In: Koblinsky C, Smith N, editors. Oceanogrpahic timeseries observations. From: observing the oceans in the 21st century. Melbourne: Bureau of Meteorology; 2001. p. 259–84.

[34] Hall J, Harrison DE, Stammer D, editors. Proceedings of OceanObs'09: sustained ocean observations and information for society, Venice, Italy, 21–25 September 2009. ESA Publication; 2010. https://doi.org/10.5270/OceanObs09. WPP-306.

[35] Fischer A. A framework for ocean observing. CLIVAR Exch 2015;19(2):3–7. No. 67.

[36] A Framework for Ocean Observing. By the task team for an integrated framework for sustained ocean observing, UNESCO 2012, IOC/INF-1284. 2012. https://doi.org/10.5270/OceanObs09-FOO.

[37] Mankins JC. Technology readiness levels. In: A white paper. Washington, DC: NASA; 1995.

[38] Pearlman JS, Buttigieg PL, Simpson P, Munoz Mas C, Heslop E, Hermes J. Accessing exiting and emerging best practices for ocean observation. In: Pearlman JS, Simpson P, Mas CM, Heslop E, Hermes J, editors. OCEANS 2017 – Anchorage. 2017. ISBN: 978-0-6929-4690-9. p. 1–7. Anchorage (AK, USA).

[39] OECD. The ocean economy in 2030. Paris: OECD Publishing; 2016. https://doi.org/10.1787/9789264251724-en.

[40] GEO. Oceans and society: blue planet. 2017 – 2019. Implementation Plan, available at: https://geoblueplanet.org/wp-content/uploads/2016/10/Blue_Planet_Impt_Plan_Sept21_2016.pdf.

[41] G7 science ministerial declaration on Ocean Observing in Italy. 2017. Available at: http://www.g7italy.it/sites/default/files/documents/ANNEX%201_WG%20Future%20of%20the%20Seas%20and%20Oceans.pdf.

[42] Harscoat V, Pouliquen S, AtlantOS WP7 partners. Data management handbook AtlantOS. 2016. https://doi.org/10.13155/48139.

[43] Roemmich D, et al. Integrating the ocean observing system: mobile platforms. 2009. Available at: http://www.oceanobs09.net/proceedings/pp/4A1-Roemmich-OceanObs09.pp.33.pdf.

[44] IODE. IODE is… 2016. Available at: http://www.iode.org/index.php?option=com_content&view=article&id=385&Itemid=34.

[45] Hoerling M, Schubert S. Oceans and drought. 2009. Available at: http://www.oceanobs09.net/proceedings/pp/5A2-Hoerling-OceanObs09.pp.22.pdf.

[46] Kite-Powell HL, Colgan CS. Estimating the economic benefits of regional ocean observing systems. NOPP, Marine Policy Centre, Woods Hole Oceanographic Institute; 2005. 35 p.

[47] Kruse J, Crompvoets J, Pearlman F. GEOValue, the socioeconomic value of geospatial information. CRC Press; 2017. ISBN: 9781498774512. 332 p.

[48] Flemming. SEPRISE Sustained, Efficient Production of Required Information and Services within Europe - SEPRISE SOCIO-ECONOMIC ANALYSIS: SCOPING REPORT A report for EuroGOOS within the SEPRISE Project, Report funded by the Sixth European Commission Framework Programme. 2007.

[49] IOOS. The ocean enterprise – a study of US business activity in ocean measurement, observation and forecasting, prepared by ERISS corporation the Maritime Alliance. 2017. https://ioos.noaa.gov/wp-content/uploads/2017/04/oceanenterprise_feb2017_secure.pdf.

[50] Nolan G. EuroGOOS - How have selected ocean observatories been evaluated so far? Presentation for AtlantOS/OECD workshop, Kiel. June 2016.

Further Reading

[1] UN. Summary of the first global integrated marine assessment, United Nations General Assembly. July 22, 2015. A/70/112.

Glossary

ADCP Acoustic Doppler Current Profiler
AODN Australian Ocean Data Network
AUV Autonomous Underwater Vehicles
CLIVAR Climate and Ocean: Variability, Predictability, and Change, is one of four core projects of the World Climate Research Program
CTD Instruments measuring Conductivity, Temperature, and Depth
EMODnet The European Marine Observation and Data Network
ENSO El Niño–Southern Ocean
EOVs Essential Ocean Variables
FGGE First Global GARP Experiment

FOO Framework for Ocean Observing, www.oceanobs09.net/foo

GARP The Global Atmospheric Research Program

GCOS Global Climate Observing System; Implementation plan: https://ane4bf-datap1.s3-eu-west-1.amazonaws.com/wmocms/s3fs-public/programme/brochure/GCOS-200_OnlineVersion.pdf?PlowENiCc1RGh9ReoeAoGBT0QhnJYm6_

GEO Group on Earth Observation

GEO-BON Group on Earth Observations Biodiversity Observation Network

GEOSS Global Earth Observing System of Systems

GFOFS Joint Global Ocean Flux Study

GHRSST GODAE High-Resolution Sea Surface Temperature Consortium, www.ghrsst.org

GODAE The Global Ocean Data Assimilation Experiment

GOOS The Global Ocean Observing System, www.goosocean.org

GO-SHIP The Global Ocean Ship-based Hydrographic Investigations Program, www.go-ship.org

GPS Global Positioning System

ICES The International Council for the Exploration of the Sea

ICSU The International Council for Science

INTAROS Integrated Arctic Observation System

IOC Intergovernmental Oceanographic Commission

IOCCP International Ocean Carbon Coordination Project

IODE The IOC International Oceanographic Data and Information Exchange

IOGOOS The Indian Ocean Global Ocean Observing System

IOOS Integrated Ocean Observing System

IPCC Intergovernmental Panel on Climate Change

JCOMM The Joint WMO–IOC Technical Commission for Oceanography and Marine Metrology

MSFD Marine Strategy Framework Directive, 2008: http://data.europa.eu/eli/dir/2008/56/oj and 2017 revised Commission Decision on Good Environmental Status: http://data.europa.eu/eli/dec/2017/848/oj and amended Annex III of the Directive: http://data.europa.eu/eli/dir/2017/845/oj

NetCDF Network Common Data Format

OceanObs International conference taking place every 10 years evaluating major observational programs

OceanSITES A worldwide system of deep-water reference stations, www.oceansites.org

OECD Organization for Economic Co-operation and Development

OOI Ocean observation initiative

pCO$_2$ Partial pressure of carbon dioxide

PICES North Pacific Marine Science Organization

SAON Sustaining Arctic Observing Networks

SeaDatanet Pan-European Infrastructure for Ocean & Marine Data Management

SDG Sustainable Development Goal; no 14 "Life below Water," https://sustainabledevelopment.un.org/sdg14

SOCCOM Southern Ocean Carbon and Climate Observations and Modeling

SOOS Southern Ocean Observing Systems

SST Sea surface temperature

TOGA Tropical Ocean Global Atmosphere

TAO Tropical Atmosphere Ocean, http://www.pmel.noaa.gov/tao/global/global.html

TPOS2020 The Tropical Pacific Observing System

TRITON Triangle Trans-Ocean Buoy Network

UN United Nations (UN Agenda 2030), http://www.un.org/ga/search/view_doc.asp?symbol=A/RES/70/1&referer=http://www.un.org/sustainabledevelopment/development-agenda/&Lang=E

UNEP UN Environment Program

UNESCO United Nations Educational, Scientific, and Cultural Organization

WCRP World Climate Research Program

WHP The WOCE Hydrographic Program

WMO The World Meteorological Organization

WOCE World Ocean Circulation Experiment (1990–2002)

XBT eXpendable BathyThermograph

2

Ocean In Situ Sensors: New Developments in Biogeochemistry Sensors

2.1

An Autonomous Optical Sensor for High Accuracy pH Measurement

Hervé Precheur[1], Eric Delory[2]

[1]Sensorlab, Gran Canaria, Spain; [2]Oceanic Platform of the Canary Islands (PLOCAN), Telde, Spain

O U T L I N E

2.1.1 INTRODUCTION

Ocean acidification is known to affect the growth of entire groups of species that are basic to the trophic chain, such as some zooplankton and shellfish larvae, and as such is an increasing threat for the entire ecosystem [1]. Ocean acidity is also directly related to the absorption of carbon dioxide (CO_2), a major greenhouse gas and one of the key variables of the carbon cycle equation in the ocean [2]. As the concentration of CO_2 in the atmosphere has increased with the advent of the industrial age, the ocean continues to play an important role in absorbing it and reducing the impact on global warming, yet at the cost of increasing ocean acidity. There are, therefore, several reasons why monitoring acidity in the ocean is essential, and several initiatives are now in place to ensure pH measurements are made on a continuous basis. For example, the Global Carbon Observing System (GCOS) provides guidance for the coordination of regional carbon-observing systems (e.g., Integrated Carbon Observation System [ICOS]), whether terrestrial, atmospheric, or oceanic, acidity being an essential variable. Tracking the decrease of ocean pH requires high-precision measurements, with sensitivity of the order of a thousandth of a pH unit (mpH) [3]. Reaching such precision is highly challenging, in particular because the stability and accuracy of most sensing techniques depend on the accurate measurement of other variables (e.g., temperature, as illustrated later). When it comes to the development of autonomous systems, additional constraints are added, like compactness, low energy consumption, and low maintenance. This chapter is engineering focused, and the authors recommend consulting the extensive scientific literature for more background on ocean carbon system science, e.g., Refs. [1–3]. The objective of the chapter is to focus on a practical solution based on optics that has been deployed successfully on moorings in the Atlantic and the Mediterranean. The system presented in this chapter has demonstrated its reliability as a high-accuracy solution for open-ocean monitoring. A detailed description of concepts of operation is provided, followed by results and analysis of future work, such as the emerging need for multifunctional capabilities.

2.1.2 CONCEPT OF OPERATION

The general system operation is based on m-cresol purple reagent addition to the water sample. When the reagent is added to the sample, its color is modified depending on the original pH value of the sample. However, adding a reagent to the sample modifies the pH value of the original sample. This requires a correction. A relatively simple solution is performing several additions and performing a least squares procedure. This procedure will allow the calculation of the original pH value, before the reagent was added.

The m-cresol reagent was selected because the measurement range it allows (7–8.5) matches well with the pH found in seawater. The system uses optical filters to determine each wavelength peak, so it is not straightforward to use reagents requiring different wavelengths.

2.1.2.1 Optical Chain

The optical subsystem consists of a wide-band light-emitting diode (LED) light (OSLON® Solid-State Lighting [SSL] family), controlled via a constant-current source.

FIGURE 2.1.1   Photodetectors assembly.

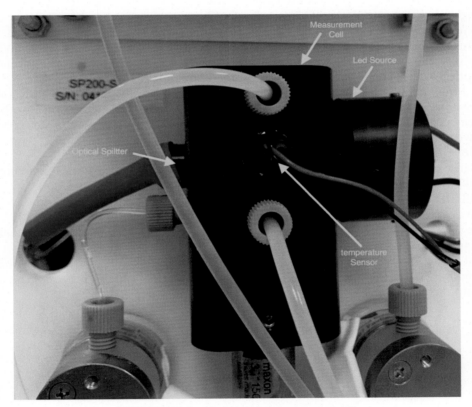

FIGURE 2.1.2   Optical chain, measurement cell.

The wide-band light source is guided into the measurement cell (Fig. 2.1.2), in which the water sample is mixed with the reagent. The light then passes through the mix of water and reagent. The spectral absorbance of the fluid will be measured using an array of custommade optical filters, photodetectors (Fig. 2.1.1), and digitizers.

2.1.2.2 Absorbance Measurement

The pH calculation is performed measuring two absorbance peaks of the m-cresol purple reagent, one at 434 nm and the second at 578 nm. Two additional absorbances are measured, one is the 487-nm isosbestic point (the wavelength at which a reagent does not change with pH, but only depends on the amount of reagent added to the solution). The last absorbance that needs to be measured is at 730 nm, at which the m-cresol purple does not react and is used to verify the baseline.

So finally, we have totally four absorbance wavelengths required to calculate pH

434 nm: First wavelength peak for the m-cresol purple
487 nm: Isosbestic value for the m-cresol purple
578 nm: Second wavelength peak for the m-cresol purple
730 nm: Wavelength for baseline monitoring

The equation used to calculate the absorbance is the following:

$$A_\lambda = -\log_{10}\left(\frac{S_\lambda - D_\lambda}{R_\lambda - D_\lambda}\right).$$

In which: D = Dark measured signal at wavelength λ. It is the output value of the photodetector, when the light source is turned off. This dark intensity is usually due to noise in the photodetector chain. R = Reference measured signal at wavelength λ. It is the output value of the photodetector, with the light source turned on and stabilized, and the initial water sample before adding any reagent. S = Sample measured signal at wavelength λ. It is the output value of the photo detector, with the light source turned on and stabilized after the reagent has been added to the sample.

To get precise and stable absorbance measurements it is important to optimize the optical chain to receive the highest possible light in the photodetector. This is important because a large absorbance implies that the light that reaches the photodetectors has been attenuated by a big factor.

There are various parameters that can be adjusted to reach this goal. The first and most obvious is getting the strongest possible luminous flux from the LED source. The second is trying to optimize the optical transmission path, to minimize the optical losses. The last parameter is the integration time of the photodetector.

The most critical wavelengths are 434 and 578 nm, because they are directly used for pH calculation. The m-cresol purple allows measuring pH values between 7 and 8.5. A pH of 7 will have a strong 434-nm absorbance peak, but the 578-nm absorbance peak will be very weak. A pH of 8.5 will show the opposite trend, a strong 578-nm absorbance peak and a very weak 434-nm absorbance peak. Typical absorbance values for 434 and 578-nm wavelengths vary from 0.1 up to 2 absorbance units (AUs; Fig. 2.1.3).

Stability in the absorbance value is critical; ideally, the overall absorbance variations due to noise should be held below 5 mAUs (0.005 AUs), to obtain good pH determination.

The analog-to-digital converters (ADCs) used to digitize the photodetector outputs are 16 bits wide, which quantize the signal into a digital value that ranges from 0 to 65,536 steps.

The light source is adjusted so that the photodetector receives the maximum light possible, usually between 50,000 and 60,000 counts (some margin is necessary to avoid saturating the ADC).

The typical total noise present at the photodetector is in the order of ±5 counts Root-Mean-Square (RMS).

As can be seen in Fig. 2.1.3, absorbances greater than 1 can lead to important errors (>5 mAUs) when the photodetector reference output counts are below 45,000.

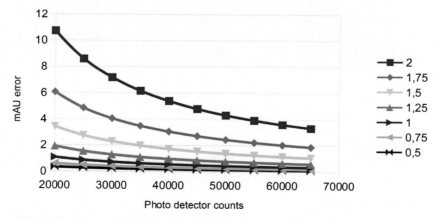

FIGURE 2.1.3   Error in mAUs for absorbances from 0.5 to 2 for reference counts from 20,000 to 65,000.

2.1.2.3 Light Source

The light source used is a wideband LED, that covers roughly from 420 to 750 nm (see Fig. 2.1.5 and Table 2.1.1, for response values).

It can be clearly seen that, although it is a wideband spectrum, the amplitude in the 434–730 nm band has an important variation.

The relative amplitudes in this particular case are presented in Table 2.1.1.

In the case of the pH sensor, the most critical wavelengths are the m-cresol purple absorbance peaks at 434 and 578 nm. For a given luminous flux, the 578 nm photodetector will receive twice as much light as the 434 nm photodetector (Table 2.1.1).

The 487-nm wavelength is the isosbestic point; the typical absorbances are relatively low, always below 1 AU. It is acceptable then to have relatively low count numbers for the reference value.

The 730-nm wavelength is used for baseline correction and will not show any absorbance. The reference counts can be relatively low, as the absorbances will always be in a very tight range around ±0.05 AUs. The errors introduced here by low reference counts are negligible.

The amount of light received by the photodetector depends on the luminous flux, as well as in the optical losses.

As we have seen previously, there is a relatively important imbalance in the luminous flux, going from a 5% relative flux for the 730-nm wavelength, to a 97% for the 578-nm wavelength.

2.1.2.4 PhotoDetectors

The photodetector model used is a TSL2581 from AMS Technologies. The integration time for each photodetector can be configured independently in a range between 173 and 688 ms. Doubling the integration time will double the number of counts that will be read by the ADC, for the same received luminous flux.

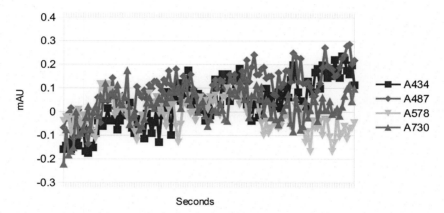

FIGURE 2.1.4 Stability of reference absorbances in a 120 s period, 60,000 reference counts, and 40 mA LED current.

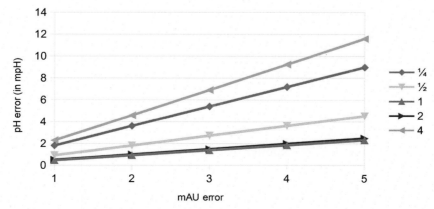

FIGURE 2.1.5 pH error in milli pH (mpHs) units, versus the error in mAUs for different absorbance ratios, from 1/4th to 4.

TABLE 2.1.1 Light Source Relative Amplitude per Wavelength

Wavelength (nm)	Relative Amplitude (%)
434	47
487	32
578	97
730	5

The system automatically balances the counts read by the photodetectors digitizers, configuring different integration times for each photodetector, and by adjusting the LED output luminous flux.

The optical path is configured before each new pH measurement. Once the measure cell has been cleaned and filled with sample water, the system starts an optical configuration procedure.

The luminous flux of the LED source is configured to reach the 60,000-count value using the minimum integration time for the 578 nm photodetector.

Once the luminous flux has been set, all the other photodetector integration times are calculated to reach 60,000 counts.

Then the system stores the count values as the reference for all the absorbances that will be measured during the pH measurement process.

Once the reference value has been stored, it is crucial that the LED source, as well as the photodetectors, is stable to <1 mAU during the full-measure process. The typical measure process takes less than 1 min.

From Fig. 2.1.4, it can be seen that the reference absorbances are stable at less than ±0.5 mAUs for a 2 min period.

2.1.2.5 pH Calculation

The equation used to calculate a pH value is the following:

$$pH = pK_2 + \log_{10}\left(\frac{A_{578}/A_{434} - \varepsilon_1\left(HI^-\right)/\varepsilon_2\left(HI^-\right)}{\varepsilon_1\left(I^{2-}\right)/\varepsilon_2\left(HI^-\right) - \left(A_{578}/A_{434}\right)\varepsilon_2\left(I^{2-}\right)/\varepsilon_2\left(HI^-\right)}\right)$$

In which: A_{578}/A_{434} Is the ratio between the absorbances at $\lambda = 578$ nm and $\lambda = 434$ nm.

The 730 nm absorbance is used to correct the absorbance of all the other wavelengths, as it is considered constant. So in fact, the 578/434 absorbance ratio can be better described as:

$$\frac{A_{578} - A_{730}}{A_{434} - A_{730}}$$

The terms:

$$\varepsilon_1\left(HI^-\right)/\varepsilon_2\left(HI^-\right) = 0.00691$$
$$\varepsilon_1\left(I^{2-}\right)/\varepsilon_2\left(HI^-\right) = 2.222$$
$$\varepsilon_2\left(I^{2-}\right)/\varepsilon_2\left(HI^-\right) = 0.1331$$

are the extinction coefficients.

The extinction coefficients are constants that are dependent on the reagent used and the optical characteristics of the measurement system. It is important to note that the different m-cresol purple manufacturers provide slightly different reagents; this means that the extinction coefficients are different for each m-cresol purple brand.

And the term:

$$pK_2 = \frac{1245.69}{T/K} + 3.8275 + 0.0021\left(35 - S\right)$$

in which: T/K = Temperature in Kelvin, S = Salinity.

As stated previously, the absorbance calculation has to be accurate to obtain a good pH calculation (Fig. 2.1.5). The following graph shows the theoretical error with 1–5 mAUs error, for absorbance ratios of ¼ to 4.

It can be seen that absorbance errors of 5 mAUs can lead to pH errors of more than 8 mpHs units, when the absorbance ratios are high.

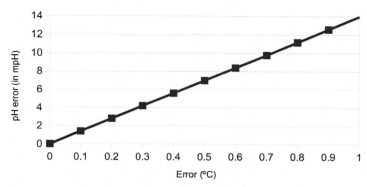

FIGURE 2.1.6   pH error (in mpHs) versus temperature measurement error (°C).

2.1.2.5.1 Temperature Dependance

The pK_2 term is strongly dependent on temperature. The following graph shows the error in mpH versus temperature.

As can be seen in Fig. 2.1.6, a 0.2°C error leads to an error of around 3 mpH units. An error of 0.5°C leads to an error of 7 mpH units.

A 0.5°C error may seem relatively high in a static environment, but when measuring sample waters with a different temperature than the measurement cell, it is easy to reach temperature differences of even some degrees. It is important then to minimize the temperature difference between the water sample and the measurement cell.

The temperature sensor inside the cell allows recording the temperature inside the measurement cell.

This works well for static conditions, such as in a submarine sensor. However, if the sample temperature is different by more than 2–3°C from the measurement cell, the temperature sensor may not be able to calculate the real temperature within the cell, as rate of temperature change within the measurement cell is difficult to determine.

2.1.2.6 Measurement Process

The core of the pH sensor is the measurement cell, in which the dye is injected, mixed with the water sample, and stirred. There is a temperature sensor inside the cell, which is required to measure the temperature of the sample.

The measurement process is as follows:

The first step is cleaning. The input and output solenoid valves are first opened, and the pump starts to operate. This purges the measurement cell and fills it with the water sample to be measured. The cleaning process time is user configurable, and will depend on the deployment conditions.

All system tubing is made of ethylene/tetrafluoroethylene copolymer (ETFE), which has proven quite effective against biofouling.

Then the injection process starts. The dye container is kept at the outside pressure (the same as the water sample). This allows the dye to be easily injected into the measurement cell, as pumping creates a low-pressure condition inside the cell.

The dye solenoid valve is opened for a short period (the order of magnitude is in the tens of milliseconds) and the reagent is injected into the measurement cell (as the pressure of the cell is lower than the sample).

Then the sample is stirred, and the first pH value is calculated.

Then a second addition is performed, the sample is stirred, and the second pH value is calculated.

This process is repeated several times, the best results are obtained when performing between 5 and 10 additions. The exact amount of dye injection in each step is not critical, as long as the whole set of additions have enough dynamic range (typically isosbestic values ranging from 0.1 to 0.5).

Each pH value will have a different temperature, as power dissipated by all of the measurement system (electronics, solenoid valves, pump, and stirrer) will tend to elevate gradually the sample temperature by typically 0.5°C.

Once all the pH values have been calculated, each pH point needs to be normalized to the same temperature. This is performed using the equation:

$$\text{pH}i_{T0} = \text{pH}i_{Ti} - 1.582E^{-2}\,(T_i - T_0)$$

(cf. Millero et al., 2007: $dpH/dT = -1.582E^{-2}$) All temperatures are in °C.

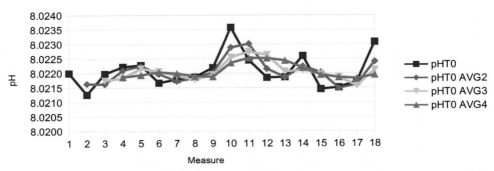

FIGURE 2.1.7   Short-term precision with different averaging values.

The reference temperature is taken at the start of the pH measurement process, as this is supposed to be the initial temperature.

Once all the pH values have been normalized to the same temperature, what we obtain is a vector of pH values. Each pH value has been modified by the amount of reagent used to calculate each point.

The last step consists of obtaining the real pH value, removing the reagent effect. This is performed with a least squares fit, representing each calculated pH value on the Y axis, and the amount of reagent used on the X axis. The previous chart shows 10 pH values measured with isosbestic values between 0.14 and 0.45, corresponding to 10 reagent additions.

The real pH value with the effect of the reagent removed is determined when x = 0. In this particular case, the pH value is 8.06164.

2.1.2.7 Short-Term Stability

Fig. 2.1.7 displays a series of 18 pH values, measured every minute. The original data from the sensor is the pHT0 line, and the other lines are rolling averages of two, three, and four samples.

It can be seen that the short-term stability with the raw data out of the sensor is better than 3 mpHs peak to peak. Using a rolling average of two samples, the stability is better than 2 mpHs, and averaging four samples, the stability is 1 mpH peak to peak.

2.1.2.8 Operation Power and Data Storage

The typical average power consumption during the measurement process is 1.8 W at 12 V. The typical energy required to perform a measurement is 28 mWHs.

The system can be switched to hibernation mode, in which the power consumption is reduced to less than 1 mW at 12 V. The system can be programmed to perform periodic measurements autonomously, and kept in hibernation between measurements.

The storage memory is an industrial-grade microstorage device (SD) card, with 1 Gigabite (GB) capacity. This allows more than 50.000 measurements.

2.1.2.9 Sensor Status and Assembly Performance

The previous methods have been integrated in a sensor assembly that is now available commercially (see Fig. 2.1.8). The design allows reaching the same accuracy achievable in the lab, but autonomously deployed in the ocean, allowing long-term measurements without human intervention. Its construction guarantees precise and stable measurements in open waters with heavy swell. The system includes a real-time clock allowing autonomous operation, as well as a high-precision internal temperature sensor. It can be connected to an external conductivity/temperature sensor for deployments where salinity variations occur (Fig. 2.1.10). The system can be equipped with an internal rechargeable Li-ion battery allowing fully autonomous operation for periods as long as one full year. The complete measurement process takes less than 1 min. The sensor has been successfully deployed on several platforms, including a wave glider (Fig. 2.1.9), which had inherent limitations in payload volume and weight. Comparison with other high-accuracy sensor systems on the market reveals one true competitor in terms of accuracy, also based on reagent, from Sunburst Sensors (Table 2.1.2). A nonreagent transistor-based method (Seafet) shows excellent response time, with yet one-tenth of the precision obtained with the Sensorlab SP200 SM and Submersible Autonomous Moored

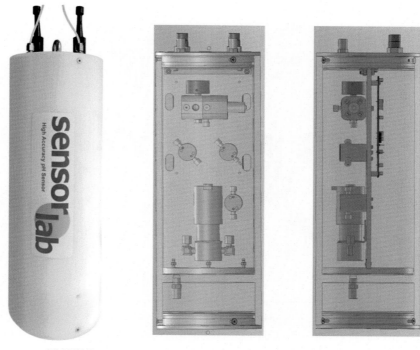

FIGURE 2.1.8   Sensorlab SP200-SM pH sensor with cutaway views.

FIGURE 2.1.9   pH sensor integrated below a MARUM SV3 Liquid Robotic wave glider, off PLOCAN facilities.

Instrument (SAMI)-pH from Sunburst Sensors. Although the Sensorlab system shows higher response time, the latter is qualified for deeper waters at the time of writing.

2.1.3 FUTURE DIRECTIONS

With an increasing number of ocean processes and variables to be monitored, the need to optimize resources leads to the design and development of multifunctional sensor systems that can be deployed on a diversity of platforms, with a preference on those platforms that have proven cost-effective, such as vessels of opportunity and surface or underwater autonomous vehicles. This will imply greater levels of integration and miniaturization, as well as the

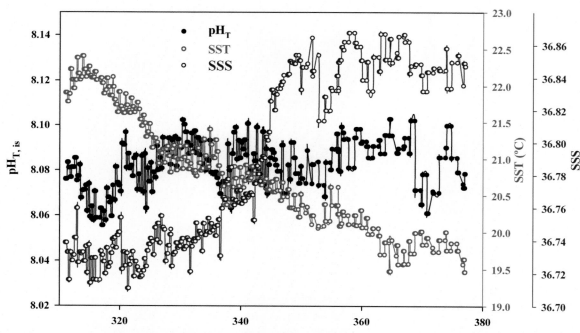

FIGURE 2.1.10   Example of pH measurement in the field with varying temperature and salinity. x-axis in days and y-axis temperature-normalized pH (Sensorlab & ULPGC - González-Dávila, M).

TABLE 2.1.2   Information Collected From Manufacturers Websites

	SP200 SM	SAMI-pH	Seafet
Manufacturer	Sensorlab	Sunburst sensors	Seabird
pH range	7–8.5	7–9	6.5–9
Salinity Range (PSU)	30–37	25–40[b]	20–40
Response time	55 s	3 min	100 ms (10 Hz)
Accuracy	0.005	0.003	0.05
Precision (pH units)	0.002	<0.001	0.004
Typical indicator life (reagent)	5000 measurements	5600 measurements[c]	Not applicable
Long-term drift	<0.005 over 1 year	<0.001 over 6 months	0.003 over 1 month
Temperature sensor accuracy	±0.1°C (−5 to +35°C)	0.1, ±0.01°C	NP
Dimensions (housing length, diameter)	52, 15 cm	55, 15.2 cm	55, 11.4 cm
Weight in air/seawater	5/−1.4 kg	7.6/1.1 kg	5.4/0.1 kg
Host interface	RS232, USB, OGC PUCK[a]	RS232	RS232, USB
Average measurement power	1.8 W (@12 V)	NP	340–400 mW
Depth	10 m	600 m	50 m

NP, not provided. SeaFet is different from the other two in that it uses a Field-Effect Transistor (FET) as sensing element.
[a]*Implemented during the NeXOS project.*
[b]*Extended range possible.*
[c]*Estimated from manufacturer's page.*

capability to pair different sensor systems in single pressure vessels, in which electromechanical components and capabilities can be shared and different data can be simultaneously acquired and transmitted for solving complex questions. This is a typical need in ocean carbon cycle and carbonate chemistry studies, in which the monitoring of pH and pCO_2 or alkalinity are required to get a complete understanding of carbon-related processes in the ocean. Such a development took place in the framework of the NeXOS project, in which pH and pCO_2 measurements could be paired on ferry routes and from a surface autonomous vehicle [4]. Higher technology readiness (technology readiness level [TRL] 9) is yet to be reached to see such systems in sustained operations, and, as for many new technologies, this will also depend on the demand for multifunctional sensors. System integrators are also likely to propose modular solutions when different sensor types can be mounted on common control systems, as this has happened in the field of physical oceanography. Although pairing pH and pCO_2 measurements was possible on an autonomous surface vehicle in NeXOS, integrating an alkalinity sensor to reduce uncertainty is not yet possible at this stage. This constitutes an important path for future innovations, including the consequent reduction of power and size.

Glossary

GCOS Global Climate Observing System.
GCOS Essential Variables (Oceans) Surface: Sea-surface temperature, Sea-surface salinity, Sea level, Sea state, Sea ice, Surface current, Ocean color, Carbon dioxide partial pressure, Ocean acidity, Phytoplankton.
ICOS The Integrated Carbon Observation System RI is a pan-European Research Infrastructure that provides harmonized and high-precision scientific data on carbon cycle and greenhouse gas budget and perturbations.

References

[1] Wang M, Jeong C-B, Lee YH, Lee J-S. Effects of ocean acidification on copepods. Aquat Toxicol 2018;196:17–24. https://doi.org/10.1016/j.aquatox.2018.01.004.
[2] Turley C, Eby M, Ridgwell AJ, Schmidt DN, Findlay HS, Brownlee C, et al. The societal challenge of ocean acidification. Mar Pollut Bull 2010;60(6):787–92. https://doi.org/10.1016/j.marpolbul.2010.05.006.
[3] Santana-Casiano JM, González-Dávila M, Rueda M-J, Llinás O, González-Dávila E-F. The interannual variability of oceanic CO_2 parameters in the northeast Atlantic subtropical gyre at the ESTOC site. Global Biogeochem Cycles 2007;21(1). https://doi.org/10.1029/2006GB002788.
[4] Memè S, Delory E, Felgines M, Pearlman J, Pearlman F, del Rio J, et al., editors. NeXOS: next generation, cost-effective, compact, multifunctional web enabled ocean sensor systems. OCEANS 2017-Anchorage, vol. 2017. September 2017. p. 18–21. http://ieeexplore.ieee.org/document/8232104/.

CHAPTER

2.2

Challenges and Applications of Underwater Mass Spectrometry

R. Timothy Short[1], Ryan J. Bell[2], Gottfried P.G. Kibelka[3], Strawn K. Toler[1]

[1]Center for Security and Survivability, SRI International, St. Petersburg, FL, United States; [2]Owner Beaver Creek Analytical, LLC, Lafayette, CO, United States; [3]CMS Field Products, OI Analytical, Pelham, AL, United States

2.2.1 INTRODUCTION

Among the techniques used in modern elemental and molecular analysis, none surpasses mass spectrometry (MS) in analytical power, particularly with regard to selectivity and sensitivity. MS allows the identification and quantification of chemical compounds based on the atomic composition of the molecules and their charge state. Because the measurement is not based on specific molecular features like functional groups, MS is the technique of choice for the analytical characterization of unknown samples.

Among the suite of analytical capabilities provided by MS are ultratrace-to-major-constituent analysis, comprehensive elemental and isotopic quantification, and identification and structural elucidation of compounds ranging from small molecules to large biomolecules. However, no single configuration of a mass spectrometer (MS)[1] and sample interface provides all of these analytical features. Each of the various combinations of MS type, ionization method, and sample interface, applies only to certain analyses and, in some cases, is completely inappropriate for others.

Interest in the development of portable MS systems for in situ analysis arose from the desire to take the demonstrated analytical capabilities of MS characterizations into the field [1]. One of the most important advantages of an in situ instrument is the high frequency with which data can be collected, allowing data fields of high temporal

[1] In this chapter, MS is used interchangeably for both mass spectrometry and mass spectrometer.

resolution and high spatial resolution for a mobile georeferenced instrument. In situ MS provides the additional advantage of simultaneously creating these high-frequency data sets for a large number of analytes. Having this type of information in near real-time also enables intelligent adaptive sampling strategies that may significantly reduce the expense associated with traditional sampling techniques. The development and use of portable MS systems have seen a dramatic increase in the last several decades, largely due to the advent of smaller, more rugged high-vacuum pumps, e.g., turbomolecular pumps that include a molecular drag-pump stage to allow use of small diaphragm and scroll pumps as backing pumps. The Harsh Environment Mass Spectrometry (HEMS) Workshop series began in 1999 to facilitate communication among scientists and engineers working on the development and use of portable MS systems, and this website contains an overview of the diverse applications that have been developed over the last 15 years [2].

Underwater MS (UMS) is a specialized subfield of HEMS, and shares many of the requirements associated with all portable MS systems, such as small size, weight, and power (SWaP), ruggedness, and reliability. However, UMS faces unique challenges, most notably, the high-pressure environment of the deep ocean. Analytes must be transported from a dissolved state at potentially several hundred times atmospheric pressure to the high vacuum of the MS analyzer. The development of UMS systems began in the mid-to late-1990s at three independent universities: the Technical University of Hamburg–Harburg in Germany [3], and the Massachusetts Institute of Technology [4] and the University of South Florida [5] in the United States [6]. Although the specifics of the designs varied, all of these groups employed a membrane interface to introduce analytes into the MS housing through a process called pervaporation [7,8], which is the primary basis for membrane introduction mass spectrometry (MIMS) [9–11]. MIMS has been demonstrated as an effective method for analyzing dissolved, volatile, relatively nonpolar analytes below approximately 200 atomic mass units (amu), such as light-stable gases (e.g., oxygen, argon, and carbon dioxide), and volatile organic compounds (VOCs) (e.g., light hydrocarbons, trihalomethanes, and monocyclic aromatics). To our knowledge, all operational UMS instruments use a membrane interface for sample introduction. As a result, the primary focus of UMS deployments has been the detection, quantification, and mapping of these classes of compounds detectable by MIMS.

2.2.2 UNDERWATER MASS SPECTROSCOPY (UMS) CHALLENGES

This section focuses on specific challenges for implementing UMS and obtaining the most useful data from deployed UMS systems

- Design considerations related to packaging and deploying MSs in an underwater environment
- Factors affecting the interpretation of MS data to identify analytes and obtain quantitative calibrated results
- Difficulties encountered when taking these laboratory instruments into the harsh seawater environment to extract meaningful information

2.2.2.1 Instrument Design Challenges

Mass spectrometers provide a wealth of information at the cost of being more complex and larger than other oceanographic sensors; therefore, SWaP optimization is critical to allow deployment in a wide variety of environments and on as many platforms as possible. However, weight is not as great a concern for UMS systems, which require only neutral buoyancy in seawater (Fig. 2.2.1).

All MSs use an electromagnetic field or a combination of electromagnetic fields to manipulate charged atoms or molecules (ions) in the gas phase. Consequently, analytes must be ionized if they are introduced into the MS as neutral species, as with MIMS. An ever-expanding variety of ionization methods are available for MS, but all UMS mass analyzers use electron impact (EI) ionization using a hot filament to generate electrons [12]. Ensuring that the analyzer's electromagnetic field is the main force acting on the ions requires shielding external fields [13] and minimizing or excluding background gas molecules that would interact and collide with the analyte ions. Therefore, all MSs operate at a reduced gas pressure (i.e., vacuum), and almost all MSs require a pressure more than a million times lower than that of ambient air [14]. The two main obstacles for every UMS are thus (1) how to maintain vacuum inside the MS while it is underwater, and (2) how to transport analytes from the ocean environment, often hundreds of times above ambient atmospheric pressure (1 bar), into the vacuum of the MS.

UMS instruments must have all components encased within a subsea housing that will withstand seawater at high pressure. The UMS components inside the pressure vessel include

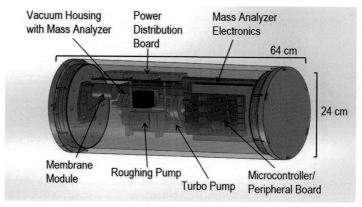

FIGURE 2.2.1 Three-dimensional model of a underwater MS identifying the major components. This model is based on the current version designed and used by SRI International (SRI).

- A vacuum chamber for an MS comprising an ion source, a mass analyzer, and a charged-particle detector
- Control electronics for the MS and peripheral systems
- Pumps to maintain a vacuum inside the vacuum chamber
- A sample introduction system

The most common materials used for the subsea vessel are aluminum, titanium, or stainless steel, with most UMS systems using subsea housings made from anodized aluminum. Titanium is typically the best choice for deployments of more than 3000 m and long-term deployments due to its relative tensile strength-to-weight ratio and its resistance to the very corrosive seawater environment.

Most UMS systems use a linear-quadrupole mass filter for the mass analyzer [15–18], which uses a combination of radio frequency (RF) and direct current (DC) voltages on four parallel rods to separate ions according to their mass (or more specifically, their mass-to-charge ratio, but for EI ionization [at 70 eV energy], the charge usually equals one) as shown in Fig. 2.2.2. Linear quadrupoles have the benefits of being well characterized, rugged, versatile, and commercially available. Their weight, size, and analytical performance are also a good fit for most inwater MIMS applications. Cycloidal MSs, which use superimposed magnetic and electric fields to separate ions according to their masses, have been used by some groups [19,20] and offer reduced power consumption over the linear quadrupoles. Ion traps have been used as mass analyzers in some cases [17,21] and can offer some advantages in sensitivity and mass range. However, when membrane interfaces are used to introduce analytes, an analyzer with a mass range above 300 amu does not usually provide an advantage due to the slow diffusion of larger molecules in the membrane material.

In Fig. 2.2.2, the water sample is pumped through the membrane capillary, through which the dissolved analytes permeate and evaporate into the analyzer vacuum housing, where they are ionized by the electron beam (red dotted line), filtered by their mass-to-charge ratio by the linear quadrupole, and ultimately detected by an electron multiplier or Faraday cup detector. Although Fig. 2.2.2 illustrates a flow-through membrane interface, most UMSs that are subjected to increased hydrostatic pressure use a flow-over or sheet-membrane configuration supported by a porous frit.

The typical power consumption of UMS instruments lies in the 50–100 W range. The greatest single power draw comes from the vacuum pumps. Maintaining a high vacuum inside the vacuum chamber while admitting as much sample as possible to increase sensitivity requires a high-vacuum pump with a sufficient pumping speed. In most cases, a small turbomolecular pump with a molecular drag stage is combined with a diaphragm pump. The power for this pump combination depends on

- The make and model of the pump
- The size of the vacuum chamber in which the MS operates
- The required pressure, usually 10^{-5} Torr, to guarantee that ions can pass through the mass filter without interacting with other ions or residual gas
- The pervaporation rate through the membrane inlet
- The work the pump has to perform to compress the exhaust

A small mass filter offers a short flight path for the target ions, reducing the probability for the ion to collide with residual gas molecules and allowing higher pressures. The pervaporation rate is, in the first approximation, a

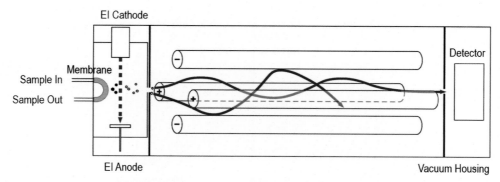

FIGURE 2.2.2   A schematic representation of a membrane introduction mass spectrometry system with a flow-through membrane interface and a linear quadrupole mass filter that employs electron impact (EI) ionization as the mass analyzer. The *solid blue line* denotes the mass-to-charge ratio that has a stable trajectory; the ratio shown by the *solid red line* does not have a stable trajectory.

function of the surface area, thickness of the membrane inlet, and the temperature of the sample. Vacuum pumps in a carefully designed UMS can be run as low as 20 W and even lower in applications that are not time critical and allow a much reduced pervaporation rate.

Over the time of a deployment, the power requirements will slowly increase. As UMS systems operate in a hermetically sealed subsea vessel, all sample that is admitted into the vacuum chamber for analysis will be exhausted by the vacuum pumps into the subsea housing and result in increased pressure inside the housing. The vacuum pumps, working against an increasing pressure, will first compensate by using more power (sometimes up to 50 W) until the pressure inside the housing exceeds pump specifications resulting in pump failure. Because most of the vapor transported into the MS will be water, a desiccant can be used to absorb a significant part of the pump exhaust, reducing the pressure inside the vessel and preventing condensation. The analytical performance of the UMS system will be impacted long before the pumps fail; however, careful monitoring of the power consumption of the pumps and the internal humidity of the housing will allow system checks that ensure system performance.

Alternatively, exhausted gases can be avoided using entrainment pumps (nonevaporative getter and ion pumps). This significantly reduces power consumption, but, ultimately, the buildup of introduced sample in the entrainment pump will also limit deployment times. Power-saving strategies that involve partial shutdown of the system require warmup and equilibration times and often destabilize the analytical result. Nonetheless, assuming continuous operation, weeks-to-months deployment times are possible when exhaust gas is well managed and the pervaporation rate is low [20].

Sample introduction by MIMS typically uses a hydrophobic membrane material such as polydimethylsiloxane (PDMS) [22], polytetrafluoroethylene (PTFE) [23], or composite membranes [24,25]. Using a membrane interface significantly restricts the amount of water vapor introduced into the vacuum system and often enhances the concentration of analytes from the dissolved phase to the gas phase [26]. The primary difference between the membrane interface used in UMS instruments and those used in a laboratory environment is that the UMS membrane cannot be porous and must be mechanically supported to withstand pressure differentials of many bars.

The design of a deep-water MIMS interface varies for different instruments, but in all cases, either a sheet or capillary membrane is supported by an inert porous structure such as a frit (Fig. 2.2.3). The pervaporation of target molecules through the membrane and into the vacuum of the MS depends on the physical and chemical properties of the membrane, the target molecule, and the matrix. Instrumental sensitivity and response time is maximized by reducing the thickness of the diffusion-controlled boundary layer at the membrane surface [27]. Designs using sheet membranes [18,20], flow-through capillaries [17], and flow-over capillaries [15,28] have been developed and optimized to increase sample velocity at the membrane surface to reduce the boundary layer.

Depending on the target compound and the desired spatial and temporal resolution of the analytical result, sampling parameters need to be adjusted. Gaseous components like oxygen, argon, and methane have a high-diffusion rate through the membrane material, even at temperatures commonly found below the thermocline, and should be sampled at a high sample flow to ensure that only fresh, nondepleted sample is in contact with the membrane. Because larger organic molecules like hydrocarbons diffuse more slowly through the membrane material, it is advantageous to heat the aqueous sample to simultaneously reduce the solubility of the organic compound in water and increase their diffusion rate through the membrane. Warming the sample reduces the risk of increasing the sample boundary layer and, therefore, allows a slower sample flow rate. Keeping a constant sample flow alongside the membrane surface is essential for the quantitation of the MS signals. To vastly improve the accuracy of the quantitative

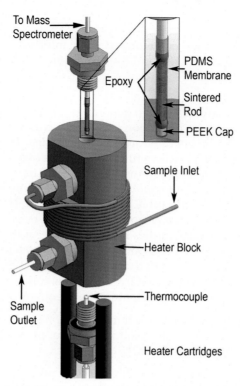

To Mass Spectrometer

PDMS Membrane

Epoxy

Sintered Rod

PEEK Cap

Sample Inlet

Heater Block

Thermocouple

Sample Outlet

Heater Cartridges

FIGURE 2.2.3 A flow-over membrane interface design that implements temperature regulation. Water is pumped through the stainless steel tubing wrapped around a heater block from the sample inlet, through the center of the heater block to flow over the capillary membrane (mechanically supported by a sintered rod to withstand increased hydrostatic pressure), and to the sample outlet. Analytes enter the vacuum of the mass analyzer by pervaporation through the capillary membrane and diffusing through the sintered rod to the hollow stainless tube at the top of the figure. The polyether ether ketone (PEEK) cap prevents the evaporated analytes from exiting through the sintered rod in that direction. A thermocouple measures the heater block temperature, and the underwater MS microcontroller controls the current to the heater cartridges to maintain the block at a constant temperature.

result, in particular for nongaseous compounds, the sample also needs to be temperature controlled, but heating the sample to a constant temperature can quickly result in an overwhelming power draw. A sample flow of 2 mL/min and a temperature increase of 30°C requires a heater with a minimum of 4.2 W. Increasing the flow to 15 mL/min requires a heater with at least 31.4 W.

UMS systems can be deployed as soon as the operational vacuum inside the vacuum chamber of the MS has been reached, but to accurately quantify results, the vacuum level and temperature of the MS components should equilibrate, ideally for 24 h before deployment. To maintain uninterrupted operation, it is best to have two power inputs (typically 24 volts direct current [VDC]) to seamlessly switch between facility or shipboard power and the power of the deployment platform. UMS instruments can be powered through a tether from a power source to a platform or, for greater flexibility, powered by on-platform batteries. However, batteries add to the cost, weight, and size of the system, and can become the limiting factor for deployment duration. A tether or umbilical from a ship or through a cabled observatory offers the advantage of a constant power supply that can deliver hundreds of watts to the instrument and offer a high-bandwidth data connection that allows receiving data in real time. In an ideal situation, the real-time data can be used to increase the value of the collected data by fine tuning or adapting sampling strategies. In addition, to triggering the use of traditional sampling methods, irregular or surprising results can be verified by increasing sample frequencies, resampling, varying the sampling parameters, or by fine-tuning the ionization, mass separation, or detection parameters of the MS. The disadvantages of an umbilical are the drag the tether creates in water, the overall limited reach of a tethered system, and the requirement for a surface-support vessel or cabled observatory infrastructure.

As with all oceanographic instruments, corrosion and biofouling can be a serious problem for extended deployments, and even short deployments in highly productive environments. Fouling of the external subsea vessel is not a major concern for performance of UMS instruments, but fouling of the fluidic lines can alter results. Periodic flushing of the lines with acidified seawater or biocides can help reduce fouling.

2.2.2.2 Data Challenges

This section focuses on the specific challenges encountered when interpreting UMS data and covers the difficulties of compound identification, methods of calibration, and the importance of mass resolution.

2.2.2.2.1 Compound Identification

Although the major strengths of MS analyses are dynamic range, selectivity, and sensitivity, compound identification is a major challenge for MSs with low mass resolution (e.g., unit resolution in which only nominal integer masses are separated). Most MS instruments can be operated in different modes. The two most typical modes for a linear quadrupole mass filter are scan mode, which produces full mass spectra over a selected mass range (e.g., 4–130 mass-to-charge ratio [m/z] taking 10–100s), and selected ion-monitoring (SIM) mode, which monitors only preselected ions (e.g., 20–40 ions taking <10s). The main disadvantage of the scan mode is the extended time it takes to scan through every mass with a sufficient signal-to-noise ratio (SNR) for each analyte. This may be acceptable when the MS is stationary or analyzing a largely homogenous water bodies, but many oceanographic phenomena of study dictate a high-temporal or spatial data resolution; such studies require the use of the faster SIM mode. Two advantages of the scan mode are (1) unknown or unanticipated compounds can be discovered, and (2) compounds can potentially be identified by comparing the acquired spectra with databases (e.g., Ref. [29]). However, the identification of compounds by MS alone is difficult, especially for small VOCs and gaseous compounds that create very few fragment ions and rather unspecific mass spectra. A more confident identification requires one of the following approaches:

- Preseparation of compounds by chromatographic techniques [30].
- High-resolution MS, ideally with exact mass values to four decimal places.
- Tandem (MS/MS) or multiple stages of MS (MS^n) [12].

Although, in principle, all of these techniques are achievable for an underwater system, the added complexity, timing, and SWaP severely limit their application. Miniaturized ion trap MS instruments and their intrinsic MS/MS and MS^n capability [31] offer the potential for enhanced VOC identification but are challenged by space-charge effects due to the abundance of water vapor ions, which can inhibit the detection of low-mass ions [32].

For most UMS analyses, in particular on moving platforms, the SIM mode provides the most information in the least amount of time. SIM analyses select a list of ions of specific masses along with a detector integration time (dwell time) for each mass, and the MS analyzer steps through these masses in sequence and repeats the sequence until instructed to stop. In most deployments, appropriate SIM masses and dwell times are compiled for a predefined list of target analytes. For a linear quadrupole mass filter, for example, a list of 45 masses can be analyzed in 5–10s, which is a practical sampling frequency given the response times of MIMS instruments. If the list is compiled carefully, each mass will have a dwell time that achieves a reasonable SNR for the analysis of a broad suite of analytes, each with their respective desired detection limits.

During the ionization process, the EI electron beam (typically 70eV) imparts excessive energy into the molecules, causing chemical bonds to break and yielding several ion masses for each analyte. This fragmentation can provide information regarding the molecular structure of the analytes, but the fragmentation process often leads to contributions from several analytes to the same mass peaks. Overlapping and interfering peaks from compounds other than the one of interest is often termed chemical noise and can make compound identification difficult. Further confusing spectral interpretation can be the creation of doubly charged ions (e.g., Ar^{+2} appears at "mass" 20 instead of 40).

Increasing confidence in compound identification requires consideration of the sample introduction method and the sample environment. Using a membrane interface restricts the types of compounds that enter the MS and simplifies the analysis and compound identification by rejecting dissolved ions and limiting the analytes to relatively nonpolar compounds that are volatile enough to evaporate on the vacuum side of the membrane. The deployment environment imposes further restrictions on the list of candidate molecules. For example, it is unlikely to observe easily reduced compounds, such as sulfides, in an oxidative environment.

Although the membrane essentially introduces all analytes simultaneously into the MS analyzer, different compounds have slightly different diffusion times through the membrane; therefore, the response times also provide clues to compound identification. The membrane inlet of a UMS usually operates in a constant flow mode with no defined starting or injection point. In most applications, the UMS is deployed so that the sensor moves through a water column sampling water of changing composition. A rapid change in signal intensity of lower mass ions, which are easily identified as the molecular peak of a gas, indicates that the UMS moved from one layer or current of the nonhomogeneous water column into a different layer. A pronounced change in gas concentration signals for ions that are normally attributed to light gases but shows a delayed permeation compared to their typical response time indicates that the ions are not from light gas molecules, but created from the fragmentation of larger molecules. With the response–time relationship in mind, ions can be selected that minimize chemical noise.

2.2.2.2.2 *Calibration*

Calibrations can be performed by monitoring quantitative ions for each analyte of interest with the analysis of standard seawater samples. Dissolved gas standards can be created by equilibrating seawater of known temperature and salinity with gas mixtures in which the partial pressure of each analyte is known, and VOC standards can be created through the serial dilution of stock solutions of each VOC dissolved in methanol at a known concentration [15,33]. The accuracy of the calibration can be improved through independent gas chromatography (GC) analysis of the prepared standards [34]. MSs are designed to produce a linear detector response over a wide concentration range (typically four to five orders of magnitude). Thus, many analyte concentrations can then be determined by linear regressions, which determine the instrumental background and sensitivity to a particular analyte. If all the interferences are accounted for, the electronic or instrument background can be determined by measurements at masses unlikely to have any ion contribution from the chemical background (e.g., a mass of five).

In the high water-vapor environment of an MIMS instrument, the EI ion source is likely to generate chemical noise during water–filament reactions and accidental chemical ionization. As a result, preferable methods for determining baselines for aqueous samples include equilibrating the sample with a low-mass gas (such as He); sampling from boiling water (condensed inline); or, if the membrane geometry is sufficiently restrictive, by turning off the sample pump, which is particularly useful in the field [35]. The water content inside the MS will be governed by membrane properties, membrane conditions (primarily temperature), and vacuum conditions (e.g., outgassing from the region around the warm filament). For this reason, a good practice leaves the EI filament on for several hours (preferably overnight) before a baseline reading is taken, which also provides sufficient time for residual gases to be pumped out of the system. It has also been demonstrated that using cold traps or other means to reduce water vapor in the vacuum system can reduce interference for some analytes and improve detection limits [36].

2.2.2.2.3 *Mass Resolution*

Mass resolution, or the ability to resolve adjacent masses, is another factor to consider in quantification. Most small mass analyzers have only unit–mass resolution, which often allows a small overlap of adjacent mass peaks (usually 5%–10%). Consequently, if mass peaks are not baseline resolved, large peaks can contribute to the intensity of an adjacent low-intensity mass peak and lead to interference, such as the contribution from mass-peak 16 arising from water vapor and molecular oxygen to mass-peak 15, complicating the quantification of methane [36]. Methods to subtract contributions of this type are often difficult to implement reliably in the field due to subtle changes in instrumental background and mass resolution that result from changing field conditions. The challenges of relating laboratory calibrations to field data are summarized in the following section.

2.2.2.3 Field Deployment Challenges

Obtaining and properly interpreting data from a field deployment requires careful planning and the collection of other environmental data. A major challenge is the application of the calibration parameters obtained under controlled laboratory conditions to data collected in changing environmental field conditions. Temperature, salinity, and hydrostatic pressure during the deployment must be considered. Calibrations require the use of solutions with salinities similar to those of the deployment sites; consequently, salinity corrections should be considered for estuarine work. Corrections resulting from transitioning from freshwater to seawater are on the order of 5% [37].

Oceanographic temperatures can vary significantly, especially during depth profiles. Temperature changes inside the subsea housing can cause temperature changes of critical electronic components, but the effect of temperature on membrane permeability is a higher concern. Membrane permeability changes significantly with respect to temperature [11,38], but this effect can be handled to some degree using internal standards. At the expense of power consumption, UMS users often opt to regulate the temperature of incoming seawater by heating to simplify calibrations. Perhaps more importantly, a warmed sample can help achieve faster response times for many VOCs. The strong relationship between membrane permeability and temperature dictates the use of a well-designed temperature controller for accurate low-noise calibrations.

Hydrostatic pressure results in the compression of PDMS and a change in permeability [15]. Empirical correction coefficients describe a reduction in permeation of light gases (through loss of voids used for permeation) and an increase in permeation for many VOCs (through molar volume effects) under hydrostatic pressure. The use of laboratory-determined correction coefficients is complicated by hysteresis and a co-dependence on membrane temperature; thus, researchers have generally relied on the use of internal standards and in situ external standards to apply pressure corrections. Successful internal standard corrections have been demonstrated using argon, which is predicted by temperature and salinity measurements [37], or oxygen measured by a dedicated dissolved oxygen sensor (DO) [20].

Because sensitivity is affected by so many parameters, a good practice is to run analytical standards regularly during a field deployment. If time permits, a daily calibration for each analyte is preferred, although when constrained by time, performing pre- and posttrip calibrations may be sufficient depending on a user's required level of precision and accuracy. A simplified daily check is performing a baseline determination by turning off the sample pump followed by the analysis of Ar using air-equilibrated seawater.

The latency of the time response of UMS measurements is another major challenge for interpreting field UMS data. Because the diffusion-based transport of analytes through the membrane delays their introduction into the MS, membrane-based sensors have inherently slower response times than those equipped with direct sample inlets. Although this delayed response is inconsequential for stationary sensors that monitor slowly varying signatures, sensors mounted on vehicles that encounter highly transient signals (e.g., dynamic chemical plumes) may necessitate further characterization. If the residence time of the sensor in the chemical plume is shorter than the time it takes to reach steady-state flow through the membrane, it is difficult to discern both the spatial extent of the plume and the concentration profile of the target analyte in the plume [39]. For field analyses using MIMS systems, experimental conditions are difficult to control and, at times, certain parameters normally required for quantitative analyses are completely unknown. However, mass spectral time-series analysis of an analyte's leading edge slope upon entry into the plume can lead to reasonable estimates of plume concentration in certain cases [39]. In addition, Janfelt et al. [33] derived a simple method to determine chemical concentrations in plumes that are so narrow that they are only briefly sampled by a moving MIMS. Despite these advances, more in situ work is required to verify these algorithms and develop true deconvolution algorithms in which the instrument response function for each analyte is characterized and then deconvolved from the field data to yield the true analyte distribution.

2.2.3 UNDERWATER MASS SPECTROMETRY APPLICATIONS

Applications of UMS have become quite diverse [6], and we anticipate more growth as the use of UMS systems continues to proliferate. Although a comprehensive review of applications is beyond the scope of this chapter, this section presents several examples based on the authors' prior work.

Vertical profiles, which are among the most common types of oceanographic measurements, are obtained by casting a deployment frame with instrumentation from a research vessel (R/V) using a winch with a cable that allows communication and a limited amount of power for sensors. In addition, to bottles for sample collection, the deployment frames typically house a suite of sensors that include a conductivity, temperature, depth sensor (CTD), DO sensor, and a fluorescence sensor. Data generated from sample collection provide a patchy picture of the water column that can be better characterized by deploying a UMS instrument. The UMS instrument can provide measurements of all the dissolved atmospheric gases and nonpolar VOCs in a depth resolution comparable to that produced by sensors in the CTD suite. For example, plots a through d in Fig. 2.2.1 show vertical profiles for methane (CH_4), nitrogen (N_2), oxygen (O_2), and carbon dioxide (CO_2) obtained by a UMS instrument (R/V Pelican, Mississippi Canyon Block 118 [MC118], Gulf of Mexico, November 2007). Also plotted are the saturation dissolved-gas profiles derived from their known atmospheric partial pressures and CTD measurements (RBR Ltd. Ottawa, Ontario, Canada). For comparison, the oxygen profile plots from the UMS and a DO sensor (Aanderaa Data Instruments, Bergen, Norway) are in good agreement (Fig. 2.2.1C). The inverse behavior of the oxygen and carbon dioxide profiles is typical for biological activity (i.e., the competition between photosynthesis and respiration). Measurable methane is missing in the water column, despite the presence of methane hydrates on the sea floor at MC118 (Fig. 2.2.1A). The ability to measure all atmospheric gases simultaneously with UMS provides a much more complete picture of processes in the water column than provided only by the CTD sensor suite and collected samples. Data shown were collected during a cruise sponsored by the Gulf of Mexico Hydrates Research Consortium that extensively studied the site [40].

A followon cruise on the R/V Pelican to MC118 in June 2010, 31 months after the first UMS deployment in that location, detected increased dissolved methane concentrations (Fig. 2.2.4E and F). Fig. 2.2.4E shows UMS dissolved methane concentrations up to 1 μmol/kg at depths of approximately 600 and 800 m. MC118 is located approximately nine miles northeast of MC252, the site of the April 2010 Deepwater Horizon blowout. The Macondo Well was not capped until July 2010 and was releasing high concentrations of oil and natural gas into the Gulf of Mexico at the time of the UMS deployment. Fig. 2.2.4F shows methane concentrations determined by GC from discrete water samples collected by rosette casts after the UMS cast. Guided by the previous in situ UMS measurements, more samples were

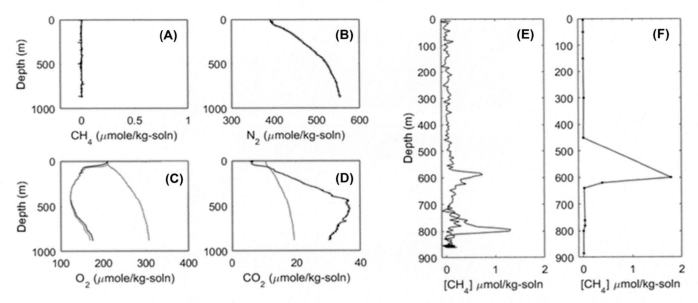

FIGURE 2.2.4   Vertical profile concentration data of dissolved gases at MC118 in the Gulf of Mexico: (A–D) November 2007, Underwater MS (UMS) data in black, calculated gas saturation profiles in blue, and O_2 data from a dissolved oxygen sensor in red; (E) June 2010, UMS data for CH_4 plotted for both upcast and downcast; (F) CH_4 concentrations determined from samples collected following the UMS cast. *Provided by Prof. Jeff Chanton at Florida State University.*

collected for GC analyses around the depths of 600 and 800 m to enhance the chance of finding samples with elevated concentrations of methane. As a result, one collection produced a sample with close to 2 μmol/kg near 600-m depth. Rosette casts prior to the UMS cast were largely unsuccessful in capturing samples with elevated methane concentrations, a clear indication for the necessity of higher sampling rates and the advantage of using UMS for both in situ measurements and to guide sampling strategies for further analyses.

Moored time series with UMS sampling rates for all relevant gases can provide valuable information on diurnal biogeochemical cycles [20] and capture highly transient events. Buoy technology has evolved so that power budgets can support at least limited duration deployments of UMS instruments for time-series measurements, and cabled underwater stations can support much longer deployments. For example, Fig. 2.2.5 shows time-series and sediment depth-profile data for dissolved gas concentrations in sediment pore water in the South Atlantic Bight. Long-term time-series measurements on cabled observatories, such as the Ocean Observing Initiative's network [42] are also becoming possible.

Use of manned submersibles, unmanned remotely operated vehicles (ROVs), and autonomous surface and underwater vehicles enables collection of data with high spatial resolutions over larger areas or volumes than that allowed by traditional vertical profiling from *R/Vs*. Tethered ROVs can provide sufficient power to deploy a UMS system and have the advantage of high-bandwidth real-time communication with the UMS and other instrumentation. The ROV range is limited, however, due to the tether constraints.

Autonomous underwater vehicles (AUVs) offer the ability to perform large area surveys, and some medium-sized AUVs, such as the Bluefin 12 (Bluefin Robotics, Quincy, MA) have sufficient power budgets to deploy UMS systems. However, the bandwidth of acoustic modems used while the AUV is underwater is typically not sufficient for real-time communication of all UMS data during the deployment. Remotely controlled and autonomous surface vehicles offer extended survey ranges and sufficient bandwidth through RF links, but only provide in situ sensing information about the near-surface water. Fig. 2.2.6A and B, demonstrate the ability to create two-dimensional chemical maps from UMS measurements using a guided unmanned surface vehicle in Bayboro Harbor, St. Petersburg, FL. The m/z 91 distribution was attributed to toluene from local motorboat exhaust, but the source of the m/z 62 distribution, although believed to be from dimethyl sulfide, was not conclusively identified [28].

Since the Deepwater Horizon blowout in 2010, many UMS deployments have focused on the detection of hydrocarbons. Interest in UMS sensing of light hydrocarbons is motivated by the ability to differentiate among hydrocarbons (methane, ethane, propane, and butane). This differentiation can provide valuable information regarding the source of an inwater oil or natural gas leak and the source of a natural hydrocarbon seep (thermogenic vs. biogenic) [43]. The power of AUV-based UMS analyses was demonstrated during a survey downstream from the Deepwater Horizon leaking Macondo Well, in which relatively high concentrations of light hydrocarbons and other VOCs were detected at depths of around 1100 m [44].

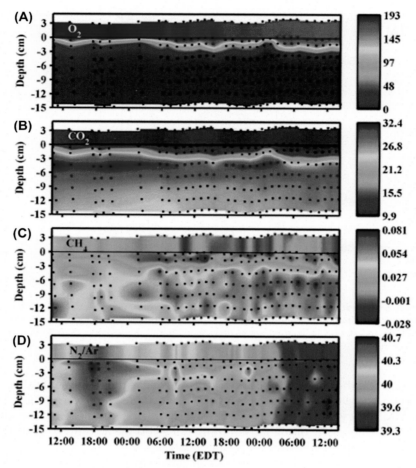

FIGURE 2.2.5   Depth–time contours of calibrated Underwater MS data in sediment pore water: (A) oxygen concentration (μmol/kg); (B) carbon dioxide concentration (μmol/kg); (C) methane concentration (μmol/kg); (D) nitrogen–argon mole ratio (no units). The horizontal line at 0 cm represents the sediment–water interface as determined by time-lapse photography. The "•" markers represent sampling locations in time and depth. *Reprinted with permission from Bell RJ, Savidge WB, Toler SK, Byrne RH, Short RT. In situ determination of porewater gases by underwater flow-through membrane inlet mass spectrometry. Limnol Oceanogr Methods 2012;10:117–28.* https://doi.org/10.4319/lom.2012.10.117; Fig. 5.

ROVs equipped with an instrument suite, including a UMS system, have been used to study deep-water natural hydrocarbon seeps (Fig. 2.2.7). For these deployments, the ROV was equipped with a high-definition camera, a multibeam sonar, a CTD, a DO sensor, a fluorometer, and a UMS instrument [45]. A sample line was installed to rapidly pump water from the tip of the ROV's starboard robotic arm to the UMS sample inlet to accurately control and observe the location of sampling. UMS data were continuously collected during multibeam surveys of hydrocarbon seep sites in the Gulf of Mexico. Exemplary data for georeferenced dissolved methane relative concentrations are superimposed on the bathymetric map (Fig. 2.2.7). The locations of the strongest seeps can be observed from spots of high-intensity methane (Fig. 2.2.7, plotted in red). Creation of three-dimensional maps of the methane concentrations in the water column surrounding the seeps are also possible [36,46]. Other scientific UMS applications include studies of methane-rich, deep-water brine pools [18,47] and carbon dioxide-accumulating subsea pools [48].

The latest version of the UMS instrument designed at SRI (Fig. 2.2.1), can be mounted in an AUV, along with a multibeam sonar, fluorometer, and CT sensor to compose an AUV-based hydrocarbon-sensing package (Fig. 2.2.8A). Preliminary deployments (Fig. 2.2.8C) have shown promise for an instrument suite of this type for exploring large-area hydrocarbon seep fields or inspecting underwater oil and gas assets to detect and characterize potential leaks.

2.2.4 FUTURE

Future UMS challenges include the further reduction of SWaP to enable longer deployments on persistent stationary and mobile platforms, the development of alternative sample introduction interfaces to allow the analysis of a broader range of chemical species, and the development of deconvolution algorithms to yield true analyte

FIGURE 2.2.6   Global Positioning Systems (GPS) track of underwater MS deployment on a guided surface vehicle (GSV) in Bayboro Harbor: (A) m/z 91 signal intensity mapped over track of the GSV; (B) m/z 62 signal intensity mapped over track of the GSV. *Reprinted with permission from Wenner PG, Bell RJ, van Ameron FHW, Toler SK, Edkins JE, Hall ML, Koehn K, Short RT, Byrne RH. Environmental chemical mapping using an underwater mass spectrometer. Trends Anal Chem 2004;23:288–95. https://doi.org/10.1016/S0165-9936(04)00404-2; Fig. 7.*

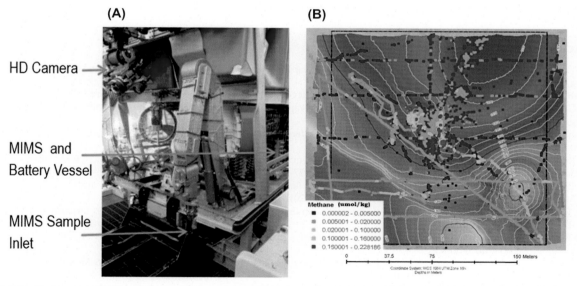

FIGURE 2.2.7   Underwater MS (UMS) deployment on an remotely operated vehicle (ROV) for surveys of hydrocarbon seep surveys: (A) SRI's UMS instrument mounted on a working class ROV along with other survey equipment, the UMS sample inlet was mounted on the end of the left robotic arm; (B) georeferenced UMS methane concentration data plotted over a high-resolution bathymetric map of a natural hydrocarbon seep area in the Gulf of Mexico. *MIMS, membrane introduction mass spectrometry. Images courtesy of SRI International, 2017.*

distribution in near-real time [6]. New types of small, lower-power MSs are continually being introduced, as both commercial products by companies and prototype designs by research groups, which have potential for incorporation into UMS systems. Improvements in membrane interface design and the use of novel membrane materials have the potential to improve UMS response times. New direct-sample introduction methods need to be explored to improve response times, increase the mass range of UMS systems, and broaden the types of analytes that can be

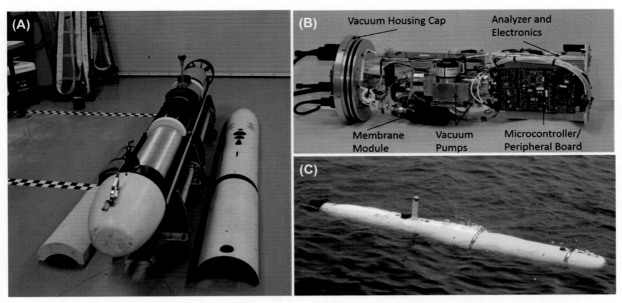

FIGURE 2.2.8   SRI's underwater MS (UMS) instrument deployed on a Bluefin 12 autonomous underwater vehicle (AUV): (A) the UMS mounted inside SRI's Bluefin 12 AUV in the lab; (B) SRI's UMS instrument without the aluminum subsea housing; and (C) the AUV underway before diving for a survey. *Images courtesy of SRI International, 2017.*

addressed to include more-polar and less-volatile compounds. Advanced instrumentation and/or data analytics are needed to improve the accuracy and speed of providing calibrated data (i.e., analyte concentrations in real time). As longer-term deployments become feasible, methods to inhibit biofouling of the instruments, particularly the sample lines and sample introduction interfaces, will become essential. Finally, a relatively new area of interest and research may push UMS system designs to an entirely new level to help explore extraterrestrial ocean worlds, such as Europa and Enceladus, moons of Jupiter and Saturn, respectively.

References

[1] Synder DT, Pulliam CJ, Ouyang Z, Cooks RJ. Miniature and fieldable mass spectrometers: recent advances. Anal Chem 2015;88:2–29. https://doi.org/10.1021/acs.analchem.5b03070.

[2] Workshop on harsh-environment mass spectrometry. 2016. Retrieved from: http://www.hems-workshop.org/.

[3] Matz G, Kibelka G. Detection of chemicals in the sea, technical report 01RA96029. Hamburg-Harburg: Technical University; 2001. https://doi.org/10.13140/RG.2.1.4597.0967.

[4] Hemond H, Camilli R. NEREUS: engineering concept for an underwater mass spectrometer. Trends Anal Chem 2002;21:526–33. https://doi.org/10.1016/S0165-9936(02)00113-9.

[5] Short RT, Fries DP, Toler SK, Lembke CE, Byrne RH. Development of an underwater mass-spectrometry system for in situ chemical analysis. Meas Sci Technol 1999;10:1195–201. https://doi.org/10.1088/0957-0233/10/12/311.

[6] Chua EJ, Savidge WB, Short RT, Cardenas-Valencia AM, Fulweiler RW. A review of the emerging field of underwater mass spectrometry. Front Mar Sci 2016;2(209):1–24. https://doi.org/10.3389/fmars.2016.00209.

[7] Baker RW, Cussler EL, Eykamp W, Koros WJ, Riley RL, Strahman H. Membrane separation systems. Department of Energy; 1990. ER/30133–H1.

[8] Noble RD, Stern SA, editors. Membrane separations tehcnology – principles and applications, vol. 2. Elsevier; 1995. ISBN: 9780444816337 (hardcover), 9780080536187 (ebook).

[9] Hoch G, Kok B. A mass spectrometer inlet system for sampling gases dissolved in liquid phases. Arch Biochem Biophys 1963;101:160–70. https://doi.org/10.1016/0003-9861(63)90546-0.

[10] Johnson RC, Cooks RG, Allen TM, Cisper ME, Hemberger PH. Membrane introduction mass spectrometry: trends and applications. Mass Spectrom Rev 2000;2000(19):1–37. https://doi.org/10.1002/jms.3447.

[11] LaPack MA, Tou JC, Enke CG. Membrane mass spectrometry for the direct trace analysis of volatile organic compounds in air and water. Anal Chem 1990;62:1265–71. https://doi.org/10.1021/ac00212a013.

[12] Ekman R, Silberring J, Westman-Brinkmalm AM, Kraj A. Mass spectrometry: instrumentation, interpretation and applications. Hoboken (New Jersey): John Wiley & Sons, Inc.; 2009.

[13] Bell RJ, Davey NG, Martinsen M, Short RT, Gill CG, Krogh ET. The effect of the Earth's and stray magnetic fields on mobile mass spectrometer systems. J Am Soc Mass Spectrom 2014;26(2):201–11. https://doi.org/10.1007/s13361-014-1027-4.

[14] Thomson JJ. Conduction of electricity through gases. 2nd ed. Cambridge: University Press; 1906.

[15] Bell RJ, Short RT, Van Amerom FHW, Byrne RH. Calibration of an in situ membrane inlet mass spectrometer for measurements of dissolved gases and volatile organics in seawater. Environ Sci Technol 2007;41:8123–8. https://doi.org/10.1021/es070905d.

[16] McMurtry G, Wiltshire JC, Bossuyt A. Hydrocarbon seep monitoring using in situ deep-sea mass spectrometry. In: Proceedings oceans 2005, vol. 1. Brest: Marine Tech. Soc. and IEEE; 2005. p. 395–400. https://doi.org/10.1109/OCEANSE.2005.1511747.

[17] Short RT, Fries DP, Kerr ML, Lembke CE, Toler SK, Wenner PG, Byrne RH. Underwater mass spectrometers for in situ chemical analysis of the hydrosphere. J Am Soc Mass Spectrom 2001;12:676–82. https://doi.org/10.1016/S1044-0305(01)00246-X.

[18] Wankel SD, Joye SB, Samarkin VA, Shah SR, Friederich G, Melas-Kyriazi J, Girguis PR. New constraints on methane fluxes and rates of anaerobic methane oxidation in a Gulf of Mexico brine pool via in situ mass spectrometry. Deep Sea Res Part II 2010;57:2022–9. https://doi.org/10.1016/j.dsr2.2010.05.009.

[19] Camilli R, Hemond HF. NEREUS/Kemonaut, a mobile autonomous underwater mass spectrometer. Trends Anal Chem 2004;23:307–13. https://doi.org/10.1016/S0165-9936(04)00408-X.

[20] Camilli R, Duryea A. Characterizing spatial and temporal variability of dissolved gases in aquatic environments with in situ mass spectrometry. Environ Sci Technol 2009;43:5014–21. https://doi.org/10.1021/es803717d.

[21] McMurtry G, Kolotyrkina IY, Lee JS, Kim KH. Underwater mass spectrometers for in situ monitoring of dissolved gases and volatile organic compounds in deep ocean and coastal environments. In: Proc. MTS/IEEE oceans 2012. 2012. p. 1–6. https://doi.org/10.1109/OCEANS-Yeosu.2012.6263597. Virginia Beach (VA).

[22] Silva ACB, Agusti R, Dalmazio I, Windmoller D, Lago RM. MIMS evaluation of pervaporation processes. Phys Chem 1999;1:2501–4. https://doi.org/10.1039/A808730J.

[23] Maden AJ, Hayward MJ. Sheet materials for use as membranes in membrane introduction mass spectrometry. Anal Chem 1999;68:1805–11. https://doi.org/10.1021/ac9509216.

[24] Alberici R, Sparrapan R, Eberlin MN, Windmoller D, Augusti R. Polyetherimide silicone: a 10 µm ultrathin composite membrane for faster and more sensitive membrane introduction mass spectrometry analysis. Anal Commun 1999;36:221–3. https://doi.org/10.1039/A902185J.

[25] Miranda LD, Bell RJ, Short RT, van Amerom FHW, Byrne RH. The influence of hydrostatic pressure on gas diffusion in polymer and nano-composite membranes: application to membrane inlet mass spectrometry. J Membr Sci 2011;385–386:49–56. https://doi.org/10.1016/j.memsci.2011.09.009.

[26] Pawliszyn J, editor. Handbook of solid phase microextraction. Elsevier Science; 2011. ISBN: 9780123914491.

[27] Sysoev A. A mathematical model for kinetic study of analyte permeation from both liquid and gas phases through hollow fiber membranes into vacuum. Anal Chem 2000;72(17):4221–9. https://doi.org/10.1021/ac991388n.

[28] Wenner PG, Bell RJ, van Ameron FHW, Toler SK, Edkins JE, Hall ML, Koehn K, Short RT, Byrne RH. Environmental chemical mapping using an underwater mass spectrometer. Trends Anal Chem 2004;23:288–95. https://doi.org/10.1016/S0165-9936(04)00404-2.

[29] NIST Mass Spec Data Center, Stein SE. Mass spectra. In: Linstrom PJ, Mallard WG, editors. NIST chemistry WebBook, NIST standard reference database number 69. Gaithersburg (MD): National Institute of Standards and Technology; 2017. p. 20899. http://webbook.nist.gov.

[30] Matz G, Kibelka G, Dahl J, Lennemann F. Experimental study on solvent-less sample preparation methods: membrane extraction with a sorbent interface, thermal membrane desorption application and purge-and-trap. J Chromatogr A 1999;830(2):365–76. https://doi.org/10.1016/S0021-9673(98)00853-X.

[31] March RE, Todd JF. Quadrupole ion trap mass spectrometry. 2nd ed. John Wiley & Sons, Inc.; 2005.

[32] Kibelka GPG, Short RT, Toler SK, Edkins JE, Byrne RH. Field-deployed underwater mass spectrometers for investigations of transient chemical systems. Talanta 2004;64:961–9. https://doi.org/10.1016/j.talanta.2004.04.028.

[33] Janfelt C, Lauritsen FR, Toler SK, Bell RJ, Short RT. Method for quantification of chemicals in a pollution plume using a moving membrane-based sensor exemplified by mass spectrometry. Anal Chem 2007;79:5336–42. https://doi.org/10.1021/ac070408f.

[34] Schlueter M, Gentz T. Application of membrane inlet mass spectrometry for online and in situ analysis of methane in aquatic environments. J Am Soc Mass Spectrom 2008;19:1395. https://doi.org/10.1016/j.jasms.2008.07.021.

[35] Bell RJ. Development and deployment of an underwater mass spectrometer for quantitative measurements of dissolved gases (Graduate Theses and Dissertations) 2009http://scholarcommons.usf.edu/etd/1849.

[36] Gentz T, Schlueter M. Underwater cryotrap-membrane inlet system (CT-MIS) for improved in situ analysis of gases. Limnol Oceanogr Methods 2012;10:317–28. https://doi.org/10.4319/lom.2012.10.317.

[37] Bell RJ, Short RT, Byrne RH. In situ determination of total dissolved inorganic carbon by underwater membrane introduction mass spectrometry. Limnol Oceanogr Methods 2011;9:164–75. https://doi.org/10.4319/lom.2011.9.164.

[38] Baker RW. Membrane technology and applications. 2nd ed. John Wiley & Sons Ltd; 2004. ISBN: 0-470-85445-6.

[39] Short RT, Toler SK, Kibelka GPG, Rueda Roa DT, Bell RJ, Byrne RH. Detection and quantification of chemical plumes using a portable underwater membrane introduction mass spectrometer. Trends Anal Chem 2006;25:637–46. https://doi.org/10.1016/j.trac.2006.05.002.

[40] Mississippi Mineral Resources Institute; 2012. Retrieved from: http://www.mmri.olemiss.edu/Home/programs/GoMHRC.aspx.

[41] Bell RJ, Savidge WB, Toler SK, Byrne RH, Short RT. In situ determination of porewater gases by underwater flow-through membrane inlet mass spectrometry. Limnol Oceanogr Methods 2012;10:117–28. https://doi.org/10.4319/lom.2012.10.117.

[42] Ocean Observatories Initiative; 2016. Retrieved from: http://oceanobservatories.org/.

[43] Sackett WM. Use of hydrocarbon sniffing in offshore exploration. J Geochem Explor 1977;7:243–54.

[44] Camilli R, Reddy CM, Yoerger DR, Van Mooy BAS, Jakuba MV, Kinsey JC, McIntyre CP, Sylva SP, Maloney JV. Tracking hydrocarbon plume transport and biodegradation at deepwater horizon. Science 2010;330:201–4. https://doi.org/10.1126/science.

[45] Carragher PD, Ross A, Roach E, Trefry C, Talukder A, Stalvies C. In: Natural seepage systems at Biloxi & Dauphin Domes & Mars Mud volcano, North East Mississippi Canyon Protraction Area, Gulf of Mexico, Offshore Technology Conference. 2013. https://doi.org/10.4043/24191-MS.

[46] Gentz T, Damm E, von Deimling JS, Mau S, McGinnis DF, Schlueter M. A water column study of methane around gas flares located at the West Spitsbergen continental margin. Cont Shelf Res 2013;72:107–18. https://doi.org/10.1016/j.csr.2013.07.013.

[47] Wankel SD, Germanovich LN, Lilley MD, Genc G, diPerna CJ, Bradley AS, Olson EJ, Girguis PR. Influence of subsurface biosphere on geochemical fluxes from diffuse hydrothermal fluids. Nat Geosci 2011;4:461–8. https://doi.org/10.1038/ngeo1183.

[48] Camilli R, Nomikou P, Escartin J, Ridao P, Mallios A, Kilias SP, Argyraki A, The Caldera Science Team. The Kallisti Limnes, carbon-dioxide accumulating subsea pools. Sci Rep 2015;5. https://doi.org/10.1038/srep12152.

CHAPTER

2.3

Nutrients Electrochemical Sensors

Carole Barus[1], D. Chen Legrand[1], Ivan Romanytsia[1], Justyna Jońca[1], Nicolas Striebig[2], Benoit Jugeau[3], Arnaud David[3], Maria Valladares[4,5], P. Muñoz Parra[5], Marcel Ramos[4,5,6], BORIS Dewitte[1,4,5,6], Veronique Garçon[1]

[1]Laboratoire d'Etudes en Géophysique et Océanographie Spatiales, UMR 5566, Université de Toulouse, CNRS, CNES, IRD, UPS Toulouse Cedex 9, France; [2]Observatoire Midi-Pyrénées, Toulouse Cedex 9, France; [3]nke Instrumentation, Hennebont, France; [4]Center for Advanced Studies in Arid Zones (CEAZA), La Serena, Chile; [5]Facultad de Ciencias del Mar, Departamento de Biología Marina, Universidad Católica del Norte, Coquimbo, Chile; [6]Millennium Nucleus for Ecology and Sustainable Management of Oceanic Islands (ESMOI), Coquimbo, Chile

2.3.1 INTRODUCTION

Nutrient cycles (nitrate, phosphate, and silicate) participate in carbon dioxide (CO_2) sequestration in the ocean and are linked with the global carbon cycle. Observing their concentration in the open and coastal ocean will allow us to better understand the major biogeochemical cycles and assess the consequences on the ocean health of anthropogenic nutrient releases [1–12]. With very high spatiotemporal data, climate models could be improved to accurately predict the ocean response to future climate-forcing events. The development of sensors that operate autonomously in situ that can be used on different platforms and underwater vehicles is a challenge. When other desired attributes such as interference free, reagentless, low energy consumption, good accuracy, and robustness are added to the requirements, the development constitutes an immense challenge [13].

The most widely used method to detect nutrients is based on colorimetric detection using traditional, discrete shipboard sampling technique, and onboard analyses [14–16]. Over the past decade, significant progress has been made in the development of in situ nutrient sensors and a few are commercially available to measure nitrate, phosphate, and silicate [16–19]. Wet chemical techniques with spectroscopic/colorimetric detection use liquid reagent additions that need to be changed regularly. Reagents' stability in time is also an unsolved issue. Furthermore, colorimetric methods suffer from interferences and refractive index effects [20].

Ultraviolet spectroscopy measurements at 220 nm are also used to detect nitrate ions in seawater [21]. In situ nitrate sensors such as the In Situ Ultraviolet Spectrophotometer (ISUS) or the Submersible Ultraviolet Nitrate Sensor (SUNA) are implementable on underwater vehicles such as Array for Real-Time Geostrophic Oceanography (ARGO) floats. However, these sensors have a high limit of detection (LOD) of 1.5 μmol/L and they suffer from strong chemical interferences such as bromide, carbonates, and Colored Dissolved Organic Matter (CDOM) [22,23].

An attractive alternative is to use electrochemistry as an innovative tool to detect nitrate, phosphate, and silicate without any liquid reagent addition. Nitrate ions being electroactive species can be detected directly by electrochemical methods [24–28]. To measure nitrate in the concentration range found in the open ocean, we proposed to use a modified working electrode with metallic nanoparticle deposition. Bimetallic electrodes showed indeed a better sensitivity toward NO_3^- detection as the signal is enhanced due to higher specific electrode surface area [14,29,30].

On the contrary, phosphate (PO_4^{3-}) and silicate ($Si[OH]_4$) are nonelectroactive species and cannot be detected directly by electrochemistry. However, the oxidation of metallic molybdenum electrodes in specially designed electrochemical cells using membrane technologies allows forming in situ the phospho- and silicomolybdic complexes detectable on gold working electrodes [31–34]. This work presents recent developments and innovations to detect nitrate, phosphate, and silicate by electrochemistry. The technology readiness levels (TRL) are presently different for the three nutrients.

In the case of nitrate, the feasibility of using modified electrodes can detect low concentrations of NO_3^- using square wave voltammetry measurements (SWV). The optimization of the silver nanoparticles deposition on gold working electrode is a crucial parameter to reach an LOD competitive with the existing in situ sensors [35]. TRL level reached so far is 3.

Phosphate and silicate sensors are more advanced, i.e., TRL 5/6 for phosphate and TRL 7/8 for silicate. For both nutrients, the optimization of all parameters and the selection of the most suitable electrochemical method to be used are made with open cells. Then, methods are transferred to small-volume prototypes developed in the laboratory. The mechanical design of the electrochemical cell(s) has to be optimized to get the highest possible signal.

In the case of phosphate detection, because the concentration is smaller than the one for silicate in the open ocean, SWV measurements are used to detect the phosphomolybdic complex. The characterization of the signal depending on the frequency used will be discussed. The validated/optimized prototype design and its characterization in a relevant environment using artificial seawater will be presented.

Optimizations (mechanical design and electrochemical parameters) have been made with the silicate sensor and led to the production of the very first in situ sensor. This sensor has been deployed for the first time on a mooring off shore Coquimbo, Chile, and results obtained will be presented and discussed in this chapter. Thanks to its small size, it was also implemented on a PROVOR profiling float (also see chapter 5.7) and deployed in the Mediterranean Sea, off shore Villefranche-sur-Mer where data were transmitted real time via satellite.

2.3.2 EXPERIMENTAL SECTION

2.3.2.1 Chemicals

All solutions were prepared in Milli-Q water (Millipore Milli-Q water system) and stored into plastic containers (polymethylpentene or polypropylene); glass is not used to avoid silicate contamination in solution. The electrolyte was either sodium chloride solution, pH$\approx$4.5 (NaCl supplied by Merck), at 34.5 g/L ([Cl$^-$] = 0.6 mol/L), artificial seawater, pH$\approx$7 made with 30.36 g of sodium chloride ([Cl$^-$] = 0.52 mol/L), 6.74 g of magnesium sulfate heptahydrate (MgSO$_4$, 7H$_2$O, Van Waters and Rogers [VWR]) and 0.16 g of sodium hydrogenocarbonate (NaHCO$_3$), or real seawater. **Nitrate** standard solutions were prepared with potassium nitrate (KNO$_3$, 99.995% purity) from Alfa Aesar. **Phosphate** standard solutions were prepared with potassium dihydrogen phosphate (KH$_2$PO$_4$, Merck). **Silicate** standard solutions were prepared either with sodium hexafluorosilicate (Na$_2$SiF$_6$, from Carlo Erba) or from certified standard solution of sodium silicate (Na$_2$SiO$_3$, H$_2$O, from Alfa Aesar) at 1001 $\pm$ 5 μg/mL. All metals were purchased from GoodFellow (molybdenum, silver, gold, platinum, titanium).

2.3.2.2 Material and Electrochemical Cells

All the experiments in the laboratory using open cells and small-volume prototypes were performed at room temperature under atmospheric conditions with either a PGSTAT-128N potentiostat/galvanostat (Metrohm$^®$) or a μAutolab III potentiostat (Metrohm$^®$) controlled by NOVA software.

A conventional three-electrodes system was used for all experiments with gold **working electrodes** (modified with silver nanoparticles [AgNPs] for nitrate measurements) made with either gold commercial electrode or gold wire inserted into epoxy resin. Prior to experiments, gold electrodes were first polished with aluminum oxide (0.3 μm diameter). Then they were electrochemically cleaned in 0.5 mol/L sulfuric acid solution (prepared from a 98% H$_2$SO$_4$ solution supplied by Merck) by polarizing the electrode 10 s at +2 V and 10 s at −2 V to form O$_2$ and H$_2$ bubbles, respectively, at the electrode surface. Then, cyclic voltammograms were recorded between E$_1$ = 0 V and E$_2$ = 1.5 V at 100 mV/s until reproducible cycles were obtained.

In the case of nitrate detection, the electrodepositions of AgNPs onto gold substrates were performed by chronoamperometry at E = −0.2 V optimized potential in an electrolyte solution containing silver ions at [Ag$^+$] = 0.17 mmol/L (AgNO$_3$ from VWR Chemicals). The end of electrolysis was controlled by the electric charge (Q in Coulombs [C]). The resulting Au/AgNP-modified electrodes were then rinsed with Milli-Q water and ready to use.

The **reference electrode** was a silver–silver chloride (Ag/AgCl) either a commercial electrode (with [Cl$^-$] = 3 mol/L, Metrohm$^®$) or made with silver wire potted into epoxy resin covered with silver chloride layer and immersed into chloride solution ([Cl$^-$] concentration depending on the electrolyte/seawater being used).

Platinum or titanium grids acted as **counter electrodes**. The design/geometry of the **molybdenum electrodes** used to form the reagents in the case of silicate and phosphate detections was adapted to the corresponding cells.

2.3.3 RESULTS AND DISCUSSION

2.3.3.1 Nitrate

The deposition of silver nanoparticles on gold electrode surfaces ($\varnothing$ = 3 mm) was performed using chronoamperometry at E = −0.2 V/Ag/AgCl/KCl (3 mol/L) and the end of electrolysis was controlled by the charge (Q) in

Coulomb (C) (and not by the time). To deposit the optimized quantity of silver and, thus, to obtain the highest possible nitrate signal by improving the electrocatalytic properties of the bimetallic Au/AgNPs electrode, the Volcano plot ($I_{peak}=f(|Q|)$) presented on Fig. 2.3.1 has been established. The peak intensities corresponding to the nitrate reduction at $E=-1.0\,V$ are measured on the cyclic voltammograms presented on Fig. 2.3.1, insert (forward cycles only), recorded on gold-modified electrode (Au/AgNPs) with different charges. The two peaks at -0.4 and $-0.6\,V$ correspond to the characteristic reduction of O_2 in two steps, first into H_2O_2, then into H_2O observed on the gold electrode [36]. For all charges applied, the gold electrodes are, therefore, not totally covered with silver.

As expected, the signal first increased when the quantity of silver deposited increased (from -14 to $-38\,\mu C$) as the effective surface area of the electrode increased. Then, the signal decreased from -60 to $-75\,\mu C$ due to the Ostwald ripening mechanism widely known in the field of crystallization. It is defined by the dissolution of smaller particles that will grow in larger ones [37–39]. The effective surface area of Au/AgNPs, therefore, decreases and the electrode tends to behave as a bare electrode [40]. According to the Sabatier principle, the maximum catalytic activity should be for a surface partially covered with AgNPs and before the Ostwald ripening process starts, which can be determined on the Volcano plot at the intersection of the two characteristic straight lines obtained [40,41]. The optimum charge found to form the modified Au/AgNPs electrode and to detect NO_3^- is $-56.3\,\mu C$. Taking into account the experimental errors to build the Volcano plot and determine the optimum condition, the charge chosen for silver nanoparticle deposition is $-52\,\mu C$ corresponding to 4.6×10^{15} Ag atoms/cm^2.

Now to go further on the detection of low concentrations of nitrate using the optimized Au/AgNPs ($Q=-52\,\mu C$) electrode, the SWV method is proposed to detect nitrate (see SWV principle in the Appendix section) [42–44]. The SWV parameters have been optimized to obtain the least noisy possible signal using classical frequency of $f=100\,Hz$, an amplitude of $E_{SW}=60\,mV$, and a staircase potential $E_{step}=5\,mV$. The square wave voltammograms obtained with Au/AgNPs ($\varnothing=3\,mm$, $Q=-52\,\mu C$) electrode for nitrate concentrations between 0.39 and 50 $\mu mol/L$ are presented on Fig. 2.3.2. As previously observed on cyclic voltammograms, three peaks are observed at -0.25, -0.6, and $-1.0\,V$ corresponding to the reduction of O_2 in two steps into H_2O_2 then into H_2O on gold surface and finally the reduction of NO_3^- into NH_4^+, respectively [36].

The peak intensities for each nitrate concentration have been measured using NOVA software, and the corresponding calibration curve is presented on Fig. 2.3.2, insert. Good linear behavior is obtained between 0.39 $\mu mol/L$ up to 50 $\mu mol/L$ with a LOD of 49 nmol/L corresponding to the smallest concentration measured that did not fit on the calibration curve. The linear behavior covers the whole concentration range found in the open ocean.

To use these Au/AgNP electrodes in a nitrate sensor for marine applications, their stability with time has to be evaluated, particularly the stability of the AgNPs. To evaluate the stability of a modified electrode Au/AgNPs ($\varnothing=3\,mm$, $Q=-52\,\mu C$), the same electrode was continuously stored in sodium chloride solution and used to detect regularly nitrate at $[NO_3^-]=25\,\mu mol/L$ using SWV during 625 h (~26 days). The peak intensity corresponding to nitrate reduction remained at 95% of the initial intensity peak ($t=0$) after 26 days showing good stability of the system and its potential as an electrode for a marine sensor.

2.3.3.2 Phosphate

The procedure for in situ reagent formation and phosphate detection without silicate interference has been previously described by Jońca et al. [33]. To detect phosphate without liquid-reagent addition, molybdenum electrodes are oxidized at 2 V to produce molybdates and protons (Eq. 2.3.1), necessary to form the phosphomolybdic complex (Eq. 2.3.2), which is detected on gold working electrode [32].

$$Mo+4H_2O \rightarrow MoO_4^{2-}+8H^+ +6e^- \tag{2.3.1}$$

$$PO_4^{3-}+12MoO_4^{2-}+24H^+ \rightarrow \left[PMo_{12}^{(V)}O_{40}\right]^{3-}+12H_2O \tag{2.3.2}$$

To avoid silicate interference, the electrochemical cell is divided in three compartments. In the first one, a primary molybdenum electrode is oxidized (Eq. 2.31) and thanks to a very thin (30 μm) proton-exchange membrane (Functional Membranes and Plant Technology [FuMaTech], fumapem® F-930), protons formed across the membrane to concentrate into the second compartment reaching $pH=1$. In this second compartment, a second molybdenum electrode is oxidized to obtain the ratio $H^+/MoO_4^{2-}=70$ required to form only the phosphomolybdic complex free from silicate interference. To avoid the reduction of protons formed, the counter electrode is isolated behind a nonproton exchange membrane (N117 Du Pont™ Nafion® perfluorosulfonic acid [PFSA] Membranes, 180 μm thickness).

Previous work performed by Jońca et al. [32,33] used hydrodynamic chronoamperometry method to detect the phosphomolybdic complex on a rotating-disk working electrode to force and control the convection. This method

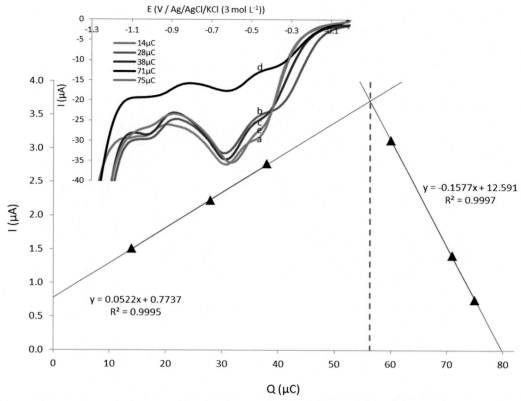

FIGURE 2.3.1 Volcano plot $\left(I_{peak}\left(NO_3^-\right)=f\left(|Q|\right)\right)$ corresponding to the cyclic voltammograms (insert) obtained in NaCl solution (34.5 g/L) containing 0.1 mmol/L NO_3^- using Au/AgNPs electrodes with AgNPs electrodeposition charges: Q = −14 μC (a–Orange), Q = −28 μC (b–Red), Q = −38 μC (c–Blue), Q = −71 μC (d–Black) and Q = −75 μC (e–Green). *Adapted from Chen Legrand D, Barus C, Garçon V. Square wave voltammetry measurements of low concentrations of nitrate using Au/AgNPs electrode in chloride solutions. Electroanalysis 2017;29:2882–87.* https://doi.org/10.1002/elan.201700447.

cannot be used with the in situ sensors as it will be impossible to stir the solution, because the volume of the cells will be too small as a miniaturization of the sensor is desired. Besides, mechanical stirring costs energy. Therefore, to detect phosphate in the concentration range found in the ocean, SWV is proposed as this method provides an enhanced sensitivity as compared to cyclic voltammetry or chronoamperometry as previously mentioned [45–48].

Several step potentials and amplitudes have been tested to obtain the smoothest and highest signal possible (results not shown). The amplitude has been fixed at $E_{SW}=25\,mV$ and the step potential at $E_{step}=1\,mV$. Two frequencies were compared i.e., f = 250 and f = 2.5 Hz corresponding to scan rates 250 and 2.5 mV/s, respectively.

The example of square wave voltammograms obtained at f = 2.5 Hz for different phosphate concentrations between 0.502 and 6.38 μmol/L is presented on Fig. 2.3.3. In the potential window between 0.40 and 0.05 V used, two peaks are obtained around 0.30 and 0.15 V corresponding to the reduction of the phosphomolybdic complex. The molybdenum $Mo^{(VI)}$ changes its oxidation state to form a mix of $Mo^{(V)}/Mo^{(IV)}$ complex by exchanging two, then three electrons according to the following Eqs. (2.3.3) and (2.3.4) [49].

$$\left[PMo_{12}^{(VI)}O_{40}\right]^{3-}+2H^++2e^-\xrightarrow{(E=0.3\,V)}\left[H_2PMo_2^{(V)}Mo_{10}^{(VI)}O_{40}\right]^{3-} \qquad (2.3.3)$$

$$\left[H_2PMo_1^{(V)}Mo_{10}^{(VI)}O_{40}\right]^{3-}+3H^++3e^-\xrightarrow{(E=0.15\,V)}\left[H_5PMo_5^{(V)}Mo_7^{(VI)}O_{40}\right]^{3-} \qquad (2.3.4)$$

For the second reduction (around 0.15 V), a shift of peak potential toward more cathodic potential when phosphate concentration increases is clearly observed and linearly correlated with the decimal logarithm of the phosphate concentration as presented on Fig. 2.3.3, insert. According to Nernst equation (Eq. 2.3.5), the slope of the $E_{peak}=f(log_{10}[PO_4^{3-}])$ curve is proportional to $\dfrac{2.3RT}{nF}\approx\dfrac{0.059}{n}$ (at 298K) with R the gas constant, T the absolute temperature, n the number of electrons exchanged, and F the Faraday constant. E is electrode potential, E^θ is the

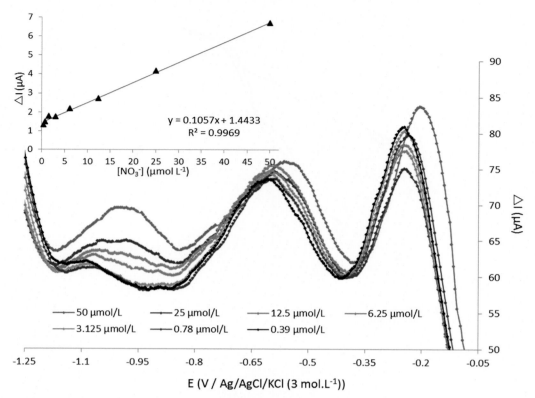

FIGURE 2.3.2   Square wave voltammograms at $f = 100\,Hz$, $E_{SW} = 60\,mV$, $E_{step} = 5\,mV$ on Au/AgNPs ($\emptyset = 3\,mm/Q = -52\,\mu C$) in $34.5\,g/L$ NaCl solution containing $[NO_3^-] = 0.39$ (black), 0.78 (purple), 3.125 (light blue), 6.25 (orange), 12.5 (green), 25 (dark blue), and $50\,\mu mol/L$ (pink)— Insert: corresponding calibration curve $\Delta I = f([NO_3^-])$ for peak around $E = -1.0\,V$. *Adapted from Chen Legrand D, Barus C, Garçon V. Square wave voltammetry measurements of low concentrations of nitrate using Au/AgNPs electrode in chloride solutions. Electroanalysis 2017;29:2882–87.* https://doi. org/10.1002/elan.201700447.\

standard potential, C_{Ox} and C_{Red} are the concentrations of the oxidant and the reducer, respectively, and x is the distance from the electrode surface [42]. In the experimental conditions of Fig. 2.3.3, insert, n = 2.95, which is very close to the theoretical value n of three electrons exchanged (Eq. 2.3.4).

$$E = E^\theta + 2.3\,\frac{RT}{nF}\log_{10}\left[\frac{(C_{Ox})_{x=0}}{(C_{Red})_{x=0}}\right]$$

(2.3.5)

For this second reduction at $E_{peak} = 0.15\,V$, the peak intensity (ΔI) also increases with the phosphate concentration in the solution. The corresponding calibration curve obtained at $f = 2.5\,Hz$ is presented in Fig. 2.3.4 (left) and compared with results obtained at $f = 250\,Hz$ (Fig. 2.3.4, right) for both peak potentials (i.e., 0.15 (▲) and 0.3 V (△)).

For both frequencies, no linear regression is observed on the curve $\Delta I = f([PO_4^{3-}])$ obtained for the first reduction at $E_{peak} = 0.3\,V$ (Fig. 2.3.4, △) indicating maybe the instability of the formed complex. The signal even decreases for the highest concentration of phosphate at $f = 250\,Hz$.

At $E_{peak} = 0.15\,V$, using low frequency ($f = 2.5\,Hz$) allows achieving a linear calibration curve for the whole concentration range of phosphate (from 0.5 to $6.38\,\mu mol/L$), whereas a saturation appears over $1.06\,\mu mol/L$ at high frequency. Indeed, at $f = 250\,Hz$, the signal is linear with the concentration of phosphate only between 0.07 and $1.06\,\mu mol/L$ (Fig. 2.3.4, right).

Similar behaviors with signal saturation or two different regimes of linear behavior for phosphate concentration determination have been observed in references [50–52] for different electrochemical methods. Kircher and Crouch [54] and later Maeda et al. [53] and Jońca et al. [55] showed the formation of different types of phosphomolybdic complexes depending on phosphate and proton concentrations and formation of polymer structures [53–55]. The hypothesis of polymers formation for the highest phosphate concentrations and their possible adsorption on the electrode surface [56–58] might explain the obtained results. These conditions could cause disturbance of the diffusion and/or make the electronic transfer harder. Furthermore, Carpenter et al. [59], demonstrated that the electron transfer is not completely mass-transport controlled. Indeed, a homogenous chemical reaction, probably protonation, occurs before the reduction of the phosphomolybdic complex and slows down the electron transfer [59]. At 250 Hz, the scan rate is too fast, explaining the saturation of the

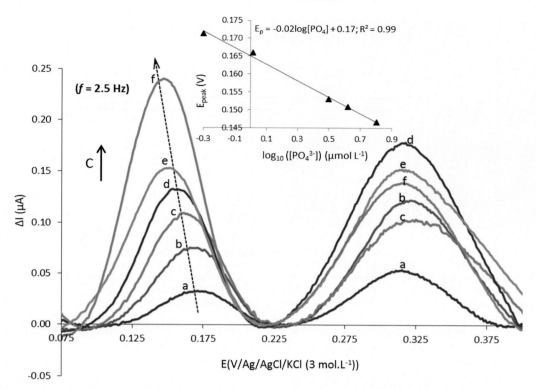

FIGURE 2.3.3 Square wave voltammograms obtained at $f = 2.5\,Hz$ ($E_{SW} = 25\,mV$, $E_{step} = 1\,mV$) on gold electrode ($\varnothing = 3\,mm$) obtained in NaCl solution (34.5 g/L) containing 0.502 (a), 1.04 (b), 2.09 (c), 3.14 (d), 4.19 (e), and 6.38 (f) $\mu mol\,L^{-1}$ of PO_4^{3-}—Insert: peak potential (reduction around 0.15 V) as a function of the decimal logarithm of phosphate concentration $E_{peak} = f(\log_{10}[PO_4^{3-}])$. *Adapted from Barus C, Romanytsia I, Striebig N, Garçon V. Toward an in situ phosphate sensor in seawater using Square Wave Voltammetry. Talanta 2016;160:417–24.* https://doi.org/10.1016/j.talanta.2016.07.057.

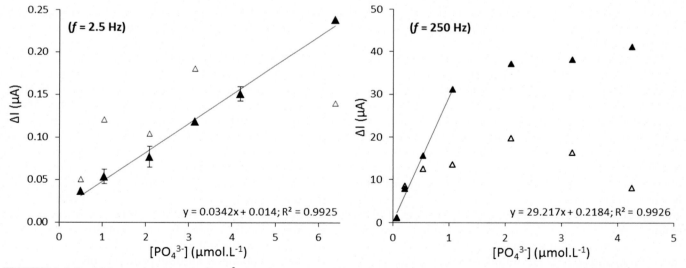

FIGURE 2.3.4 Calibration curves $\Delta I = f([PO_4^{3-}])$ for peaks at $E = 0.3\,V$ ($\triangle$) and for peaks at $E = 0.15\,V$ ($\blacktriangle$) extracted from square wave voltammograms obtained at $f = 2.5\,Hz$ (left) and $f = 250\,Hz$ (right) ($E_{SW} = 25\,mV$, $E_{step} = 1\,mV$). *Adapted from Barus C, Romanytsia I, Striebig N, Garçon V. Toward an in situ phosphate sensor in seawater using square wave voltammetry. Talanta 2016;160:417–24.* https://doi.org/10.1016/j.talanta.2016.07.057.

signal at the highest concentrations. On the contrary, at 2.5 Hz, the low scan rate (i.e., 2.5 mV/s) used provided enough time to either the protonation prestep, or the complex diffusion and/or the electron transfer through polymer structures, to occur.

As expected, the limit of quantification is higher at $f = 2.5\,Hz$: 0.5 μmol/L instead of 0.07 μmol/L at $f = 250\,Hz$. Voltammograms can be recorded at both frequencies and depending on the oceanic area under study, either the

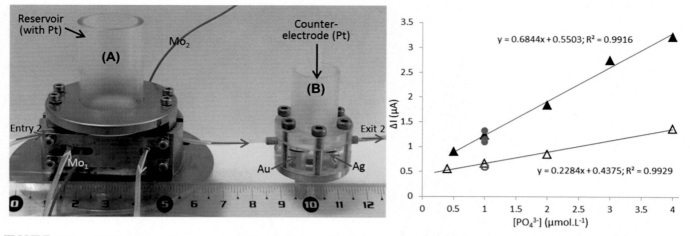

FIGURE 2.3.5   (left) Phosphate laboratory prototype with (A): complexation compartment and (B): detection compartment— (right) Calibration curves $\Delta I = f([PO_4^{3-}])$ from SWV (at $f = 250\,Hz$, $E_{SW} = 25\,mV$, $E_{step} = 1\,mV$) recorded after 5 min of complexation waiting time in artificial seawater for peak at $E = 0.27\,V$ ($\triangle$) and peak at $E = 0.15\,V$ ($\blacktriangle$). In red: ΔI obtained for artificial seawater containing $1\,\mu mol/L$ $(PO_4^{3-}) + 50\,\mu mol/L$ of $Si(OH)_4$ in the same experimental conditions for peak at $E = 0.27\,V$ ($\bigcirc$) and for peak at $E = 0.15\,V$ ($\bullet$). *Adapted from Barus C, Romanytsia I, Striebig N, Garçon V. Toward an in situ phosphate sensor in seawater using Square Wave Voltammetry. Talanta 2016;160:417–24. https://doi.org/10.1016/j.talanta.2016.07.057.*

upper ocean where nanomolar phosphate concentrations are observed, Eastern Boundary Upwelling Systems, or the deep ocean, where the highest phosphate concentrations are found, the suitable linear calibration will be used to determine phosphate concentrations.

All parameters determined for SWV with the open cell had to be adapted for the small-volume prototype designed in the laboratory presented in Fig. 2.3.5, left, especially the charges for molybdenum oxidations. As the oxidation of molybdenum does not follow Faraday's law (coformation of molybdenum oxides: MoO_2, MoO_3…) [55], the theoretical charges for both molybdenum oxidations cannot be calculated and they have been determined experimentally. Because the solution is not stirred during molybdenum oxidations, contrary to the open cell, the homogenization of the solution will only be due to diffusion and might take time. As the geometry of the cells and the electrode orientations have a strong influence on the reagent diffusion and complex formation, the complexation cell (Fig. 2.3.5, left-A) is separated from the detection cell (Fig. 2.3.5, left-B), allowing a decrease in the homogenization time of the solution by using larger molybdenum electrode surfaces [49]. Therefore, the gold and silver electrodes are moved in another cell: the detection cell (B).

The phosphomolybdic complex formed is transferred into the detection cell (B) using a solenoid Lee–Co® pump injecting $50\,\mu L$ of solution at each impulsion. Square wave voltammograms using the same parameters as previously used in the open cell, i.e., $E_{SW} = 25\,mV$, $E_{step} = 1\,mV$, and $f = 250\,Hz$, were recorded for various concentrations of phosphate dissolved in artificial seawater, after only 5 min of complexation waiting time and four pump impulsions ($200\,\mu L$) to transfer the phosphomolybdic complex formed into the detection cell (B). The corresponding ΔI for both peaks were analyzed and measured with NOVA software, and the calibration curves obtained between 0.4 and $4\,\mu mol/L$ of phosphate are presented on Fig. 2.3.5, right. Good linearities with correct regression coefficients were obtained for both peaks at this frequency over the whole concentration range. Saturation for the highest concentrations of phosphate does not appear here as was the case previously with the open cell using $f = 250\,Hz$. The first reduction at $0.27\,V$ is even linear with phosphate concentration in these experimental conditions. The complexation waiting time of 5 min was enough to detect and measure low concentration of phosphomolybdic complex at $0.4\,\mu mol/L$ and because the reaction of complexation is probably not 100% complete, it prevents the signal saturation for highest concentrations. Moreover, the short complexation waiting time prevents as well any variation of pH in the cell (A) (leakage through the Nafion® membrane) and so tends to stabilize the complex, explaining the linearity obtained for both peak potentials.

To check the absence of silicate interference, artificial seawater containing $1\,\mu mol/L$ of phosphate and $50\,\mu mol/L$ of silicate has also been recorded using the same experimental and electrochemical conditions. The results obtained are reported on the calibration curves in Fig. 2.3.5, right (red dots). The same peak intensities are obtained for solution with and without silicate for both peaks proving the non-interference of silicate while measuring phosphate concentration.

Considering the good results obtained with this small volume prototype, in terms of limit of quantification and response time (complexation waiting time), the mechanical design of the in situ immersible phosphate sensor was directly adapted from this prototype design. The in situ phosphate sensor is being assembled and in situ electronics integrating the SWV is currently under development.

FIGURE 2.3.6 (A) Silicate electrochemical sensor, (B) stainless steel cage equipped with the silicate sensor (bottom) and its battery pack (top) ready to be deployed at 55 m depth on a mooring offshore Coquimbo, Chile, at Talcaruca Point; and (C) silicate sensor implemented on PROVOR float before its deployment in the Mediterranean Sea offshore Villefranche-sur-Mer.

2.3.3.3 Silicate

Like previously for phosphate detection, silicate also needs to be complexed with MoO_4^{2-} at acidic pH and, therefore, the counter electrode is isolated behind a Nafion® membrane to avoid the reduction of the H^+ formed during the oxidation of molybdenum electrode (Eq. 2.3.1) in the complexation cell. In the case of silicate, only one molybdenum electrode is required [31,34]. The silicomolybdic complex signal is detected by cyclic voltammetry in the detection cell.

The in situ immersible silicate sensor presented in Fig. 2.3.6-A is an anodized aluminum cylindrical sensor of 2.2 kg in air, 90-mm diameter, and 250-mm height without the connector. The solenoid pump is placed into a reservoir equipped with a membrane and filled with dielectric oil to be in equipressure with seawater. To transfer the silicomolydic complex formed in the complexation cell into the detection cell without diluting it, the flow rate of the pump has to be slow enough to be close to a laminar flow. The ideal flow rate ≈3.3 μL/s is obtained by increasing the back pressure of the pump using a diameter of 0.8-mm internal diameter tube at the entrance of the pump (inlet) and a 0.3-mm internal diameter tube with a substantial length (≈50 cm) at the exit of the pump toward the cell [60]. The electronics were inserted into a dry compartment behind the pump compartment. The in situ electronics has been previously characterized and compared with a commercial potentiostat Metrohm® [60]. The electrochemical cells (in polyether ether ketone [PEEK] pieces) with all electrodes are on the top of the sensor. The molybdenum electrode for reagents formation is placed at the bottom of the complexation cell (376 μL). The silver-wire reference electrode and the titanium-grid counter electrode are behind the

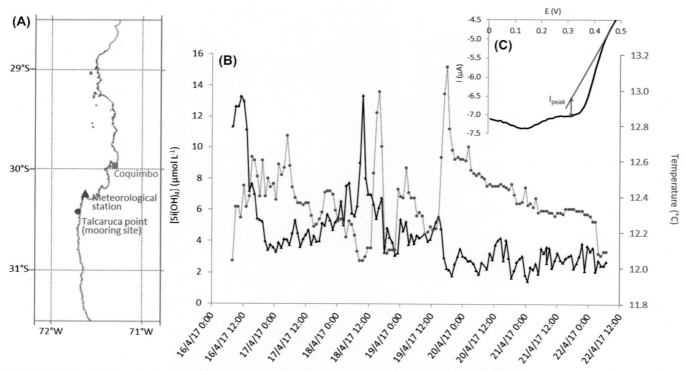

FIGURE 2.3.7   (A) Map of sensor deployment area with ● mooring at Talcaruca point location and ▲ meteorological station location (Punta Lengua de Vaca of Center for Advanced Studies in Arid Zones [CEAZA]—(B) Time series of silicate concentrations and in situ temperatures (recorded at 55-m depth between April 16 and 22 of 2017 (local time: UTC-4h)—(C) Cyclic voltammogram (forward cycle) at 100mV/s and its tangent *Adapted from Barus C, Chen Legrand D, Striebig N, Jugeau B, David A, Valladares M, et al. First deployment and validation of in situ silicate electrochemical sensor in seawater. Front Mar Sci 2018;5:60.* https://doi.org/10.3389/fmars.2018.00060.

Nafion® membrane, directly in contact with the open ocean. Inside the detection cell (94 µL), the three electrodes ($\varnothing$ = 2 mm each): i.e., gold working electrode, silver reference electrode, and platinum counter electrode are assembled in the same PEEK piece with epoxy resin. All the electrodes are plugged and can be easily recovered for reconditioning, making this sensor very handy. The sensor is also equipped with pressure and temperature sensors measuring the pressure at the beginning and the end of seawater sampling and the temperature when the sensor measures the silicate concentration. The housing has been validated up to 60bars corresponding to 600m depth using a pressure column. The cyclic voltammogram recorded on gold working electrode by the electronics is posttreated. The first peak intensity at around 0.35 V corresponding to silicate concentration is measured from the highest intensity (peak, in absolute value) to the interception of the tangent drawn before (the highest potential) the peak, continuously to the baseline. An example of peak intensity measurement is shown in Fig. 2.3.7C.

The in situ silicate electrochemical sensor connected to its battery pack was assembled on a stainless steel cage (Fig. 2.3.6B) and deployed at 55-m depth at Talcaruca point, offshore Coquimbo, Chile, (30° 27.2′ S, 71° 42.7′ W) between April 16 and April 29, 2017 (Fig. 2.3.7A). This observation site is the most intense center of upwelling of the central–northern zone of Chile [61,62]. The complexation time chosen for this deployment was 30min. The whole procedure took 62min, so the silicate concentration (peak intensity) was measured almost every hour. The evolution of silicate concentration and in situ seawater temperature measured by the sensor at 55-m depth through time (local time, UTC-4h is presented in Fig. 2.3.7B. The mean hourly value of surface wind speed from the nearby meteorological station (see location on the map in Fig. 2.3.7A, blue triangle) recorded between April 15 and 30 of 2017 is shown in Fig. 2.3.8. The black rectangle highlights the period of silicate sensor deployment.

When the sensor was deployed, a discrete sample was taken to validate the sensor measurement. The sample was analyzed in triplicate after deployment by colorimetric method at Universidad Católica del Norte and gave a silicate concentration of 11.4 ± 0.1 µmol/L [63]. The first cyclic voltammogram recorded by the sensor at the same time and the same depth as the discrete sample is presented in Fig. 2.3.7C as well as the tangent drawn to measure the peak intensity. It gives a silicate concentration of 11.3 µmol/L, which is in good agreement with the reference value. The

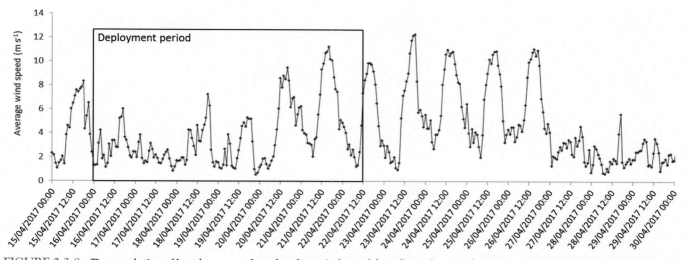

FIGURE 2.3.8 Time evolution of hourly mean value of surface wind speed from Punta Lengua de Vaca meteorological station. *Black rectangle* indicates the period of sensor deployment. *Adapted from Barus C, Chen Legrand D, Striebig N, Jugeau B, David A, Valladares M, et al. First deployment and validation of in situ silicate electrochemical sensor in seawater. Front Mar Sci 2018;5:60.* https://doi.org/10.3389/fmars.2018.00060.

process to calculate the real concentration of silicate using the calibration made before deployment and considering the complexation time was previously reported [60].

The first peak of silicate concentration recorded on April 16 (Fig. 2.3.7B) could be interpreted as resulting from upwelling favorable wind peak on April 15 that could either uplift silicate-rich waters or produce vertical mixing (Figs. 2.3.7B and 2.3.8). Then, the increase of the local temperature at 55-m depth is well correlated with the silicate concentration decrease to ~3.5 µmol/L. This result is in agreement with the sea-surface temperature (SST) observations from Multiscale Ultrahigh Resolution (MUR) shown in Fig. 2.3.9. The April 16 SST map shows a lower temperature (silicate-rich seawater) at Talcaruca Point (● black dots on the maps) than on the SST maps from April 17 and 18. The intense silicate peak recorded on April 18 (~13 µmol/L) (Fig. 2.3.7B) is probably due to eddy advection of silicate-rich seawater because the upwelling favorable winds do not exhibit a clear enhancement and the SST is experiencing a slight increase. Restratification processes are also possible for explaining the SST warm peak on April 18 [64].

Then, the evolution of SST between April 19 and 29 of 2017 (Fig. 2.3.9) shows the establishment and progression of the upwelling toward Talcaruca Point correlated with strong winds recorded between April 20 and 27 (Fig. 2.3.8). Results obtained with the electrochemical sensor show also the decrease of the temperature at 55-m depth from April 20 and a weak increase of silicate concentration centered around 2.8 µmol/L (Fig. 2.3.7B). The memory card of the sensor was full on April 22 and did not save (erased) the following data measured by the sensor. The sensor was originally supposed to be recovered on April 21 but the wind increased drastically the day before, and the weather and sea-state conditions prevailing during several days made impossible the recovery before April 29. A second discrete sample was taken when the sensor was recovered and analyzed in triplicate, giving concentration of 6.7 ± 0.2 µmol/L. As expected, the silicate concentration probably continued to increase after April 22 due to the upwelling progression (Fig. 2.3.9).

After being deployed on the mooring in Chile, the sensor has been directly adapted and implemented on a PROVOR float (Fig. 2.3.6C). Adding a new sensor on the float meets some challenges (e.g., see Chapter 5.7). One of them was the float behavior with large and heavy structure. Studies were carried out to reach a configuration compliant with float-navigation capabilities. Some trials have been done to adjust buoyancy by adding syntactic foam (in pink on Fig. 2.3.6C) to compensate weight modification and check all navigation, data acquisition, and communication phases. Special effort has been made to design the holding and provide easy access to all connectors on the sensor. It can be easily removed from the holding without disassembling the entire system. The Exchange Multi Application Protocol (EMAP) protocol, a tool developed during the SenseOCEAN project to define a standard protocol for communication with the platform, was implemented to communicate with the sensor. The float was deployed off Villefranche-sur-Mer in the Mediterranean Sea on Wednesday May 10, 2017 from the Sagitta III, National Center for Scientific Research, France (CNRS)—Technical Division National of the Institute of Sciences of the Universe (DT INSU) ship and drifted until Monday May 15, 2017 afternoon when it was recovered. The float remained close to the shore throughout the deployment period due to wind and current that pushed the float toward

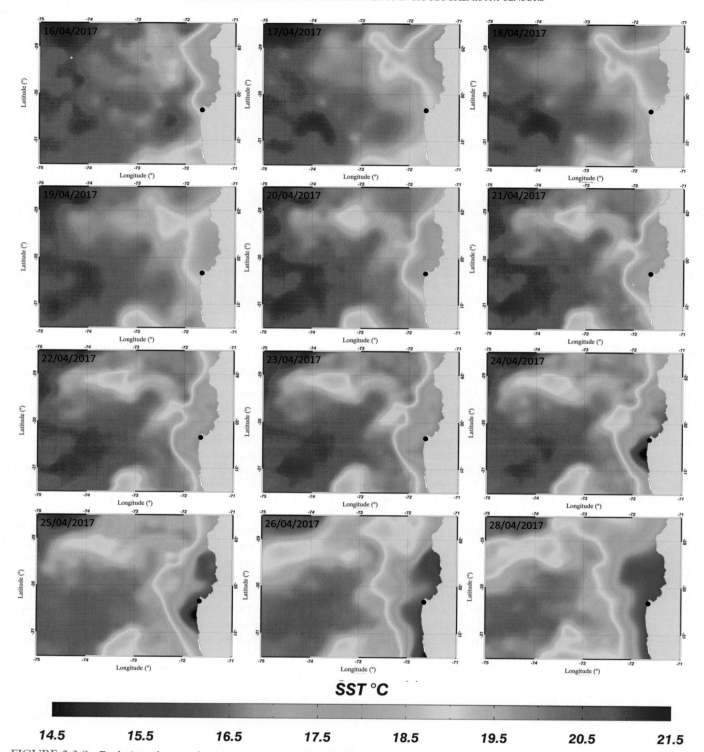

SST °C

14.5   15.5   16.5   17.5   18.5   19.5   20.5   21.5

FIGURE 2.3.9   Evolution of sea-surface temperature from the Multiscale ultrahigh resolution (0.01 degrees) satellite product between April 16 and 28 of 2017— ● indicates the mooring position. *Adapted from Barus C, Chen Legrand D, Striebig N, Jugeau B, David A, Valladares M, et al. First deployment and validation of in situ silicate electrochemical sensor in seawater. Front Mar Sci 2018;5:60. https://doi.org/10.3389/fmars.2018.00060.*

the coast, instead of offshore as we were expecting prior to deployment. Therefore, no data deeper than 160 m was obtained. Nevertheless, the sensor measured silicate concentration when the float was at parking depth. Data were sent through satellite when the float reached the surface. An example of data recorded at 14°C and 160-m depth gives a silicate concentration of $[Si(OH)_4] = 5.94\,\mu mol/L$ [60], which is close to previously published historical silicate concentrations from this area, depth, and season [65].

2.3.4 CONCLUSIONS

In this chapter, "Nutrients electrochemical sensors," we highlighted and proved the great innovative potential of electrochemical methods to develop in situ, autonomous sensors to measure nutrient concentrations without any liquid-reagent addition, interference free, and with sensitivity allowing coverage of the whole concentration range observed in the open ocean for the three major macronutrients.

The results presented show a sensitive method for nitrate detection in seawater using a very simple three-electrode electrochemical system. The optimization of nanoparticle deposition and electrochemical method chosen allowed to detect nitrate in the concentration range found in the open ocean (0.39–50 μmol/L) with a LOD of ~50 nmol/L, which is very promising and competitive with the existing nitrate sensors and detection technologies. To increase sensitivity, a larger electrode surface could be used. In addition, a comparison with other bimetallic couples should be tried, as copper showed very good electrocatalytic properties for nitrate detection [27] or platinum substrate instead of gold for example. Then the next step will be to produce the in situ submersible electrochemical nitrate sensor. The electronics using SWV measurements will be directly adapted from the phosphate sensor also using SWV detection.

Electrochemical detection of phosphate using SWV at different frequencies highlighted the complexity of phosphomolybdic complex reduction. However, even if the mechanism of its reduction is not yet fully elucidated, we managed to measure phosphate concentration in seawater, free of silicate interference, and without any liquid-reagent addition using simple molybdenum oxidations and partitioned electrochemical cells. A good linearity was obtained in the concentration range between 0.5 and 4 μmol/L using small-volume laboratory prototype. The first in situ electrochemical phosphate sensor including electronics (with SWV measurements) is being assembled and will be soon characterized and deployed for the first time in the open ocean.

The development made in recent years to detect silicate in seawater led to the very first in situ autonomous silicate electrochemical sensor, which represents a significant increase in Technological Readiness Level (TRL) and an important innovation in nutrient sensing. Its dimensions and energy consumption (25 mAh for 1 sample per h configuration) allow its deployment on moorings as well as its implementation on underwater vehicles such as a PROVOR float. Both deployments made were "a world premiere" and provided satisfactory results as compared to the reference sample and literature, proving the potential of our silicate sensor. After a complete characterization of its life-time and accuracy statistics, it will be ready for commercialization by 2020.

An effort has to be done on the posttreatment of the recorded electrochemical signals. The future development of an algorithm able to automatically integrate the peak intensities and directly convert them into in situ nutrient concentrations will considerably increase the performance of our sensors, making them easier to use for nonelectrochemists.

Glossary

CDOM Colored dissolved organic matter
CEAZA Center for advanced studies in arid zones
EMAP Exchange multi application protocol
ISUS In situ ultraviolet spectrophotometer
LOD Limit of detection
LOQ Limit of quantification
MUR Multiscale ultrahigh resolution
PEEK Polyether ether ketone
SST Sea-surface temperature
SUNA Submersible ultraviolet nitrate sensor
SWV Square wave voltammetry
TRL Technology readiness level

Acknowledgments

Nitrate detection work using Au/AgNPs was supported by the MIACTIS project: "Microsystèmes intégrés pour l'analyse des composés en traces in situ" (Integrated microsystems for the analyses of in situ trace compounds) as part of the overall project "Instrumentation and Environmental Sensors" funded by the RTRA-STAE: Thematic network of the Foundation for scientific cooperation "Sciences et Technologies pour l'Aéronautique et l'Espace" (Sciences and Technologies for Aeronautic and Space) in Toulouse, France.

Contracts of Dr. Dancheng Chen Legrand and Dr. Ivan Romanytsia as well as the development on silicate and phosphate sensors were supported by the SenseOCEAN project: Marine sensors for the 21st Century funded by the European Commission Seventh Framework Programme (FP7) Oceans of Tomorrow program under Grant Agreement n°614141 (http://www.senseocean.eu/).

The authors would like to thank Marcel Belot from the Groupe d'Instrumentation Scientifique of the Observatory Midi-Pyrénées, Toulouse, France for mechanical manufacturing and assistance and Joël Sudre from LEGOS laboratory for his help on the SST MUR plots.

The authors also thank the crew of the scientific vessel Stella Maris II, Coquimbo, Chile and Cécile Guieu, Laure Mousseau, Antoine Poteau from the Laboratoire d'Océanographie de Villefranche-sur-Mer UMR7093 and Jean-Yves Carval, captain of the SAGITTA 3 vessel of the Observatoire

Océanologique de Villefranche-sur-Mer for their help and their involvement during the PROVOR float deployment. IRD is also thanked for financial support. We are grateful to the Center for Advanced Studies in Arid Zones (CEAZA) for providing the wind data from the meteorological station of Punta Lengua de Vaca and assistance for processing the data.

References

[1] DeMaster DJ. The supply and accumulation of silica in the marine environment. Geochim Cosmochim Acta 1981;45(10):1715–32. https://doi.org/10.1016/0016-7037(81)90006-5.

[2] Dugdale RC, Wilkerson FP, Minas HJ. The role of silicate pump in driving new production. Deep Sea Res Part I 1995;42(5):697–719. https://doi.org/10.1016/0967-0637(95)00015-X.

[3] Dugdale RC, Wilkerson FP. Silicate regulation of new production in the equatorial Pacific upwelling. Nature 1998;391:270–3. https://doi.org/10.1038/34630.

[4] Mann DG. The species concept in diatoms. Phycologia 1999;38(6):437–95. https://doi.org/10.2216/i0031-8884-38-6-437.1.

[5] Falkowski PG, Barber RT, Smetacek V. Biogeochemical controls and feedback on ocean primary production. Science 1999;281:200–6.

[6] Yool A, Tyrrell T. Role of diatoms in regulating the ocean's silicon cycle. Global Biogeochem Cycles 2003;17(4):1103–24. https://doi.org/10.1029/2002GB002018.

[7] Arrigo KR. Marine microorganisms and global nutrient cycles. Nature 2005;437:349–55. https://doi.org/10.1038/nature04159.

[8] Ward BB, Capone DG, Zehr JP. What's new in the nitrogen cycle? Oceanography 2007;20(2):101–9. https://doi.org/10.5670/oceanog.2007.53.

[9] Matear RJ, Wang YP, Lenton A. Land and ocean nutrient and carbon cycle interactions. Curr Opin Environ Sustain 2010;2(4):258–63. https://doi.org/10.1016/j.cosust.2010.05.009.

[10] Tréguer PJ, De La Rocha CL. The world ocean silica cycle. Annu Rev Mar Sci 2013;5:477–501. https://doi.org/10.1146/annurev-marine-121211-172346.

[11] Frings PJ, Clymans W, Fontorbe G, De La Rocha C, Conley DJ. The continental Si cycle and its impact on the ocean Si isotope budget. Chem Geol 2016;425:12–36. https://doi.org/10.1016/j.chemgeo.2016.01.020.

[12] Bristow LA, Mohr W, Ahmerkamp S, Kuypers MMM. Nutrients that limit growth in the ocean. Curr Biol 2017;27(11):R474–8. https://doi.org/10.1016/j.cub.2017.03.030.

[13] COCA Working group. The Collaborative on Oceanography and Chemical Analysis (COCA) and suggestions for future instrumental analysis methods in chemical oceanography. Virtural special issue in marine chemistry. 2014. http://media.journals.elsevier.com/content/files/cocameeting-30202026.pdf.

[14] Moorcroft MJ, Davis J, Compton RG. Detection and determination of nitrate and nitrite: a review. Talanta 2001;54(5):785–803. https://doi.org/10.1016/S0039-9140(01)00323-X.

[15] Patey MD, Rijkenberg MJA, Statham PJ, Stinchcombe MC, Achterberg EP, Mowlem M. Determination of nitrate and phosphate in seawater at nanomolar concentrations. Trends Anal Chem 2008;27(2):169–82. https://doi.org/10.1016/j.trac.2007.12.006.

[16] Ma J, Adornato L, Byrne RH, Yuan D. Determination of nanomolar levels of nutrients in seawater. Trends Anal Chem 2014;60:1–15. https://doi.org/10.1016/j.trac.2014.04.013.

[17] Thouron D, Vuillemin R, Philippon X, Lourenço A, Provost C, Cruzado A, et al. An autonomous nutrient analyser for oceanic long-term *in situ* biogeochemical monitoring. Anal Chem 2003;75(11):2601–9. https://doi.org/10.1021/ac020696+.

[18] Legiret FE, Sieben VJ, Woodward EM, Abi Kaed Bey SK, Mowlem MC, Connelly DP, et al. A high performance microfluidic analyser for phosphate measurements in marine waters using the vanadomolybdate method. Talanta 2013;116:382–7. https://doi.org/10.1016/j.talanta.2013.05.004.

[19] Grand MM, Clinton-Bailey GS, Beaton AD, Schaap AM, Johengen TH, Tamburri MN, et al. A lab-on-chip phosphate analyser for long-term *in situ* monitoring at fixed observatories: optimization and performance evaluation in estuarine and oligotrophic coastal waters. Front Mar Sci 2017;255(4):1–16. https://doi.org/10.3389/fmars.2017.00255.

[20] McKelvie ID, Peat DMW, Matthews GP, Worsfold PJ. Elimination of the Schlieren effect in the determination of reactive phosphorus in estuarine waters by flow-injection analysis. Anal Chim Acta 1997;351:265–71.

[21] Finch MS, Hydes DJ, Clayson CH, Weigl B, Dakin J, Gwilliam P. A low power ultra violet spectrophotometer for measurement of nitrate in seawater: introduction, calibration and initial sea trials. Anal Chim Acta 1998;377:167–77.

[22] Johnson KS, Coletti LJ. *In situ* ultraviolet spectrophotometry for high resolution and long-term monitoring of nitrate, bromide and bisulfide in the ocean. Deep Sea Res Part I 2002;49:1291–305.

[23] Johnson KS, Coletti LJ, Jannasch HW, Sakamoto CM, Swift DD, Riser SC. Long-term nitrate measurements in the ocean using the *in situ* ultraviolet spectrophotometer: sensor integration into the APEX profiling float. J of Atmos Ocean Technol 2013;30:1854–66. https://doi.org/10.1175/JTECH-D-12-00221.1.

[24] Carpenter NG, Pletcher D. Amperometric method for the determination of nitrate in water. Anal Chim Acta 1995;317:287–93.

[25] Koparal AS, Ogutveren UB. Removal of nitrate from water by electroreduction and electrocoagulation. J Hazard Mater 2002;89(1):83–94. https://doi.org/10.1016/S0304-3894(01)00301-6.

[26] Kim D, Goldberg IB, Judy JW. Chronocoulometric determination of nitrate on silver electrode and sodium hydroxide electrolyte. Analyst 2007;132:350–7. https://doi.org/10.1039/b614854a.

[27] Hafezi B, Majidi MR. A sensitive and fast electrochemical sensor based on copper nanostructures for nitrate determination in foodstuffs and mineral waters. Anal Methods 2013;5:3552–6. https://doi.org/10.1039/C3AY26598F.

[28] Bui MN, Brockgreitens J, Ahmed S, Abbas A. Dual detection of nitrate and mercury in water using disposable electrochemical sensors. Biosens Bioelectron 2016;85:280–6. https://doi.org/10.1016/j.bios.2016.05.017.

[29] Davis J, Moorcroft MJ, Wilkins SJ, Compton RG, Cardosi MF. Electrochemical detection of nitrate and nitrite at a copper modified electrode. Analyst 2000;125:737–42. https://doi.org/10.1039/A909762G.

[30] Prussse U, Vorlop KD. Supported bimetallic palladium catalysts for water-phase nitrate reduction. J Mol Catal A Chem 2001;173:313–28.

[31] Lacombe M, Garçon V, Thouron D, Le Bris N, Comtat M. Silicate electrochemical measurements in seawater: chemical and analytical aspects towards a reagentless sensor. Talanta 2008;77:744–50. https://doi.org/10.1016/j.talanta.2008.07.023.

[32] Jońca J, León FV, Thouron D, Paulmier A, Graco M, Garçon V. Phosphate determination in seawater: toward an autonomous electrochemical method. Talanta 2011;7:161–7. https://doi.org/10.1016/j.talanta.2011.09.056.

[33] Jońca J, Giraud W, Barus C, Comtat M, Striebig N, Thouron D, et al. Reagentless and silicate interference free electrochemical phosphate determination in seawater. Electrochim Acta 2013;88:165–9. https://doi.org/10.1016/j.talanta.2016.07.057.

[34] Aguilar D, Barus C, Giraud W, Calas E, Vanhove E, Laborde A, et al. Silicon-based electrochemical microdevices for silicate detection in seawater. Sensor s and Actuators B 2015;211:116–24. https://doi.org/10.1016/j.snb.2015.01.066.

[35] Majidi MR, Asadpour-Zeynali K, Hafezi B. Fabrication of nanostructured copper thin films at disposable pencil graphite electrode and its application to elecrocatalytic reduction of nitrate. Int J Electrochem Sci 2011;6:162–70.

[36] Chen Legrand D, Barus C, Garçon V. Square wave voltammetry measurements of low concentrations of nitrate using Au/AgNPs electrode in chloride solutions. Electroanalysis 2017;29:2882–7. https://doi.org/10.1002/elan.201700447.

[37] Serizawa N, Katayama Y, Miura T. Ag(I)/Ag electrode reaction in amide-type room-temperature ionic liquids. Electrochim Acta 2010;56:346–51. https://doi.org/10.1016/j.electacta.2010.08.072.

[38] Redmond PL, Hallock AJ, Brus LE. Electrochemical Ostwald ripening of colloidal Ag particles on conductive substrates. Nano Lett 2005;5(1):131–5. https://doi.org/10.1021/nl048204r.

[39] Wang F, Han R, Liu G, Chen H, Ren T, Yang H, et al. Construction of polydopamine/silver nanoparticles multilayer film for hydrogen peroxide detection. J Electroanal Chem 2013;706:102–7. https://doi.org/10.1016/j.jelechem.2013.08.008.

[40] Calle-Vallejo F, Huagn M, Henry JB, Koper MTM, Bandarenka AS. Theoretical desing and experimental implementation of Ag/Au electrodes for the electrochemical reduction of nitrate. Phys Chem Chem Phys 2013;15:3196–202. https://doi.org/10.1039/c2cp44620k.

[41] Laursen AB, Varela AS, Dionigi F, Fanchiu H, Miller C, Trinhammer OL, et al. Electrochemical hydrogen evolution: Sabatier's principle and the Volcano plot. J Chem Educ 2012;89(12):1595–9. https://doi.org/10.1021/ed200818t.

[42] Mirčeski V, Komorsky-Lovrić Š, Lovrić M. In: Scholz F, editor. Square-wave voltammetry: theory and application. Springer; 2007.

[43] Mirceski V, Laborda E, Guziejewski D, Compton RG. New approach to electrode kinetic measurements in square-wave voltammetry: amplitude-based quasireversible maximum. Anal Chem 2013;85(11):5586–94. https://doi.org/10.1021/ac4008573.

[44] Lovric M, Jadresko D. Theory of square-wave voltammetry of quasireversible electrode reactions using an inverse scan direction. Electrochim Acta 2010;55:948–51. https://doi.org/10.1016/j.electacta.2009.09.043.

[45] Bard A, Faulkner LR. Electrochemical methods: fundamentals and applications. 2nd ed. John Wiley and Sons; 2001.

[46] Ramaley L, Krause Jr MS. Theory of square wave voltammetry. Anal Chem 1969;41(11):1362–5. https://doi.org/10.1021/ac60280a005.

[47] Osteryoung J. Voltammetry for the future. Acc Chem Res 1993;26(3):77–83. https://doi.org/10.1021/ar00027a001.

[48] Osteryoung JG, Osteryoung RA. Square wave voltammetry. Anal Chem 1985;57(1). https://doi.org/10.1021/ac00279a004. 101A–10A.

[49] Barus C, Romanytsia I, Striebig N, Garçon V. Toward an in situ phosphate sensor in seawater using square wave voltammetry. Talanta 2016;160:417–24. https://doi.org/10.1016/j.talanta.2016.07.057.

[50] Bai Y, Tong J, Wang J, Bian C, Xia S. Electrochemical microsensor based on gold nanoparticles modified electrode for total phosphorus determinations in water. IET Nanobiotechnol 2014;8(1):31–6. https://doi.org/10.1049/iet-nbt.2013.0041.

[51] Kolliopoulos AV, Kampouris DK, Banks CE. Rapid and portable electrochemical quantification of phosphorus. Anal Chem 2015;87(8):4269–74. https://doi.org/10.1021/ac504602a.

[52] Fogg AG, Bsebsu NK, Birch BJ. Differential-pulse anodic voltammetric determination of phosphate, silicate, arsenate and germinate as β-heteropolymolybdates at a stationary glassy-carbon electrode. Talanta 1981;28(7):473–6.

[53] Maeda K, Himeno S, Osakai T, Saito A, Hori T. A voltammetric study of Keggin-type heteropolymolybdate anions. J Electroanal Chem 1994;364:149–54.

[54] Kircher CC, Crouch SR. Kinetics of the formation and decomposition of 12-molybdophosphate. Anal Chem 1983;55(2):242–8. https://doi.org/10.1021/ac00253a016.

[55] Jońca J, Barus C, Giraud W, Thouron D, Garcon V. Electrochemical behaviour of isopoly- and heteropolyoxomolybdates formed during anodic oxidation of molybdenum in seawater. Int J Electrochem Sci 2012;7:7325–48.

[56] Rong C, Anson FC. Spontaneous adsorption of heteropolytungstates and heteropolymolybdates on the surfaces of solid electrodes and the electrocatalytic activity of the adsorbed anions. Inorg Chim Acta 1996;242:11–6. https://doi.org/10.1016/0020-1693(95)04843-X.

[57] Wang B, Dong S. Electrochemical study of isopoly- and heteropoly-oxometallates film modified microelectrodes-VI. Preparation and redox properties of 12-molybdophosphoric acid and 12-molybdosilicic acid modified carbon fiber microelectrodes. Electrochim Acta 1996;41(6):895–902. https://doi.org/10.1016/0013-4686(95)00383-5.

[58] Choi S, Kim J. Adsorption properties of Keggin-type polyoxometalates on carbon based electrode surface ant their electrocatalitic activities. Bull Korean Chem Soc 2009;30(4):810–6.

[59] Carpenter NG, Hodgson AWE, Pletcher D. Microelectrode procedures for the determination of silicate and phosphate in waters – fundamental studies. Electroanalysis 1997;9(17):1311–7. https://doi.org/10.1002/elan.1140091703.

[60] Barus C, Chen Legrand D, Striebig N, Jugeau B, David A, Valladares M, et al. First deployment and validation of in situ silicate electrochemical sensor in seawater. Front Mar Sci 2018;5:60. https://doi.org/10.3389/fmars.2018.00060.

[61] Strub P, Mesías J, Montecino V, Rutllant J, Salinas S. Coastal ocean circulation off western South America. In: Robinson A, Brink K, editors. The sea, the global coastal ocean, vol. 11. New York: Wiley; 1998. p. 272–313.

[62] Thiel M, Macaya EC, Acuña E, Arntz WE, Bastias H, Brokordt K, et al. The Humboldt current system of Northern and Central Chile. Oceanographic processes, ecological interactions and socioeconomic feedback. Oceanogr Mar Biol Ann Rev 2007;45:195–344.

[63] Grasshoff K, Kremling K, Ehrhardt M, editors. Methods of seawater analysis. Weinheim: Wiley-VCM; 1999. p. 353–64.

[64] Renault L, Dewitte B, Falvey M, Garreaud R, Echevin V, Bonjean F. Impact of atmospheric coastal jets on SST off central Chile from satellite observations (2000-2007). J Geophys Res 2009;114(C08006):1–22. https://doi.org/10.1029/2008JC005083.

[65] Pasqueron de Fommervault O, Migon C, D'Ortenzio F, Ribera d'Alcalà M, Coppola L. Temporal variability of nutrient concentrations in the northwestern Mediterranean sea (DYFAMED time-series station). Deep Sea Res, Part I 2015;100:1–12. https://doi.org/10.1002/2015JC011103.

APPENDIX: SQUARE WAVE VOLTAMMETRY PRINCIPLE

The main advantage of pulse techniques, such as Square Wave Voltammetry, is to provide data of high quality and enhanced sensitivity compared to cyclic voltammetry or chronoamperometry. Short-term square-shaped potential pulses, combined with staircase potential ramp (E_{step}) and current sampling, allow the decrease of the contribution of capacitive current [42–44]. Capacitive current (I_c) corresponds to the charging/discharging of electrical double-layer capacitance. It is observed when the potential is changing, or when the distribution of ions is changing near the electrode/electrolyte interface that acts like a capacitor. Capacitive current does not involve any chemical reaction (or charge transfer).

On the contrary, faradaic current (I_f) is causing (or caused by) a charge transfer occurring at the electrode surfaces (chemical reaction). The electric current corresponds to the reduction or oxidation of chemical species in the cell, which involves the transfer of one or several electron(s) across the interface between a metallic electrode surface and a solution phase (electrolyte).

Therefore, to obtain the highest sensitivity (corresponding to the electron transfer) it is important to reduce the effect of the charging current and only measure the contribution of the faradaic current. Because the faradaic current is proportional to $I_f \propto t^{-1/2}$, and the capacitive current is proportional to $I_c \propto e^{-t/RC}$ in which t is time, R the solution resistance, and C is the double-layer capacitance, the faradaic current decreases more slowly than the capacitive current. Therefore, by sampling the current at the end of each potential pulse (I_1 for the forward pulse and I_2 for the backward pulse), the capacitive current can be considered negligible and only the faradaic contribution is measured. The differential current between both measurements ($\Delta I = I_1 - I_2$) is plotted versus the potential staircase: E_{step}. The potential pulses are characterized by the frequency (f in Hz) and the amplitude of the pulse E_{SW} in volts (V). The peak obtained has a Gaussian form in which the peak height is directly proportional to the concentration of the species in solution.

C H A P T E R

2.4

Microfluidics-Based Sensors: A Lab on a Chip

Matthew Mowlem, Allison Schaap, Alex Beaton
National Oceanography Center, United Kingdom

2.4.1 INTRODUCTION

What constitutes Lab-on-chip (LOC) technology is poorly defined, but typically involves microfluidic structures, such as channels or wells, with dimensions less than a few hundred μm and integrating more than one laboratory process or component on a single substrate. It is closely allied to, if not indistinct from, micro Total Analysis System(s) (μTAS) [1], which were introduced in the late 1980s/early 1990s and are microfabricated [2–5] systems integrating all elements required to transduce chemical signals into data. In contrast to the very large number of laboratory applications of microfluidics and LOC [6–12], oceanographic sensor systems must be self-reliant [13,14], integrating all reagent supply, mixing, transduction, control, and data handling functions in a robust package. As such, an in situ LOC system to date typically comprises a "chip" element with varying levels of functionality integrated with pumps, sometimes valves, as well as reagent stores, optical or electrical transducers, and embedded electronics/software for control of the sensor and for handling the resulting data. It is usual for the system to be placed into a housing for protection from the seawater. This can be pressure resistant, which limits the maximum deployment depth, and is costly. Alternatively, housings can be pressure balanced: the ambient (environmental) pressure is communicated inside the housing. This is often achieved by filling the housing with a fluid, such as oil, and including some compliance in the housing, for example by including a bellows or diaphragm that enables expansion and contraction of the balancing fluid.

Despite there being relatively few examples of successful in situ deployment in oceans or seas, this technology addresses many of the challenges facing ocean in situ sensors.

A particular advantage is the ability of LOC to use a wide variety of high-performance analytical chemistry techniques or assays (e.g., Refs. [15,16]) that have long provenance in the laboratory, are proven at environmentally relevant concentrations, and address targets that are typically dilute (pM–μM per L). This can include the use of concentration or sample manipulation prior to analytical determinations (e.g., Refs. [17–21]). They also benefit from innovations in analytical techniques, and adaptations and improvements of existing techniques to better access new or existing analytical targets. It is also possible to carry standard and blank solutions and to analyze these during the deployment to enable in situ calibration. This enables LOC to measure with excellent metrology a number of chemical and biology/biodiversity Essential Ocean Variables (EOVs[1], [22]) and Essential Biodiversity Variables (EBVs [23,24]). Further parameters of interest such as pollutants, pathogens, and microplastics are also tractable, and achieving measurement with LOC is a global research activity.

LOC sensors are also potentially low cost ($5–15 k) or at least have a favorably comparable cost to existing physical and chemical sensors (typically $10–30 K). This is important in enabling scaleup and deployment in large numbers to address the chronic undersampling of the ocean environment in both space and time.

The small physical dimensions of fluidic paths used by LOC technology results in small volumes of both sample and reagent being required for each analysis—typically a few μL. Even in systems that require extensive flushing, total waste production can be limited to much less than 1 mL per sample, at least an order of magnitude less than the comparable macrofluidic systems. Macrofluidic systems, again, are poorly defined, but generally use flow paths greater than a few hundred μm, frequently achieve flows than are nonlaminar, or are actively mixed and often are constructed from tubing and discrete components with limited integration. The latter can lead to significant poorly flushed (dead) volume and internal volume resulting in significant fluidic and, hence, power consumption for each analysis. In contrast, in microfluidics, the use of such small volumes minimizes the consumption of reagents, standards, and blanks as well as power consumption. Scaling has a number of additional effects [6,25–27], including the reduction in the Reynolds and Péclet numbers (Re and Pe, respectively; see glossary) resulting in more laminar flow with diffusive mixing and mass transport, whereas reduction in length scale decreases time taken for diffusion to occur, for example, across a channel or flow stream. Length-scale reductions also result in reductions in thermal mass and insulation and improved heat transfer. Small dimensions of fluidic features also means that LOC systems can be of reduced overall physical size, allowing easy integration into autonomous underwater vehicles (AUVs).

However, there are a number of disadvantages of the technology, and for successful widespread take-up these must be addressed (see the following). These include:

- Robustness: Because LOC systems typically have moving parts such as pumps and/or valves, long-term reliability must be carefully managed. In addition, many users are cautious because of the perceived vulnerability to fouling or blocking of microchannels and microfluidics because of particles or biological growth. In practice, this is rarely operationally significant (see the following). In addition, the systems require a complex integration of electrical and fluidic systems that can, without careful design, lead to a network of tubes, wires, and cables each with numerous connectors that are vulnerable to damage and/or failure.
- Reagent supply: Many of the analytical techniques used in ocean LOC require a supply of reagents including buffers, reactive elements, and frequently standards and blanks to enable in situ calibration. Many reagents are, without additional measures, unstable leading to a degradation of performance over time, and this can be accelerated in, for example, high-temperature and high light intensity environments.
- Ease of use: without careful and advanced design, LOC systems can require a high level of user skill, which limits the extent of their application. For example, without careful design a user may have to:
  - make up the reagents using standard but skilled analytical techniques
  - connect reagent containers to the LOC system and ensure it is flushed and air bled from the system
  - calibrate the LOC and check functioning
  - complete a custom integration with a platform (e.g., mooring or autonomous vehicle), which typically supplies power and communications links
  - analyze raw data from the sensor converting these into measurements as well as performing quality control on these data.

---

[1] http://www.goosocean.org/index.php?option=com_content&view=article&id=14&Itemid=114.

- Time response: Without multiplexing, or use of segmented or droplet flows [28–31], the combined effects of assay kinetics (and hence reaction time) and fluidic dispersion [32], in which fluid is smeared in space and time as it flows along a channel or pipe, conspire to reduce the maximum sampling frequency, typically to a sample every 2–10 min. This may be a limitation for moving platforms, particularly when profiling vertically or in areas of high gradient. For example, to locate a rapid change of nutrient concentration, known as a nutricline [33], to within 5 m with a vehicle moving at 20 cm/s requires a sample every 25 s.

2.4.2 IN SITU FLUIDIC SENSING SYSTEMS FOR ENVIRONMENTAL APPLICATIONS SIMILAR TO LAB ON CHIP

A number of reviews present the development of in situ sensor technology, including those using reagent-based techniques and fluid-based analysis. These include sensors for ecosystems [34,35], environmental analysis [36], marine monitoring [13], natural waters [37] as well as in situ microfluidic electrochemical sensors [10] for the cryosphere [38], microalgae [39], and chemical and biological ocean sensors [14,40]. These report a number of operational in situ fluidic-based systems for nutrients, the carbonate system, and biological parameters that can be considered *macrofluidic* (i.e., in which the fluidic structures are of dimensions larger than several hundred μm, and the fluid volumes used per measurement are several mL or tens of mL rather than μL). These are of interest as they illustrate that microfluidic techniques and LOC are not the only technologies that can provide high-performance metrology of key ocean variables. However, they also serve to show where LOC offers operational or metrology advantages. Macrofluidic systems typically use tubing assembled into a sensor using connectors or joints and loosely integrated, or discrete optical, mixing, and pumping elements. This lack of integration frequently results in large numbers of fluidic and electrical connections and a complex mix of discrete components, all of which can lead to potential for greater unreliability. It is striking that very few of the macrofluidic systems present in the literature have become widely used, and this is despite many having excellent performance. Much of this is due to reticence in user communities because of perceived high price, complexity, and unreliability of the technology.

Of the macrofluidic systems presented in the literature, the osmotically pumped analyzer described by Jannasch et al. [41] has characteristics most similar to LOC technology. Originally applied to nitrate analysis but adaptable, e.g., for iron or manganese, it uses salt gradient-powered membrane-based pumps to drive fluid through 500 μm channels formed by wire embossing in a polyvinyl chloride (PVC) fluidic manifold that is sealed against a clamped vinyl sheet. The manifold includes an 8 mm × 0.5 mm optical cell for absorption measurement using a simple light-emitting diode (LED) and photodiode pair. With a flow rate of only 12 μL/h the analyzer has a 90% rise time of 30 min, somewhat worse than LOC, and an LOD of 0.1 μM and a dynamic range 0.1–20 μM. However, osmotic pump variability and even reversal of flow can be experienced due to pressure and temperature changes, and hence widespread application has not occurred.

The SubChemPac [42] is another early macrofluidic analyzer, and can be configured for nitrate, nitrite, Fe(II)/Fe (III), or phosphate. It includes a chip-like fluidic manifold, but with a separate detection cell 3 mm × 25 cm, sample/reagent flow rates of 20 and 1 mL/min, and maximum power consumption of 150 W, however, it does not achieve performance matching LOC. It does, however, enable high-frequency measurement (order 15 samples/min).

The Shipboard Environmental Data Acquisition System (SEAS) family of analyzers [43–49] can be applied to a wide range of parameters including nitrate, nitrite, phosphate, iron, copper, and all carbonate-system variables. They do not include LOC technology, but do integrate mixing, waveguide, optical fiber, and for some analytes, gas transfer in an integrated optofluidic cell, or "tube in a tube" structure. The core of this technology is a gas-permeable liquid core waveguide (LCW) formed from Teflon AF®-2400 or quartz and Teflon AF®-1600 (both with refractive index lower than water), which enables long (0.1–5 m) absorption cells and hence high sensitivity and low limits of detection (2.5 nM for nitrite). The diameter frequently used (600 μm [49]–810 μm [43]) is on the border of macro and microfluidics, but coupled with macro pumps, heaters, and a multiwavelength spectrometer results in a high-power consumption system suitable for shorter deployments.

The Robotic Analyzer for the TCO_2 System (RATS) [50] similarly, is not classical LOC but uses an LCW for pH analysis to allow low-concentration dyes to be used for pH determination while avoiding perturbation of the measured pH by the dye [50,51] or a tube within a tube gas-exchange cell (ID 1.47 mm) for TCO_2 analysis. The resultant performance for TCO_2 is excellent (precision and accuracy ±2.7 μM in the lab, ±3.6 μM in situ) but a large (but unspecified) external battery pack is required and approximately 11 mL of waste is produced per analysis.

The Submersible Autonomous Moored Instrument (SAMI) family of analyzers [52–59] is predominantly macrofluidic, though some variants do have very small internal volume from carefully constructed and jointed pipe and optical fiber networks [58,59] or use an integrated optofluidic cell (ID 0.5 mm [52], 1.4 mm [54]), and a 300 µm ID gas-exchange tube membrane for CO_2 analysis [52]. The Total Alkalinity (TA) sensor [56] uses interconnecting tubing of 157 µm ID and an optional filter tube of 76 µm ID but has a macrofluidic optical cell requiring a magnetic stirrer and flushing with 35–40 mL of sample. The pH variant [54,57], which has minimum internal volume (100 µm ID mixing coil, 610 µm × 1.72 mm optical cell), requires 50 µL of reagent, 5 mL of flushing and ~1 kJ of energy per measurement[2].

Also addressing the variables of the ocean carbonate system, the Channelized Optical System (CHANOS) [60] pH and DIC sensor and its predecessors [60–64] and benchtop variants including for TA [65] similarly, also include the use of microbore tubing. Principally, this is Teflon AF 2400 (typically 0.5 mm OD, 0.4 mm ID) for gas exchange and sometimes as an LCW for pH determinations. In pH variants problems with time variant apparent optical length of the Teflon AF 2400 LCW are obviated by using a separate commercial optofluidic "Z" cell 10 mm × 0.4 mm ID [60,61]. Despite these small fluidic cross sections, moderate flow rates, and continuous flow in operation (to produce a counterflow during gas exchange) result in relatively large reagent and sample consumption. The Dissolved inorganic carbon (DIC) sensor variant [61] has a time response on the order of 40 s, and indicator flow rate of 1.0 mL/min, resulting in 0.67 mL consumption per measurement. The sample flow rate is 4.0 mL/min or 2.7 mL per sample. The TA variant [65] uses a large optical cell and requires 20 mL sample as well as, 0.5 mL of titrant per measurement.

The NH_4-Digiscan ammonia sensor [66] is predominantly macrofluidic, but does use a gas-exchange cell (chip) with channel depths of only 0.20 and 0.10 mm separated by a 0.071-mm thick Teflon tape. During operation, ammonia gas diffuses into an acid-receiving solution and reacts with protons forming ammonium ions. Because an ammonium ion is less mobile than a proton, this results in a drop in the conductivity of the receiving solution. The device has been widely used in estuaries, has an LOD of 0.014 µM, measurement period of 17.6 min, stability exceeding 30 days, but is designed for operation to a maximum depth of 3 m.

Commercially available systems use macrofluidic designs and hence produce significant waste and have higher reagent and power consumption. They include the Wetlabs/Seabird Scientific Cycle-P Phosphate analyzer [67]; the Systea Wiz sensor for orthophosphate, ammonia, nitrite, nitrate, silicates, iron, and trace metals [68–70]; Sunburst Sensors SAMI family of carbonate parameter sensors (see previous discussion); and SensorLAB pH sensors[3] (see Chapter 2.1).

2.4.3 IN SITU MICROFLUIDICS AND LAB ON A CHIP

The earliest use of engraved microfluidic manifolds integrated with electromechanical microdispensing pumps for in situ oceanographic applications was in the "Autonomous Nutrient Analyzer In Situ" (ANAIS) [71], which can measure nitrate, silicate, and phosphate. It uses a pressure-balanced fluidic design and electronics in a pressure case rated to 100 bar (1000 m depth). Its performance figures of merit include a detection range of 0.1–40 µM nitrate, 150 µM silicate, and 5 µM phosphate, detection limit less than 0.1, 0.5, and 0.1 µM, and an accuracy of 1 (@40 µM), 1 (@80 µM), and 3% (@4 µM), respectively. ANAIS used engraved channels in cadmium (for nitrate reduction to nitrite accessible by the Griess assay), polymethyl methacrylate (PMMA) and polyether ether ketone (PEEK) which were sealed by clamping a lid/secondary layer. The manifold included mounts and sealing surfaces for the solenoid (Lee Co.) valves and, integrates the cadmium column and fluidic connections in the main manifold. A separate optofluidic flow cell is used for absorption measurements, which includes LED light sources, discrete optical (band-pass) filters, lenses, 10-mm Quartz glass windows, and two photodiode detectors (one for monitoring LED power, and the other for measuring the output of the absorption cell). The device, which includes heating, requires on the order of 900 J/nutrient measurement[4] when in continuous operation.

An updated version of this instrument, CHEMINI [72–74] similarly integrates pumps (in this instance, dual two-way peristaltic pumps) and valves (8 three port [two-way]) onto a fluidic manifold formed in PMMA by milling. The channel diameter used is 0.8 mm, bordering on macrofluidic and detection is not integrated but in a separate quartz optofluidic cell (3 cm × 1 mm ID) illuminated by polymer fiber-coupled LEDs and detected by similar fiber-coupled photodiode. The fiber optics feed through into a pressure case that houses the optoelectronic and electronic

---

[2] Estimated from battery configuration (18 D cells and 16 C cells and 2 samples/h for 1 month [57].

[3] http://www.sensorlab.es/sites/default/files/FL_SP101_SM_4.pdf.

[4] Estimated from stated 21 mAh per measurement and assuming a 12 V power supply.

components. The energy consumption is high at 8.6 kJ/measurement[5] but the system is depth rated to 600 bar (6000 m) and has an LOD of 0.3 and 0.1 µM for iron and sulfide respectively. The system has also been deployed for 6 months at 1100-m depth while exhibiting excellent reproducibility (1.07% on 25 µM in situ standard).

Many of the few in situ microfluidic/LOC sensors that have been deployed globally have been produced in Japan. Many of these have very advanced capabilities but have been deployed for only short periods and, therefore, can use materials, such as polydimethylsiloxane (PDMS) [5] which have poor stability in the long term (more than a few weeks) [75].

Variants include a sensor for quantification of adenosine triphosphate (ATP) [76–80], which performs optional noncellular ATP elimination [78] by mixing the sample with ATPase, mixing of a lysis buffer, cell lysis, mixing of lysate with luciferin–luciferase reagent, and detection of the resultant luminescence. The chip that accomplishes this is complex and consists of two PDMS layers containing channels. The lower layer has channels sealed against a glass base and the upper layer seals against the lower layer to which it is bonded. Through-hole (vias) link these layers together and also allow connection to reagent supplies and waste. The upper layer performs the mixing functions, whereas the lower layer consists of a spiral pattern of large surface area and long path length that ensures the maximum luminescence is observed by the large viewing area of the detector (photon multiplier housed together with electronics in a pressure case). The glass substrate has integrated transparent heaters and a resistive temperature sensor to maintain the reaction at 25°C. The chip is connected to off-chip valves, gear, and syringe pumps all housed in a pressure-balanced housing. Reagents are stored externally. Ingeniously, some of the reagent containers resemble a syringe (i.e., include a plunger and seal) but have fluid connections at each end. This enables a working fluid (water) to be pumped into one end of the reagent store to deliver meter reagents from the other end. This can be useful in limiting the dead volume normally experienced in pumps. Because of the luminescence assay used, the stated detection limit is excellent (0.11 pM), but no figures of merit for long-term performance or power consumption are available.

DNA analyzer technology [21,80–84], developed by the same group, similarly uses off-chip pumps and valves (12 off, four of which are two-way arranged on a separate manifold) with PDMS to form microchannels, which are sealed against a temperature-controlled glass substrate. These create a network of tubing and off-chip systems supporting a microfluidic chip (see Fig. 2.4.1). The technique of driving reagents held in syringes with a working fluid is again used as is a photon-multiplier and electronics package within a pressure case. In this case, the temperature-controlled glass substrate forms zones of different temperature for steps in a Polymerase Chain Reaction (PCR) [85] assay. Specifically sample is mixed with lysis buffer to liberate DNA into solution in a delay fluid path and DNA is then captured on glass beads all at 50°C. Subsequently, this first heater is turned off for elution with a PCR buffer (to avoid inhibition experienced with standard wash buffers). The eluent is then mixed with a PCR mix (PCR buffer, deoxynucleotide triphosphates, primers, DNA polymerase, intercalating DNA reporting dye (SYBR green, Molecular Probes, Inc.) and Tween [nonionic surfactant principally to block reagent adsorbing to the PDMS]). This mixture then passes through a number of serpentine bends that take the resultant mixture over three thermostated heated zones: 98, 50, and 72°C for DNA denaturation, annealing, and extension, respectively. The resultant amplified products pass through a fiber optic excited and interrogated fluorescence cell before archiving of a subsample in a storage tube with successive samples separated by segmentation with oil. The whole chip is flushed with "DNA away" reagent and sampled prior to repeating the process for subsequent samples. Laboratory testing and field trials (to 1471 m) validated short-term operation (five field samples analyzed) and an LOD of between 10^3 and 10^4 cells/mL. Power consumption is not reported.

This group has also produced microfluidic chemical sensors for pH [80] and manganese (II) [80,86–89]. The latest reported device for Mn^{2+} uses multiple layers of PDMS to form: the microfluidic channels and networks, as well as connections to reagents, pumps, and flow sensors, sealing layers, and a thin internal deformable layer for the formation of integrated and hence low dead volume on-chip valves and flow regulators. The on-chip valves (normally closed configuration) and flow regulator (normally open configuration) are actuated with a working fluid (methanol) moved by miniature (commercial off the shelf) electro–osmotic pumps (EOPs). The chip is operated with a single pump operating at low pressure on the waste (exit) of the chip, and the system uses the valves and flow regulator (sample line) in conjunction with flow sensors to regulate flows and mixing ratios. This is extremely space, reagent, and power efficient. However, a photomultiplier tube (PMT) detector is required for the luminescence measurement and the total power consumption of the electronics (housed in a pressure case) is 7.2 W. The LOD is 280 nM and linear range 280–500 nM, but deployment durations of only 8 h are reported.

Though not yet applied to ocean sensing, a significant locus of in situ microfluidic environmental sensors has been developed in Dublin City University and in the group of Prof. Dermot Diamond in particular. The focus has been to produce low-cost sensor systems that can be used in environmental sensor networks [90]. Sensors demonstrated include phosphate [90–97], nitrite [91,98,99], nitrate [91,100,101], ammonia [91], Chemical Oxygen Demand (COD) [102], and pH [102–104]. Their fully autonomous field-deployable systems include a microfluidic chip made from

---

[5] 20 mAh at 12 V.

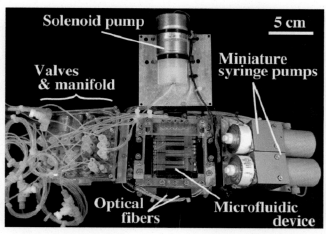

FIGURE 2.4.1 Photograph of the "IISA-Gene" in situ DNA analyzer showing modular elements connected with tubing from supplementary information for Ref. [21].

PMMA or cyclic olefin copolymers (COC) with integrated absorption cell actuated with either peristaltic [101] or syringe [104,105] pumps, check valves, and integrated with sample filter, batteries, electronics, GSM communication systems, and reagents, all housed in a robust air-filled (and hence limited depth rating) hinged-lid container. The optical cells used differ in each variant varying from 1 [92] to 3-mm [101,105] ID × 3 [92] to 28-mm [105] long. The pH measurement performance (0.04 pH precision, 0.11 pH accuracy) [104] and limits of detection achieved (phosphate 3 μM [92], nitrate 5 μM [101], nitrite 0.7 μM, and ammonia 0.8 μM [105]) are well matched to many terrestrial water applications but are not sufficient for many oceanographic applications.

Investment in biosensor research by the European Commission through the Oceans of Tomorrow (FP7-2013-OCEAN) funding also provided for the development of a number of microfluidic sensor systems.

The BRAAVOO project developed both bioreporter [106] and immunosensor [107]-based microfluidic devices with the aim of incorporation in a surface buoy. In such an arrangement the sensors would operate in air on samples automatically acquired from the near sea surface. However, publication of realization of such a setup is absent from the literature.

A demonstration of the bioreporter technology used *Escherichia coli* DH5α strain 1598 (pPR-ArsR-ABS) [106], which expresses enhanced green fluorescent protein and protein ArsR, and results in green fluorescent protein (GFP) expression in response to arsenite (As(III)). The relatively complex chip made from PDMS-included pneumatically driven hydraulically lines (filled with water) to operate on chip valves, and connections to pressurized nutrient, cleaning, and sample containers that obviate the need for additional pumps. The bioreporters are cultured on the chip and subcultures can be periodically moved into a reactor including a microfabricated PDMS filter/trap in which bioluminescence is read (using a fluorescence microscope with appropriate filters and a charge-coupled device [CCD] detector) on exposure to the sample. A response was observed for 0.47 μM arsenite (from $NaAsO_2$) within 40 min.

The immunosensor used a silicon interferometric channel waveguide with a portion of waveguiding cladding removed as a refractive index sensor. This senses the refractive index in an immobilized label-free competitive immunoassay. Here, the analyte is covalently bonded to the sensing layer and binding of target-specific antibodies decreases with increasing concentration of target in the sample. In demonstration (laboratory) experiments, nine successive measurements of Okadaic Acid concentration were performed with an LOD of 0.2 μg/L (approx. 0.2 nM), and maximum sensitivity at 1.3 μg/L (approx. 1.6 nM). Little detail of the chip is published, though a prototype is shown in [108].

The ENVIGUARD project produced microfluidic sensors similarly, for an air-filled surface buoy-based platform. Advances include the development of a macrofluidic sample-processing system to feed to each of the devices. One of these is an adaptation of the semiautomated electrochemical detection of microalgae (ALGADEC) device [109,110], which uses a multielectrode "chip" for algae species quantification using electrochemical detection. The assay uses algae species-specific 18S-DNA hybridization probes to form an immobilized sandwich for 18S-rRNA target with the upper probe labeled with digoxigenin. This label is detected by binding of a digoxigenin antibody itself labeled with the enzyme horseradish-peroxidase which promotes degradation of hydrogen peroxide in a supplied substrate. This reduction in peroxide can be detected electrochemically. This chip is operated within a macrofluidic analyzer using peristaltic pumps, valves and temperature control necessary for the assay. Detection limits for operation on purified total RNA were 6250 cells whereas interference from cellular contents in filtered lysate resulting in LOD of 25,000 cells.

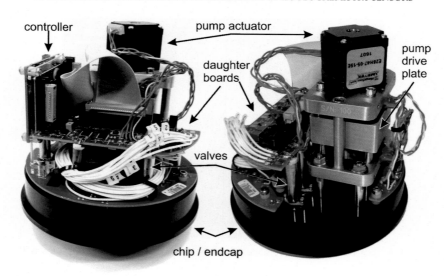

FIGURE 2.4.2   Photographs of internal views (front and back) of LOC chemical sensor. With valves and pumps mounted directly onto the chip, which is also the endcap, there are no internal connecting tubes. The electronics (controller and daughter boards) are tightly integrated.

The ENVIGUARD project also investigated the use of aptamer biosensors and microfluidic sensors using resonant nanopillars [111] for detection within a microfluidic biosensor, but results for the latter suggest that operation in aqueous phase is inferior to "dry" operation, suggesting that, at present, this method is unsuitable for in situ oceanographic sensors.

Other laboratory devices produced within this Oceans of Tomorrow program include: microfluidic chips formed from Graphene Oxide–PDMS composite for enhanced adsorption of heavy metals [112] and flame retardants [113] (SMS project), Lab-on-disc sensors for chemical pollutants and biotoxins operated within a macrofluidic system designed for use in a surface-buoy platform [114] (MARIABOX).

Also funded by the EU Oceans of Tomorrow program, the SenseOCEAN project augmented existing activity at the National Oceanography Center and the University of Southampton in the development of submersible oceanographic lab-on-chip chemical sensors. The family of sensors uses similar components and are typically pressure-balanced in design (including all electronics and optics) with a polymer microfluidic chip typically forming the end cap of the pressure-balanced housing. The chip also provides a mounting plane for electronics, off-chip pumps and valves, as well as recesses for integration of optical components to create an integrated and actuated optofluidic manifold (see Fig. 2.4.2).

The technology is based on stopped-flow analysis (whereby discrete measurements are taken after fluids are mixed and left to react for a set period), and utilizes tinted PMMA substrates to suppress stray and scattered light [115]. This is particularly useful for absorption spectroscopy and colorimetric analysis in which a range of optical path-lengths are required. Using a partially optically absorbing material at the wavelengths used for the analytical determinations enables sufficient suppression of stray and scattered light yet still allows high levels of light to be coupled into and out of the optical cell. This is optimized by thinning the PMMA, and hence reducing optical power loss in the PMMA in the region in which light is launched into or received from the cell (i.e., the absorption cell windows). This enables micromachining of axially illuminated absorption cells of varying lengths (e.g., 2.5 and 25 mm on a single chip [116]) with small cross sectional area, (typically $300\,\mu m \times 150\,\mu m$) and hence low volume in a single substrate. This results in excellent metrology performance: for example, a version of the technology targeting nitrate has a dynamic range of 0.025–$350\,\mu M$, an LOD of 25 nM and precision 7 nM. Time between measurements is typically 3–5 min (chemistry dependent) although techniques such as multiplexing [117] and more efficient pumping and valving will attempt to reduce this (see Section 2.4.4).

The chips are formed by micromilling and to further enhance optical and fluidic performance the resultant machining marks are reduced, and polishing achieved, using solvent vapor exposure prior to bonding [118]. This exposure enables solvent (typically chloroform) to condense on the surface of the chip, and due to their higher surface area-to-volume ratio machining marks absorb a higher concentration of solvent. With optimized exposure conditions the machine marks reflow and flatten. This process also softens the surface of the chip making it possible to form strong solvent-mediated bonds with further layers of PMMA with reduced risk of collapsing microchannels. This enables the manufacture of complex multilayer chips.

FIGURE 2.4.3 Photograph of LOC chemical sensor in pressure-balanced housing shown with blanking plug on one of the two electrical (power and data) connectors. Locking ring removed for clarity. No fluid connections are present (reagent/standards/input sample filter).

The chips are actuated with solenoid valves (LFNA1250325H, Lee Products Ltd., UK) and a custom-made multi-barrel syringe pump. The latter has a single drive plate actuating multiple plungers in the barrels of multiple pumps which creates accurately controlled ratios of flow determined by the cross-sectional area of the plunger/barrel. This is essential for high-performance colorimetric assays in which, for example, error-inducing variable dilution of the sample with reagent is obviated. These pumps and valves are mounted on the side of the chip enclosed within the pressure balanced housing and are bathed in oil. Electronics also included within the housing. They control the valves, pumps, and optical sources (LEDs) while also reading the outputs from optical detectors (typically photodiodes), thermistors or conductivity cells and deal with data logging, processing, communication and interfacing with users, platforms and wider ocean observing systems. The electronics, currently based on the SAM4L microcontroller (ARM, UK) have been custom designed and use components selected and extensively tested for inherent pressure tolerance. A flexible diaphragm is included in the pressure-balanced case to compensate for oil thermal expansion and compression under pressure. This pressure balanced design results in their being no need for expensive pressure cases or pressure differential tolerant penetrators or bulkhead connectors. The pressure balanced cases are formed from low-cost stock polymer pipe (see Fig. 2.4.3) and are tested to 6000 m depth equivalent pressure.

Reagents and standards are connected to the endcap. Most recently connecting tubes have been replaced by leak free push fit fluidic connectors and a replaceable reagent cartridge. Prior to this bags were connected via tubes and ¼″ 28 UNC fluidic connectors (e.g., see Fig. 2.4.4) and stored in a protective cylindrical case. The move to a reagent cassette reduces the user skill required to swap out fluids and enables users to rapidly setup or replenish sensors in the field.

To date the LOC systems that have been deployed have addressed nitrite [116,119,120], nitrate [116,119,121], phosphate [122–124], pH [125,126] and high-concentration Fe [127] with the deepest deployment[6] of 4800 m and the longest unattended deployment[7] of 1 year. During these deployments, a range of fouling environments

---

[6] Andrew Morris, NOC, personal communication.

[7] FixO3 mooring, Fram Strait, Alfred Wegener Institute Helmholtz Center for Polar and Marine Research, Alex Beaton personal communication.

(A)   (B)

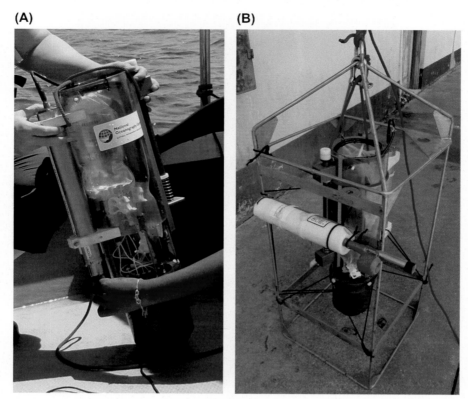

FIGURE 2.4.4   Photographs of deployment configurations (A) packaged to replace a Niskin bottle in a standard CTD rosette; (B) housed together with CTD on a mooring frame.

FIGURE 2.4.5   Photograph showing macrofouling on top of an LOC nitrate sensor after deployment in Chesapeake Bay, the device function was unaffected.

has been experienced (see Fig. 2.4.5). Despite expectations, fouling has not been a measureable problem in the vast majority of deployments. Only in one test in an extremely turbid river, as part of the Alliance for Coastal Technology's Nutrient Sensor Challenge, did filters clog with particles. We have only observed fouling affecting measurements in one deployment of the pH sensor due to growth on the inlet filter in a sunlit coastal deployment (Socratis Loucaides personal communication). Both issues are being resolved by redesigning the inlet filter (larger filters areas in high-turbidity environments, and applying biofilm control) or by using multiple filters, which are sequentially used.

The sensors consume 1–1.8 W when operating, consume approximately 300 J/measurement and have an extremely low-power sleep mode. In addition, the SAM4L microcontroller-based electronics and associated software allow ready integration with a wide range of ocean going vehicles and observing systems. A standard

(A)   (B)

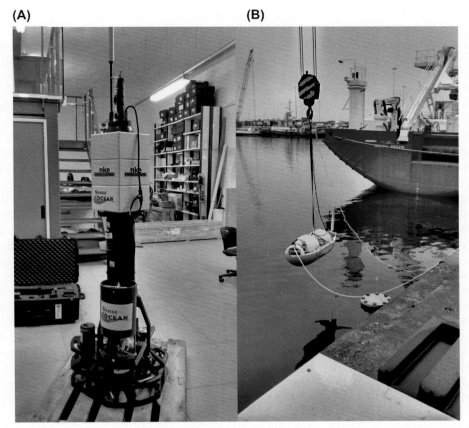

FIGURE 2.4.6   Photographs of LOC integration with autonomous observing systems (A) NKE Provor profiling float (also see Chapter 5.7); (B) autonomous underwater vehicle Autosub long range (ALR, a.k.a. BoatyMcBoatface).

method of interfacing and for managing data has been implemented[8]. Examples of where this has been applied include integration of the sensors with profiling floats as part of the SenseOCEAN project and AUVs as part of the European Union Horizon 2020 (EU H2020) Strategies for Environmental Monitoring of Marine Carbon Capture and Storage (STEMM-CCS) project and UK Energy Technologies Institute Carbon Capture and Storage– Measuring, Monitoring, and Verification (CCS-MMV) project (see Fig. 2.4.6). In both the latter projects, lab-on-chip sensors are used to examine the carbonate system and nutrients to enable identification of a Carbon Capture and Storage (CCS) reservoir emissions versus natural variability and biological activity by perturbation of seawater stoichiometry.

2.4.4 DEVELOPING IMPACT AND TAKE-UP OF LAB-ON-CHIP TECHNOLOGY

Despite the advances made in the development of microfluidic and LOC technologies, a considerable gap remains between current capability and that which should be achievable by the technology. Because of the technology's ability to measure with high performance and potentially low cost so many of the priority parameters and EOVs, it is well placed to fill the gap in current measurement capability to address these internationally agreed priorities. However, at the time of writing, none of the commercially available reagent-based chemical or biological sensors use lab-on-chip technology, and current in situ oceanographic lab-on-chip technology remains the preserve of a few laboratories in just a few countries, with the majority of activity focused in the United Kingdom and Japan.

Addressing this shortfall in real-world applications and hence realized impact requires technological improvements and innovations as well as acceleration of commercialization of the technology. The latter is essential as

[8] http://www.senseocean.eu/senseocean/sites/senseocean/files/documents/Deliverable%20D7.8%20Policy%20Document%20 Sensor%20Development%20for%20the%20Ocean%20of%20Tomorrow_r.pdf.

research labs do not have the capacity to manufacture large numbers of instruments to support global field programs. For example, the laboratories at the National Oceanography Center currently manufacture and deploy circa 60 sensors per year with many requests for international collaborations going unmet. An increase in production is not feasible without partnering, and indeed the existing throughput places pressures on the ability to develop new instruments. Transfer of the technology to industry is underway and should help address these problems. A similar model will be required for other developers of in situ ocean chemical sensors to achieve widespread impact.

The technical improvements and innovations that would increase impact and take-up of LOC technology include: improved longevity/robustness and ease of use; reduced cost; development of sensors for further parameters; and improved time response.

Improved longevity reduces cost of ownership and deployment ultimately reducing the cost per data point, but also providing benefits in terms of time overhead for owners. At present, only a handful of lab-on-chip sensors (earlier) have been deployed for durations greater than a few weeks. However, reagent-based sensors have been deployed for longer durations, and pioneering long-term LOC deployments show that it is feasible to produce sensors with sufficient reliability and robustness (see earlier). This requires reliability engineering of the sensors from the outset including choice and careful examination of materials, and engineering out expected failure modes.

In addition to the long-term reliability of the sensor, the principal limitations on deployment duration are reagent lifetime and energy storage. The benefits of size scaling associated with LOC mean that LOC sensors are already energy efficient and further improvement is likely [37]. However, reagent lifetime, including the stability of onboard standards and blanks, remains a challenge common to all reagent-based sensors. Although some assays, such as pH determination with spectrophotometric dyes [125,126,128–130], are stable for years, others have much shorter useable lifetimes in deployments, for example the orthophthalaldehyde ammonia assay [131]. Stabilization of assays and adjustment of recipes are, therefore, an important area of research. Results can be surprising; for example, it has been shown that the stability of reagents for the molecular assay NASBA (for analysis of RNA) can be extended from hours in the laboratory to months using dehydration and stabilization with sugars [132,133]. Deployment conditions can also alleviate this problem; for example, the Griess assay [134] for nitrate typically has an operational working lifetime of 3–6 months, but this was extended to over a year in the cold dark conditions experienced on the FixO3 mooring in the Fram Strait [135].

The ease of use of in situ ocean lab-on-chip sensors is improving with increased numbers of deployments and efforts to commercialize them. An increased emphasis on operational performance in the long term as required by industry and users, rather than publishable novelty (as often tolerated by academic inventors) will also be necessary. Particular challenges include: the supply and exchange of reagents that can be addressed with reagent cartridges and automated loading/quality control or calibration routines; eased integration that can be addressed with plug-and-play functionality and the use of common-interface standards; the development of user-friendly software and data processing back-end; and extended service intervals.

Although some LOC ocean sensors are already lower cost than other commercial sensor technologies, further cost reduction is required if they are to be deployed in the large numbers required to achieve global synoptic assessment of key biogeochemical parameters/EOVs. Here, simplification, and use of low-cost components is a key tool, though significant gains can be made through scaling with increased production volumes. An example of possible simplification is the use of pressure-balanced design removing the need for expensive pressure cases and connectors, or the use of droplet microfluidic platforms that require only a single actuator that operates the rotor of a miniature peristaltic pump operating on multiple reagents simultaneously [30]. An alternative strategy could be to use on-chip pumps and valves in stopped-flow architectures [117,136].

Lab-on-chip technology is well placed to address a number of unaddressed EOVs, EBVs, and parameters required by wider user groups including nutrients, the (inorganic) carbonate system, trace metals, particulate matter, organic carbon, phytoplankton and zooplankton biomass and diversity, and possibly further EBVs on abundance and distribution using eDNA approaches [137–145]. For chemical sensors this will require the development of improved sensors for ammonia, silicate, trace iron, total alkalinity, dissolved inorganic carbon, particulate matter, organic nutrients, and total organic carbon. These are all tractable with adaptation of existing reagent-based in situ LOC technology and assays. In addition, there is opportunity for LOC technology to better address the measurement of parameters that are currently addressed including nitrate, phosphate, pCO2 and pH.

For biological parameters (biomass, diversity EBVs, but also productivity and functioning), there are a number of technologies that could be applied. These include the development of microcytometry [146–157]. This technology passes a sample containing particles through an analysis region in which cells are individually counted and measured electrically and optically for size, shape, and optical properties (for example photosynthetic pigment fluorescence or scatter). These data can be used or combined to size, count, and discriminate phytoplankton and other particles present in natural samples at high rate (thousands of particles per second). This could enable estimation of microbial biomass, abundance, diversity and productivity. The challenge here is to realize a rugged deployable system at reasonable cost, and to demonstrate sufficient particle discrimination for operational applications.

Alternatively, nucleic acid analysis is a leading technology for essential biodiversity variable (EBV) measurement not least because it has demonstrated utility in laboratory studies and macrofluidic implementations [158–161]. It is also attractive because it can address abundance, diversity and function for microbes (by direct water analysis) and higher organisms through eDNA approaches. The challenge is to improve the robustness, longevity and performance of existing in situ LOC ocean sensors (see earlier). An intriguing possibility is the use of microfluidic modules within existing macrofluidic analyzers to increase the number and quality of measurements possible.

In some applications, such as profiling on rapidly moving vehicles, or those surveying particularly dynamic environments time response of current LOC systems (typically minutes) is a limitation. Further improving time response may improve energy efficiency if sensors can be operated for shorter periods. There are a number of technological solutions to this requirement including the application of droplet microfluidics [28–31]. Here, a sample and/or mixtures of sample and reagent is encapsulated into a stream of droplets, typically segmented by a hydrophobic oil phase including surfactant(s). Droplets may be formed at high rates up to the kHz range and prevent transfer of material between them. This is an effective preventative measure against dispersion [32], which otherwise blurs samples and reduces effective time response in continuous flow systems. Dispersion, the smearing out of fluid along the main long axis of a channel, is a direct result of the parabolic flow profile characteristic of unsegmented flow inside microfluidic channels, where the fluid near the walls moves much more slowly than the fluid in the center of the channel. Dispersion also has a negative effect on fluidic consumption, particularly as it exacerbates the need for flushing between samples. In segmented flow systems mixing is promoted within droplets as flow circulates within them in response to drag forces and their defined boundaries. Another advantage is that droplets maintain separation almost indefinitely and hence temporal information can be preserved and high sample frequency achieved even with assays with long reaction times. However, there are disadvantages: droplets are small in size (typically <1 mm) and thus requiring stretching or aggregation to enable the long optical path lengths required for ocean-relevant performance in many colorimetric assays, and the oils/segmenting phases used are gas permeable, which can be problematic in some assays. These issues are being addressed by current research efforts.

An alternative strategy for improved time response is to use multiplexed stop flow [117] in which dispersion is avoided by incubating reactions while the fluid is stationary. Temporal throughput is maintained by having multiple reaction or analysis channels that are sequentially loaded with sample and reagent. This approach requires large numbers of valves. Using off chip valves is problematic because of cost and because they have significant internal dead volumes and elasticity resulting in difficulty in achieving stopped flow in multiplexed systems. This suggests that on chip valve solutions will be required [117,136].

As applications of LOC sensors broaden to include wider parameters, including pollutants and toxins of regulatory and industrial interest, LOC biosensor and immunosensor technologies are likely to play a more significant role. They will be particularly adapted to quantification of low-concentration analytes such as organic pollutants for which there are existing recognition molecules or for which assays can be readily produced. The challenge for in situ ocean applications is to improve longevity and robustness while accommodating or reducing the complexity of the assay so that automation can be achieved.

Although the genesis of in situ ocean LOC and microfluidic sensors has been motivated by applications in oceanography, there is increasing interest in wider applications. These include: chemical and biological analysis in aquaculture, shellfisheries and fisheries waters. This is driving development of low-cost chemical sensors and nucleic acid and biosensors for pathogens, harmful algal bloom species and pollutants. There are also applications in direct human health protection, particularly as the technologies are transferred into coastal and inland waters. Here, the robustness and metrology performance of ocean LOC may be useful in quantification of natural pathogens and manmade toxins.

2.4.5 CONCLUSIONS

Building on a long history of in situ reagent-based ocean sensor development and deployment and significant research activity in microfluidic and LOC technology outside the oceanographic and environmental monitoring communities, in situ ocean sensors using microfluidics and LOC are on the cusp of widespread take-up and impact in operational settings. The technology is well suited to measurement of hard to measure and frequently dilute or rare chemical and biological parameters, many of which are priorities for the international community as reflected by their status as EOVs and EBVs. There are examples in which microfluidic LOC sensors are able to make measurements of higher quality than existing sensor technologies and of sufficient quality for operational oceanography and marine science. These include measurements of nutrients such as nitrate and phosphate as well as the carbonate system, trace metals, and, at least in demonstration deployments, nucleic acids. Further, numerous sensors and assays are in development with the aim of transfer from the laboratory to the operational setting. This includes sensors for organic pollutants and toxins detected with biosensors. However, there remain challenges to achieving the full potential of this technology principally achieving commercialization to enable widespread use which itself requires improved sensor longevity/robustness and ease of use; reduced cost; and in some cases improved time response. Addressing these challenges remains the focus of global research efforts.

Glossary

μM Micromoles, a unit of chemical concentration equal to a mole of the molecule of interest per L of fluid. Many ocean chemical parameters are in the range of micromoles down to picomoles.

μTAS Micro total analysis system, a class of technologies introduced in the late 1980s/early 1990s consisting of microfabricated devices integrating all elements required to transduce chemical signals into data.

ANAIS Autonomous nutrient analyzer in situ, an autonomous microfluidic platform for measuring nitrate, phosphate, and silicate concentrations in the ocean.

ATP Adenosine triphosphate, an energy-carrying organic chemical found in all lifeforms.

AUV Autonomous underwater vehicle, a class of robotic subsea systems that includes propelled devices and buoyancy-driven gliders. AUVs can be used as deployment platforms for small oceanographic sensors and can provide power and communications capabilities and can reduce the cost of sensor deployments by reducing the need for research ship travel.

CCS Carbon capture and storage, a method of reducing the concentration of carbon in the atmosphere by capturing and storing anthropogenically produced CO_2.

COC Cyclic olefin copolymer, a class of polymers used for microfluidic devices.

COD Chemical oxygen demand, a measurement of the amount of oxygen consumed by the chemical reactions occurring in fixed volume of water.

DIC Dissolved inorganic carbon, the sum of the inorganic carbon species dissolved in a volume of water. DIC is one of the four chemical parameters which are used to describe the oceanographic carbonate system.

EBV Essential biodiversity variables, a set of high-priority target parameters for quantifying and managing changes in biodiversity. The EBVs are selected and defined by the Group on Earth Observations.

EOP Electroosmotic pump, a pump which applies an electric field to an uncharged fluid in contact with a charged surface to generate flow.

EOV Essential ocean variables, a set of high-priority target parameters for ocean measurement. The EOVs are selected and defined by the Global Ocean Observing System.

GSM Global System for Mobile Communications, an international standard developed for cellular network communications.

ID/OD Inner diameter/outer diameter (for tubing).

LED Light-emitting diode, a low-cost light source that can be used for optical measurements of chemical reactions. LEDs typically output a wider range of wavelengths than lasers but are significantly lower in cost.

LOC Lab on chip.

LOD Limit of detection, a parameter used to quantify the performance of sensors. Although exact definitions vary, the LOD generally represents the lowest quantity of a substance that a sensor can reliably distinguish from a blank and is typically defined as three times the standard deviation of the baseline noise.

LCW Liquid core waveguide, a sensing mechanism in which the liquid analyte is passed through a tube or channel which is designed such that the analyte itself acts as an optical waveguide, thus confining and guiding the interrogating light. This can be accomplished, for example, by using a tubing material with a lower index of refraction as that of the liquid analyte.

NASBA Nucleic acid sequence-based amplification, a method of amplifying RNA signals at a constant temperature.

PCR Polymerase chain reaction, a method of massively amplifying a DNA sequence.

PDMS Polydimethylsiloxane, a transparent silicone-based elastomer often used for microfluidics prototyping.

Pe The Péclet number, a dimensionless number used in fluid mechanics to provide a size-independent description of transport phenomenon in a system. The Péclet number is the ratio of advective transport (i.e., transport by bulk fluid flow) to diffusive transport. In microfluidic systems, Pe is typically very low, indicating a relative dominance of diffusion over bulk fluid movement. The Péclet number is defined as $Pe = UX/D$, in which U is the average fluid velocity, X a representative length of the flow system (e.g., the width of a rectangular channel), and D the diffusion constant of the species of interest.

PEEK Polyether ether ketone, a thermoplastic polymer.

PMMA Poly(methyl methacrylate), a rigid, often transparent polymer which can be used for micromachining. A thermoplastic, PMMA can be milled or molded and is also known by various trade names include acrylic, Plexiglas, or Perspex.

PMT Photomultiplier tube, an optical detector capable of measuring extremely low levels of light.

RATS Robotic analyzer for the TCO_2 system.

Re The Reynolds number, an important dimensionless number used in fluid mechanics to provide a size-independent description of fluid behavior in a given system. The Reynolds number is a ratio of the inertial forces to the viscous forces in a fluid; a high Re (Re > 2000) indicates turbulent flow and a low Re indicates laminar flow. In microfluidics systems, Re < 100 is typical and Re ≈ 1 is common. The Reynolds number is defined as $Re = \rho UX/\mu$, in which ρ is the fluid density, U the average velocity in the channel, X a representative length of the flow system (e.g., the diameter of a tube), and μ the fluid dynamic viscosity.

References

[1] Manz A, Graber N, Widmer HM. Miniaturized total chemical-analysis systems - a novel concept for chemical sensing. Sens Actuators B Chem 1990;1(1–6):244–8.

[2] Becker H, Gartner C. Polymer microfabrication methods for microfluidic analytical applications. Electrophoresis 2000;21(1):12–26.

[3] Yetisen AK, Akram MS, Lowe CR. Paper-based microfluidic point-of-care diagnostic devices. Lab Chip 2013;13(12):2210–51.

[4] Weibel DB, DiLuzio WR, Whitesides GM. Microfabrication meets microbiology. Nat Rev Microbiol 2007;5(3):209–18.

[5] Jo BH, et al. Three-dimensional micro-channel fabrication in polydimethylsiloxane (PDMS) elastomer. J Microelectromech Syst 2000;9(1):76–81.

[6] Ohno K, Tachikawa K, Manz A. Microfluidics: Applications for analytical purposes in chemistry and biochemistry. Electrophoresis 2008;29(22):4443–53.

[7] Akyazi T, Basabe-Desmonts L, Benito-Lopez F. Review on microfluidic paper-based analytical devices towards commercialisation. Anal Chim Acta 2018;1001:1–17.

[8] Auroux PA, et al. Micro total analysis systems. 2. Analytical standard operations and applications. Anal Chem 2002;74(12):2637–52.

[9] Castro ER, Manz A. Present state of microchip electrophoresis: state of the art and routine applications. J Chromatogr A 2015;1382:66–85.

[10] Kudr J, et al. Microfluidic electrochemical devices for pollution analysis-a review. Sens Actuators B Chem 2017;246:578–90.

[11] Zhang J, et al. Fundamentals and applications of inertial microfluidics: a review. Lab Chip 2016;16(1):10–34.

[12] Murphy TW, et al. Recent advances in the use of microfluidic technologies for single cell analysis. Analyst 2018;143(1):60–80.

[13] Mills G, Fones G. A review of in situ methods and sensors for monitoring the marine environment. Sens Rev 2012;32(1):17–28.

[14] Daly KL, et al. Chemical and biological sensors for time-series research: current status and new directions. Mar Technol Soc J 2004;38(2):121–43.

[15] Patey MD, et al. Interferences in the analysis of nanomolar concentrations of nitrate and phosphate in oceanic waters. Anal Chim Acta 2010;673(2):109–16.

[16] Patey MD, et al. Determination of nitrate and phosphate in seawater at nanomolar concentrations. TrAC Trends Anal Chem 2008;27(2):169–82.

[17] Floor GH, et al. Combined uncertainty estimation for the determination of the dissolved iron amount content in seawater using flow injection with chemiluminescence detection. Limnol Oceanogr Methods 2015;13(12):673–86.

[18] Lohan MC, Aguilar-Islas AM, Bruland KW. Direct determination of iron in acidified (pH 1.7) seawater samples by flow injection analysis with catalytic spectrophotometric detection: application and intercomparison. Limnol Oceanogr Methods 2006;4:164–71.

[19] Shelley RU, et al. Determination of total dissolved cobalt in UV-irradiated seawater using flow injection with chemiluminescence detection. Limnol Oceanogr Methods 2010;8:352–62.

[20] Worsfold PJ, et al. Determination of dissolved iron in seawater: a historical review. Mar Chem 2014;166:25–35.

[21] Fukuba T, et al. Integrated in situ genetic analyzer for microbiology in extreme environments. RSC Adv 2011;1(8):1567–73.

[22] Lindstrom E, et al. A framework for ocean observing. UNESCO; 2012.

[23] Kissling WD, et al. Building essential biodiversity variables (EBVs) of species distribution and abundance at a global scale. Biol Rev 2018;93(1):600–25.

[24] Pereira HM, et al. Essential biodiversity variables. Science 2013;339(6117):277–8.

[25] Kamholz AE, Yager P. Theoretical analysis of molecular diffusion in pressure-driven laminar flow in microfluidic channels. Biophys J 2001;80(1):155–60.

[26] Kamholz AE, Yager P. Molecular diffusive scaling laws in pressure-driven microfluidic channels: deviation from one-dimensional Einstein approximations. Sens Actuators B Chem 2002;82(1):117–21.

[27] Janasek D, Franzke J, Manz A. Scaling and the design of miniaturized chemical-analysis systems. Nature 2006;442(7101):374–80.

[28] Teh SY, et al. Droplet microfluidics. Lab Chip 2008;8(2):198–220.

[29] Chong ZZ, et al. Active droplet generation in microfluidics. Lab Chip 2016;16(1):35–58.

[30] Nightingale AM, et al. Phased peristaltic micropumping for continuous sampling and hardcoded droplet generation. Lab Chip 2017;17(6):1149–57.

[31] Shang LR, Cheng Y, Zhao YJ. Emerging droplet microfluidics. Chem Rev 2017;117(12):7964–8040.

[32] Stroock AD, et al. Chaotic mixer for microchannels. Science 2002;295(5555):647–51.

[33] McLaughlin FA, Carmack EC. Deepening of the nutricline and chlorophyll maximum in the Canada Basin interior, 2003–2009. Geophys Res Lett 2010;37(24).

[34] Kroger S, Law RJ. Sensing the sea. Trends Biotechnol 2005;23(5):250–6.

[35] Kroger S, Law RJ. Biosensors for marine applications - we all need the sea, but does the sea need biosensors? Biosens Bioelectron 2005;20(10):1903–13.

[36] Jokerst JC, Emory JM, Henry CS. Advances in microfluidics for environmental analysis. Analyst 2012;137(1):24–34.

[37] Nightingale AM, Beaton AD, Mowlem MC. Trends in microfluidic systems for in situ chemical analysis of natural waters. Sens Actuators B Chem 2015;221:1398–405.

[38] Bagshaw EA, et al. Chemical sensors for in situ data collection in the cryosphere. TrAC Trends Anal Chem 2016;82:348–57.

[39] Tsaloglou MN. In: Tsaloglou MN, editor. Microfluidics and in situ sensors for microalgae. Microalgae: current research and applications. Wymondham: Caister Academic Press; 2016. p. 133–51.

[40] Degrandpre MD, Bellerby RGJ. Chemical sensors in marine science. Oceanus 1995;38(1):30–2.

[41] Jannasch HW, Johnson KS, Sakamoto CM. Submersible, osmotically pumped analyzers for continuous determination of nitrate in-situ. Anal Chem 1994;66(20):3352–61.

[42] Hanson AK. In: A new in situ chemical analyzer for mapping coastal nutrient distributions in real time. Oceans 2000 Mts/Ieee - where Marine Science and Technology Meet, vols. 1–3, Conference Proceedings. IEEE; 2000. p. 1975–82.

[43] Steimle ET, Kaltenbacher EA, Byrne RH. In situ nitrite measurements using a compact spectrophotometric analysis system. Mar Chem 2002;77(4):255–62.

[44] Callahan MR, Rose JB, Byrne RH. Long pathlength absorbance spectroscopy: trace copper analysis using a 4.4 m liquid core waveguide. Talanta 2002;58(5):891–8.

[45] Callahan MR, Kaltenbacher EA, Byrne RH. In-situ measurements of Cu in an estuarine environment using a portable spectrophotometric analysis system. Environ Sci Technol 2004;38(2):587–93.

[46] Byrne RH, Kaltenbacher EA. Construction and intensive field testing of SEAS-II sensors for trace element, nutrient and CO_2 system analyses. Office of Naval Research; 2003. Grant ref. #N00014-03-1-0612.

[47] Byrne RH, et al. Spectrophotometric measurement of total inorganic carbon in aqueous solutions using a liquid core waveguide. Anal Chim Acta 2002;451(2):221–9.

[48] Adornato LR, et al. High-resolution in situ analysis of nitrate and phosphate in the oligotrophic ocean. Environ Sci Technol 2007;41(11):4045–52.

[49] Liu XW, et al. Spectrophotometric measurements of pH in-situ: laboratory and field evaluations of instrumental performance. Environ Sci Technol 2006;40(16):5036–44.

[50] Sayles FL, Eck C. An autonomous instrument for time series analysis of TCO2 from oceanographic moorings. Deep Sea Res Oceanogr Res Pap 2009;56(9):1590–603.

[51] Chierici M, Fransson A, Anderson LG. Influence of m-cresol purple indicator additions on the pH of seawater samples: correction factors evaluated from a chemical speciation model. Mar Chem 1999;65(3):281–90.

[52] Degrandpre MD, et al. In-situ measurements of seawater pCO_2. Limnol Oceanogr 1995;40(5):969–75.

[53] Gray SEC, et al. Applications of in situ pH measurements for inorganic carbon calculations. Mar Chem 2011;125(1–4):82–90.

[54] Seidel MP, DeGrandpre MD, Dickson AG. A sensor for in situ indicator-based measurements of seawater pH. Mar Chem 2008;109(1–2):18–28.

[55] Spaulding R, DeGrandpre M, Harris K. Autonomous pH and pCO(2) Measurements in marine environments Quantifying the inorganic carbon system with in-situ Sami technology. Sea Technol 2011;52(2). 15-+.

[56] Spaulding RS, et al. Autonomous in situ measurements of seawater alkalinity. Environ Sci Technol 2014;48(16):9573–81.

[57] Martz TR, et al. A submersible autonomous sensor for spectrophotometric pH measurements of natural waters. Anal Chem 2003;75(8):1844–50.

[58] Degrandpre MD. A renewable-reagent fiber optic sensor for ocean pCO_2. In: Lieberman RA, editor. Chemical, biochemical, and environmental fiber sensors Iii, vol. 1587. 1992. p. 60–6.

[59] Degrandpre MD. Measurement of seawater pCO_2 using a renewable-reagent fiber optic sensor with colorimetric detection. Anal Chem 1993;65(4):331–7.

[60] Wang ZA, et al. In situ sensor technology for simultaneous spectrophotometric measurements of seawater total dissolved inorganic carbon and pH. Environ Sci Technol 2015;49(7):4441–9.

[61] Wang ZA, Chu SN, Hoering KA. High-frequency spectrophotometric measurements of total dissolved inorganic carbon in seawater. Environ Sci Technol 2013;47(14):7840–7.

[62] Wang ZHA, et al. Simultaneous spectrophotometric flow-through measurements of pH, carbon dioxide fugacity, and total inorganic carbon in seawater. Anal Chim Acta 2007;596(1):23–36.

[63] Wang ZA, et al. A long pathlength liquid-core waveguide sensor for real-time pCO(2) measurements at sea. Mar Chem 2003;84(1–2):73–84.

[64] Wang Z, et al. A long pathlength spectrophotometric pCO_2 sensor using a gas-permeable liquid-core waveguide. Talanta 2002;57(1):69–80.

[65] Li QL, et al. Automated spectrophotometric analyzer for rapid single-point titration of seawater total alkalinity. Environ Sci Technol 2013;47(19):11139–46.

[66] Plant JN, et al. NH_4-Digiscan: an in situ and laboratory ammonium analyzer for estuarine, coastal, and shelf waters. Limnol Oceanogr Methods 2009;7(2):144–56.

[67] Barnard AH, et al. Real-time and long-term monitoring of phosphate using the in-situ CYCLE sensor. In: Oceans 2009, vols. 1–3. 2009. p. 1698–703.

[68] Moscetta P, et al. Instrumentation for continuous monitoring in marine environments. In: Oceans 2009, vols. 1–3. 2009. 737-+.

[69] Vuillemin R, et al. Continuous Nutrient automated monitoring on the Mediterranean Sea using in situ flow analyser. In: Oceans 2009, vols. 1–3. 2009. 517-+.

[70] Bodini S, et al. Automated micro Loop Flow Reactor technology to measure nutrients in coastal water: state of the art and field application. In: Oceans 2015-Genova. 2015.

[71] Thouron D, et al. An autonomous nutrient analyzer for oceanic long-term in situ biogeochemical monitoring. Anal Chem 2003;75(11):2601–9.

[72] Sarradin PM, et al. EXtreme ecosystem studies in the deep OCEan: Technological developments. In: Chung JS, et al., editor. Proceedings of the fourteenth. 2004. p. 738–45.

[73] Vuillemin R, et al. CHEMINI: a new in situ CHEmical MINIaturized analyzer. Deep Sea Res Part I Oceanogr Res Pap 2009;56(8):1391–9.

[74] Laes-Huon A, et al. Long-term in situ survey of reactive iron concentrations at the EMSO-Azores observatory. IEEE J Ocean Eng 2016;41(4):744–52.

[75] Graiver D, Farminer KW, Narayan R. A review of the fate and effects of silicones in the environment. J Polym Environ 2003;11(4):129–36.

[76] Hanatani K, et al. Development of in situ microbial ATP analyzer and internal standard calibration method. In: 2015 IEEE underwater technology (Ut). 2015.

[77] Fukuba T, et al. A microfluidic in situ analyzer for ATP quantification in ocean environments. Lab Chip 2011;11(20):3508–15.

[78] Aoki Y, et al. Development of "IISA-ATP" system for in situ microbial activity assessment in deep-sea environment. In: Oceans 2008-Mts/Ieee Kobe Techno-Ocean, vols. 1–3. 2008. p. 548–51.

[79] Fukuba T, et al. Development of an integrated in situ analyzer for quantitative analysis of microbial ATP in aquatic environments. In: 2007 symposium on underwater technology and Workshop on scientific use of submarine cables and related technologies, vols. 1 and 2. 2007. p. 240–4.

[80] Fuji T, Fukuba T. Microfluidics-based in situ biological and chemical sensing - towards integrated and real-time measurement in Deep Sea. In: 2007 symposium on underwater technology and Workshop on scientific use of submarine cables and related technologies, vols. 1 and 2. IEEE; 2007. p. 210.

[81] Fukuba T, et al. Development of miniaturized in situ analysis devices for biological and chemical oceanography. In: 2005 3rd IEEE/EMBS special topic conference on microtechnology in medicine and biology. 2005. p. 56–9.

[82] Matsunaga M, et al. Microfabricated devices for DNA extraction toward realization of deep-sea in situ gene analysis. In: Oceans '04 Mts/Ieee Techno-ocean '04, Vols. 1-2, Conference Proceedings, vols. 1–4. 2004. p. 89–94.

[83] Fukuba T, et al. Microfabricated flow-through device for DNA amplification - towards in situ gene analysis. Chem Eng J 2004;101(1–3):151–6.

[84] Fukuba T, et al. Microfabricated flow-through PCR device for underwater microbiological study. In: Proceedings of the 2002 international symposium on underwater technology. 2002. p. 101–5.

[85] Saiki R, et al. Enzymatic amplification of beta-globin genomic sequences and restriction site analysis for diagnosis of sickle cell anemia. Science 1985;230(4732):1350–4.

[86] Takagi N, et al. Development of microfabricated in situ chemical analyzer. In: Oceans '04 Mts/Ieee Techno-ocean '04, Vols. 1-2, Conference Proceedings, vols. 1–4. 2004. p. 83–8.

[87] Provin C, et al. Development of integrated in situ analyzer for manganese (IISA-Mn) in deep sea environment. In: 2007 Symposium on underwater technology and Workshop on scientific use of submarine cables and related technologies, vols. 1 and 2. 2007. 658-+.

[88] Provin C, et al. Integrated in situ analyzer for manganese (IISA-Mn) for deep sea environment. In: Oceans 2008-Mts/Ieee Kobe Techno-ocean, vols. 1–3. 2008. p. 208–12.

[89] Provin C, et al. An integrated microfluidic system for manganese anomaly detection based on chemiluminescence: description and practical use to discover hydrothermal plumes near the Okinawa trough. IEEE J Ocean Eng 2013;38(1):178–85.

[90] Diamond D, et al. Wireless sensor networks and chemo-/biosensing. Chem Rev 2008;108(2):652–79.

[91] Cogan D, et al. Next generation autonomous chemical sensors for environmental monitoring. In: Lekkas TD, editor. Proceedings of the 13th international conference on environmental science and technology. 2013.

[92] Slater C, et al. Validation of a fully autonomous phosphate analyser based on a microfluidic lab-on-a-chip. Water Sci Technol 2010;61(7):1811–8.

[93] Cleary J, et al. An autonomous microfluidic sensor for phosphate: on-site analysis of treated wastewater. IEEE Sensor J 2008;8(5–6):508–15.

[94] Slater C, et al. Autonomous field-deployable device for the measurement of phosphate in natural water. In: VoDinh VT, Lieberman RA, Gauglitz G, editors. Advanced environmental, chemical, and biological sensing technologies. 2007.

[95] McGraw CM, et al. Autonomous microfluidic system for phosphate detection. Talanta 2007;71(3):1180–5.

[96] Hayes J, et al. Intelligent environmental sensing with a phosphate monitoring system and online resources. In: Simos TE, Maroulis G, editors. Computation in modern science and engineering, vol. 2. 2007. p. 1216–9. Pts a and B.

[97] Cleary J, et al. Field-deployable microfluidic sensor for phosphate in natural waters. In: 2007 IEEE sensors, vols. 1–3. 2007. p. 1001–4.

[98] Czugala M, et al. CMAS: fully integrated portable centrifugal microfluidic analysis system for on-site colorimetric analysis. RSC Adv 2013;3(36):15928–38.

[99] Czugala M, et al. Portable integrated microfluidic analytical platform for the monitoring and detection of nitrite. Talanta 2013;116:997–1004.

[100] Cogan D, et al. Integrated flow analysis platform for the direct detection of nitrate in water using a simplified chromotropic acid method. Anal Methods 2013;5(18):4798–804.

[101] Cogan D, et al. Development of a low cost microfluidic sensor for the direct determination of nitrate using chromotropic acid in natural waters. Anal Methods 2015;7(13):5396–405.

[102] Cleary J, et al. Microfluidic analyser for pH in water and wastewater. In: Lekkas TD, editor. Proceedings of the 13th international conference on environmental science and technology. 2013.

[103] Czugala M, et al. Optical sensing system based on wireless paired emitter detector diode device and ionogels for lab-on-a-disc water quality analysis. Lab Chip 2012;12(23):5069–78.

[104] Sansalvador I, et al. Autonomous reagent-based microfluidic pH sensor platform. Sens Actuators B Chem 2016;225:369–76.

[105] Cogan D, et al. The development of an autonomous sensing platform for the monitoring of ammonia in water using a simplified Berthelot method. Anal Methods 2014;6(19):7606–14.

[106] Buffi N, et al. An automated microreactor for semi-continuous biosensor measurements. Lab Chip 2016;16(8):1383–92.

[107] Chocarro-Ruiz B, et al. Nanophotonic interferometric immunosensors for label-free and real-time monitoring of chemical contaminants in marine environment. In: VoDinh T, Lieberman RA, editors. Advanced environmental, chemical, and biological sensing technologies Xiv. 2017.

[108] Chocarro-Ruiz B, et al. Nanophotonic label-free biosensors for environmental monitoring. Curr Opin Biotechnol 2017;45:175–83.

[109] Diercks-Horn S, et al. The ALGADEC device: a semi-automated rRNA biosensor for the detection of toxic algae. Harmful Algae 2011;10(4):395–401.

[110] Diercks S, Metfies K, Medlin LK. Development and adaptation of a multiprobe biosensor for the use in a semi-automated device for the detection of toxic algae. Biosens Bioelectron 2008;23(10):1527–33.

[111] Hernandez AL, et al. How the surrounding environment affects the biosensing performance of resonant nanopillars arrays: under dry conditions or immersed in fluid. Sens Actuators B Chem 2018;259:956–62.

[112] Chalupniak A, Merkoci A. Graphene oxide-poly(dinnethylsiloxane)-based lab-on-a-chip platform for heavy-metals preconcentration and electrochemical detection. ACS Appl Mater Interfaces 2017;9(51):44766–75.

[113] Chalupniak A, Merkoci A. Toward integrated detection and graphene-based removal of contaminants in a lab-on-a-chip platform. Nano Res 2017;10(7):2296–310.

[114] Bonasso M, et al. MariaBox: first prototype of a novel instrument to observe natural and chemical pollutants in seawater. In: OCEANS 2017-Aberdeen. 2017.

[115] Floquet CFA, et al. Nanomolar detection with high sensitivity microfluidic absorption cells manufactured in tinted PMMA for chemical analysis. Talanta 2011;84(1):235–9.

[116] Beaton AD, et al. Lab-on-chip measurement of nitrate and nitrite for in situ analysis of natural waters. Environ Sci Technol 2012;46(17):9548–56.

[117] Ogilvie IRG, et al. Temporal optimization of microfluidic colorimetric sensors by use of multiplexed stop-flow architecture. Anal Chem 2011;83(12):4814–21.

[118] Ogilvie IRG, et al. Reduction of surface roughness for optical quality microfluidic devices in PMMA and COC. J Micromech Microeng 2010;20(6).

[119] Yucel M, et al. Nitrate and nitrite variability at the seafloor of an oxygen minimum zone revealed by a novel microfluidic in-situ chemical sensor. PLoS One 2015;10(7).

[120] Beaton AD, et al. An automated microfluidic colourimetric sensor applied in situ to determine nitrite concentration. Sens Actuators B Chem 2011;156(2):1009–14.

[121] Beaton AD, et al. High-resolution in situ measurement of nitrate in runoff from the Greenland ice sheet. Environ Sci Technol 2017;51(21):12518–27.

[122] Clinton-Bailey GS, et al. A lab-on-a-chip analyzer for in situ measurement of soluble reactive phosphate: improved phosphate blue assay and application to fluvial monitoring. Environ Sci Technol 2017;51(17):9989–95.

[123] Legiret FE, et al. A high performance microfluidic analyser for phosphate measurements in marine waters using the vanadomolybdate method. Talanta 2013;116:382–7.

[124] Grand MM, et al. A lab-on-chip phosphate analyzer for long-term in situ monitoring at fixed observatories: optimization and performance Evaluation in Estuarine and oligotrophic coastal waters. Front Mar Sci 2017;4(255).

[125] Rerolle VMC, et al. Development of a colorimetric microfluidic pH sensor for autonomous seawater measurements. Anal Chim Acta 2013;786:124–31.

[126] Rerolle VMC, et al. Seawater-pH measurements for ocean-acidification observations. TrAC Trends Anal Chem 2012;40:146–57.

[127] Milani A, et al. Development and application of a microfluidic in-situ analyzer for dissolved Fe and Mn in natural waters. Talanta 2015;136:15–22.

[128] Papadimitriou S, et al. The measurement of pH in saline and hypersaline media at sub-zero temperatures: characterization of Tris buffers. Mar Chem 2016;184:11–20.

[129] Loucaides S, et al. Characterization of meta-cresol purple for spectrophotometric pH measurements in saline and hypersaline media at sub-zero temperatures. Sci Rep 2017;7.

[130] Papadimitriou S, et al. The stoichiometric dissociation constants of carbonic acid in seawater brines from 298 to 267 K. Geochem Cosmochim Acta 2018;220:55–70.

[131] Bey S, et al. A high-resolution analyser for the measurement of ammonium in oligotrophic seawater. Ocean Dyn 2011;61(10):1555–65.

[132] Tsaloglou MN, et al. Real-time isothermal RNA amplification of toxic marine microalgae using preserved reagents on an integrated microfluidic platform. Analyst 2013;138(2):593–602.

[133] Loukas CM, et al. Detection and quantification of the toxic microalgae Karenia brevis using lab on a chip mRNA sequence-based amplification. J Microbiol Methods 2017;139:189–95.

[134] Griess P. Bemerkungen zu der Abhandlung der HH. Weselsky und Benedikt „Ueber einige Azoverbindungen". Berichte der deutschen chemischen Gesellschaft 1879;12(1):426–8.

[135] Polarstern Cruise report PS99.2. 2016.

[136] Ogilvie IRG, et al. Chemically resistant microfluidic valves from Viton (R) membranes bonded to COC and PMMA. Lab Chip 2011;11(14):2455–9.

[137] Foote AD, et al. Investigating the potential use of environmental DNA (eDNA) for genetic monitoring of marine mammals. PLoS One 2012;7(8).

[138] Thomsen PF, et al. Detection of a diverse marine fish fauna using environmental DNA from seawater samples. PLoS One 2012;7(8).

[139] Valentini A, et al. Next-generation monitoring of aquatic biodiversity using environmental DNA metabarcoding. Mol Ecol 2016;25(4):929–42.

[140] Bakker J, et al. Environmental DNA reveals tropical shark diversity in contrasting levels of anthropogenic impact. Sci Rep 2017;7.

[141] Minamoto T, et al. Environmental DNA reflects spatial and temporal jellyfish distribution. PLoS One 2017;12(2).

[142] Morard R, et al. Planktonic foraminifera-derived environmental DNA extracted from abyssal sediments preserves patterns of plankton macroecology. Biogeosciences 2017;14(11):2741–54.

[143] Stat M, et al. Ecosystem biomonitoring with eDNA: metabarcoding across the tree of life in a tropical marine environment. Sci Rep 2017;7.

[144] Stoeckle MY, Soboleva L, Charlop-Powers Z. Aquatic environmental DNA detects seasonal fish abundance and habitat preference in an urban estuary. PLoS One 2017;12(4).

[145] Salter I. Seasonal variability in the persistence of dissolved environmental DNA (eDNA) in a marine system: the role of microbial nutrient limitation. PLoS One 2018;13(2).

[146] Benazzi G, et al. Discrimination and analysis of phytoplankton using a microfluidic cytometer. IET Nanobiotechnol 2007;1(6):94–101.

[147] Ateya DA, et al. The good, the bad, and the tiny: a review of microflow cytometry. Anal Bioanal Chem 2008;391(5):1485–98.

[148] Golden JP, et al. Multi-wavelength microflow cytometer using groove-generated sheath flow. Lab Chip 2009;9(13):1942–50.

[149] Ligler FS, et al. Microflow cytometer. In: Fauchet PM, editor. Frontiers in pathogen detection: from nanosensors to systems. 2009.

[150] Ligler FS, et al. A multiwavelength microflow cytometer. NRL Rev 1988;2009:152–4.

[151] Barat D, et al. Design, simulation and characterisation of integrated optics for a microfabricated flow cytometer. Optic Commun 2010;283(9):1987–92.

[152] Golden JP, et al. A microflow cytometer on a chip. In: Fauchet PM, Miller BL, editors. Frontiers in pathogen detection: from nanosensors to systems. 2010.

[153] Kim JS, Ligler FS. Microflow cytometer. Microflow Cytom 2010:1–379.

[154] Hashemi N, et al. Microflow cytometer for optical analysis of phytoplankton. Biosens Bioelectron 2011;26(11):4263–9.

[155] Hashemi N, et al. Optofluidic characterization of marine algae using a microflow cytometer. Biomicrofluidics 2011;5(3).

[156] Barat D, et al. Simultaneous high speed optical and impedance analysis of single particles with a microfluidic cytometer. Lab Chip 2012;12(1):118–26.

[157] Golden JP, et al. A microflow cytometer for optical analysis of phytoplankton. In: Miller BL, Fauchet PM, editors. Frontiers in biological detection: from nanosensors to systems Iv. 2012.

[158] Scholin C, et al. Remote detection of marine microbes, small invertebrates, harmful algae, and biotoxins using the environmental sample processor (ESP). Oceanography 2009;22(2):158–67.

[159] Ryan J, et al. Harmful phytoplankton ecology studies using an autonomous molecular analytical and ocean observing network. Limnol Oceanogr 2011;56(4):1255–72.

[160] Harvey JBJ, et al. Robotic sampling, in situ monitoring and molecular detection of marine zooplankton. J Exp Mar Biol Ecol 2012;413:60–70.

[161] Pargett DM, et al. Development of a mobile ecogenomic sensor. In: Oceans 2015-Mts/IEEE Washington. New York: IEEE; 2015.

Ocean In Situ Sensors: New Developments in Biological Sensors

Plankton Needs and Methods

Jochen Wollschläger[1,2], Klas Ove Möller[2]

[1]Institute for Chemistry and Biology of the Marine Environment, Carl von Ossietzky University of Oldenburg, Wilhelmshaven, Germany; [2]Institute of Coastal Research, Helmholtz-Zentrum Geesthacht, Centre for Materials and Coastal Research, Geesthacht, Germany

3.1.1 INTRODUCTION

The term plankton describes all water-living organisms the inherent movement abilities for which are small compared to the movement of the surrounding water. Strictly speaking, this also includes comprehensively large organisms like jellyfish, but most planktonic organisms are of relatively small size (micrometers to millimeters). Phytoplankton as the "plant" part of the plankton community comprises a large variety of photosynthetically active organisms, both prokaryotic and eukaryotic, which as primary producers transform inorganic carbon into biomass using light as their source of energy. Although they constitute only 1% of global primary-producer biomass, these organisms are responsible for approximately 50% of global carbon fixation and oxygen production [1] and, hence, mediate carbon flux from the atmosphere to the oceans. The heterotrophic zooplankton, the "animal" part of plankton, in contrast, are the primary consumers of phytoplankton, and in turn a major food source for larger organisms. Thus, zooplankton link phytoplankton production to higher trophic levels (e.g., fish), the members of which are relevant for human economy and nutrition. In total, plankton constitute the base of the aquatic food web, and due to this crucial ecological position, investigation and monitoring of its dynamics in terms of biomass and community structure is a topic of ongoing importance. Plankton populations can provide an indicator of the status of the marine environment [2,3]. Thus, especially in the light of climate change, ocean acidification, introduction of foreign species, and various types of pollution due to human activities, potential changes in the plankton have to be detected and evaluated to estimate potential consequences for marine ecosystem services [4].

However, compared to terrestrial ecosystems, the aquatic, and especially the vast marine environment, is substantially less-well-sampled. The reason is lower accessibility, associated with relatively high costs due to, e.g., ship time. Furthermore, with laboratory-based analysis of discrete samples by traditional approaches like microscopy, or modern, but sophisticated, methods like molecular biology, often only a relatively small number of samples can be taken and processed in a given time. In the light of the high spatiotemporal variability of the marine environment, phenomena like plankton patchiness, migratory patterns, and temporal dynamics might not be adequately resolved by these means, which also hinders understanding of underlying processes. Furthermore, plankton samples are usually processed after the cruise and thus require preservation that might introduce additional sources of bias [3]. Thus, in situ methods that allow the investigation of plankton in its natural environment and that are capable of high-frequency measurements are necessary to close gaps in data acquisition [4,5]. However, this chapter will not only consider submersible instruments (measuring in situ in a strict sense), but include also benchtop instruments that are capable of analyzing a continuously pumped water stream in high frequency (quasi-in situ measurements). Although there has been some progress in the automation of molecular methods and in their adaptation for in situ measurements [6–9] and acoustic methods have a long standing in the observation of plankton and fish [10], most in situ measurements of plankton are nowadays based on optics. For this reason, they will be the focus of this chapter.

Optical methods are valuable tools because they are comparatively inexpensive, convenient to use, automatable, and allow high measurement frequency. Of course, there often exists a trade-off between these advantages and the level of detail in the information the sensors can provide. Nevertheless, optical sensors can tackle the plankton community on various levels of detail, from bulk biomass over functional groups (size classes, phytoplankton pigment groups) to specific species. Thus, these measurements can be considered as valuable additions to traditional analyses of discrete samples.

In principle, there are two approaches for investigating plankton in situ by optical means: The first is that plankton is to a certain degree optically active, either by changing the inherent optical properties of the water (absorption, scattering), or by showing fluorescence. Thus, by measuring these parameters, information about plankton biomass and—to a certain degree—plankton composition can be obtained. However, these methods are primarily applicable to phytoplankton, which are (due to various photosynthetic pigments) more optically active than zooplankton. The second approach targets plankton on a more individual level by imaging, taking advantage of the continuously improving possibilities of image analysis. For both approaches, a variety of instruments are available today ([11,12,13,75] and references therein). See also the publication of Moore et al. [12] for a list of manufacturers. The working principles of the most common optical approaches are summarized in the following.

3.1.2 METHODS TAKING ADVANTAGE OF OPTICAL PROPERTIES OF PLANKTON

3.1.2.1 Measurements of the Inherent Optical Properties of Water

When a light beam passes through a water body, its intensity is reduced as the result of two processes: absorption and scattering [14]. They are inherent optical properties (IOPs) of the water determined by the water itself and all its different constituents like chromophoric dissolved organic matter (CDOM), phytoplankton cells with their various pigments, nonpigmented plankton, or suspended sediments. The sum of both processes determines the total attenuation of the light beam.

Absorption takes place if the energy of a respective photon corresponds to one of the resonance levels of the molecules the light is interacting with [15]. In that case, its energy is transferred to that molecule, whereas the photon itself is eliminated; what results is a detectable loss of light. Because the photon's energy depends on its wavelength and it has to match a certain resonance level, absorption is a wavelength-specific process that is linked to the structure of the absorbing molecule. Thus, different substances have characteristic absorption spectra. Pure water, for example, shows high absorption in the red part of the visible spectrum [16], which is the reason why the clear open ocean appears blue to the human eye. However, depending on the type and concentration of other constituents present in the water, the absorption spectrum will vary considerably.

As stated before, not only absorption, but also scattering, contributes to the observable reduction in light intensity when light passes a water sample. In this form of interaction, the energy of the photon is not, or only partly, reduced (elastic or inelastic scattering), but its direction of movement is altered. The magnitude and angles of scattering are not uniformly distributed, but determined by the volume-scattering function (VSF) and depend on size, shape, concentration, and composition of the particles present [15]. The observable loss of light is caused in this case by photons that are scattered out of the field of view of the detector. Detectors are commonly a set of photomultipliers or photodiodes (channels), which simply measure the number of photons arriving. Depending on the number of channels available, the measurements can be performed at single or multiple wavelengths, or even in high resolution over the whole spectrum (hyperspectral). The latter is advantageous, because it provides maximum spectral information. With respect to the measurement, the differentiation between the IOPs is realized by the geometrical setup of the instrument with respect to position of the light source and the detector, as well as by the design of the measurement cuvette.

Transmissometers or *attenuation meters* consist of a collimated light source opposite a detector, whereas the space between is filled with the sample. They measure the combined light loss caused by absorption and scattering and give, therefore, a general measure of water clarity. The contribution of absorption to the signal depends on the wavelength(s) chosen for the measurement. For example, the use of green light minimizes the influence of chlorophyll-*a* (chl-*a*) absorption on light attenuation. This allows flexibility in the measurements with respect to their purpose, but still a specific monitoring of plankton is not possible, because, for example, the instrument cannot differentiate between living and nonliving material. In clear waters where plankton is nearly the only source of absorption and scattering, the attenuation signal will be a good proxy of plankton abundance, but in waters with high sediment loads, this relationship will be much less direct.

Scattering meters (or *scatterometers*) are designed to measure predominantly the light redirected by particles present in a water sample. Commonly, the detector of scattering meters is arranged in an angle between 90 and 180 degrees to the light beam, thus the respective instruments measure a proportion of the light that is scattered in the backward direction. From the measurement at the specific angle, total backscattering, or even the total magnitude of the VSF, can be calculated. Some instruments also measure more than one angle of the VSF to reduce calculation errors. Biases due to absorption are reduced by the use of short path lengths, wavelengths for which absorption is generally low (e.g., in the green region of the spectrum), and empirical correction factors. The principle of *turbidity meters* (which are frequently used for monitoring purposes) is similar to that of scattering meters. Here, the light scattered in an angle of 90 degrees is measured and expressed in units of a standard used for calibration (e.g., formazine) instead of scattering coefficients. All instruments measuring scattered light give information about the particle load present in the water. However, as for the attenuation meters, a good correlation to plankton biomass is only obtained when it is the only source of scattering. However, some instruments (e.g., Laser In Situ Scattering and Transmissiometry [LISST], Sequoia Scientific, United States) take advantage of laser diffraction measurements, evaluating the scattering patterns of laser light on particles. This at least provides information about the size structure of the scattering particles.

Absorption meters have to overcome two main problems to obtain accurate measurements: first, the difficulty to separate the light loss of absorption from the light loss due to scattering on particles present in the sample, and second the problem that the concentration of absorbing materials in natural waters can be relatively low, leading to a low signal-to-noise ratio. There are different approaches to overcome these problems: the first is the use of a reflective tube of sufficient length (tens of centimeters), as is realized in the ac-9 or ac-s instrument (Western Environmental Testing Laboratory [WETLABS], United States, Fig. 3.1.1). The basic setup is comparable to an attenuation meter (see earlier), but due to the reflectivity of the tube containing the water sample, also the light scattered in the forward

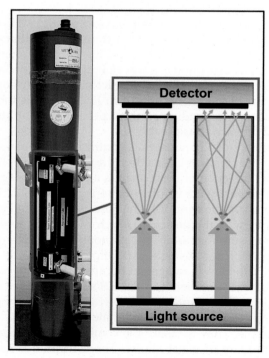

FIGURE 3.1.1   Measurement principle of the ac-9/ac-s instrument. For explanation see text.

direction is caught by the detector. Because this is the majority of the scattered light, the light loss measured is approximately only caused by the absorption of the water sample. In fact, the ac-s instruments measures both absorption and attenuation coefficients in parallel using one reflective and one nonreflective tube fed by the same water source (Fig. 3.1.1, schematic on the right). That also offers the possibility of obtaining scattering coefficients of the water sample by subtracting the absorption from the attenuation. Nevertheless, absorption coefficients obtained using reflective tubes have to be corrected for remaining biases [17,18]. A second approach is the use of integrating cavities [19,20]. These instruments use a diffuse-light field for the absorption measurement, which is realized inside the cavity due to multiple reflections of the measurement light on the highly reflective walls. Sample water is filled directly in the cavity and absorption can be measured with negligible influence of scattering [20], because scattering on particles cannot make the light field more diffuse. Thus, all light intensity loss measured is due to absorption in these instruments. Furthermore, due to the multiple reflections of the light on the cavity walls, its photons run several times through the water sample, which enhances the chance of them being absorbed by a molecule. Thus, even if the cavity is comparatively compact, the effective optical path length is sufficiently long to enable highly sensitive measurements. Typically, the cavity has a diameter of approximately 10 cm, but this can be varied depending on the expected concentration of absorbing substances in the water. However, it has to be taken into account that the optical path length in these instruments is a function of cavity reflectivity, which can change due to contamination of the cavity wall. Therefore, the instruments require periodic careful calibration before use. On the basis of these integrating cavity setups for the analysis of discrete water samples, integrating cavity instruments for continuous flow-through usage have also been developed ([21–24]; Fig. 3.1.2). Despite being comparatively difficult to measure, absorption coefficient measurements provide valuable insights in the composition of the investigated water, especially when obtained hyperspectrally. Certain coefficients can be related to the chl-*a* concentration in the water [25–27] and

FIGURE 3.1.2   Examples of integrating cavity instruments for the measurement of absorption coefficient spectra. (A and B) The hyperspectral absorption sensor (Helmholtz–Zentrum Geesthacht, Germany), a custom-built setup allowing automated flow through measurements. (C) **O**nline hyperspectral integrating cavity **a**bsorption mete**r** (OSCAR) (TriOS GmbH, Germany), a compact, submersible integrating cavity sensor. Both were developed in the course of the EU-project NeXOS. *(C) Image courtesy of TriOS GmbH.*

FIGURE 3.1.3 Particulate absorption coefficient spectra of different groups of phytoplankton. Spectra are normalized by the total sum of the absorption.

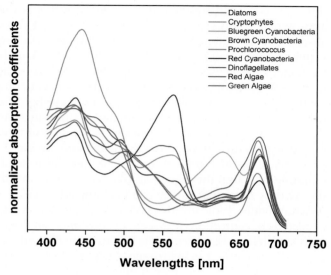

provide, therefore, an optical proxy for phytoplankton biomass, whereas certain features in the absorption spectra can be related to the—often group-specific—pigmentation of phytoplankton cells (Fig. 3.1.3). A comparison of these signatures has the potential of discriminating between phytoplankton types present in different water samples, or identifying groups or species with distinct pigments [28,29].

3.1.2.2 Measurement of Fluorescence

Another important and widely used optical parameter used for investigating water constituents is the fluorescence that is emitted from molecules after excitation by light of specific wavelength. Fluorescence is a process that is closely related to absorption: When a molecule absorbs a photon of an appropriate wavelength, the received energy lifts it to a higher, but unstable, energetic state. When the molecule eventually returns to its lower state, the previously absorbed energy is emitted in form of heat or in form of a photon with lower energy—and thus longer wavelength—than the photon initially absorbed (Stokes Shift). This emitted light is called fluorescence and it is proportional to the concentration of the fluorophore present. Like any other form of light, it can be quantified by photomultipliers or photodiodes.

Several chemical substances present in the water exhibit fluorescence, but the most important in context of plankton research is chl-*a*, the major pigment of photosynthesis. Thus, the measurement of in situ fluorescence is primarily used for estimating the concentration of chl-*a* in the water, and thus phytoplankton biomass. It is a well-studied process, and several reviews are available dealing with chl-*a* fluorescence in much detail (e.g., Refs. [30,31]). Generally, chl-*a* fluorescence represents the part of light that is absorbed by the phytoplankton cell, but which cannot be used in the photochemical reactions of photosynthesis. It is, therefore, a protection mechanism to dissipate energy that would otherwise inflict damage to the pigments, lipid membranes, or other molecules involved in the photosynthetic process. The basic principle of measuring chl-*a* fluorescence is relatively simple: a sample containing phytoplankton cells is illuminated typically with blue light, because the absorption maximum of chlorophyll is at approximately 430 nm [32]. This can be achieved, e.g., by a blue light-emitting diode (LED), or with white light and a blue filter. The detection of chl-*a* fluorescence occurs commonly at wavelengths around 680 nm. The photodiode or photomultiplier used for this purpose commonly includes a red filter to avoid additional input from sources other than chl-*a* fluorescence. However, there are various deviations from this design [33]. Due to the simplicity of this approach, there were early attempts using such measurements also in a continuous, quasi-in situ manner [34]. Nowadays, a wide variety of real in situ instruments is available from different manufacturers ([12]; examples shown in Fig. 3.1.4). In the course of the technical progress with respect to energy consumption and miniaturization of components, modern in situ fluorometers can be deployed on nearly every kind of operating platform [35]. Even normal smartphones can be converted into simple fluorometers [36], and although they cannot be considered as in situ instruments in the strict sense, they allow chl-*a* fluorescence measurements even in remote locations, without the need of specialized

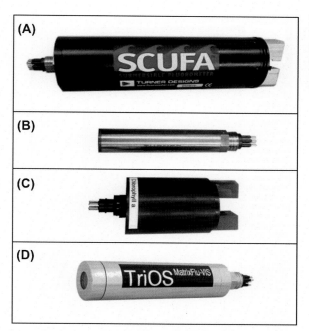

FIGURE 3.1.4 Examples of in situ fluorescence sensors. (A) Self-Contained Underwater Fluorescence Apparatus (SCUFA) submersible chl-*a* fluorometer (Turner Designs, United States; no longer manufactured). (B) Cyclops chl-*a* sensor (Turner Designs, United States). (C) Seapoint chl-*a* sensor (Seapoint Sensors, United States). (D) MatrixFlu VIS multifluorescence sensor developed in the course of the EU-project NeXOS (TriOS GmbH, Germany).

equipment. However, it has to be always remembered that chl-*a* fluorescence is only a proxy for chl-*a* concentration, and that this in turn is only a proxy for phytoplankton biomass. There are several variables influencing the relationships of these parameters [37,38], thus interpretation of fluorescence measurements is not always straightforward.

In addition to monitoring phytoplankton biomass, there are also attempts of using fluorescence measurements to obtain phytoplankton taxonomical information [35,39,40]. They take advantage of the fact that normally phytoplankton cells contain not only chl-*a*, but also accessory pigments that do not fluoresce under in vivo conditions, but transfer their absorbed energy to chl-*a*. The corresponding instruments use multiple wavelengths of different color instead of monochromatic blue light for excitation of chl-*a* fluorescence. By measuring the intensity of chl-*a* fluorescence in dependence of the wavelength of the light used for excitation, chl-*a* excitation–emission spectra can be created. These spectra vary with pigment composition of the phytoplankton investigated and can thus be used to create fluorescence fingerprints for phytoplankton groups characterized by specific pigment compositions. With these predefined fingerprints, the proportion of the different phytoplankton groups on the total amount of chlorophyll present in the measured sample is estimated. The advantage of these instruments is to provide temporally highly resolved estimations of phytoplankton community composition; however, due to the already mentioned variability of fluorescence, the fingerprints standing for the different groups may not be always valid for the phytoplankton present in the field [41–43].

3.1.2.3 Flow Cytometric Approaches

In contrast to the aforementioned methods that measure the bulk optical properties of a water sample, flow cytometry is an approach that obtains these properties from single cells or cell aggregates. Originally developed for medical purposes, flow cytometry is nowadays an important tool for the investigation of plankton [44]. Its basic principle is that the particles present in a water sample are passing one after another a measurement chamber in which they are illuminated with a focused light source (typically a laser beam). Detectors arranged around the chamber in specific angles measure the light scattered by the particles as well as potentially induced fluorescence (Fig. 3.1.5A). A sheath fluid (buffer solution or filtered water, depending on the type of analysis) of high velocity in which the sample fluid is injected enables the necessary separation of the particles as well as their arrangement in the focal plane of the chamber. Particles are counted based on the number of scattering events detected, and they are characterized with respect to size and type by their fluorescence and scattering properties. By these means, the discrimination is allowed of different particle and plankton populations, whereas the analysis is usually restricted to relatively small cells due to the diameter of the water jet orifice and disturbances of the water jet due to larger cells. Furthermore, the sample volume is typically in the order of μL to mL; what makes this method especially useful is the investigation of the

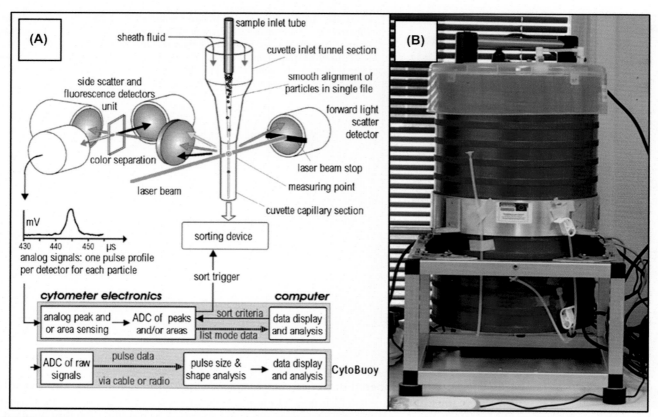

FIGURE 3.1.5   Principle (A) of flow cytometry and examples (B) of portable flow cytometers (CytoSens, CytoBuoy, The Netherlands). *(A) Modified from Dubelaar and Jonker, 2000.*

highly abundant and small cells of the pico- and nanophytoplankton (size up to 2 μm), which due to their small size are also difficult to count and identify by other means. Due to their usefulness, more and more instruments are coming on the market specifically designed for marine research (Fig. 3.1.5B). They are portable and adapted for analysis of a broader range of size classes [45–47]. Furthermore, due to technologic progress in camera technology, flow cytometers designed for the analysis of larger cells are increasingly equipped with image-taking capabilities, which offer an even more detailed view on plankton populations on the single-cell level. These instruments are described in more detail in the next part of the chapter.

3.1.3 CAMERA INSTRUMENTS AND IMAGING ANALYSES

The origin of imaging systems was not only a response to the increased availability of camera hardware and electronic components. It was also the recognition of limitations of traditional sampling instruments and the demand to accelerate processing of plankton samples, obtain in situ information on the taxonomic composition and distribution on small spatiotemporal scales, as well as to quantify fragile organisms and aggregates. To overcome the spatial limitation, acoustical sampler [48] and nonimaging optical samplers, like the Optical Plankton Counter [49] and Laser Optical Plankton Counter [50], have been developed. These instruments provide high-resolution data on biomass and particle size, but not on the taxonomic composition of plankton. However, recent advances in imaging technology have led to the development of new optical-imaging devices. This chapter provides a short background and overview of plankton imaging and its potential.

3.1.3.1 In Situ Camera Instruments

Due to the inability of traditional sampling methods to sample certain organisms, aggregates, and particles quantitatively, the reality of the planktonic world was long-time bound to the reality of net designs [51]. Potential factors

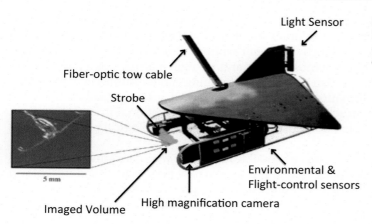

Light Sensor

Fiber-optic tow cable

Strobe

5 mm

Imaged Volume

High magnification camera

Environmental &
Flight-control sensors

FIGURE 3.1.6 Schematic drawing of the real-time Video Plankton Recorder II mounted below a depressor v-fin imaging a copepod while being towed behind a research vessel. *Image courtesy of Carin Ashjian, Woods Hole Oceanographic Institution (WHOI).*

influencing this traditional-sample reality are extrusion of zooplankton from nets, clogging of the mesh, and active avoidance of the sampling gear [52]. The resulting net selectivity can lead to biases in abundance estimates of organisms and losses of fragile and smaller species [53]. Furthermore, manual identification of preserved samples is a labor-intensive and time-consuming process, which generally involves subsampling, counting, and identifying individuals under the microscope by human experts. Having the plankton sample in a digital format allows to speed up the classification procedure, and concurrent advances in image processing [54] as well as pattern recognition [55] of plankton have made it possible to automatically determine coarse taxonomic composition and, in some cases, even quantify to species level.

Initially, optical in situ samplers were usually deployed in combination with traditional plankton net systems mounted to the nets frame [56]. Additionally, these early systems were often limited in their quantitative value due to a poorly defined imaged volume [52]. However, in the past three decades various standalone optical plankton samplers, capable of imaging and quantifying plankton, have been developed. The Video Plankton Recorder (VPR, Fig. 3.1.6), an underwater camera system towed by a research vessel developed by Davis et al. [57] was one of the forerunners [58]. It was the first plankton-sampling device that was able to automatically identify and count phytoplankton, zooplankton, and marine snow aggregates in situ and quantitatively map their abundance and distribution patterns with high resolution in real time [59]. The VPR was followed by a suite of different other camera systems, including the Underwater Vision Profiler (UVP, [60]), the ZOOplankton VISualization system (ZOOVIS, [61]), the Lightframe On-sight Keyspecies Investigate System (LOKI, [62]), the Shadow Image Particle Profiling Evaluation Recorder (SIPPER, [63]), and the In Situ Ichthyoplankton Imaging System (ISIIS, [64]), to name the most prominent. Most of the commercially available camera systems use dark- or bright-field illumination to create an image [65]. Technical limitations in size and resolution of imaging sensors mean that plankton-imaging systems necessarily have much smaller sample volumes than plankton nets. Nonetheless, because plankton nets typically undersample the number of certain organisms and aggregates as mentioned previously, imaging systems can provide a more realistic estimate of their abundance [84]. Some imaging systems such as the VPR have relatively small image volumes of a few to tens of milliliters per image depending on their calibration. Therefore, it has to be taken into account that low-abundance species (e.g., fish larvae) are potentially undersampled. However, generally all these instruments provide a reliable spatial resolution along the instrument's towpath on scales of centimeters and greater. The focus of these instruments has been mostly on imaging organisms in the size range of mesoplankton (0.2–20 mm) and marine-snow aggregates as they typically have to deal with the trade-off between size of the sampling volume and the resulting image resolution due to a confined number of camera pixels. However, some imaging systems have been developed, which are able to resolve phytoplankton in the size range from 10 to 100 μm including many diatoms and dinoflagellates (e.g., Imaging Flowcytobot, [13]). Additionally, holographic imaging instruments have been developed to resolve three-dimensional positions and identification of plankton and particles in the size range from nano-to mesoplankton from a large volume of water [66]. Although most of these instruments are towed behind research vessels or mounted on other research platforms (e.g., remotely operated vehicles), current approaches also include stationary camera systems for long-time monitoring and equipping autonomous vehicles or gliders with camera systems (e.g., Guard1, Corgnati et al. [85]). Furthermore, camera systems are frequently used in combination with other instruments (e.g., acoustics, [67]).

All of these camera systems are usually equipped with environmental sensors and, hence, allow the measurement of hydrographic parameters on scales that can be directly related to the organisms and thereby provide insights into the subtle relationships between hydrography and species distributions. Another advantage of cameras is that noninvasive sampling obtains quantitative data on gelatinous zooplankton (e.g., hydromedusae and appendicularians), colonial phytoplankton (e.g., chain forming diatoms), and other fragile particulate matter and aggregations (e.g., marine snow and aggregated organisms) in their natural environment without damaging them or destroying the association [68]. Camera systems have been used in numerous studies and various marine ecosystems during recent years to, e.g., develop microscale prey fields for larval fish to include in foraging-model studies [69], to resolve how copepod swimming behavior contributes to the formation of fine-scale layers [70], and especially to describe zooplankton distributions in relation to hydrographic structures (e.g., Refs. [71,72]). Additional information from images includes the undisturbed, natural orientation of plankton organisms and even physiological parameters like egg-production estimations [73].

These studies provided unique information on scales, which are not possible to retrieve with conventional collecting equipment, and, hence showed the potential of optical sampling techniques to give new perspectives and insights on the distribution, composition, and interactions in the plankton community as well as on whole-ecosystem dynamics.

3.1.3.2 Benchtop Imaging Systems

Simultaneously with in situ camera systems, the development of benchtop-imaging systems has comparably advanced, and imaging and processing of preserved plankton samples in the laboratory has become a popular method. Recent progress in scanner technology and hardware allow direct scanning, and many commercial scanners are capable of producing high-quality images of plankton. Additionally, specialized laboratory instruments have also been invented. Zooscan [74] is an instrument that uses a scanner sensor with a custom-built lighting system and a watertight scanning chamber. Zooplankton samples can be poured into the chamber, digitized at high resolution, and recovered without damage. The contents of these images can be processed and classified using image-classification software. This method offers fast-processing times and the advantage of a permanent record of the plankton sample, but generally suffer from reduced resolution relative to the original planktonic organisms, which may limit the level of taxonomic detail that can be obtained [58]. Additionally, flow-through imaging systems like the FlowCam [75] and the benchtop-VPR have been developed, in which the sample is pumped through a flow cell and imaged. Therefore, these systems can also be directly connected to the research vessel's seawater system or other pumped sensor systems on board (e.g., FerryBox, [76]). The potential benefits from surveying plankton with in situ imaging systems, analyzing net, or pump samples via image-processing techniques results in a considerable quantity of images, which both sampling approaches produce.

3.1.3.3 Image Classification

Optical-imaging samplers have recently become widely used in plankton ecology, whereas imaging-analysis methods have so far lagged behind image-acquisition rates [55]. However, the large amount of visual data coming from in situ plankton samplers, benchtop systems, and cabled underwater observatories clearly requires an automated approach. Furthermore, the fatiguing process of manually classifying vast amounts of images is prone to bias and can introduce errors [77]. Automated methods for analysis and recognition of plankton images have been developed, which are capable of real-time processing of incoming image data into major taxonomic groups. However, the limited accuracy of these methods and low-quality images due to turbid water conditions, as discussed later, can sometimes still require significant manual postprocessing to correct the automatically generated results [78].

As a first step, automatic focus detection is used to avoid visual inspection of all images. Each image frame containing infocus plankton or particles are identified, and these objects, so called regions of interest are then isolated using several image-processing steps (e.g., binarization, segmentation, and connectivity routines) and afterward written to disk [59]. For underwater imagery analysis, this visual content recognition is one of the most important and challenging tasks. This may result from variable illumination, scales, orientation, and nonrigid deformations [79]. Furthermore, plankton as a group is very heterogeneous, which varies in size and morphology and is surrounded by a medium containing a variety of nonliving targets such as marine snow, sediment particles, and bubbles. Larger organisms or aggregated organisms such as siphonophores and gelatinous zooplankton might exceed the size of the imaged volume and, hence, only parts of the organism are apparent on the image. Once the images are created, certain features are extracted from an often manually sorted subset of images, which is used to build and train the classifier [78].

Although automatic classification of plankton is a challenging task, there has been considerable progress in the field of machine vision. Artificial neural networks, support vector machines, and decision trees have been introduced (e.g., Refs. [54,80]) and automated systems using texture, shape, and other image features are currently capable of correctly classifying plankton images with an accuracy of 70–80% for 10–20 taxonomic categories [58]. Even further accuracy can be achieved using a dual classification method including automatic correction [78] or by using an ensemble of classifiers [81]. Additionally, recently introduced deep-learning algorithms have already significantly influenced the field of machine learning and are a promising tool for further advancement [82]. Most of our progress in the field of plankton imaging has been the result of individuals or small groups working toward development of unique instruments, which led to development of many different customized image-classification solutions, e.g., Visual Plankton [59] and Zooimage [58,74]. The effectiveness of existing imaging systems could be dramatically enhanced when flexible software toolboxes become available to the oceanographic community, which are capable of handling images from different in situ and laboratory imaging systems, provide a variety of classification algorithms, and classifying images with a high accuracy (e.g., EcoTaxa, [83]).

3.1.4 CONCLUSIONS

This chapter gave a brief overview about the different approaches available for investigating and monitoring phyto- and zooplankton, which both play major roles in the marine food web and many biogeochemical cycles. The focus was put on optical in situ sensors or flow-through instruments because they are—despite the advances in in situ molecular biology and the availability of other (e.g., acoustic) methods—still the most widespread type of instruments used in this respect. They meet requirements like high measurement frequency, robustness, cost-effectiveness, and versatility with respect to the used platform. Furthermore, many of them are based on relatively simple principles known for decades, and thus they have been optimized with respect to size, power consumption, and reliability. For other, more sophisticated sensors (e.g., automated microscopes or ones using integrating cavities), these optimizations have to be done in the future. However, there is still a demand for increasing the operation time of the sensors, as this decreases operational costs and allows the sensors to be mounted in unattended, automated measurement platforms like, e.g., gliders, floats, or moored platforms in remote locations. A big issue in this respect was and still is the prevention of biofouling, which is one of the biggest problems in the long-term deployment of optical sensors.

Another important aspect to consider is that many optical approaches target primarily plankton biomass. This limits the understanding of plankton dynamics with respect to, e.g., seasonal succession. A relatively small, nevertheless increasing number of more-sophisticated instruments provide additional information about the type or even species of (phyto)plankton that is present. However, the evaluation and interpretation of the obtained measurements (e.g., multi-spectral fluorescence or absorption spectra) in this respect is not straightforward, but requires some scientific expertise and is subject to biases, which makes these sensors and their data more difficult to use in an operational manner.

With camera instruments, it is often possible to obtain information on the same level of detail as provided by conventional microscopy. Thus, these instruments can provide the required additional taxonomical information that is needed to understand the dynamics in the plankton in more detail. However, their application is still limited by their comparatively high costs and the sheer amount of data that these instruments provide makes the automatic classification challenging. Thus, in general, with increasing amount and complexity of data, emphasis must be put also on enhancing the capabilities of data analysis and data quality control to improve data reliability. The development and application of common standards for calibration, operation, data quality control, and data processing would be a great advance in this respect.

Acknowledgments

This work was partly funded by the EU-project NeXOS ("Next generation, Cost-effective, Compact, Multifunctional Web Enabled Ocean Sensor Systems Empowering Marine, Maritime and Fisheries Management", Grant agreement no: 614102) and JERICO- NEXT (Joint European Research Infrastructure network for Coastal Observatory – Novel European eXpertise for coastal observaTories", Grant agreement no: 654410).

References

[1] Field CB, Behrenfeld MJ, Randerson JT, Falkowski P, Keeling RF, Shertz S, Falkowski PG, Barber RT, Smetacek V, Lindeman RL, Thompson MV, Randerson JT, Malmström CM, Field CB, Tynan CT, McNaughton SJ, Behrenfeld M, Falkowski P, Morel A, Behrenfeld MJ, Falkowski PG, Balch WM, Byrne CF, Berthon J-F, Morel A, Ruimy A, Dedieu G, Saugier B, Antoine D, Andre JM, Morel A, Balch W, Potter CS, Malmström CM, Randerson JT, Thompson MV, Fung IY, Conway T, Field CB, Feldman G, Bishop JKB, Rossow WB, Smith SV, Woodwell GM, Martin JH, Fitzwater SE, Gordon RM, Martin JH, Gordon RM, Fitzwater SE, Schimel DS, Vitousek PM, Mooney HA, Lubchenco J, Melillo JM, Shaver GR, Schimel DS, Braswell RH, Schimel DS, Liner E, Moore B, Myneni RB, Keeling CD, Tucker CJ, Asrar G, Nemani RR, Ciais P, Tans P, Trolier M, White JWC, Francey RJ. Primary production of the biosphere: integrating terrestrial and oceanic components. Science 1998;281:237–40.

[2] Devlin M, Best M, Coates D, Bresnan E, O'Boyle S, Park R, Silke J, Cusack C, Skeats J. Establishing boundary classes for the classification of UK marine waters using phytoplankton communities. Mar Pollut Bull 2007;55:91–103.

[3] Racault MF, Platt T, Sathyendranath S, Agirbas E, Martinez Vicente V, Brewin R. Plankton indicators and ocean observing systems: support to the marine ecosystem state assessment. J Plankton Res 2014;36:621–9.

[4] Zielinski O, Busch JA, Cembella AD, Daly KL, Engelbrektsson J, Hannides AK, Schmidt H. Detecting marine hazardous substances and organisms: sensors for pollutants, toxins, and pathogens. Ocean Sci 2009;5:329–49.

[5] Delory E, Castro A, Waldmann C, Rolin JF, Woerther P, Gille J, Del Rio J, Zielinski O, Golmen L, Hareide NR, Pearlman J. NeXOS development plans in ocean optics, acoustics and observing systems interoperability. Sensor Systems for a Changing Ocean (SSCO). IEEE; 2014. p. 1–3. https://doi.org/10.1109/SSCO.2014.7000382.

[6] Diercks-Horn S, Metfies K, Jäckel S, Medlin LK. The ALGADEC-device: a semi-automated rRNA biosensor for th detection of toxic algae. Harmful Algae 2011;10:395–401.

[7] Medlin LK. Mini review: molecular techniques for identification and characterization of marine biodiversity. Ann Mar Biol Res 2016;3(2). https://www.jscimedcentral.com/MarineBiology/marinebiology-3-1015.pdf.

[8] Scholin C, Everlove C, Harris A, Alvarado N, Birch J, Greenfield D, Vrijenhoek R, Mikulski C, Jones K, Doucette G, Jensen S, Roman B, Pargett D, Marin RI, Preston C, Jones W, Feldman J. Remote detection of marine microbes, small invertebrates, harmful algae, and biotoxins using the Environmental Sample Processor (ESP). Oceanography 2009;22:158–67.

[9] Ussler, iii W, Preston C, Tavormina P, Pargett D, Jensen S, Roman B, Marin, iii R, Shah SR, Girguis PR, Birch JM, Orphan V, Scholin C. Autonomous application of quantitative PCR in the deep sea: in situ surveys of aerobic methanotrophs using the deep-sea environmental sample processor. Environ Sci Technol 2013;47(16):9339–46.

[10] Jonsson P, Sillitoe I, Dushaw B, Nystuen J, Heltne J. Observing using sound and light – a short review of underwater acoustic and video-based methods. Ocean Sci Discuss 2009;6:819–70.

[11] Davis CS, Gallager SM, Marra M, Stewart WK. Rapid visualization of plankton abundance and taxonomic composition using the Video Plankton Recorder. In: Deep sea research Part II: topical studies in oceanography, vol. 43. 1996. p. 1947–70. Issues 7–8, 1996.

[12] Moore C, Barnard A, Fietzek P, Lewis M, Sosik H, White S, Zielinski O. Optical tools for ocean monitoring and research. Ocean Sci 2009;5:661–84.

[13] Olson RJ, Sosik HM. A submersible imaging-in-flow instrument to analyze nano- and microplankton: imaging FlowCytobot. Limnol Oceanogr Methods 2007;5:195–203.

[14] Mobley CD. Light and water: radiative transfer in natural waters. Academic press; 1994.

[15] Kirk JTO. Light & photosynthesis in aquatic ecosystems. 2nd ed. Cambridge: Cambridge University Press; 1994.

[16] Pope RM, Fry ES. Absorption spectrum (380 -700 nm) of pure water. II. Integrating cavity measurements. Appl Opt 1997;36:8710–23.

[17] Röttgers R, McKee D, Woźniak SB. Evaluation of scatter corrections for ac-9 absorption measurements in coastal waters. Methods Oceanogr 2013;7:21–39.

[18] Zaneveld JRV, Kitchen JC, Moore C. In: Ackleson SG, editor. Scattering error correction of reflecting tube absorption meters. Bergen: Ocean Optics XII. Proc. SPIE; 1994. p. 44–55.

[19] Fry ES, Kattawar GW, Pope RM. Integrating cavity absorption meter. Appl Opt 1992;31:2055–65.

[20] Röttgers R, Schönfeld W, Kipp P-R, Doerffer R. Practical test of a point-source integrating cavity absorption meter: the performance of different collector assemblies. Appl Opt 2005;44:5549–60.

[21] Dana DR, Maffione RA. A new hyperspectral spherical-cavity absorption meter. In: Poster presentation ocean sciences 2006. 2006. Honolulu (Hawaii).

[22] Gray DJ, Kattawar GW, Fry ES. Design and analysis of a flow-through integrating cavity absorption meter. Appl Opt 2006;45:8990–8.

[23] Musser JA, Fry ES, Gray DJ. Flow-through integrating cavity absorption meter - experimental results. Appl Opt 2009;48:3596–602.

[24] Wollschläger J, Voß D, Zielinski O, Petersen W. In situ observations of biological and environmental parameters by means of optics-development of next-generation ocean sensors with special focus on an integrating cavity approach. IEEE J Ocean Eng 2016;41:753–62.

[25] Bricaud A, Morel A, Babin M, Allali K, Claustre H. Variations of light absorption by suspended particles with chlorophyll a concentration in oceanic (case 1) waters: analysis and implications for bio-optical models. J Geophys Res 1998;103(C13):31033–44.

[26] Roesler CS, Barnard AH. Optical proxy for phytoplankton biomass in the absence of photophysiology: rethinking the absorption line height. Methods Oceanogr 2013;7:79–94.

[27] Wollschläger J, Röttgers R, Petersen W, Wiltshire KH. Performance of absorption coefficient measurements for the in situ determination of chlorophyll-a and total suspended matter. J Exp Mar Bio Ecol 2014;453:138–47.

[28] Kirkpatrick GJ, Millie DF, Moline MA, Schofield O. Optical discrimination of a phytoplankton species in natural mixed populations. Limnol Oceanogr 2000;45:467–71.

[29] Millie DF, Schofield OME, Kirkpatrick GJ, Johnsen G, Evens TJ. Using absorbance and fluorescence spectra to discriminate microalgae. Eur J Phycol 2002;37:313–22.

[30] Krause GH, Weis E. Chlorophyll fluorescence and photosynthesis: the basics. Annu Rev Plant Physiol Plant Mol Bioi 1991;42:313–49.

[31] Maxwell K, Johnson GN. Chlorophyll fluorescence–a practical guide. J Exp Bot 2000;51:659–68.

[32] Bidigare RR, Ondrusek ME, Morrow JH, Kiefer DA. In vivo absorption of algal pigments. In: SPIE 1302. 1990.

[33] Zeng L, Li D. Development of in situ sensors for chlorophyll concentration measurement. J Sens 2015. Article ID e903509.

[34] Lorenzen CJ. A method for the continuous measurement of in vivo chlorophyll concentration. Deep Sea Res Oceanogr Abstr 1966;13:223–7.

[35] Pearlman J, Zielinski O. A new generation of optical systems for ocean monitoring - matrix fluorescence for multifunctional ocean sensing. Sea Technol 2017;2:30–3.

[36] Friedrichs A, Busch JA, van der Woerd HJ, Zielinski O. SmartFluo: a method and affordable adapter to measure chlorophyll-a fluorescence with smartphones. Sensors 2017;17(4).

[37] Falkowski P, Kiefer DA. Chlorophyll a fluorescence in phytoplankton: relationship to photosynthesis and biomass. J Plankton Res 1985;7:715–31.

[38] Loftus ME, Seliger HH. Some limitations of the in vivo fluorescence technique. Chesap Sci 1975;16:79–92.

[39] Beutler M, Wiltshire KH, Meyer B, Moldaenke C, Lüring C, Meyerhöfer M, Hansen UP, Dau H. A fluorometric method for the differentiation of algal populations in vivo and in situ. Photosynth Res 2002;72:39–53.

[40] Yoshida M, Horiuchi T, Nagasawa Y. In Situ Multi- Excitation Chlorophyll Fluorometer for Phytoplankton Measurements. Waikoloa, HI: Oceans'11 MTS/IEEE Kona; 2011. p. 1–4.

[41] MacIntyre HL, Lawrence E, Richardson TL. Taxonomic discrimination of phytoplankton by spectral fluorescence. In: Suggett DJ, Borowitzka MA, Prasil O, editors. Chlorophyll a fluorescence in aquatic sciences: methods and applications. Netherlands: Springer; 2010. p. 129–69.

[42] Seppälä J, Balode M. The use of spectral fluorescence methods to detect changes in the phytoplankton community. Hydrobiologia 1998;363:207–17.

[43] Soohoo JB, Kiefer DA, Collins DJ, Mcdermid IS. In vivo fluorescence excitation and absorption spectra of marine phytoplankton: I. Taxonomic characteristics and responses to photoadaptation. J Plankton Res 1986;8:197–214.

[44] Wang Y, Hammes F, De Roy K, Verstraete W, Boon N. Past, present and future applications of flow cytometry in aquatic microbiology. Trends Biotechnol 2010;28:416–24.

[45] Dubelaar GBJ, Gerritzen PL. Cytobuoy: a step forward towards using flow cytometry in operational oceanography. Sci Mar 2000;64:255–65.

[46] Dubelaar GBJ, Groenewegen AC, Stokdijk W, Van Den Engh GJ, Visser JWM. Optical plankton analyser: a flow cytometer for plankton analysis, ii: specifications. Cytometry 1989;10:529–39.

[47] Olson RJ, Shalapyonok A, Sosik HM. An automated submersible flow cytometer for analyzing pico- and nanophytoplankton: FlowCytobot. Deep Res Part I Oceanogr Res Pap 2003;50:301–15.

[48] Holliday DV, Pieper RE, Kleppel GS. Determination of zooplankton size and distribution with multifrequency acoustic technology. Cons Int Explor Mer 1989;46(1):52–61.

[49] Herman AW. Simultaneous measurement of zooplankton and light attenuance with a new optical plankton counter. Cont Shelf Res 1988;8:205–21.

[50] Herman B, Beanlands E, Phillips F. The next generation of optical plankton counter: the laser-OPC. J Plankton Res 2004;26(10):1135–45.

[51] Reeve MR. Zooplankton - the connecting link: a historical perspective. In: Rothschild BJ, editor. Toward a theory on biological-physical interactions in the world ocean. Kluwer Academic Publ.; 1988. p. 501–12.

[52] Wiebe PH, Benfield MC. From the Hensen net toward four-dimensional biological oceanography. Prog Oceanogr 2003;56(1):7–136.

[53] Gallienne CP, Robins DB. Is Oithona the most important copepod in the world's oceans? J Plankton Res 2001;23(12):1421–32.

[54] Tang XO, Stewart WK, Vincent L, Huang H, Marra M, Gallager SM, Davis CS. Automatic plankton image recognition. Artif Intell Rev 1998;12(1):77–199.

[55] Hu Q, Davis CS. Automatic plankton image recognition with co-occurrence matrics and support vector machine. Mar Ecol Prog Ser 2005;295:21–31.

[56] Ortner PB, Hill LC, Edgerton HE. In situ silhouette photography of Gulf Streamzooplankton. Deep Sea Res Part I 1981;28:1569–76.

[57] Davis CS, Gallager SM, Solow AR. Microaggregations of oceanic plankton observed by towed video microscopy. Science 1992;257:230–2.

[58] Benfield MC, Grosjean P, Culverhouse PF, et al. RAPID research on automated plankton identification. Oceanography 2007;20:172–87.

[59] Davis CS, Hu Q, Gallager SM, Tang X, Ashjian CJ. Real-time observation of taxa specific plankton distributions: an optical sampling method. Mar Ecol Prog Ser 2004;284:77–96.

[60] Picheral M, Guidi L, Stemmann L, Karl DM, Iddaoud G, Gorsky G. The Underwater Vision Profiler 5: an advanced instrument for high spatial resolution studies of particle size spectra and zooplankton. Limnol Oceanogr Methods 2010;8:462–73.

[61] Bi H, Cook S, Yu H, Benfield MC, Houde ED. Deployment of an imaging system to investigate fine-scale spatial distribution of early life stages of the ctenophore *Mnemiopsis leidyi* in Chesapeake Bay. J Plankton Res 2012;35(2):270–80.

[62] Schulz J, Barz K, Ayon P, Lüdtke A, Zielinski O, Mengedoht D, Hirche HJ. Imaging of plankton specimens with the light-frame on-sight Keyspecies investigation (LOKI) system. J Eur Opt Soc Rap Public 2010;5(10017):1–9.

[63] Samson S, Hopkins T, Remsen A, Langebrake L, Sutton T, Patten J. A system for high-resolution zooplankton imaging. IEEE J Ocean Eng 2001;26:671–6.

[64] Cowen RK, Guigand CM. In situ Ichthyoplankton Imaging System (ISIIS): system design and preliminary results. Limnol Oceanogr Methods 2008;6:126–32.

[65] Kapitza HG, Microscopy from the very beginning, Translated by Kohl M. Carl Zeiss: Oberkochen; 1994. 1:40.

[66] Malkiel E, Abras JN, Widder EA, Katz J. On the spatial distribution and nearest neighbor distance between particles in the water column determined from in situ holographic measurements. J Plankton Res 2006;28:149–70.

[67] Wiebe PH, Stanton TK, Greene CH, Benfield MC. BIOMAPER-ii: an integrated instrument platform for coupled biological and physical measurements in coastal and oceanic regimes. IEEE J Ocean Eng 2002;27(3):700–16.

[68] Benfield MC, Davis CS, Gallager SM. Estimating the in situ orientation of *Calanus finmarchicus* on georges bank using the video plankton recorder. Plankton Biol Ecol 2000;47(1):69–72.

[69] Lough RG, Broughton EA. Development of micro-scale frequency distributions of plankton for inclusion in foraging models of larval fish, results from a Video Plankton Recorder. J Plankton Res 2007;29(1):7–17.

[70] Gallager SM, Yamazaki H, Davis CS. Contribution of fine-scale vertical structure and swimming behavior to formation of plankton layers on Georges Bank. Mar Ecol Prog Ser 2004;267:27–43.

[71] Jacobsen HP, Norrbin MF. Fine-scale layer of hydromedusae is revealed by video plankton recorder (VPR) in a semi-enclosed bay in northern Norway. Mar Ecol Prog Ser 2009;380:129–35.

[72] Möller KO, St John M, Temming A, Floeter J, Sell AF, Herrmann JP, Möllmann C. Marine snow, zooplankton and thin layers: indications of a trophic link from small-scale sampling with the Video Plankton Recorder. Mar Ecol Prog Ser 2012;468:57–69.

[73] Möller KO, Schmidt JO, St John M, Temming A, Diekmann R, Peters J, Floeter J, Sell AF, Herrmann JP, Möllmann C. Effects of climate-induced habitat changes on a key zooplankton species 2015;37(3): 530–41.

[74] Grosjean P, Picheral M, Warembourg C, Gorsky G. Enumeration, measurement, and identification of net zooplankton samples using the ZOOSCAN digital imaging system. ICES J Mar Sci 2004;61:518–25.

[75] Sieracki CK, Sieracki ME, Yentsch CS. An imaging-in-flow system for automated analysis of marine microplankton. Mar Ecol Prog Ser 1998;168:285–96.

[76] Petersen W, Schroeder F, Bockelmann FD. FerryBox - application of continuous water quality observations along transects in the North Sea. Ocean Dyn 2011;61(10):1541–54.

[77] Culverhouse PF, Williams R, Reguera B, Herry V, Gonzales-Gil S. Do experts make mistakes? A comparison of human and machine identification of dinoflagellates. Mar Ecol Prog Ser 2003;247:17–25.

[78] Hu Q, Davis CS. Accurate automatic quantification of taxa-specific plankton abundance using dual classification with correction. Mar Ecol Prog Ser 2006;306:51–61.

[79] Py O, Hong H, Zhongzhi S. Plankton classification with deep convolutional neural networks. In: Information technology, networking, electronic and automation control conference. IEEE; 2016. p. 132–6.

[80] Blaschko MB, Holness G, Mattar MA, Lisin D, Utgoff PE, Hanson AR, Schultz H, Riseman EM, Sieracki ME, Balch WM, Tupper B. Automatic in situ identification of plankton. In: Application of computer vision, WACV/MOTIONS'05, vol. 1. 2005. p. 79–86.

[81] Ellen J, Li H, Ohman MD. Quantifying California current plankton samples with efficient machine learning techniques. In: OCEANS'15 MTS/IEEE Washington, vol. 15. IEEE; 2015. p. 1–9.

[82] Qin H, Li X, Yang Z, Shang M. When underwater imagery analysis meets deep learning: a solution at the age of big visual data. In: OCEANS 2015-MTS/IEEE. 2015. p. 1–5.

[83] Picheral M, Colin S, Irisson J-O. EcoTaxa, a tool for the taxonomic classification of images. 2017. http://ecotaxa.obs-vlfr.fr.

[84] Davis CS, Thwaites F, Gallager SM, Hu Q. A three-axis fast-tow digital Video Plankton Recorder for rapid surveys of plankton taxa and hydrography. Limnol Oceanogr Methods 2005;3:59–74.

[85] Corgnati L, Marini S, Mazzei L, Ottaviani E, Aliani S, Conversi A, Griffa A. Looking inside the Ocean: toward an autonomous imaging system for monitoring gelatinous zooplankton. Sensors 2016;16(12):2124. https://doi.org/10.3390/s16122124.

[86] Dubelaar GBJ, Jonker RR. Flow cytometry as a tool for the study of phytoplankton. Sci 2000;64:136–156.

CHAPTER

3.2

Surface Plasmon Resonance sensors for oceanography

Florent Colas

Ifremer, REM/RDT/LDCM, F-29280 Plouzané, France

3.2.1 INTRODUCTION

For the last few decades, the use of Surface Plasmon Resonance (SPR) sensors has spread over many areas. SPR is an optical phenomenon that can be exploited to measure the changes of refractive index as small as 10^{-7} in close proximity of a surface. Such high sensitivity is even capable of detecting a few picograms/mm^2 of molecules that get adsorbed on the surface. SPR sensors are then often presented as optical scales.

SPR systems have attributes that make them very useful for ocean applications. They are sensitive to any change of refractive index. They can be turned into a specific sensor thanks to the surface functionalization. A layer specific to a molecule is deposited on the surface. Then only the target compound can reach the surface. SPR systems are sensitive to the amount of molecules on the surface, which makes them quantitative.

SPR sensors are very versatile. Any kind of molecules or microorganisms can be detected specifically as long as a specific layer can be synthesized. As will be discussed in detail later in this chapter, SPR sensors have been developed for the detection and assay of molecules such as methane, carbon dioxide, metallic ions, polycyclic aromatic hydrocarbons (PAHs), pesticides, phycotoxin, nucleic acid, and microorganisms. The potential of SPR sensors is great, which explains why they have spread so widely and so quickly during the last few decades.

However, SPR sensors are not often used in operational oceanography despite the tremendous need for specific, sensitive, and quantitative assay. To our knowledge, three in situ SPR systems have been reported: two of them as refractometer [1,2] and one as biosensor of domoic acid [3], a phycotoxin responsible for Amnesic Shellfish Poisoning (ASP). These works demonstrated that all the technologies required for in situ SPR sensors are ready, and many new sensors can now emerge.

In this chapter, the principle of SPR sensors will be exposed. Then, some examples of SPR sensors relevant to oceanography will be reviewed to illustrate the great capability of this technology. Last, the three in situ systems will be detailed.

3.2.2 PRINCIPLE OF SURFACE PLASMON RESONANCE SENSORS

3.2.2.1 Extrinsic Sensors

An SPR sensor is an extrinsic sensor, the principle for which is illustrated on Fig. 3.2.1. It is generally composed of two parts: a specific layer and a transducer. The former reacts with the sample, whereas the latter detects the induced change and produces a signal.

The transduction of SPR sensors is based on an optical phenomenon. SPR sensors are intrinsically sensitive to small refractive index changes at the surface of a metal film. They can detect changes as small as 3×10^{-9} for the most sensitive configuration [4]. However, a detection limit of 10^{-7} is more common [5].

This particular behavior was turned into refractive index sensor for coastal and deep applications [1,2]. However, its main success was to serve as a transducer for biosensors. In this special case of an extrinsic sensor, the specific layer is made of biological molecules. The main idea is to use the very high specificity of some biomolecules toward a compound to be detected. They can be antibodies (Abs), nucleic acid, peptides, enzymes… Then, combining the very high sensitivity of SPR and specificity of biomolecule, biosensors are capable of assaying traces of chemicals.

3.2.2.2 Surface Plasmon Resonance Transduction

3.2.2.2.1 Principle of Surface Plasmon Resonance

A SPR is a special resonance phenomenon that can take place in metals such as silver, copper, gold, and aluminum. The electron density of the conduction band oscillates. SPR can be generated when light interacts with a nanostructure such as a nanoparticle, a thin film, or a grating. The most commonly spread configuration is based on a thin film of metal of few tens of nanometers thickness deposited on top of a dielectric material, generally a prism. This is the so-called Kretschmann configuration [6]. Gold is often preferred over the others because of its chemical stability. In addition, it can be easily functionalized because it can create strong bond with –SH, or–NH_2 moieties. Indeed, biomolecules that possess these functional groups can be attached to the metal without complex chemical reactions.

Let us consider a prism covered by a thin film of gold of typically 50 nm thickness. The sample circulates in the flow cell at the other side of the metallic film (Fig. 3.2.2). A beam of light of wavelength λ illuminates the prism/metal interface at an angle of incidence θ. It gets reflected onto the gold film and goes out of the prism. For a particular wavelength and incidence angle, noted λ_{SPR} and θ_{SPR} hereafter, the incoming photons interact with the conduction electrons that enter into resonance. The incoming power is then transferred to the metal, which leads to a strong absorption and a then a minimum of reflected light power.

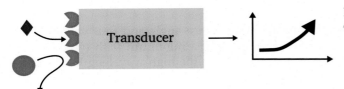

FIGURE 3.2.1 Extrinsic SPR sensor composed of a specific layer deposited on a transducer.

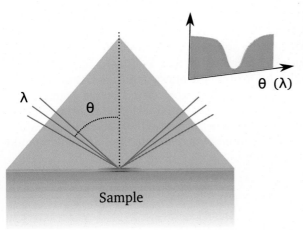

FIGURE 3.2.2 Kretschmann configuration. A thin film of noble metal is deposited on a prism. The sample is in contact with the thin film. A beam of light of angle of incidence θ and wavelength λ illuminates the metal/prism interface. At the angle corresponding to the resonance wavelength, the light is absorbed. The reflectance presents a dip.

FIGURE 3.2.3 Reflectance of a collimated and polychromatic beam when the refractive index of the sample varies (A) and the corresponding resonance wavelength (λ_{SPR}) with regard to n (B).

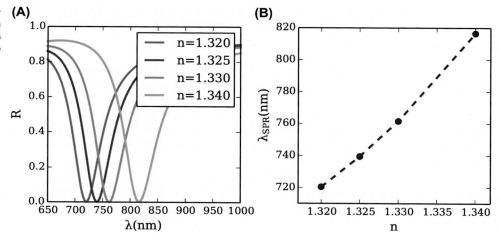

FIGURE 3.2.4 Reflectance of a collimated and polychromatic beam as a thin film of biomolecule grows on the surface from a thickness (e) of 0–10 nm (A) and the corresponding resonance wavelength (λ_{SPR}) with regard to e (B).

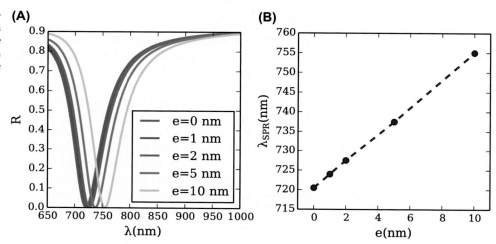

Graphs of the reflectance (R) with regard to the wavelength for a fixed incidence angle and four values of the refractive index of the sample (n) are superimposed on Fig. 3.2.3A. They present a minimum with typical shape called a *dip*, which corresponds to the SPR. As n increases, the dip is translated to the high wavelengths. The resonance wavelength, λ_{SPR}, with regard to n is plotted on Fig. 3.2.3B. It shows a typical linear increase with the refractive index of the sample.

The SPR leads to an evanescent wave propagating at the interface over a distance of a few microns. Its penetration depth is typically a few hundreds of nanometers. This evanescent wave probes the sample medium and only the events that occur at the vicinity of the metal film (typically less than 100 nm) produce a dip shift. Let us consider the previous configuration. Instead of changing the refractive index of the solution, a biomolecule layer is formed as the solution is injected. Its thickness is noted e. Its refractive index is taken equal to 1.46 [7]. On Fig. 3.2.4A, the reflectance with regard to the wavelength for a fixed incidence angle and for five thickness values of biomolecules is superimposed. As e increases, the dip shifts toward longer wavelengths. In the range considered in this example, the resonance wavelength changes linearly with the molecule layer thickness (Fig. 3.2.4B) and thus to the amount of molecule absorbed on the surface. It is this property that is exploited in SPR sensors, which are often seen as an optical scale.

Figs. 3.2.3 and 3.2.4 illustrate the case of a fixed incidence angle and a scanning wavelength. Similar results would have been obtained if considering a fixed wavelength and a scanning angle.

3.2.2.2.2 Surface Plasmon Resonance Transducer Configurations

Typically, four configurations exist [7] (Fig. 3.2.5):

- angular interrogation: the light wavelength is fixed and θ_{SPR} is measured,
- wavelength interrogation: the incidence angle is fixed and λ_{SPR} is measured,

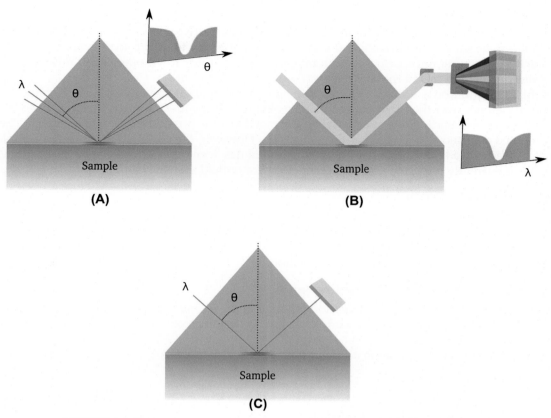

FIGURE 3.2.5   Angular (A), spectral (B), and (C) intensity interrogation configurations.

- intensity interrogation: the incidence angle and the wavelength are fixed and the reflected intensity is measured,
- phase interrogation: when the light generates SPR, a step-phase shift occurs, which is measured.

3.2.2.2.2.1 ANGULAR INTERROGATION

The angular interrogation is the most commonly used configuration. Commercial systems such as the Biacore (GE Healthcare), or the Spreeta (Texas Instruments) instruments are based on this principle. Several transducer technologies were developed to measure the resonance angle. The more straightforward way consists in scanning the incident angle with a rotating mirror and registering the minimum of reflectivity with a detector [8]. This configuration works well, but for some applications such as onfield sensors that require very high robustness, a nonscanning configuration has to be preferred. For instance, Melendez and co-worker suggested using a diverging beam of light and a photodiode array to measure the angle of minimum reflectivity in real time without any moving part in a very compact system [9]. From this architecture, the Spreeta system was born. Texas Instruments commercializes the modules that can be integrated into different systems. The SPIRIT (Surface Plasmon Instrumentation for the Rapid Identification of Toxins) instrument that is now commercialized by the Seattle technologies integrates up to 24 of these modules into a robust packaging for onfield deployment [10].

3.2.2.2.2.2 WAVELENGTH INTERROGATION

The basic configuration of a wavelength interrogation system uses a white light source and a spectrometer to measure λ_{SPR}. Although it requires more complex instrumentation than the angular interrogation, the wavelength interrogation can exploit optical fibers to improve the compactness of the sensitive part as reported by Cahill and co-workers [11]. The prism can then be deported from the source and spectrometer, which can be very convenient for some configurations. In 2016, Colas and co-workers reported an in situ SPR system utilizing this configuration [3].

The wavelength interrogation can be implemented with optical fibers. The prism is replaced by an optical fiber, the cladding for which has been removed, and a thin film of gold deposited on the core.

To improve the compactness and cost efficiency, Sereda and co-workers replaced the white light by a set of five LEDs [12]. Then, the reflected power at each of the wavelengths is measured by a photodiode. The dip is then reconstructed by fitting the data with a pseudo-Lorentzian profile.

3.2.2.2.2.3 INTENSITY INTERROGATION

The intensity interrogation can be very highly compact and cost effective. However, to reach a high sensitivity, the incident angle has to be set very accurately and it is generally necessary to utilize some scanning part to set up the instrument, such as a rotating mirror. Such design may lead to sensitivity to vibrations and misalignment and then is not appropriate for field or in situ robust systems.

However, this configuration is utilized in the SPR imaging technology. A camera replaces the photodiode. A map of the reflectance changes over the substrate can then be obtained. If the metal layer is functionalized with different spots, each of them corresponding to a specific layer, the metal layer is turned into a real biochip [13]. Several components can then be detected and assayed at the same time. However, the configuration is more complex than the previous ones. As far as we know, no onfield or in situ system has been reported. It seemed that putting in parallel several SPR cells is currently preferred over this technology for onfield systems [10].

3.2.2.2.2.4 PHASE INTERROGATION

Phase interrogation was proven the most sensitive system. A resolution of 3×10^{-9} was demonstrated [4]. However, such a sensitive measurement requires very high constraints. The refractive indices of several materials change with temperature. A thermal regulation has to be as good as to guarantee temperature drift lower than 10^{-4}°C, which is prohibitive in the case of onfield or in situ instrumentation.

3.2.2.2.3 Sensitivity, Detection Limit, and Resolution

The transducer is sensitive to the refractive index change in the vicinity of the metal film. Two cases have to be considered. First, the refractive index of the sample varies. Second, molecules get adsorbed on the metallic surface and impact the refractive index.

The refractive index sensitivity, S, of the different transducers is defined as:

$$S = \frac{\partial M}{\partial n} \tag{3.2.1}$$

in which M is the measurand. It can be λ, θ, intensity, or phase.

The unit of S is generally expressed as the unit of the measurand per refractive index unit (RIU).

The value of S depends on the metal, the glass material, the thin-film thickness, and the wavelength, which have to be optimized for a particular detection configuration [7,8].

The resolution (r) of a transducer is defined as the smallest refractive index change that can be detected. Let δM the smallest measurand value that can be detected. The resolution is given by:

$$r = S\delta M \tag{3.2.2}$$

This parameter is proportional to S but also to δM, which depends on the detection chain. Both have to be optimized for a special application. Generally, δM is considered as three times the standard deviation of the signal as the running buffer circulates in the flow cell. A resolution of 10^{-6} is achieved for standard sensors and of 10^{-7} for well-optimized sensors [14].

SPR transducers are often used to measure the quantity of molecules that gets adsorbed on the surface. In this case, the SPR sensor is sensitive to the surface mass concentration of the molecule (noted Γ). It can be shown that when the molecular layer to detect is much thinner than the penetration depth of the evanescent wave (of the order of 100 nm):

$$\Delta \Gamma = \frac{Lz}{\frac{\partial n}{\partial C} S} \Delta M \tag{3.2.3}$$

in which $\Delta \Gamma$ and ΔM are the change of the surface mass concentration of the molecule and of the measurand, respectively, and $\partial n / \partial C$ is the derivative of the refractive index with regard to the mass concentration (in mg/mL) of the compound in solution. Lz is the penetration depth of the surface plasmon in the sample. It is typically a few hundreds of nanometers.

$\Delta \Gamma$ is then inversely proportional to the molecular mass of the molecule. The SPR signal is then lower for small molecules.

The smallest amount of a molecule that can be detected (called Limit Of Detection [LOD] hereafter) scales as $1/S$. S is then an interesting parameter for comparing transducers. However, δM and Lz are of course to take into account.

3.2.2.2.4 *Algorithm for Dip Detection*

The determination of λ_{SPR} or θ_{SPR} can be done via different algorithms. Fitting the data at the vicinity of the minimum by a polynomial of degree from 2 to 5 can lead to a resolution of 10^{-7} [15–17]. Other algorithms can be used such as the centroid algorithm, which consists in calculating the abscissa of the barycenter of the dip area [18], spectral decomposition [19], or dynamic baseline correction [20].

3.2.2.3 Surface Plasmon Resonance as a Molecular Sensor

Most SPR applications consist in molecular sensors: a layer specific to an analyte is deposited on the metallic surface and the SPR transducer measures the refractive index changes as the target compounds in solution reaches the surface. An appropriate calibration protocol enables one to deduce the concentration of the analyte in solution.

The kinetics of the SPR signal is called a sensorgram. A typical SPR measurement consists in five steps:

1. At t=0, the running buffer circulates into the flow cell,
2. When the sample is injected, the analytes start reacting with the specific layer, and the SPR signal starts increasing. The specific layer saturates as the analytes react with it. The signal reaches a plateau,
3. The flow cell is rinsed with the running buffer: there is a slow decrease of the signal because molecules that are not strongly adsorbed at the surface are removed,
4. The sensitive layer can generally be regenerated by the injection of a regeneration solution: The signal then decreases or increases very fast as the regenerating buffer is injected into the flow cell because of a refractive index change between the running buffer and the regeneration solution,
5. The flow cell is rinsed with the running buffer: the SPR signal gets back to its initial value.

A typical sensorgram is showed on Fig. 3.2.6.

The kinetics depends on the amount of analyte that reacts with the specific layer. The total shift or the whole kinetics can be exploited. When working with low concentrations, the saturation cannot be reached within a reasonable lapse of time. The total shift with a given period or the slope at the beginning of the analyte reaction can be related to the target compound in solution after an appropriate calibration process.

To measure an analyte concentration, the whole process has to be carried out even though the saturation is not reached. This cycle then takes several minutes. SPR molecular sensors are then not appropriate for measurements that require response times smaller than about 5 min.

3.2.2.4 Different Detection Formats

The typical detection strategy is to graft the specific layer on the transducer and to detect the analyte bindings. Such a detection format is called *direct assay*. When dealing with a low molecular weight molecule (typically less than 10 kDa), this approach may not be sensitive enough. There are three other assay formats: sandwich, competitive, and inhibition assay (Fig. 3.2.7).

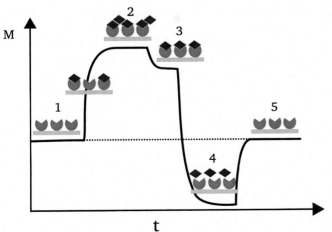

FIGURE 3.2.6 Kinetics of a typical surface plasmon resonance detection. The number corresponds to the steps described as follows.

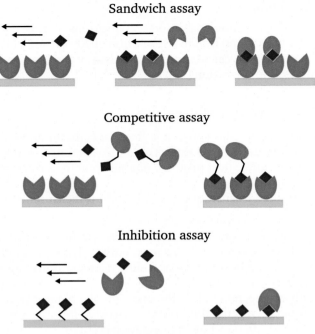

FIGURE 3.2.7   Assay format.

3.2.2.4.1 *Sandwich Assay*

The surface is functionalized by a molecular layer specific to the target molecule. First, the sample is injected and the target molecules bind to the receptor. Then, another molecule that is specific to and heavier than the target compound is injected. The signal is then amplified by the second injection.

3.2.2.4.2 *Competitive Assay*

The surface is functionalized by a molecular layer specific to the target molecule. The sample is mixed with a solution containing the analyte conjugated to a larger molecule such as Bovine Serum Albumin (BSA) or casein before injection. The signal mainly comes from the binding of the conjugated analyte. The SPR signal is then inversely proportional to the concentration of the target molecule in the sample.

3.2.2.4.3 *Inhibition Assay*

The surface is functionalized with the target molecule. The sample is mixed with the specific receptor. The mixture is injected in the flow cell. The SPR signal is then proportional to the amount of receptor that did not bind to the analyte in the sample. It is then inversely proportional to the concentration of the analyte in the sample.

Generally, the direct assay is preferred when fast measurement and straightforward development is required. Indeed, it only requires the injection of the sample, some rinsing, and the regeneration of the chip. However, it is not adapted to low molecular weight (typically lower than 10 kDa) molecules.

Inhibition and competition assays can overcome this limitation but at the cost of more complex development. Indeed, the target molecule has to be grafted on a surface in the case of inhibition format or conjugated with a bigger molecule in the case of competition assay. In both cases, it must remain accessible to the receptor. The latter is generally easier.

The inhibition assay requires that the reaction of the target molecule with the receptor reaches the equilibrium before the injection in the SPR flow cell. This is not necessary for the competition assay, which is then faster.

3.2.2.5 Specific Layer and Functionalization

The specific layer can be of several kinds. Biosensors utilize biomolecules such as protein (Abs, enzyme…), small peptide sequences, nucleic acid, etc.

Protein is a heterogeneous group of molecules with some particular functionalities. The common denominator is that they are made of sequences of amino acids. Abs are a relatively homogeneous class of protein that has been extensively used as the recognition element of the specific layer. Abs can bind with a high affinity and specificity to a particular molecule. They are relatively large molecules in the case of Immunoglobulin G (IgG; about 150 kDa). They can serve as the specific layer or the second molecular recognition element in indirect assays.

The immobilization of proteins can be done by using the chemical moieties of the amino acids. For instance, lysine contains an NH_2 function, glycine SH that can bind with the gold surface.

Proteins are generally large molecules. However, some short sequences of amino acid, forming a small peptide, can present some recognition capability. For instance, NH_2–Gly–Gly–His–COOH and NH_2–$(His)_6$–COOH can bind selectivity to Ni^{2+} and Cu^{2+} [20].

Oligonucleotides are natural receptors when DNA or RNA sequences are to be detected. The specific layer is then the complementary sequence. However, a special class of oligonucleotide is particularly interesting: the aptamer. They are nucleic acid sequences selected from pools of random-sequence oligonucleotides to bind to a target molecule. They can achieve an affinity and specificity toward molecules comparable to Abs. Two of their advantages are a lower cost and a higher stability than Abs. They have been used for several sensing applications [21] such as microcystin [22] and saxitoxin [23] detection.

Other molecular sensors employ polymers. The specificity can come from encapsulated molecules or from special properties of the polymer [24]. Molecular Imprinted Polymer (MIP) is a special class of specific polymer. The polymer is synthesized to leave small cavities with affinity toward a template molecule, which is the target molecule when dealing with a sensor. MIP can be engineered for several kinds of compounds. Two of its advantages are the low cost and the high stability compared to Abs.

These specific layers can be immobilized on the metallic surface by several means. This step is crucial. The specific molecule has to remain active after being immobilized. The recognition part has to remain accessible so that the receptor can bind to the target molecule. In addition, the surface must be specific and not enabling nonspecific adsorption. This is assessed by specificity tests. The layer needs to be stable over time and measurement cycles. These points must be carefully addressed during the functionalization optimization process, which has to be adapted to each layer and each application.

3.2.2.6 Fluidic System

Generally, the flow cell and the flow rates are chosen to get the flow in a laminar regime. The fluid then circulates at the same velocity over the whole sensitive area.

The Reynolds number R is a dimensionless parameter that can be used to characterize the flow. It can be calculated from the density (ϱ) and the viscosity (η) of the liquid, the height of the flow cell (h), and the flow rate (ϕ):

$$R = \frac{\rho\phi}{\eta h} \tag{3.2.4}$$

The flow is laminar when R is less than 2100. This condition implies that in the case of water at 20°C, $\phi/h < 2100$. The typically the height of a flow cell is of few tens of μm. If $h = 50$ μm, the flow rate has to be less than 6 mL/min.

In the case of molecular sensors, the SPR signal is sensitive to the amount of molecule that reacts with the receptor. It depends on the ability of the molecules to reach the surface and then on the diffusion of the analytes to the surface. This phenomenon can be described by the first Fick's law, which states that the diffusion rate of a molecule depends on its concentration gradient. Then, the diffusion rate of an analyte depends on its binding rate with the specific layer. In the case of an SPR sensor, the flow rate has to be high enough to get a strong concentration gradient at the vicinity of the metallic surface within a reasonable time [25].

It is also worth noting that the diffusion rate of a molecule scales as η. The kinetics of the response of the sensor then depends on the temperature.

SPR sensors then require a high control of the sample flow on the sensitive area, which is a strong constraint for in situ instrumentation. However, some recent developments of an in situ instrument utilizing microfluidics [17,18] shows that this can be overcome with appropriate engineering.

3.2.3 EXAMPLES OF SURFACE PLASMON RESONANCE SENSOR FOR THE MARINE ENVIRONMENT

In the following section, some examples of laboratory or infield SPR sensors that are potentially interesting for marine applications are illustrated. This review is not exhaustive and just underlines the very high versatility of SPR instrumentation and how it could be useful for in situ sensors.

3.2.3.1 Dissolved Gases

Methane and carbon dioxide are gases of major interest as they both contribute to the greenhouse gas effect. However, their emission by aquatic ecosystems cannot be measured accurately and continuously over a long period of time, which creates a strong need for new sensors.

In 2008, Boulart and co-workers reported a sensor capable of assaying dissolved methane in the 0–400 nM range with a detection limit of 0.2 nM [26] in laboratory. The transducer consisted of an Spreeta module and the specific layer in a polydimethylsiloxane (PDMS) matrix permeable to CH_4 in which molecules of cryptophane-A were incorporated [27]. This compound can encapsulate methane and presents a strong and reversible affinity toward this dissolved gas. This sensor was first tested in a laboratory and then used during a cruise in the Baltic Sea. The analyses were performed on board on pumped seawater. The sensor enabled a real-time profile of the dissolved-methane concentration.

An SPR carbon dioxide sensor was recently reported [28]. The specific layer consisted of a hydrophobic amidine (N,N,N'-tributylpentane-amidine) incorporated in a polymer matrix permeable to CO_2: ethyl cellulose. More qualification of the sensor is required however; this approach enabled then to assay the pCO_2 in laboratory with an LOD of 7 milliatmospheres (matm).

3.2.3.2 Trace Metal

To measure trace metals, the main difficulty is to get a receptor specific to one element. Most receptors are sensitive to a family of elements. In 2004, Wu and co-workers detected Cd, Zn, and Ni in a buffer using rabbit metallothionein as a receptor [29]. This protein can bind to these metallic elements. The sensor could only detect these elements at concentrations down to 0.1 µg/mL, which is not low enough to be turned into an environmental sensor. In 2005, Forzani and co-workers demonstrated that functionalizing the surface with short peptide sequences could lead to a selective detection of Cu^{2+} or Ni^{2+} in the sub nanomolar (nM) range [48]. Ock and co-worker proposed a polymeric layer consisting of a polyvinylchloride–polyvinylacetate–polyvinylalcohol (PVC–PVAc–PVA) copolymer film in which they incorporated a squarilyum molecule [30]. The dye has the ability to bind specifically to Cu^{2+}, which changes its absorption properties. This impacts not only the real part but also the imaginary part of the refractive index change when the ions bind to the dye. The method enabled them to reach very high sensitivity. They could detect Cu^{2+} at concentrations as low as a few picomolar (pM). They also showed that their sensor was not significantly sensitive to other ions such as: Ag^+, K^+, Na^+, Zn^{2+}, Mg^{2+}, Ca^{2+}, Co^{2+}, Li^+, Cd^{2+}, Hg^{2+}, and Pb^{2+}. However, the sensors were only used with buffer spiked with the different ions and more investigation is needed.

3.2.3.3 Pollutants

SPR biosensors were developed for detecting many different pollutants. To detect small molecules, most of the assays were based on inhibition format with Abs. With this approach, pesticides such as atrazine [31] and simazine [32] were detected at concentrations as low as 0.05 and 0.11 ng/mL respectively. Phenolic compounds such as bisphenol A were detected in the 0.05–1 ng/mL [33]. Polycyclic Aromatic Hydrocarbons (PAHs) such as the very dangerous benzo-[A]-pyrene could be detected at concentrations as low as 0.01 ng/mL [34]. PolyChlorinated Biphenyls (PCBs) were assayed at concentrations as low as 0.1 ng/mL by direct assay. To achieve such high sensitivity the gold surface was functionalized with cytochrome c, which changes its conformation when exposed to PCBs [35].

3.2.3.4 Seafood

The traditional analytical methods to assay phycotoxins are liquid chromatography with tandem mass spectroscopy or high-performance liquid chromatography with fluorescence detection. These techniques are very accurate but require skilled operators, are long, and relatively expensive. In addition, thesetechnologies are not adapted to infield or in situ systems. Several approaches have been reported to detect phycotoxins with SPR sensors.

For instance, Loterzio et al. developed a sensor system using MIP for assaying domoic acid (DA) in the 2–3000 ng/mL range. To detect such low concentrations of DA, the detection format was a competitive assay with DA conjugated to HorseRadish Peroxidase (HRP) enzyme [24]. Although the sensor demonstrated some interesting properties, it was only characterized with the running buffer spiked with DA. Yu et al. reported an inhibition assay using monoclonal Abs capable of detecting DA in clam extracts [36]. They achieved a limit of detection as low as 0.1 ng/mL.

Recently, SPR biosensors capable of multidetection were reported. Campbell and co-workers used a laboratory microfluidic device containing 16 measurement channels for detecting four phycotoxins (saxitoxin family, domoic acid, and okadaic acid) simultaneously diluted in a buffer in the ng/mL range [37]. To achieve such a high sensitivity, the detection was based on an inhibition assay taking advantage of monoclonal Abs. Two years later the system enabled McNamee et al. to assay the toxins in cell extracts from 256 seawater samples coming from European waters [38].

These few examples showed how SPR biosensors can detect very small molecules in very different matrices.

3.2.3.5 Organism Detection

The identification and numbering of organisms is particularly important for ecological studies. Automated imaging systems coupled to machine-learning algorithms [39] have been proven very efficient for micro- or mesozooplankton. However, many species cannot be discriminated on morphological criteria only. Biosensors have then been developed to identify and count specifically targeted species.

There are currently two main approaches for detecting a species: nucleic acid or whole-cell sensors. The first consists in extracting the nucleic acid, which can be DNA and RNA, and detecting the specific sequence by SPR. In 1999, Kai and co-worker showed that they could detect and assay the polymerase chain reaction product of a specific gene of *Escherichia coli* by SPR [40]. However, other technologies such as (quantitative) Polymerase Chain Reaction (PCR) were proven efficient for in situ species detection [41].

The second approach requires the synthesis of a receptor specific to the whole cell. This has been carried out for different species of *E. coli*, which is currently targeted in bath water. Many studies have been published on this approach. For instance, Torun and co-worker reported a biosensor functionalized by Abs specific to the whole cell capable of detecting few tens of colony-forming units (CFU)/mL [42]. Recently, Abadian and co-workers applied SPR imaging to study *E. coli* biofilm and showed that they could detect an individual colony [43]. In 2010, Dreanno et al. reported the SPR imaging detection of HAB species *Alexandrium minutum*. The SPR chip was functionalized with Abs specific to the cell [44].

3.2.4 IN SITU SURFACE PLASMON RESONANCE SENSORS

3.2.4.1 Refractive Index Sensors

The new equation of state (TEOS-10) can be used to determine all the thermodynamic properties of seawater. It requires the knowledge of absolute salinity (S_A). This quantity is defined as the mass fraction of dissolved material in seawater [45]. The introduction of S_A in TEOS-10 instead of the practical salinity (noted S_P) defined in the Practical Salinity Scale 1978 (PSS-78) comes from the necessity to take into account nonionic compounds that change the density of seawater and S_A but not S_P. Up until now, the absolute salinity can be calculated from traditional measurements of conductivity using a location-specific algorithm [46].

New sensors that would measure accurately the absolute salinity are necessary for the accurate application of TEOS-10. In 2007, in Italy, the Scientific Committee on Oceanic Research (SCOR) and International Association for the Physical Sciences of the Oceans (IAPSO) working group 127 dedicated to the Thermodynamics and Equation of State of Seawater identify the optical sensors measuring the refractive index of seawater as good candidates. The resolution of refractive index measurements as well as the corresponding uncertainties of theoretical formulas are required to be 10^{-6} at atmospheric pressure, and 3×10^{-6} at high pressures, corresponding to 4×10^{-6} and 10^{-5} in density, respectively.

The first underwater SPR sensor for refractive index measurement was reported by Diaz–Herrera and co-workers [1]. They developed an underwater sensor for coastal application using optical fibers. The system utilizes monomode optical fibers polished and covered by multilayer thin films [47]. It relies on intensity interrogation: an LED as a monochromatic source and a photodiode for measuring the reflectance changes. They reported a refractive index resolution of about 3×10^{-5}. Unfortunately, only one paper reported the use of this system, and the data were not related to oceanographic quantities.

A few years later, Kim and co-workers [2] reported an in situ SPR sensor for characterizing the dissolved organic carbon (DOC) concentration in coastal areas. The main idea is that both the refractive index and the conductivity of seawater depend on the DOC and ionic concentrations but differently. By combining these two techniques in coastal areas, they evidenced areas where practical salinity was constant but the refractive index changed. They attribute this special water property to DOC.

In their paper, they utilized spectral interrogation through optical fibers. The sensitivity of the system was estimated to 1600 nm/RIU and the resolution to 2×10^{-4}. Its performance is relatively low compared to prism-based sensors. However, as the optical-fiber probe is very small, it responds very fast to temperature changes.

Both studies showed that SPR sensors can be very promising for salinity measurement but fundamental metrology work needs to be carried out on how to relate the refractive index to the fundamental properties of seawater.

3.2.4.2 Biotoxin Biosensor

As mentioned earlier, in 2016, Colas and co-workers reported an underwater SPR biosensor for the detection of dissolved DA [3]. The transducer was based on spectral interrogation. Unlike the two previous systems, optical fibers only bring the light from the light source to the prism and from the prism to the spectrometer. Indeed, the optical-fiber systems are generally less sensitive than the prism-based systems because of some polarization issues in the optical fiber. The sensor is composed of three parts (Fig. 3.2.8): the optode containing the prism and some optics, a container containing the electronics, an embedded computer, the spectrometer, the light source, and a fluidic module previously used for deep-water flow-injection analysis [49]. The sensor is controlled from a boat using Ethernet protocol for communication with an embedded computer. It was designed as a remote lab.

The sensitivity of the transducer is estimated in situ to be about 5000 nm/RIU. The detection protocol relies on an inhibition assay with monoclonal Abs for achieving a very low LOD. It was estimated in situ to 0.1 ng/mL by injecting calibration solutions of known DA concentration. The sensor can be regenerated by injecting a NaOH solution.

Typically, to perform an assay, the sample was pumped through a 0.2 μm mesh filter. It was incubated with the Abs for 15 min. Then the mixture was injected in the SPR flow cell for 15 min. It was then rinsed with phosphate-buffered saline (PBS) for 10 min, regenerated by injecting NaOH for 5 min, and rinsing with PBS again. The whole measurement process took about 45 min.

An example of sensorgram acquired with this system is shown in Fig. 3.2.9. It represents two successive measurements. Two samples containing no DA and 0.5 ng/mL were assayed. They yielded a total shift of the SPR

FIGURE 3.2.8   The optode in front of the electronic container.

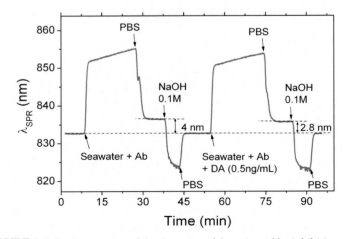

FIGURE 3.2.9   Sensorgram of the detection of domoic acid by inhibition assay.

wavelength of 4 and 2.5 nm, respectively. Then, an appropriate calibration procedure leads to a quantitative assay of the toxin.

As far as we know, this SPR biosensor is the first reported, but there should be many others. SPR is a very versatile technique that has been demonstrated to be very efficient for assaying several compounds of interest to the oceanographer and marine biologist communities. The fast development of SPR sensors these last few decades has overcome most of the technological issues. Now everything is ready for a fast development of in situ SPR sensors.

3.2.5 CONCLUSIONS

SPR sensors have widely spread in many application areas as they are specific, sensitive, quantitative, and do not require labeling of the molecules (label-free). Any kind of target can be detected, from small molecules such as methane, carbon dioxide, metallic ions, PAHs, pesticides, phycotoxin, nucleic acid up to microorganisms. SPR sensors present a great opportunity for oceanography applications.

To our knowledge, three in situ SPR sensors have been reported. Two of them were refractometers. The refractive index of seawater is directly related to the density of seawater. Optical sensors are then considered as good candidates for the measurement of the absolute salinity that is required in the recent TEOS-10. The other in situ SPR system is a biosensor, capable of assaying domoic acid, a small molecule responsible for ASP, in the 0.1–2 ng/mL range.

These works demonstrated that SPR technology is now ready for the development of new sensors to address many scientific questions.

References

[1] Díaz-Herrera N, Esteban O, Navarrete MC, Haitre ML, González-Cano A. In situ salinity measurements in seawater with a fibre-optic probe. Meas Sci Technol 2006;17(8):2227.

[2] Kim Y-C, Cramer JA, Booksh KS. Investigation of a fiber optic surface plasmon spectroscopy in conjunction with conductivity as an in situ method for simultaneously monitoring changes in dissolved organic carbon and salinity in coastal waters. Analyst 2011;136(20):4350–6.

[3] Colas F, Crassous M-P, Laurent S, Litaker RW, Rinnert E, Le Gall E, et al. A surface plasmon resonance system for the underwater detection of domoic acid: domoic acid detection using SPR. Limnol Oceanogr Meth 2016;14(7):456–65.

[4] Li Y-C, Chang Y-F, Su L-C, Chou C. Differential-phase surface plasmon resonance biosensor. Anal Chem 2008;80(14):5590–5.

[5] Piliarik M, Homola J. SPR sensor instrumentation. In: Homola J, editor. Surface plasmon resonance based sensors. Berlin, Heidelberg: Springer Berlin Heidelberg; 2006.

[6] Kretschmann E, Raether H. Radiative decay of nonradiative surface plasmons excited by light. Z Naturforsch 1968;A23(12):2135–6.

[7] Maillart E. Développement d'un système optique d'imagerie en résonance de plasmons de surface pour l'analyse simultanée de multiples interactions biomoleculaires en temps réel. [Orsay]: Université Paris XI; 2004.

[8] Homola J, Piliarik M. Surface plasmon resonance (SPR) sensors. In: Homola J, editor. Surface plasmon resonance based sensors. Berlin, Heidelberg: Springer Berlin Heidelberg; 2006. p. 45–67. [Internet].

[9] Melendez J, Carr R, Bartholomew DU, Kukanskis K, Elkind J, Yee S, et al. A commercial solution for surface plasmon sensing. Sensor Actuator B Chem 1996;35(1–3):212–6.

[10] Chinowsky TM, Soelberg SD, Baker P, Swanson NR, Kauffman P, Mactutis A, et al. Portable 24-analyte surface plasmon resonance instruments for rapid, versatile biodetection. Biosens Bioelectron 2007;22(9–10):2268–75.

[11] Cahill CP, Johnston KS, Yee SS. A surface plasmon resonance sensor probe based on retro-reflection. Sensor Actuators B Chem 1997;45(2):161–6.

[12] Sereda A, Moreau J, Boulade M, Olivéro A, Canva M, Maillart E. Compact 5-LEDs illumination system for multi-spectral surface plasmon resonance sensing. Sensor Actuators B Chem 2015;209:208–11.

[13] Maillart E, Brengel-Pesce K, Capela D, Roget A, Livache T, Canva M, et al. Versatile analysis of multiple macromolecular interactions by SPR imaging: application to p53 and DNA interaction. Oncogene 2004;23(32):5543–50.

[14] Piliarik M, Homola J. Surface plasmon resonance (SPR) sensors: approaching their limits? Opt Express 2009;17(19):16505.

[15] Bardin F, Bellemain A, Roger G, Canva M. Surface plasmon resonance spectro-imaging sensor for biomolecular surface interaction characterization. Biosens Bioelectron 2009;24(7):2100–5.

[16] Bolduc O, Live L, Masson J. High-resolution surface plasmon resonance sensors based on a dove prism. Talanta 2009;77(5):1680–7.

[17] Piliarik M, Vala M, Tichý I, Homola J. Compact and low-cost biosensor based on novel approach to spectroscopy of surface plasmons. Biosens Bioelectron 2009;24(12):3430–5.

[18] Hu J, Zhao X. An improved centroid algorithm for a surface plasmon resonance bioanalyzer using microprocessors. In: IEEE eXpress Conference publishing. 2009. p. 1–4. Available from: http://ieeexplore.ieee.org/document/5230137/.

[19] Nenninger GG, Piliarik M, Homola J. Data analysis for optical sensors based on spectroscopy of surface plasmons. Meas Sci Technol 2002;13(12):2038–46.

[20] Thirstrup C, Zong W. Data analysis for surface plasmon resonance sensors using dynamic baseline algorithm. Sensor Actuators B Chem 2005;106(2):796–802.

[21] Ruscito A, DeRosa MC. Small-molecule binding aptamers: selection strategies, characterization, and applications. Front Chem May 10, 2016;4. [Internet]. Available from: http://journal.frontiersin.org/Article/10.3389/fchem.2016.00014/abstract.

[22] Ng A, Chinnappan R, Eissa S, Liu H, Tlili C, Zourob M. Selection, characterization, and biosensing application of high affinity congener-specific microcystin-targeting aptamers. Environ Sci Technol 2012;46(19):10697–703.

[23] Handy SM, Yakes BJ, DeGrasse JA, Campbell K, Elliott CT, Kanyuck KM, et al. First report of the use of a saxitoxin–protein conjugate to develop a DNA aptamer to a small molecule toxin. Toxicon 2013;61:30–7.

[24] Lotierzo M, Henry OY, Piletsky S, Tothill I, Cullen D, Kania M, et al. Surface plasmon resonance sensor for domoic acid based on grafted imprinted polymer. Biosens Bioelectron 2004;20(2):145–52.

[25] Navratilova I, Myszka DG. Investigating biomolecular interactions and binding properties using SPR biosensors. In: Homola J, editor. Surface plasmon resonance based sensors. Berlin, Heidelberg: Springer Berlin Heidelberg; 2006. p. 155–76.

[26] Boulart C, Mowlem MC, Connelly DP, Dutasta J-P, German CR. A novel, low-cost, high performance dissolved methane sensor for aqueous environments. Opt Express 2008;16(17):12607.

[27] Brotin T, Dutasta J-P. Cryptophanes and their complexes—present and future. Chem Rev January 14, 2009;109(1):88–130.

[28] Lang T, Hirsch T, Fenzl C, Brandl F, Wolfbeis OS. Surface plasmon resonance sensor for dissolved and gaseous carbon dioxide. Anal Chem 2012;84(21):9085–8.

[29] Wu C-M, Lin L-Y. Immobilization of metallothionein as a sensitive biosensor chip for the detection of metal ions by surface plasmon resonance. Biosens Bioelectron 2004;20(4):864–71.

[30] Ock K, Jang G, Roh Y, Kim S, Kim J, Koh K. Optical detection of Cu2+ ion using a SQ-dye containing polymeric thin-film on Au surface. Microchem J 2001;70(3):301–5.

[31] Minunni M, Mascini M. Detection of pesticide in drinking water using real-time biospecific interaction analysis (BIA). Anal Lett 1993;26(7):1441–60.

[32] Harris R, Luff B, Wilkinson J, Piehler J, Brecht A, Gauglitz G, et al. Integrated optical surface plasmon resonance immunoprobe for simazine detection. Biosens Bioelectron 1999;14(4):377–86.

[33] Hegnerová K, Piliarik M, Šteinbachová M, Flegelová Z, Černohorská H, Homola J. Detection of bisphenol A using a novel surface plasmon resonance biosensor. Anal Bioanal Chem 2010;398(5):1963–6.

[34] Gobi KV, Miura N. Highly sensitive and interference-free simultaneous detection of two polycyclic aromatic hydrocarbons at parts-per-trillion levels using a surface plasmon resonance immunosensor. Sensor Actuators B Chem 2004;103(1–2):265–71.

[35] Hong S, Kang T, Oh S, Moon J, Choi I, Choi K, et al. Label-free sensitive optical detection of polychlorinated biphenyl (PCB) in an aqueous solution based on surface plasmon resonance measurements. Sensor Actuators B Chem 2008;134(1):300–6.

[36] Yu Q, Chen S, Taylor AD, Homola J, Hock B, Jiang S. Detection of low-molecular-weight domoic acid using surface plasmon resonance sensor. Sensor Actuators B Chem 2005;107(1):193–201.

[37] Campbell K, McGrath T, Sjölander S, Hanson T, Tidare M, Jansson Ö, et al. Use of a novel micro-fluidic device to create arrays for multiplex analysis of large and small molecular weight compounds by surface plasmon resonance. Biosens Bioelectron 2011;26(6):3029–36.

[38] McNamee SE, Elliott CT, Delahaut P, Campbell K. Multiplex biotoxin surface plasmon resonance method for marine biotoxins in algal and seawater samples. Environ Sci Pollut Res 2013;20(10):6794–807.

[39] Grosjean P, Picheral M, Warembourg C, Gorsky G. Enumeration, measurement, and identification of net zooplankton samples using the ZOOSCAN digital imaging system. ICES J Mar Sci 2004;61(4):518–25.

[40] Kai E, Sawata S, Ikebukuro K, Iida T, Honda T, Karube I. Detection of PCR products in solution using surface plasmon resonance. Anal Chem 1999;71(4):796–800.

[41] Preston CM, Harris A, Ryan JP, Roman B, Marin R, Jensen S, et al. Underwater application of quantitative PCR on an ocean mooring. Z. Zhao (Ed.) PLoS One 2011;6(8):e22522.

[42] Torun Ö, Hakkı Boyacı İ, Temür E, Tamer U. Comparison of sensing strategies in SPR biosensor for rapid and sensitive enumeration of bacteria. Biosens Bioelectron 2012;37(1):53–60.

[43] Abadian PN, Tandogan N, Jamieson JJ, Goluch ED. Using surface plasmon resonance imaging to study bacterial biofilms. Biomicrofluidics 2014;8(2):021804.

[44] Dreanno C, Milgram S, Gas F, Colas F, Livache T, Compère C. Direct detection of marine toxic algae by SPR imaging. Biosensors May 16, 2012. Cancun.

[45] Millero FJ, Feistel R, Wright DG, McDougall TJ. The composition of standard seawater and the definition of the reference-composition salinity scale. Deep Sea Res Part I Oceanogr Res Pap 2008;55(1):50–72.

[46] McDougall TJ, Jackett DR, Millero FJ, Pawlowicz R, Barker PM. A global algorithm for estimating Absolute Salinity. Ocean Sci 2012;8(6):1123–34.

[47] Esteban Ó, Cruz-Navarrete M, González-Cano A, Bernabeu E. Measurement of the degree of salinity of water with a fiber-optic sensor. Appl Opt 1999;38(25):5267–71.

[48] Forzani ES, Zhang H, Chen W, Tao N. Detection of heavy metal ions in drinking water using a high-resolution differential surface plasmon resonance sensor. Environ Sci Technol 2005;39(5):1257–62.

[49] Vuillemin R, Le Roux D, Dorval P, Bucas K, Sudreau JP, Hamon M, et al. CHEMINI: a new in situ CHEmical MINIaturized analyzer. Deep Sea Research Part I:. Oceanogr Res Pap 2009;56(8):1391–9.

Glossary

Immunoglobulin G (IgG) The most common type of antibodies.
Molecular Imprinted Polymer (MIP) Polymer that was synthetized to leaves cavities specific to a particular molecule.
Surface Plasmon Resonance Oscillation of the conduction band electrons of a metal at the interface with a dielectric material.
Target molecule Molecule of interest to be detected and assay.

CHAPTER

3.3

Biosensors for Aquaculture and Food Safety

Carmem-Lara Manes[1], Chrysi Laspidou[2]

[1]**Microbia Environnement, Marine Station Banyuls sur Mer, France; [2]Civil Engineering Department,
University of Thessaly, Thessaly, Greece**

3.3.1 INTRODUCTION

Toxin-producing Harmful Algal Blooms (HABs) constitute a serious threat to public health as well as to sustainable development and are responsible for socioeconomic impacts worldwide. Global costs associated with HAB impacts are difficult to access as they depend on geographic location, the frequency and intensity of blooms, the seafood-product type as well as the associated healthcare provision costs. Furthermore, these costs can vary from year to year. However, Berdalet et al. [1], and later Groeneveld et al. [2] reported an estimation of the annual economic effect of ~US$100 million in the United States back in 2006 related to public health, commercial fisheries, recreation, and tourism, and monitoring and management altogether. In Europe the direct impact on lost sales in fisheries and aquaculture associated with HABs was reported to have achieved 800 million € in 2003 and more than US$1 billion in Japan in 2006. Recently, because of the El Niño weather phenomenon, an exceptional HAB in Southern Chile caused dramatic socioeconomic impacts [3]. The local fish industry accounted for US$800 million losses (https://www.theguardian.com/environment/2016/mar/10/chiles-salmon-farms-lose-800m-as-algal-bloom-kills-millions-of-fish). Trends in the increasing frequency of HABs have been reported extensively in the specialized literature over the past three decades; however, whether actually increasing or just an awareness phenomenon, they will remain a growing threat to human activities related to the sea [4]. Therefore, understanding long-term trends and large-scale distribution patterns of harmful species is a most relevant goal, to predict whether, where, and when changes in HAB frequency and intensity are to be expected to plan effectively management and common use of the sea space.

Although HABs are generally linked with eutrophication of coastal waters and estuaries, identifying specific nutrient loads that would result in given concentrations of toxin-producing microorganisms remains a big challenge. A wide range of factors influence and/or trigger the blooms, such as climate change, existence of invasive species, local hydrology as well as socio-economic factors [5–8]. So, even though several factors are known to influence the appearance of algal blooms, such as nutrient load, water temperature, and water stratification, and despite intense work in the field of microalgale ecology, environmental conditions triggering the appearance of a toxic bloom are not well characterized. Thus, prediction and control of bloom occurrence is a difficult task at present and the development of systems to obtain relevant real-time measurementsimperative [9].

HABs result from the high concentrations of toxin-producing microorganisms in both pelagic and benthic aquatic systems. The most common types of diseases produced by several species are paralytic shellfish poisoning (PSP), diarrheic shellfish poisoning (DSP), amnesic shellfish poisoning (ASP), neurotoxic shellfish poisoning, and toxins such as the hepatotoxic Microcystins (MCs) and the Nodularins (NOD) [10]. Some HABs are also caused by epiphytic benthic microalgae [1], with the most well-known being those causing ciguatera fish poisoning (CFP) and aerosol toxicity (AST) caused by other palytoxin derivatives produced in blooms of *Ostreopsis* spp., which, besides seafood contamination, enter the sea spray causing various degrees of respiratory tract and skin irritation [11,50].

Monitoring and time-series acquisition of HAB occurrences and associated toxins are essential not only to understand both naturally and anthropogenic-driven marine ecosystem change but to build an effective early warning system for better management of fisheries and aquaculture production sites. Even though a large number of sensors have been developed and have even reached commercialization, very few have been deployed in the marine environment. This is because: (1) the working environment is very demanding with issues of sensor fouling and high corrosivity; (2) monitoring of microorganisms and/or biotoxins in their natural environment represents an enormous scientific challenge due to the diversity of toxin-producing organisms and their temporal and spatial distribution; and (3) expertise is limited due to a small number of specialized laboratories working in this field compared to other sectors, such as medicine or pharmaceuticals [12].

To tackle this difficult task, an urgent need has emerged for innovative techniques with fast response and field-deployable sensors that can monitor algal species and toxins on site. Biosensors represent a convenient choice to monitor HABs and marine toxins in situ, due to their low cost, high sensitivity, and fast-response features. Sensor technology to detect both HABs and biotoxins has greatly evolved over the past years using distinct molecular-based assays. The applications eventually go beyond HAB monitoring and extend to new and different areas; they involve, among others, fisheries and food safety: Fisheries management and assessment of fish stock is a sector that can greatly benefit from online biosensor measurement technologies. This sector includes not only the specific fish, but other marine biota, such as mammals and birds [13]. Furthermore, assessing the impact of fisheries through measurements is critical because fishing methods, such as trawls, could lead to significant changes in the structure of the seabed, its ecology, and associated functions. Finally, because toxic algae can directly affect shellfish in multiple ways, a biosensor-based online monitoring program for toxins in seafood can greatly improve food safety presenting a great potential for applications [12]. Besides, many ocean contaminants that accumulate in fish tissues are transferred to humans through the food chain and cause possible damage to human health [14] and the Marine Strategy Framework Directive (MSFD) descriptor 9 states that contaminants in seafood must not exceed relevant standards. Naturally, HAB monitoring is directly related to food safety as well, because toxins lead to large-scale fish kills and local oxygen depletion. In this chapter we highlight examples of molecular biosensing platforms used to detect and quantify toxin-producing algal species and/or biotoxins in marine environments and/or seafood.

3.3.2 CURRENTLY AVAILABLE SENSORS AND TECHNIQUES

Biosensors are traditionally composed of two parts: a biological recognition element (usually an enzyme, antibody, receptor, or microbe) coupled to a transducer that could be chemical or physical [15]. The goal is maximum response so the biological component is placed at or near the transducer surface. Various parts could act as transducers: they could be electrochemical, such as electrodes; mass associated, such as piezoelectric crystals or surface acoustic wave devices; optical, such as optodes; and thermal, with thermistors or heat-sensitive sensors. Transducers convert the particular biological element that is being recognized into an electrical signal, resulting in detection that is specific biochemically and can be easily transferred to a computer or similar system. Therefore, existing biosensing platforms are based on DNA/RNA aptamers, Abs, and DNA/RNA probe assays, and they all incorporate biorecognition elements with a chemical or physical measurement element to allow for detection and quantification of a target analyte. The use of biosensors in the food sector includes enzyme sensors for food components, and immunosensors for pathogenic bacteria and pesticides. Their applications in the food industry include, among others, determination of residues, naturally occurring toxins, and microbial contamination [16]. The details of construction of biosensors have been described elsewhere [17–19,52]. In this section, we present a comprehensive list of sensor types used in marine environments and/or food industry, along with associated examples.

Aptasensors take advantage of target-induced conformational change of an aptamer, short oligo-DNA or -RNA molecules when binding to specific targets, generally small molecules as biotoxins, with high affinity and specificity [20]. This feature allows the specific identification of several small molecules including marine toxins as Brevetoxin-2 (NSPs), Saxitoxin (PSPs), and Okadaic acid (DSPs) (reviewed in Ref. [21]). Electrochemical aptasensors have been used to detect neurotoxin Brevetoxin-2 and Saxitoxin in spiked shellfish [22,23]. However, the complex, expensive, and time-consuming selection of functional high-affinity aptamers represents a discouraging drawback for the immediate application of this method to in situ monitoring of environmental water and seafood.

Immunosensors are based on surface antigen recognition commonly used in rapid tests for toxins. There are several commercially available immunoassays for the detection of DSP, ASP, PSP (Saxitoxin), hepatotoxic MCs and NOD (several distributors, www.abraxis.com), and Cigua-Check; Oceanit Test System, Hawaii, United States (CTX) toxin categories. Most of these are commercialized as enzyme-linked immunosorbent assays (ELISA)-based kits in a microplaque format or in lateral-flow analysis paper strip (e.g., Jellett Rapid Testing Ltd., US Food and Drug Administration [FDA] approved for PSP in seafood). Both configurations are suited for quick, inexpensive, and easy screening of algal toxins in a wide range of matrices from water bodies to flesh material. However, they present known technical limitations concerning the lower sensitivity, the lack of selectivity, and cross-reactivity when working in complex natural-water sample matrices [51]. An improved immunosensor using plasmon resonance (SPR; see also Chapter 3.2) was developed for marine biotoxins as a less laborious alternative to the official high-performance liquid chromatography (HPLC) method. The biosensor assay was validated according to International Union of Pure and Applied Chemistry (IUPAC) and Association of Official Analytical Chemists (AOAC) guidelines, for the detection of PSP toxins in shellfish with detection capability of 120 µg/kg below the EU regulatory limit of 800 µg

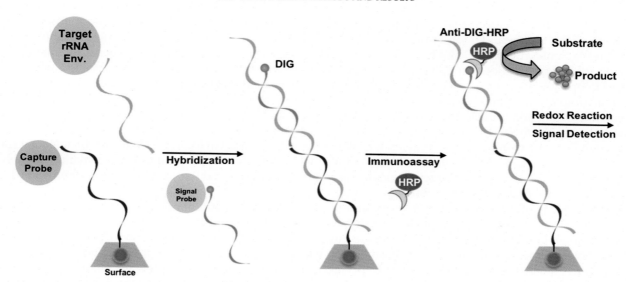

FIGURE 3.3.1   The sandwich hybridization assay process in genosensors. Schematic of nucleic acid identification based on sandwich hybridization of environmental ribosomal RNA (rRNA) with capture and signal oligonucleotide probe coupled to antibodies. Biotinylated capture probe is immobilized onto a surface, which then captures the target rRNA; this complex binds to the digoxigenin (DIG) labeled signal probe, and the antibody (anti-DIG) conjugated to horseradish peroxidase (HRP) recognizes the final hybridized complex. Signal detection occurs when HRP initiates the chemical redox reaction in the presence of an appropriate substrate.

saxitoxin equivalents (STX eq) per kg of shellfish meat [24]. Later McNamee et al. [25] adapted this biosensor format to a multitoxin SPR-based immunosensor prototype capable of detecting PSTs, DSTs, and amnesic shellfish toxins simultaneously in water samples in a few minutes, exempting sample-processing time. Although promising, the reported sensitivities of the prototype were lower than those obtained with the commercial ELISA test; besides, the SPR format does not allow for portability. Moreover, the challenge for immunosensors is their performance when working with different matrices: water, algae, or seafood, taking into account the adapted sample-processing method and time, which is critical for achieving appropriate limits of detection.

Genosensors are powerful tools for HAB species detection and quantification (for a detailed review, refer to Ref. [26]). The identification of toxic algal abundance and distribution pattern is greatly facilitated by the application species-specific ribosomal RNA/DNA probe-based assays, notably those involving quantitative polymerase chain reaction (qPCR) and sandwich hybridization methods [27]. Although qPCR has the power of high sensitivity, enabling the amplification of small quantities of target RNA or DNA, the enzymatic amplification reaction is prone to complex matrices inhibitory effects [28] in addition, to the limitation of multiplex qPCR in natural samples. On the other hand, the target sequence of interest is detected directly on the sensor platform via a sandwich hybridization, which uses RNA or DNA capture probes printed or spotted onto a support matrix, to target algal genus or species (Fig. 3.3.1). This assay format was applied to develop a multispecies RNA microarray able to identify and quantify simultaneously several HAB species on a single array ([29]; commercially available from Microbia Environnement). Due to its broad range of detection, this tool is particularly useful for large-scale monitoring of marine harmful algal species for research and modeling purposes. This microarray was successfully applied in the monitoring of a variety of HAB species, including an invasive azaspiracid shellfish poisoning toxins producer, *Azadinium* spp., in a shellfish production area in France [30]. However, due to its relatively complex experimental procedure and postanalytical processing of results this system should remain as an excellent laboratory tool for monitoring purposes and consequently not suited for field-deployed applications.

3.3.3 INNOVATION METHODS AND RESULTS

Owing to the versatility of RNA/DNA probe sandwich hybridization format, new methodological improvements resulted in the development of promising fast, sensitive, and accurate genosensors for the detection of HAB species. Recently, Orozco et al. [31] developed an electrochemical RNA genosensor array for 13 Mediterranean toxic algal species based on the concept of a semiautomated rRNA biosensor [32] with improved species specificity and minimizing potential nonspecific cross-reactions. A similar RNA genosensor was adapted to automation within the framework of the European Union Framework Program7–Ocean of Tomorrow–2013.1 Sensing toxicants in Marine waters makes Sense (EU FP7-OCEAN-2013.1 SMS) project (www.project-sms.eu). The genosensor manual assay involving species recognition through sandwich hybridization was fully transferred into a portable analyzer based on micro Loop Flow Reactor (μLFR) technology, so that the entire analytical sequence was performed by the instrument in a completely autonomous way. Besides, the overall process from sampling to data acquisition was fully automated enabling unmanned control of the procedure when deployed in a maritime observation buoy (Fig. 3.3.2A; Manes et al., unpublished). The compact integrated system comprises two interconnected modules, one for sampling and preanalytical processing and a second analytical module able to detect at least two toxin-producing algal species from the *Alexandrium* spp. and *Pseudonitzschia spp.* in situ. This opens new opportunities for automation of lab-based reactions and the quantification of target species using compact, portable, and less expensive field-deployable instrumentation suited for aquaculture monitoring.

The vast majority of actual monitoring programs are still under the constraint of in situ sampling for later laboratory analysis whether performed manually or using complex and costly instrumentation, both depending on ship time, appropriate infrastructure, and technical personnel. Therefore, it is of paramount importance the introduction of autonomous online systems fits the purpose, which will lower the associated high costs and as a consequence enable high-frequency monitoring of toxic algal species. Until now, the efforts made toward rendering instrumentation for monitoring biological information in the sea automatic, online, and real time have been dedicated to the sampling procedure and to adapted in situ sample conservation for specific downstream laboratory analysis (reviewed in Ref. [33]. An example is the development of an autonomous microbial sampler used to conserve samples in situ for further metabolic profiling of deep-sea microbial communities [34]. Currently, few instruments are able to perform

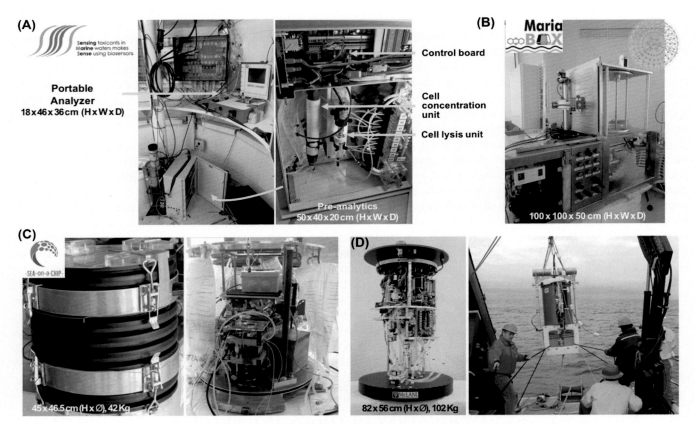

FIGURE 3.3.2 Autonomous field-deployable biosensors for harmful algal bloom (HAB) species and biotoxin detection. (A) SMS integrated system comprising an automated genosensor detecting two HAB species connected to a dedicated pre-analytic module performing sampling, cell concentration, and cells lysis, operating in a buoy on the Slovenian coast; (B) design of the multiplex lab-on-a-disk MariaBox prototype for in situ detection of four categories of biotoxins among other targets; (C) compact Sea-on-a-Chip module harboring an immunosensor for domoic acid (amnesic shellfish poisoning) among other sensors; (D) environmental sample processor analytical core for the detection of HAB species and associated toxins, and next to it the full equipment in a field deployment.

the sampling, sample processing, and analysis in situ. To date, the reference of field-deployed genosensors is still the ESP, in its third generation, developed at the Monterey Bay Aquarium Research Institute (MBARI) and commercially available at McLane Research Laboratories (Fig. 3.3.2D). The ESP, a fully molecular biology automated lab, performs molecular diagnostic tests in situ detecting several HAB species from *Pseudonitzschia* genus and its associated toxin domoic acid (ASP) through RNA hybridization, qPCR, and competitive enzyme-linked immunosorbent assay (cELISA) assays, as well as sample conservation and archiving for later laboratory testing [35,36]. Although the ESPs have been successfully applied to in situ HAB monitoring and exhibited extraordinary sea-time autonomy (45 days—[37]), still the deployment and maintenance of such instrument requires skilled professional time and hence is economically unsuitable for large-scale production and/or wide application in monitoring programs dedicated to aquaculture and food safety. New adaptations of ESP into mobile sensing platforms such as autonomous underwater vehicles (AUVs), which are under development as previously reported [38], might partially overcome those shortcomings.

Recently funded EU projects from the dedicated FP7-OCEAN-2013.1 program have been working on innovative solutions to tackle marine environment challenges through biosensing. The SMS project delivered compact automated systems harboring optical geno- and immunosensors for two toxin-producing microalgal species and domoic acid (ASP), okadaic acid (DSP), and saxitoxin (PSP) detection in seawater, suited to be used as an early warning system in aquaculture facilities. At least four related projects developed biosensors for real-time monitoring of toxin-producing algal species and/or biotoxins in the marine environment, e.g., MariaBox (www.mariabox.net) and Sea on a Chip (www.sea-on-a-chip.eu)—Fig. 3.3.2B and C, though their outcomes are yet to be disclosed. Those innovative solutions are still prototype versions and need dedicated long-term field deployment to offer coastal managers and aquaculture producers better-fit monitoring systems.

3.3.4 CHALLENGES

3.3.4.1 Forecasting Harmful Algal Blooms With Biosensors

For HAB management and mitigation to be effective, it is important to have relevant information on time so that the development, trajectory, and toxicity of bloom populations can be forecasted, even before they start developing to allow time for action. Typical development times and duration of HABs are relatively hard to define, because hydrological regimes, nutrient cycling, and meteorological conditions, through multiple complex interrelations, define the food web, and finally, the trophic state of water bodies [6]. Several reviews have been conducted on the intensification and global expansion of harmful cyanobacterial blooms in terms of abundance, geographic extent, and effects on ecosystem health, as well as factors that may be facilitating this expansion. Environmental factors such as temperature, light intensity, pH, nutrients, salinity, ultraviolet radiation, wind, trace metals, and environmental pollutants can influence the growth of the cyanobacterial species and their cyanotoxin production [39]. Cusack et al. [40] employed an HAB forecast system for Bantry Bay, southwest Ireland, with the application of a 3D physical hydrodynamic model. As a result, using direct measurements of water temperature, salinity, and density with depth produced a reliable 3-day forecast of physical hydrodynamics closely linked to toxic HAB episodes; furthermore Cusack et al. [40], concluded that signals of the upcoming HAB episode were evident up to 10 days prior to shellfish contamination in the Bay. The modeling and biosensor communities have worked toward establishing location-specific early warning capabilities and relying on models that support forecasts of bloom-development rates and toxicity [41]. HAB-relevant data might be collected by several different organizations at different water bodies and various times, but the lack of data-sharing practices make all this information go unused. In California, an integrated, statewide, HAB monitoring and alert network was created by coordinating organizations and researchers currently collecting HAB data and developing a centralized portal for the dissemination of this information, because all programs for monitoring HABs were found uncoordinated with each other [42]. To meet this challenge, HAB forecasts could be developed in a way parallel to weather forecasts: in the latter case, a large number of meteorological stations are set up around the country and collect a variety of data that when assimilated together enable the production of a meaningful forecast. In the same way, data streams for HAB-specific physicochemical and biological variables should be collected throughout the water bodies, at regular spatial and temporal granularity to produce accurate forecasts of bloom events and assess their associated risks. The fact is, HAB monitoring has fallen behind meteorological, physical, and chemical ocean observatories due in large part to the difficulty of designing, integrating, and maintaining HAB-specific instrumentation within these observatories [43].

Several types of platforms exist that could be used for a sustained in situ observation of HAB organisms and toxins, yet with varying and still unresolved integration challenges. This can be achieved by different technologies, platforms, and payload capacity. For instance, buoyancy gliders and autonomous underwater vehicles (AUVs) can carry and power smaller instruments for extended periods. As sizes get larger, the Wave Glider provides more power for deployed instrumentation and thus longer autonomous duration. Extended periods can be achieved with ships or research vessels that may deploy technologies both in the field, but also in labs on board. Moorings and buoys are also a solution for equipment deployment, even though they can only provide point-based measurements; however, identifying optimal deployment sites for such fixed-location assets requires a knowledge of regional oceanography and its regulation of HAB dynamics.

To address the HAB forecast challenges, the sensors developed should not only be limited to detecting the known algal species, but adaptable for detection of newly identified HAB species. Similarly, sensors detecting marine biotoxins in situ should be easily adaptable for detection of newly discovered biotoxins, to catch up with current trends. Naturally, because these sensors will be operated in the harsh marine environment, issues like maintenance, long-term stability, and biofouling need to be addressed [44]. To return to the aforementioned meteorological forecast analogy, to be able to accurately forecast HABs, suitable data mining, machine learning, and artificial intelligence techniques need to be applied to the vast amount of data already produced, coupled with spatiotemporal analysis of the potentially predictive value of these datasets, to improve our understanding of the processes underlying the development and spread of HABs, and possibly to identify proxies that can be used as early warning indicators [45]. These results should enable the development of early warning systems possibly based on proxies that are used as indicators of the HABs to come, which is explicitly requested by end-users. The involvement of authorities, farmers, fishermen, and the public for access to all this information regarding HABs is important, to raise awareness and build trust with the forecasting authorities. Finally, effective HAB treatment methods, such as nanotechnologies, or use of substances like H_2O_2 [8] or modified clay [46], should also be developed.

3.3.4.2 Diversity in Toxin Groups Affecting Detection

One of the challenges of aquatic algal toxin detection is that various toxin groups differ in their chemical structure. As a result, they come from different sources, act differently, and are distributed through different pathways through the trophic chain. As expected, they pose a variety of threats to human health, because they have different toxicity levels [47]. Furthermore, the appearance of new toxin groups and new toxin analogs within a group seems to have increased lately; it is also possible that this rise is artificial and is linked to the advances in detection and identification techniques. In other words, we think more toxin analogs are present nowadays, because we can actually detect and measure them.

3.3.4.3 Variety of Samples Affected by Toxins

Another challenge is the need to cope with different kinds of samples: seafood, microalgae, and water, to cover all levels of the food chain and provide reliable early warning systems. Innovative methods in aquatic toxin detection are focused on the production of fast, sensitive, multitoxin-detection techniques. Trending in this field is the aforementioned multiplex SRP-based immunosensor for ASP, DSP, and PSP toxins [25]. Assays that are easy to perform and do not require highly trained personnel and high-throughput methods would be a highly valuable tool considering the increase of the number of samples to be processed by routine testing laboratories. Innovation in analytical methods and biological methods will have to be an important goal.

3.3.4.4 R&D Focus on and Beyond the Analytical Box

The successful use of sandwich-like immunosensors in an autonomous biosensing device relies on their regeneration capacity and a sensing surface's binding property. These two parameters are critical for the reuse of the same sensor and the accuracy of the results in view of applying autonomous field-deployed biosensors for toxic algal and biotoxin monitoring. Biosensor methodology applied to aquaculture should be conceived as an independent, robust, and autoregenerative test able to integrate different automatic systems in a cost-effective way. In line with this, a promising development of a biosensor disk format able to perform several measurements on a disk and automatically replaced when fully used are being carried on in the EU-MARIne environmental in situ Assessment and monitoring tool BOX ([MARIABOX] project). Alternative configurations are being developed using functionalized magnetic particles as mobile-sensing binding surface. This allows surface regeneration and adaptation to robust fluidic-based analyzers (EU-SMS project). The use of fluid-based automated devices allows integrating and compacting the laboratory functions within a single portable analyzer, characterized by reduced analytical times and decreased reagent consumption. However, an important limitation of automated biosensing systems is sample processing prior to analysis. The challenge concerns the concentration of algal target cells and biotoxins as well as the analyte extraction as needed, to achieve appropriate limits of detection, which can be at least as elaborate as the analytical procedure itself. Usually, cell concentration techniques are based on tangential filtration or using hollow-fiber filter cartridges allowing for concentration of large volumes of seawater. Retained cells are recovered with appropriate buffer for the cell lysis step, which could be both mechanical (e.g., sonication and French press) and chemical (e.g., acidic/basic, enzymatic), releasing the analyte in the solution prior to analysis. Eventually, and depending on the application, the analyte will require further purification (e.g., affinity column separation). Mastering these preanalytic steps represents a real challenge to be tackled by a multidisciplinary approach to keep biosensor sensitivity at satisfactory levels.

The difficulties encountered in deploying automated biosensors are also mainly related to the need of proper instrument housing, which comprises expensive marinization integration subjected to market constraints of qualified suppliers. This results in complex scientific instruments that require intensive ship time and skilled personnel for its operation. One could opt for relatively economical compact online systems that can be field deployable in existing operational buoys or platforms in which the degree of automation would allow remote operation in the marine environment as exemplified previously. Still other technical challenges associated with long-term deployment in the sea environment, e.g., biofouling of any material in contact with seawater, are to be foreseen, taken into account, and surmounted.

3.3.4.5 Data Management

Data acquisition, processing, and transfer to end-users and interoperability with global sensing platforms are altogether a major challenge in online monitoring systems. Beforehand, the frequency and interval of analysis should be adapted to the target information needed, taking into account that an online monitoring system does not necessarily have to be a real-time monitoring system. The system should be configured to deliver to the end-users the necessary information–knowledge to make optimal decisions. Because the variables involved in the generation and surveillance of HABs are diverse—weather and season, physicochemical water properties, such as temperature, pH, salinity, oxygen, turbidity, suspended matter, dissolved organic matter, nutrients, etc.—and data sources (remote sensing, wireless sensor networks, Geographic Information Systems [GIS], etc.) are becoming more and more abundant as technologies mature, advanced Artificial Intelligence (AI) Techniques (neural networks and fuzzy logic) enable working with the large quantities of data in real time. As a result, we get systems that become smarter with time, as more data are produced and the AI algorithms are trained, assisting decision makers and authorities in producing timely alerts that have the potential to protect public health [45]. In the case of fish farming, real-time monitoring of toxic algal species will probably not be relevant because of the slow-growth rate of such organisms, e.g., the doubling time of *Dinophysis* spp. natural population was reported to be 3 days [48]. In fact, an online monitoring system displaying an alert system with a web-based service will allow the farmers to manage all the necessary information even remotely.

3.3.5 CONCLUSIONS

Ocean in situ sensors and biosensors are essential in addressing the needs of the aquaculture industry and associated food-safety sector, because they can provide an easy and inexpensive way of monitoring coastal and estuarine waters for toxins and other contaminants that threaten public health, as they are transferred through the food chain. Such monitoring will assist authorities in complying with the MSFD and will assist in developing public trust for the aquaculture and food-safety sectors. Various different biosensors currently exist that are based on detecting different molecules in the waterbodies and food, namely aptasensors, immunosensors, and genosensors. The field of ocean biosensors still faces important challenges: (1) forecasting HABs is a difficult task, because a long time series of much physicochemical and biological data needs to be considered in a coordinated manner; and (2) toxin groups are diverse and continuously changing, with new groups being added every so often. In addition, (3) the samples that are affected are variable, requiring adapted processing; and (4) real-time online systems need innovative energy supplies that would render them autonomous for long periods. Furthermore, (5) biosensor detection limits are usually insufficiently low to be used as an early warning system, and (6) data production from sensors may be massive, making data management and data mining a challenge as well.

References

[1] Berdalet E, Fleming LE, Gowen R, Davidson K, Hess P, Backer LC, Moore SK, Hoagland P, Enevoldsen H. Marine harmful algal blooms, human health and wellbeing: challenges and opportunities in the 21st century. J Mar Biol Assoc UK 2015;2015.

[2] Groeneveld RA, Bartelings H, Börger T, Bosello F, Buisman E, Delpiazzo E, Eboli F, Fernandes JA, Hamon KG, Hattam C, Loureiro M, Nunes PALD, Piwowarczyk J, Schasfoort FE, Simons SL, Walker AN. Economic impacts of marine ecological change: review and recent contributions of the VECTORS project on European marine waters, Estuarine. Coast Shelf Sci 2016. https://doi.org/10.1016/j.ecss.2016.04.002.

[3] Hernández C, Díaz PA, Molinet C, Seguel M. Exceptional climate anomalies and northwards expansion of paralytic shellfish poisoning outbreaks in Southern Chile. Harmful Algae News 2016;54:1–2.

[4] Zingone A, Enevoldsen H, Hallegraeff G. Are HABs and their societal impacts expanding and intensifying? A call for answers from the HAB scientific community. In: Proença LAO, Hallegraeff G, editors. Marine and Fresh-water harmful algae. Proceedings of the 17th International Conference on harmful algae. International Society for the study of harmful algae 2017. 2017.

[5] Gilbert PM, Allen JI, Bouwman AF, Brown CW, Flynn KJ, Lewitus AJ, Madden CJ. Modeling of HABs and eutrophication: status, advances, challenges. J Mar Syst 2010;83(3–4):262–75.

[6] Laspidou CS, Kofinas D, Mellios N, Latinopoulos D, Papadimitriou T. Investigation of factors affecting the trophic state of a shallow Mediterranean reconstructed lake. Ecol Eng 2017;103:154–63.

[7] Mellios N, Kofinas D, Laspidou C, Papadimitriou T. Mathematical modeling of trophic state and nutrient flows of Lake Karla using the PCLake model. Environ Process 2015;2(1):85–100.

[8] Papadimitriou T, Kormas K, Dionysiou D, Laspidou C. Using H_2O_2 treatments for the degradation of cyanobacteria and microcystins in a shallow hypertrophic reservoir. Environ Sci Pollut Res 2016;23(21):21523–35.

[9] Hallegraeff GM. Ocean climate change, phytoplankton community responses, and harmful algal blooms: a formidable predictive challenge. J Phycol 2010;46:220–35.

[10] Lassus P, Chomérat N, Hess P, Nézan E. Toxic and Harmful Microalgae of the World Ocean/Micro-algues toxiques et nuisibles de l'océan mondial. Denmark, International Society for the Study of Harmful Algae/Inergovernmental Oceanographic Commission of UNESCO; 2016. IOC Manuals and Guides, 68.

[11] Brissard C, Herrenknecht C, Séchet V, Hervé F, Pisapia F, Harcouet J, Lémée R, Chomérat N, Hess P, Amzil Z. Complex toxin profile of French Mediterranean Ostreopsis cf. ovata strains, seafood accumulation and ovatoxins prepurification. Mar Drugs 2014;12:2851–76.

[12] Kröger S, Law R. Biosensors for marine applications: we all need the sea, but does the sea need biosensors? Biosens Bioelectron 2005;20(10):1903–13.

[13] Papadimitriou T, Katsiapi M, Vlachopoulos K, Christopoulos A, Laspidou C, Moustaka-Gouni M, Kormas K. Cyanotoxins as the "common suspects" for the Dalmatian pelican (Pelecanus crispus) deaths in a Mediterranean reconstructed reservoir. Environ Pollut 2017;234:779–87.

[14] Han F, Huang X, Mahunu GK. Exploratory review on safety of edible raw fish per the hazard factors and their detection methods. Trends Food Sci Technol 2017;59:37–48.

[15] Venugopal V. Biosensors in fish production and quality control. Biosens Bioelectron 2002;17:147–57.

[16] O'Connell PJ, Sullivan CK, Guilbault GG. Biosensors for food analysis. Irish J Agric Food Res 2000;39:321–30.

[17] Turner APF. Biosensors — sense and sensitivity. Science 2000;290:1315–7.

[18] Wagner G, Guilbault GG. Food biosensor analysis. Madison (NY): Marcel Dekker; 1994.

[19] Karube I, Tamiya E. Biosensors for food industry. Food Biotechnol 1987;1:147–66.

[20] Rapini R, Marrazza G. Electrochemical aptasensors for contaminants detection in food and environment: recent advances. Bioelectrochemistry 2017;118:47–61.

[21] Pfeiffer F, Mayer G. Selection and biosensor application of aptamers for small molecules. Front Chem 2016;4:1–25.

[22] Eissa S, Siaj M, Zourob M. Aptamer-based competitive electrochemical biosensor for Brevetoxin-2. Biosens Bioelectron 2015;69:148–54.

[23] Hou L, Jiang L, Song Y, Ding Y, Zhang J, Wu X, Tang D. Amperometric aptasensor for saxitoxin using a gold electrode modified with carbon nanotubes on a self-assembled monolayer, and methylene blue as an electrochemical indicator probe. Microchim Acta 2016;183:1971–80.

[24] Campbell K, Haughey SA, Top H, Egmond H, Vilariño N, Botana LM, Elliott CT. Single laboratory validation of a surface plasmon resonance biosensor screening method for paralytic shellfish poisoning toxins. Anal Chem 2010;2010(82):2977–88.

[25] McNamee SE, Elliott CT, Delahaut P, Campbell K. Multiplex biotoxin surface plasmon resonance method for marine biotoxins in algal and seawater samples. Environ Sci Pollut Res 2013;20:6794–807.

[26] McPartlin DA, Jonathan H, Loftus JH, Crawley AO, Silke J, Murphy CS, O'Kennedy RJ. Biosensors for the monitoring of harmful algal blooms. Curr Opin Biotechnol 2017;45:164–9.

[27] McCoy GR, McNamee S, Campbell K, Elliot CT, Fleming GTA, Raine R. Monitoring a toxic bloom of Alexandrium minutum using novel microarray and multiplex plasmon resonance biosensor technology. Harmful Algae 2014;32:40–8.

[28] Hata A, Katayama A, Furumai H. Organic substances interfere with reverse transcription-quantitative PCR-based virus detection in water samples. Appl Environ Microbiol 2015;81:1585–93.

[29] McCoy GR, Touzet N, Fleming GTA, Raine R. Evolution of the MIDTAL microarray: the adaption and testing of oligonucleotide 18S and 28S rDNA probes and evaluation of subsequent microarray generations with Prymnesium spp. cultures and field samples. Environ Sci Pollut Res 2015;22:9704–16.

[30] Kegel JU, Del Amo Y, Costes L, Medlin LK. Testing a microarray to detect and monitor toxic microalgae in Arcachon Bay in France. Microarrays 2013;2:1–23.

[31] Orozco J, Villa E, Manes CL, Medlin LK, Guillebault D. Electrochemical RNA genosensors for toxic algal species: enhancing selectivity and sensitivity. Talanta 2016;161:560–6.

[32] Diercks-Horn S, Metfies K, Jackel S, Medlin LK. The ALGADEC device: a semi-automated rRNA biosensor for the detection of toxic algae. Harmful Algae 2011;10:395–401.

[33] McQuillan JS, Robidart JC. Molecular-biological sensing in aquatic environments: recent developments and emerging capabilities. Curr Opin Biotechnol 2017;45:43–50.

[34] Edgcomb VP, Taylor C, Pachiadaki MG, Honjo S, Engstrom I, Yakimov M. Comparison of Niskin vs. in situ approaches for analysis of gene expression in deep Mediterranean Sea water samples. Deep Sea Res II 2016;129:213–22.

[35] Doucette GJ, Mikulski CM, Jones KL, King KL, Greenfield DI, Marin III R, Jensen S, Roman B, Elliott CT, Scholin CA. Remote, subsurface detection of the algal toxin domoic acid onboard the Environmental Sample Processor: assay development and field trials. Harmful Algae 2009;8:880–8.

[36] Greenfield DI, Marin R, Doucette GJ, Mikulski C, Jones K, Jensen S, Roman B, Alvarado N, Feldman J, Scholin C. Field applications of the second-generation Environmental Sample Processor (ESP) for remote detection of harmful algae: 2006–2007. Limnol Oceanogr Methods 2008;6:667–79.

[37] Yamahara KM, Demir-Hilton E, Preston CM, Marin R, Pargett D, Roman B, Jensen S, Birch JM, Boehm AB, Scholin CA. Simultaneous monitoring of faecal indicators and harmful algae using an in-situ autonomous sensor. Lett Appl Microbiol 2015;61:130–8.

[38] McPartlin DA, Lochhead MJ, Connell LB, Doucette GJ, O'Kennedy RJ. Use of biosensors for the detection of marine toxins. Essays Biochem 2016;60:49–58.

[39] Neilan BA, Pearson LA, Muenchhoff J, Moffitt MC, Dittmann E. Environmental conditions that influence toxin biosynthesis in cyanobacteria. Environ Microbiol 2013;15(5):1239–53.

[40] Cusack C, Dabrowski T, Lyons K, Berry A, Westbrook G, Salas R, Duffy C, Nolan G, Silke J. Harmful algal bloom forecast system for SW Ireland. Part II: are operational oceanographic models useful in a HAB warning system. Harmful Algae 2011;53:86–101.

[41] Anderson DM, Cembella AD, Hallegraeff GM. Progress in understanding harmful algal blooms: paradigm shifts and new technologies for research, monitoring, and management. Ann Rev Mar Sci 2012;4:143–76.

[42] Kudela RM, Bickel A, Carter ML, Howard MDA, Rosenfeld L. The monitoring of harmful algal blooms through ocean observing: the development of the California harmful algal bloom monitoring and alert program. Coast Ocean Observing Syst 2015:58–75.

[43] Doucette GJ, Kudela RM. Chapter twelve—in situ and real-time identification of toxins and toxin-producing microorganisms in the environment. Compr Anal Chem 2017;78:411–43.

[44] Kalogerakis N, Arff J, Banat IM, Broch OJ, Daffonchio D, Edvardsen T, Eguiraun H, Giuliano L, Handå A, Lopez-de-Ipina K, Marigomez I, Martinez I, Oie G, Rojo F, Skjermo J, Zanaroli G, Fava F. The role of environmental biotechnology in exploring, exploiting, monitoring, preserving, protecting and decontaminating the marine environment. N Biotechnol 2015;32(1):157–67.

[45] Mellios N, Papadimitriou T, Laspidou C. Predictive modeling of microcystin concentrations in a hypertrophic lake by means of Adaptive Neuro Fuzzy Inference System (ANFIS). Eur Water 2016;55:91–103.

[46] Seger A, Hallegraeff G. Mitigating fish-killing algal blooms with PAC modified clays: efficacy for cell flocculation and ichthyotoxin adsorption. In: Proença LAO, Hallegraeff GM, editors. Marine and fresh-water harmful algae. Proceedings of the 17th International Conference on harmful algae. International Society for the study of harmful algae 2017. 2017.

[47] Vilarino N, Louzao MC, Fraga M, Rodriguez LP, Botana LM. Innovative detection methods for aquatic algal toxins and their presence in the food chain. Anal Bioanal Chem 2013;405:7719–32.

[48] Garces E, Delgado M, Camp J. Phased cell division in a natural population of Dinophysis sacculus and the in situ measurement of potential growth rate. J Plankton Res 1997;19:2067–77.

[49] Davis CS, Thwaites F, Gallager SM, Hu Q. A three-axis fast-tow digital Video Plankton Recorder for rapid surveys of plankton taxa and hydrography. Limnol Oceanogr Methods 2005;3:59–74.

[50] Ciminiello P, Dell'Aversano C, Fattorusso E, Forino M. Palytoxins: a still haunting Hawaiian curse. Phytochem Rev 2010;9(4):491–500.

[51] Zhang C, Zhang J. Current techniques for detecting and monitoring algal toxins and causative harmful algal blooms. J Environ Anal Chem 2015;2:123.

[52] Brooks SL, Higgins IJ, Newman JD, Turner APF. Biosensors for process control. Enzyme Microb Technol 1991;13:946–55.

Further Reading

[1] Ng A, Chinnappan R, Eissa S, Liu H, Tlili C, Zourob M. Selection, characterization, and biosensing application of high affinity congener-specific microcystin-targeting aptamers. Environ Sci Technol 2012;46:10697–703.

Glossary

AOAC Association of Official Analytical Chemists
ASP Amnesic shellfish poisoning
AST Aerosol toxicity
AUVs Autonomous underwater vehicles
AZP Azaspiracid shellfish poisoning
cELISA (competitive) enzyme-linked immunosorbent assay
CFP Ciguatera fish poisoning
DNA Deoxyribonucleic acid
DSP Diarrheic shellfish poisoning
ELISA Enzyme-linked immunosorbent assay
ESP Environmental sample processor
FDA Food and Drug Administration
HABs Harmful algal blooms
HPLC High performance liquid chromatography
IUPAC International Union of Pure and Applied Chemistry
LFA Lateral flow analysis
MBARI Monterey Bay Aquarium Research Institute
MCs Microcystins

MSFD Marine Strategy Framework Directive
NOD Nodularins
NSP Neurotoxic shellfish poisoning
PSP Paralytic shellfish poisoning
qPCR (quantitative) polymerase chain reaction
RNA Ribonucleic acid
SMS Sensing toxicants in marine waters makes Sense using biosensors

CHAPTER

4

Ocean In Situ Sensors Crosscutting Innovations

CHAPTER

4.1

A New Generation of Interoperable Oceanic Passive Acoustics Sensors With Embedded Processing

Eric Delory[1], Daniel Mihai Toma[2], Joaquin del Rio Fernandez[2], Pablo Cervantes[3], Pablo Ruiz[3], Simone Memè[1]

[1]Oceanic Platform of the Canary Islands (PLOCAN), Telde, Spain; [2]UPC, Universitat Politècnica de Catalunya, Vilanova i la Geltrú, Spain; [3]Centro Tecnológico Naval y del Mar (CTN), Fluente Álamo (Murcia), Spain

4.1.1 THE CHALLENGE

The European Marine Strategy Framework Directive (MSFD [1]) and the increasing need to monitor underwater noise for Environmental Impact Assessments of marine activities have increased demand for cost-effective multipurpose instrumentation. There is, as yet, no satisfactory solution available in the market. In contrast, within the terrestrial domain, new technologies have followed from new legislation. The developed instruments (sound-level meters or analyzers) embed several functionalities up to report generation, and their use requires little training, substantially reducing their operation costs. In marine applications, although regulations are now in place for underwater noise (e.g., in Europe, the MSFD Good Environmental Status (GES) Descriptor 11/Underwater Noise [2]), marine acoustic sensors and acoustic data processing remain costly. This is due to acquisition costs and the need for experts at each stage, from sensor operation through to data processing and publishing.

4.1.1.1 Applications Landscape

In addition to monitoring noise, new sensors and platforms can provide cost-efficient information for the assessment of marine mammal populations (MSFD/GES descriptor 1), as was demonstrated with T-POD porpoise detectors [3] and highlighted in Ref. [4]. There are several further potential applications of these sensors, such as the detection of fish reproduction areas (relevant to descriptor 3 of the MSFD and the European Common Fisheries Policy [CFP]) [5,6]. Others include the nonlethal mitigation of catch depredation by whales in fisheries activities [7], detection of GreenHouse Gas (GHG) seeps from pipelines and deep-sea carbon storage, gasification of methane clathrates [8], detection of low-frequency seismic events, ice cracking, ocean-basin thermometry and tomography [9], and acoustic communication.

Wind is also an important contributor to ocean ambient noise, the cause of noise being surface-wave breaking. Briefly, the higher the wind, the higher the number of wind waves and the sea state, the higher the ambient-noise level. Ambient-noise sound-pressure levels on a 500 Hz to 50 kHz bandwidth was shown to correlate well with wind speed as reported in Refs. [10,11].

Rainfall is finally one of the principal natural sources of underwater sound [12]. Sound spectra span from 1 up to 50 kHz depending on drop size. Most sounds are created by smaller and larger drops (smaller or larger than 2–3.5 mm), with 13–25 kHz and 1–10 kHz frequency ranges, respectively. Medium-size drops (1.2–2 mm) are paradoxically quieter than smaller drops (0.8–1.2 mm). Larger-drop sizes (>3.5 mm) produce loud sounds in a 1–50 kHz window [13].

To illustrate this diversity, Fig. 4.1.1 provides a spectral representation of the variety of sound sources in the ocean.

4.1.1.2 Passive Acoustics Sensor Systems

Current oceanic passive acoustic sensor systems generally include one or several transducers, signal conditioning, instrument interface with communication and control capability, and internal or external power supply. In standalone platforms, passive acoustic data are generally stored for later recovery and analysis unless (rarely) there is a direct radio frequency (RF) link to shore. Data may also be sent directly through cabling to a host, whether it is a ship or vessel of opportunity (towed array [16]) or a cabled observatory shore station [17,18]. Hydrophones can also work in array configuration, needing synchronized acquisition (a k a simultaneous sampling), which is technically challenging and further increases costs. Most ocean-monitoring platforms are still standalone with an RF link of limited bandwidth, and the growing use of glider and profiler technologies in global observation will continue to increase the use of RF transmission, mainly via costly and energy-demanding satellite links. Autonomous platforms, such as gliders, are also now being increasingly considered for the monitoring of cetaceans via passive acoustics, as in Ref. [19]. Current passive acoustic sensor systems also need to address the broad dynamic range of underwater sound intensity, which implies specific analog- and/or digital-design strategies, allowing for measurements ranging from low-level ambient noise (e.g., down to 50 dB re 1 µPa) to high-level sounds produced by active sources, whether bioacoustic or anthropogenic (up to 200 dB re 1 µPa and higher). Several innovations are presented in this regard in the following section.

With respect to sensor metadata, internet connectivity, and infrastructure interoperability, important functionalities are also desirable, such as discoverability of sensors, availability, accessibility of information, and usability through documented datasets, i.e., with corresponding metadata. Machine readability is also desirable for future machine-to-machine information communication and automated processing, implying the use of standards wherever possible at sensor, host (platform), and user interfaces.

4.1.2 INNOVATIONS

4.1.2.1 Introduction

For the development of a cost-effective and reliable solution, the aforementioned challenges must first be turned into system engineering requirements. A representative set could be:

- High dynamic range to accommodate the broad range of acoustic intensities.
- Compressed information, to accommodate bandwidth-limited platforms and costly satellite links, by means of preprocessed data, e.g., estimated sound spectra and sound-source characterization.

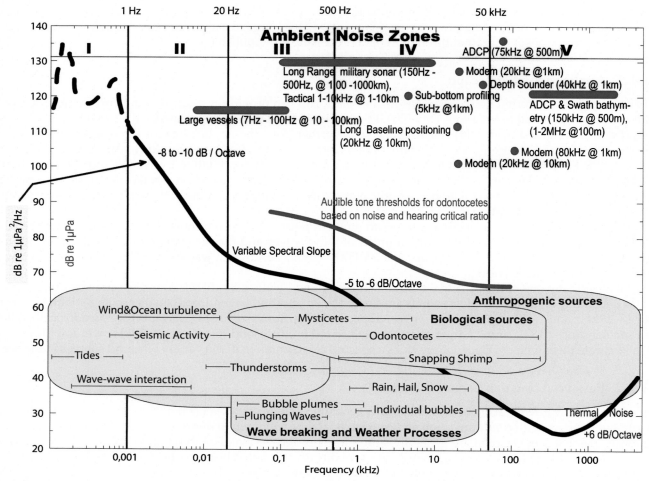

FIGURE 4.1.1   Ocean soundscape diagram with typical natural and anthropogenic sources. The two *heavy-lined curves* show (in black) a typical ocean ambient-noise power spectral density, and (in blue) source levels may reach between 190 and 210 dB re 1 micropascal (μPa) @1 m for instruments and 210–240 dB for long-range low-frequency military sonar. The *light blue boxes* (the vertical size and position for which is arbitrary) show the frequency range of natural sources [14,15].

- Autonomous platforms compatibility and small form factor to facilitate integration on small mobile platforms, such as profilers and gliders, but also fixed platforms, resulting in multiplatform capability.
- Infrastructure interoperability: communication protocols include serial and Ethernet, as well as plug-and-play capability for compatible platforms to allow for multiprotocol communication and sensor–web integration.
- Sensor networking capability for array configuration to allow simultaneous sampling, e.g., through precision timing standards for accurate phase measurement and source localization.
- Commercialization: calibration of spectral response and dynamic range, and reliability measures are performed to reach sufficient technology readiness, cost-efficiency, and accelerated commercialization for usage by a majority of users.

4.1.2.2 Technical Requirements

In systems engineering, system general requirements are translated into technical requirements. These can be summarized in table or matrix format for improved traceability and documentation. For example, 180 dB re 1 μPa was set as maximum-received level. This accounts for attenuation due to source–receiver distance, because measurements are oceanic and not meant to be performed at close range. Resolution, dynamic range, accuracy, sampling frequency, special requirements, among other requirements are provided in Table 4.1.1.

TABLE 4.1.1   Hydrophone Requirements

Resolution	Dynamic Range (DR)	Directivity	Accuracy	Sampling Frequency	Special Requirements	Other1	Other2
Fit for DR	<SS0; 50–180 dB re 1 μPa	Omnidirectional	±3 dB	100 thousand samples per second (kSps)	Embedded noise and bioacoustics statistics	Store on detection, Solid state storage	Low power for small vehicle integration ESONET Label [28]

TABLE 4.1.2   Examples of Platform Diversity and Characteristics Relevant for Sensor Payload Integration

Platforms	Glider	Profiler	Cabled Observatory (ESONET Label Used or Recom. [28])	Standalone (ESONET Label Used or Recom.)
Power available for new sensors	500 mW • Batteries voltage: 19.0–29.4 V • Can be Switched by navigation CPU	10.8 VDC (150 Ampere hour [Ah]), add battery. possible +100 Ah	48 VDC15 VDC 600 W min. (>=1 port) 200 W min. (>= 2 ports) 20 W min (>=4 ports)	48 VDC15 VDC-20 W min (>=4 ports)
Range	600–1500 km	Highly variable (drifter)	Not applicable (fixed)	Not applicable (fixed)
Deployment duration	Up to 2 months	Up to 5 years (250–300 profile cycles)	6 months–1 year	6 months–1 year
Operating depth range	700 m (850 m survival)	2000 m (survival 2200 m)	Down to thousands of m	Surface (buoy) down to thousands of m (mooring – lander)
Navigation	GPS Waypoints, Pressure Sensor, Altimeter (optional)	ARGOS, pressure sensor	Not applicable (fixed)	Not applicable (fixed)
Transmission	RF Modem, Iridium	RF Iridium (Rudics mode) 100 kB/s when surfacing	Fiber or copper	Iridium for world coverage, ORBCOMM, ARGOS 3. VHF, Wi-fi – AirMAx, GSM, GPRS, EDGE, UMTS, HSPA, WiMAX
Payload and ports size	9 L/5 kg in two sections - 4 × Puck™ size Dry and wet Virtual cylinders dia160 mm × h80 mm, Virtual cone dia 150 mm × h270 mm × dia 75 mm Front end tap and front frame modifiable	30 mm bottom – 40 mm top, if mounted on top, max 110 mm × 150 mm × 25 mm	Highly variable	Highly variable
Ports communication protocols	• 4 × ADC • 5 × Serial RS232/TTL selectable • 1 × I2C bus • 4 × Outputs TTL 3.3 V • 4 × Input TTL 3.3 V	TTL	RS232-422-485 – Eth 100 base T (copper), 1000 Nase T (copper), > if fiber	RS232-422-485 – Eth 100 base T (copper)
Other relevant sensors	Conductivity, temperature, and pressure (CTD)			

From the numerous platform constraints, we derived additional sensor requirements in Table 4.1.3, accounting for the most constraining requirements for dimensions (profiler), voltage (down to 3.3 V) and low power (<1 W).

TABLE 4.1.3   Sensor Requirements for MultiPlatform Usability Accounting for the Most Constraining Requirements in Table 4.1.2

Dimensions	Voltage	Power	Data Management	Protocol
120 mm × 30 mm Ø + connector	3.3–18 VDC (5 V nominal),	<1 W	PUSH, .wav 24 bit storage OGC-SWE formats for statistics and detection	PUCK (Plug and work) and OGC SWE, IEEE1588

4.1.3 SENSOR DEVELOPMENT

This section describes the development of a hydrophone (NeXOS A1) with multiplatform integration capability, the characteristics of which derive from the previous requirements. The description includes hardware, firmware, and interfacing. Examples of host (platform) integration on low-power autonomous mobile platforms are provided later in the chapter.

4.1.3.1 Overview

A compact and low-power board was chosen as the main Microcontroller unit (MCU) for the hydrophone, which would be consistent with allowing for the implementation of sufficient firmware capabilities to encompass interfacing, control, and digital signal-processing functionalities, in which:

- The sensor-side interface provides the communication with the onboard sensors such as temperature and clock, and the acoustic transducers after the digital conversion, which is done by the sensor.
- The "plug and play" middleware implements a standard protocol including sensor metadata traceability.
- The host-side interfacing allows for serial or Ethernet-capable platforms communication.
- A self-configuration service decodes the sensor metadata encoded in a standard format at power-on, retrieving the parameter values and starting the hydrophone sensing functions.
- Data recording (.wav file format for acoustic data), synchronization, and data transfer between internal processes is taken care of by the hydrophone MCU. Lossless compression algorithms can be programmed and compiled by user.
- User-selectable digital signal processing tasks are performed by the hydrophone MCU, such as click detection, and whistle detection algorithms may be executed sequentially on the acoustic raw data.
- Ambient noise measurements and statistics are processed based on acoustic calibration results stored on hydrophone's embedded memory.
- Standard Commands for Programmable Instruments (SCPI) encode all instrument commands.

4.1.3.2 Hardware

This section summarizes hardware characteristics of the new hydrophone, including acoustic specifications for three transducer types (D70, SQ26, and JS-B100) meant for different depth ratings. Fig. 4.1.2 provides an overview of the signal flow from transducer to control and communication layers. Electrical, mechanical, and software characteristics are then presented, leading to the final assembly in Fig. 4.1.3.

4.1.3.2.1 Calibration

Acoustic calibration of sensor assemblies was performed by the Centro Tecnológico Naval y del Mar (CTN) facilities (Spain) following the International Electrotechnical Committee (IEC) 60565:2007 standard, using reference

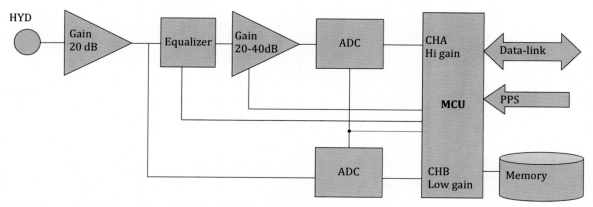

FIGURE 4.1.2   Two channels, defined as channels A and B, allow for different gain configurations and broader dynamic range, from ambient noise to high intensities. Equalizer performs whitening of natural sounds, preventing saturation at high gain.

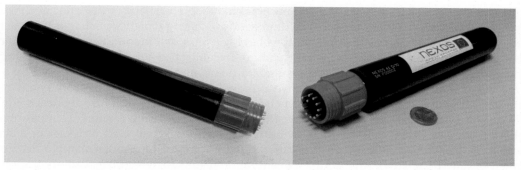

FIGURE 4.1.3   NeXOS A1 with SQ26 (left) and D70 (right) transducers, with both 12-pin European Seas Observatory Network - European Multidisciplinary Seafloor and water column Observatory (ESONET-EMSO) label connector for power, and serial and Ethernet communication.

TABLE 4.1.4   Transducer Types and Characteristics

Hydrophone Type	D/70	SQ26-01	JS-B100
Sensitivity ChA	−138/−158 dB re 1 µPa	−133.5/−153.5 dB re 1 µPa	−141/−161 dB re 1 µPa
Sensitivity ChB	−178 dB re 1 µPa	−173 dB re 1 µPa	−181 dB re 1 µPa
Frequency range (±1.5 dB)	from 1 Hz to 50 kHz	from 1.15 Hz to 28 kHz	from 1 Hz to 50 kHz
Sea noise equalizer	HP filter one pole 3.2 kHz	HP filter one pole 3.2 kHz	HP filter one pole 3.2 kHz
Beam pattern	Omnidirectional	Omnidirectional	Omnidirectional
Input equivalent noise (@5 kHz G = 60 dB)	27 dB re. $1\,\mu Pa/\sqrt{Hz}$	22.5 dB re. $1\,\mu Pa/\sqrt{Hz}$	30 dB re $1\,\mu Pa/\sqrt{Hz}$
Full BW dynamic range	136 dB	136 dB	136 dB

NeXOS A1 Electrical	Mechanical Characteristics
Supply voltage	4.6–42 Vdc.
Power consumption sleep	30 mW.
Power consumption min	350 mW.
Max consumption (full MCU speed)	1 W.
EMI protection	
Connector	SubCon MCBH12M
Communication	Serial RS232 (Serial and Ethernet: A1 hybrid)
External coat	Polyurethane.
Size	Φ34 mm × 255 mm.

Hydrophone Type	D/70	SQ26-01	JS-B100
Working depth	up to 1500 m	up to 2000 m	up to 3600 m
Weight	333 g	317 g	333 g

Software-configurable hardware settings

It is possible to set the following parameters, through software commands:

Sampling frequency:	up to 100 kSps (software configurable).
CHA sensitivity:	−138/−158 dB re 1 µPa (D70 transducer).
Equalizer (HP filter):	1 Hz/3200 Hz (one pole)
Channel selection:	A&B—A—A with Equalizer—B

instruments, a truncated cone-shaped tank, and an anechoic chamber. The shape of the truncated cone-shaped tank delays the first early reflections so that the echo-free time window during which acoustic measurements are carried out is increased, which leads to opening the possibility of having a more stable sound signal (due to the use of longer stimuli) and being able to work with lower-frequency stimuli. Calibration was performed following the free-field calibration by comparison, in which the sound pressure at a point in the sound field generated by a sound source (projector) is measured with a calibrated standard hydrophone. Then the standard hydrophone is replaced by the hydrophone under calibration. The ratio of the open-circuit voltages of the two hydrophones is equal to the ratio of their free-field sensitivities. The standard also considers the option of using a reference microphone for hydrophone calibration in air. The calibration by this method requires either a calibrated projector or a standard hydrophone or microphone. The chosen method was the calibration by comparison using a standard hydrophone (Reson TC4033) and microphone (B&K 4189). An example of calibration results for spectral sensitivity response is provided in Fig. 4.1.4.

4.1.3.2.2 Environmental Testing

Environmental testing including depth qualification is a prerequisite to sensors aimed for use in the field and undergoing severe conditions, more so if the sensor will be integrated with expensive devices like autonomous vehicles and costly operations. A failure of the sensor can not only ruin months of preparation but also have different degrees of consequences on the host platform, from damage to loss. The NeXOS project implemented a series of standard tests on all sensors developed, based on the NF X 10.812 (Class B) standard. Failures were identified with first prototypes, such as molding deformation after cycles of pressure tests beyond 300 bars, and subsequent malfunctions (the sensor recovered shape and functioning after 2–3 days under atmospheric conditions). This led to updating the mechanical design and producing a new prototype with increased reliability. The hydrophone was

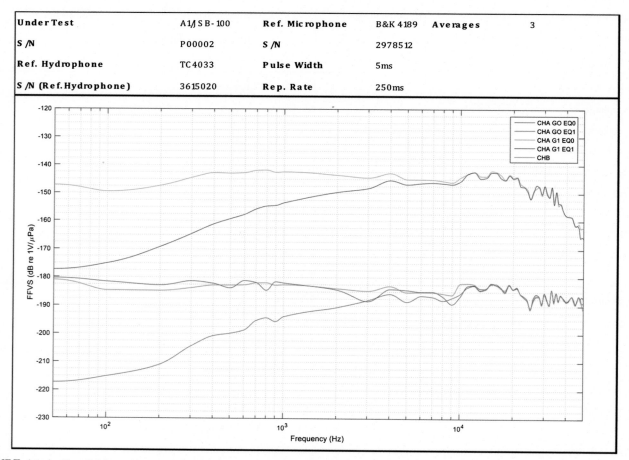

FIGURE 4.1.4   Free-field sensitivity calibration for A1 equipped with JSB-100 transducer. Sensitivity is measured for different software-configurable gain and filtering settings.

Résultats

L'hydrophone NEXOS A1 JS-B100 S/N P00002a été qualifié suivant la norme NF X 10-812 classe B pour la pression hydrostatique, le stockage en chaleur humide, le stockage en froid, la tenue aux vibrations et la tenue aux chocs mécaniques.

Chronogramme des essais en pression hydrostatique

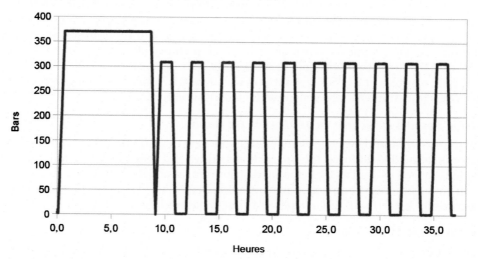

FIGURE 4.1.5   NF X 10-812 Class B series of pressure cycles and text stating that A1 qualified on environmental tests on hydrostatic pressure, hot moisture, cold storage, vibration, and mechanical shocks (December 2016).

tested again with satisfactory results for all tests, i.e., hydrostatic pressure, hot moisture, cold storage, vibration, and mechanical shocks (Fig. 4.1.5).

4.1.3.3 Preprocessing Firmware

The hydrophone implements signal-processing algorithms to provide the capabilities required to cover MSFD[2] Descriptors 1 and 11 for "Good Environmental Status," ensuring respectively that "Biodiversity is maintained" and "introduction of energy (including underwater noise) does not adversely affect the ecosystem." For Descriptor 1, three different algorithms (Click Detector, Whistle Detector, and Low-Frequency Tonal Sounds) are implemented. The first two are based on the open-source Passive Acoustic Monitoring (PAM) software PAMGuard,[3] the third on the work of Zaugg et al. [20]. This decision was taken after a study of applicable referenced passive acoustic monitoring software (see Ref. [21] for details). For Descriptor 11, four different algorithms have been developed. MSFD requirements regarding Indicators 11.2.1 and 11.1.1 have been taken into account. These requirements have been extracted from Refs. [22–24]. The purpose of these indicators is to assess the pressure on the environment by making available an overview of all low- and midfrequency impulsive-sound sources over a period of 1 yr throughout regional seas. This will provide the European Member States with an overview of the environmental pressures.

4.1.3.3.1 Click and Whistle Detectors

The click and whistle detector algorithms are based on the PAMGuard open-source code, a leading community-based software development used in the cetacean research community. These broadband detectors mainly work on specific acoustic features of the upper part of the spectrum and are fittest for detecting toothed whales in general, like dolphins, beaked whales, and sperm whales. Further details on the detector software designs are available online.[3] Most efforts for this specific functionality have focused on optimizing and porting the code to an embedded platform, while maintaining it open-source in its ported version. Functional integrity of ported algorithms was then checked by comparing algorithm behavior on well-known commercial technical computing software.

[2] European Marine Strategy Framework Directive.

[3] www.pamguard.org.

4.1.3.3.2 Low-Frequency Tonal Sounds

The algorithm is based on the technique depicted in Ref. [20], using the quantification of the prominence of spectral peaks. The result of this metric is compared with a threshold defined by the user. This metric can be computed in different ways; the entropy metric was chosen with values ranging from 0 to −1, being −1 when the spectrum is perfectly flat and 0 when there is a single peak in the spectrum. The algorithm is generally adapted to signals produced by baleen whales. The most challenging aspect of the implementation consisted in porting the algorithm to the embedded system, due to the memory limitations (MCU static random access memory [RAM] of 136 kBytes) hindering efficient (i.e., real-time) processing of the relatively long time durations necessary to detect consistent spectral peaks. Although there was no known work-around on the memory allocation issue, a performance comparison eventually showed that the algorithm was able to work satisfactorily. Performance is illustrated in Fig. 4.1.6, using slices of 93 ms, i.e., 2048 samples at 22,050 Hz sampling rate. Fig. 4.1.7 illustrates noise-level statistics for 63 and 125 Hz (L10, L90, and RMS) transmitted to the sensor web visualisation tool.

4.1.3.3.3 Impulsive Sounds Indicator in 10 Hz–10 kHz Band

The measure of impulsive sounds was implemented in this new development as a contribution to MSFD Indicator 11.1.1. The indicator is defined as follows: the proportion of days and their distribution within a calendar year over areas of a determined surface, as well as their spatial distribution, in which anthropogenic sound sources exceed levels that are likely to entail significant impact on marine animals measured as Sound Exposed Level (in dB re 1 μPa²s) or as peak sound-pressure level (in dB re 1 μPa peak) at 1 m, measured over the frequency band 10 Hz–10 kHz.

4.1.3.3.4 Trends in Third Octave Bands

The measure contributes to Indicator 11.2.1, defined as: trends in the ambient noise level within 1/3 octave bands 63 and 125 Hz (center frequency) (re 1 μPa Root Mean Square [RMS]; average noise level in these octave bands over a year) measured by observation stations and/or with the use of models if appropriate (11.2.1). Two digital filters are applied to the input signal. The filters are third-octave band filters, which comply with base–10 systems and class 0 filters according to IEC 61260 (1995) standard. Although indicator 11.2.1 requires third-octave bands 63 and 125 Hz, recommendations have been expressed to extend the range of ambient noise level to (20 Hz–20 kHz). The number of third-octave bands within this frequency range according to IEC 61260 (1995) being 30, 30 filters are applied to the input signal to obtain the sound-pressure level (SPL$_{rms}$) for each frequency band. Signal decimation stages were implemented to this aim. This was needed due to filtering implementation constraints of the processing platform, and consisted of prefiltering and decreasing the sampling frequency of a given signal; therefore, after the decimation

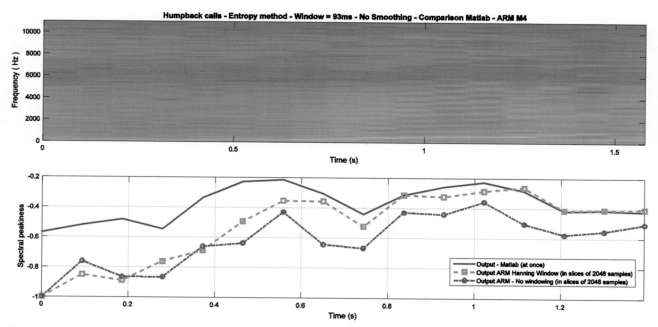

FIGURE 4.1.6 Low-frequency tonal sound-detection algorithm performance comparison between processing the whole input and the input in signal slices of 93 ms.

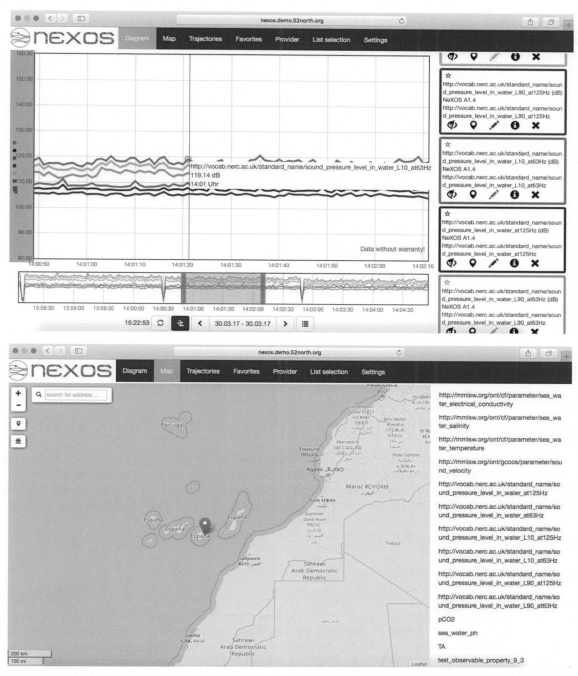

FIGURE 4.1.7   NeXOS visualization front end displaying preprocessed noise-level statistics for 63 and 125 Hz (L10, L90, and RMS) transmitted to the sensor web.

process by two different orders, two new sampling frequencies were obtained (1 and 16 kHz). Basic modules of this algorithm are the same as in the indicator 11.2.1 except for the decimation module.

4.1.3.3.5 Noise Band Monitoring

This algorithm calculates ambient noise level trends within a frequency band defined by the user. The general foundation is the same as for indicator 11.2.1. The structure defined for this algorithm is similar to indicator 11.2.1. Instead of two filters (63 and 125 Hz), the algorithm takes a user-defined filter.

4.1.3.4 Interfacing Firmware

As proposed in Refs. [25,26] for passive acoustics and further detailed in Chapter 6.2 extended to all ocean sensors, innovation is also now underway to enhance interoperability of passive acoustic information and sensors, in a similar way as what has now long been promoted by the environmental and oceanographic data management community. This results in better usability, sustainability of passive acoustic information, and improved discoverability of sensors and sensor status. By encoding sensor metadata following common standards, a diversity of users and user clients (human or machines) can also connect to self-describing services independently of the technology. Here is an example of encoding of SCPI instruction encoded in SensorML, supporting plug-and-play configuration under the Open Geospatial Consortium–Plug-and-Work (OGC-PUCK) protocol and Sensor Web Enablement (SWE) services (see Chapters 6.1 and 6.2):

```
<sml:input name="dataIn">
    <sml:DataInterface>
        <sml:data>
            <swe:DataStream>
                <swe:elementType name="command"/>
                    <swe:encoding>
                    <!-- Define the text encoding-->
                        <swe:TextEncoding tokenSeparator="," blockSeparator=""/>
                    </swe:encoding>
                    <!-- Define the command string-->
                    <swe:values>MEAS:NOISE?</swe:values>
            </swe:DataStream>
        </sml:data>
    </sml:DataInterface>
</sml:input>
```

4.1.4 INTEGRATION, VALIDATION, AND DEMONSTRATION ON VEHICLES

Sensor hardware, software, mechanical, and electrical integration has been performed on several platforms. In this section, we illustrate the integration work for the Alseamar *Sea Explorer* deep glider, the NKE PROVOR profiling float, and a wave glider from Liquid Robotics. Sensor and data interoperability solutions have been implemented via the OGC-PUCK protocol and the OGC Sensor Web Enablement for all sensor–platform pairs. In this chapter we will show how the SensorML standard, core to the interoperability framework in NeXOS, is used for encoding hydrophone metadata. See Chapters 6.1 and 6.2 of this volume for details on these solutions.

4.1.4.1 The PROVOR Float

The PROVOR float is a drifting float, developed by the French NKE company.[4] Its typical operational profile is between the surface and selected depths at slow speed. Typically, the profile is between 0 and 2000 m depth. The vertical speed is controlled by the buoyancy of the float, the horizontal speed by the ambient ocean currents. The float hosted a NeXOS A1 sensor for a demonstration in the Central Atlantic region. Before integration, the sensor has been successfully qualified according the AFNOR NF X10-812 standards to comply with the reliability constraints of a deep-vehicle sensor payload. The OGC-PUCK protocol was implemented on the profiling float firmware, allowing for automatic driver download from sensor to platform, system self-configuration, and sensor–web infrastructure integration. Fig. 4.1.8 illustrates computer-aided design (CAD) design and the first marine integration test on an operational float.

The PROVOR float equipped with the A1 hydrophone was successfully deployed off Las Palmas, Canary Islands, on Tuesday May 23, 2017, from the ship *Pisko Uno*. The float drifted until Wednesday May 24, 2017, afternoon when it was recovered. This short mission aimed to check that the "acoustic PROVOR" float was fully functional, while allowing recovery. The float was programmed to achieve parking and profiling depths of 500 m for the specific needs of in situ demonstration. A1 was located at the top of the float close to the conductivity, temperature, and pressure (CTD) probe to measure data in the same water layer. A1 hydrophone measured MSFD descriptor 11 sound characteristics every 5–6 s during float ascent to the surface, whereas SBE41 sensor measured temperature,

---

[4] http://www.nke-instrumentation.com.

FIGURE 4.1.8 NeXOS smart hydrophone (gray and black Polyurethane *rod shape*) physical integration on NKE PROVOR Float, from CAD to first deployments. *Courtesy of SMID Technology & NKE Instrumentation.*

pressure, and salinity every 2 s. The float was deployed by the boat officer using a small stair located at the back of the ship, with direct access to seawater. The map in Fig. 4.1.9 shows the float position on May 23, 2017. During the mission, the float transmitted both a technical file including Global Positioning System (GPS) position, internal parameters, a data file from the A1 hydrophone and CTD probe in comma-separated values (CSV) format. The mission ran smoothly. The PROVOR float managed its buoyancy to descend to 500 m, compliant to settings, and during the ascent acoustics data were acquired. Acquisition duration was limited to the first 100 m over the reached depth to reduce satellite communication costs. Data could be visualized on the NeXOS Sensor Web client (Fig. 4.1.9).

4.1.4.2 The Alseamar *Sea Explorer* Glider

The *Sea Explorer*[5] glides through the water column by varying its buoyancy, resulting in the potential for very long-endurance operation. As part of its vertical cycles, it resurfaces regularly to transmit information via satellite link back to the operator including its GPS position, scientific data, and internal parameters. Its modular design includes an independent payload section located at the front of the vehicle. This section can be changed rapidly according to mission requirements. It features two large compartments: a hyperbaric and a wet section. The *Sea Explorer* does not have wings or external moving parts, which facilitates easy launch and recovery operations and reduces the risk of entanglement (plastic debris, seaweed, and so on). In addition, the absence of any external moving parts limits any risks of leaks. Because of the length of the A1 sensor, the vehicle nose cone was extended and the chassis was modified (Fig. 4.1.10).

On June 16, 2017, the glider was deployed for a demonstration mission, northwest of Runde Island, Norway. It progressed out across the shallow (100–200 m depth) continental shelf out to the shelf break, down the slope (crossing a strong nearly barotropic, topographically trapped shelf-edge current, out to deep water (P2). Fig. 4.1.11 shows team, path, and sound-pressure levels transmitted near real-time from the glider. A pattern on 125 third-octave bandwidth appears as the glider in P2 approaches a high-density route. Spikes are also noted due to glider mechanical noise.

[5] https://www.alseamar-alcen.com/products/underwater-glider/seaexplorer.

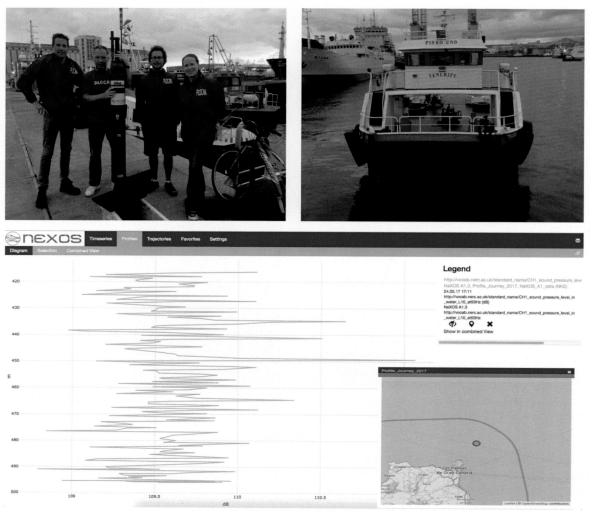

FIGURE 4.1.9　Demonstration of A1 sensor on NKE PROVOR profiler. Top left: team for preparation and deployment, from left to right: Damien Malardé (NKE), Eric Delory, and Simone Memé (Oceanic Platform of the Canary Islands [PLOCAN]), and Allison Haefner (Center for Marine Environmental Sciences, University of Bremen [MARUM]).

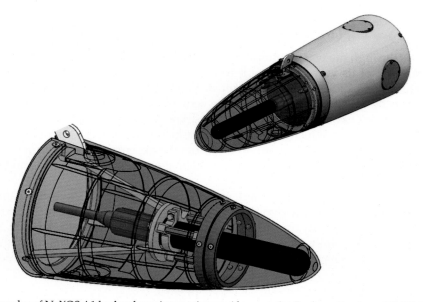

FIGURE 4.1.10　Examples of NeXOS A1 hydrophone integration on Alseamar *Sea Explorer*. *Courtesy of SMID Technology and Alseamar.*

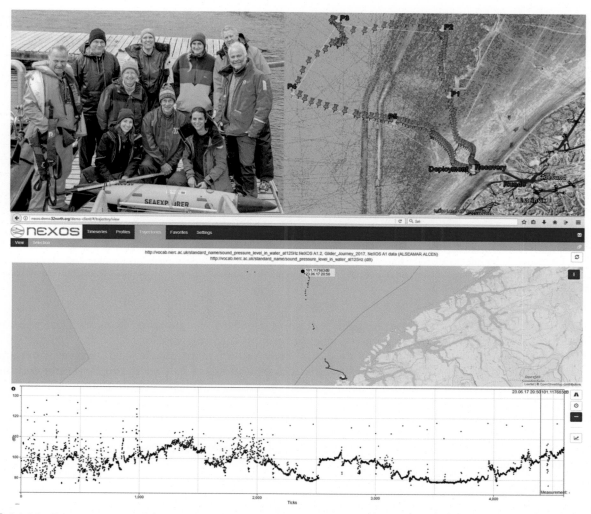

FIGURE 4.1.11   Demonstration of A1 sensor on deep glider. Top left: the team with the *Sea Explorer* glider: Virginie Del Marro (Alseamar), Céline Bonaldi (Alseamar), Karsten Kvalsund (Runde Center), Claudia Erber (Runde Center), Nils–Roar Hareide (Runde Center), Jay Pearlman (IEEE France), Françoise Pearlman (IEEE France), Tom Knudsen (Photographer), Per Eide (Photographer), Roger Kvalsund (Runde Center), Jenny Ullgren (Observer from "FjUF"— Fjernstyrte Undervassfarkoster, local project), Emily Eide (Photographer), and Arild Hareide (Photographer). Top right: glider path, off Runde Island Norway, and overlaid maritime traffic density map.[6] Bottom: sound-pressure level (dB) along one stretch of previous path.

4.1.4.3 The Liquid Robotics Wave Glider

Powered by wave and solar energy, the Wave Glider[7] is an autonomous, unmanned surface vehicle (USV) that can deliver real-time data for up to a year with no fuel. An Adaptive Modular Power System can payload power requirements even for higher-power systems (e.g., sonar) and thus support a wide array of sensors. A PLOCAN-operated wave glider was chosen to test its capability to host the acoustic sensor (A1), as well as an optical sensor (NeXOS O1 [27]) (Fig. 4.1.12).

Hardware integration concerns four items: sensor A1; data-logger; antenna; and a tow-body to maintain the sensor at sufficient distance and depth from platform and surface noise. The tow-body is integrated and supported by the float part of the Wave Glider and relies on vehicle supply for powering. For this test, it was decided to set a parallel communication and transmission path, based on Iridium Short-Burst Data (SBD), and not interfere with the vehicle firmware to simplify the integration. The tow-body includes a specific electromechanical-jacketed and

[7]https://www.liquid-robotics.com.

FIGURE 4.1.12   Wave glider integration—the four port PLOCAN blue Sensor Box prototype implements the plug-and-play protocol, controls the sensor, and communicates with the Iridium SBD modem. The vehicle manufacturer's control system was bypassed in this sensor–platform pair for facilitating rapid integration.

ballasted cable to transmit real-time data from A1. A damping system was included between float and tow-body cable to minimize stress and vibration from the vehicle movements during operation.

The Wave Glider equipped with the A1 sensor was first deployed offshore Taliarte (east coast of Gran Canaria) on Tuesday May 17, 2017, from PLOCAN-2 rubber boat in the surroundings of Taliarte harbor and operated until May 24, 2017, when it was successfully recovered. Two areas of operation were selected: off Taliarte and PLOCAN test site; and a 24/7 piloting schedule was conducted continuously during the period of operation. Several failures were identified (tow-body poor ballasting, leak in surface data-logger, GPS transmission failure) and fixed. A new mission was completed in August, when data files were recorded on a local server and converted to comply with the Sensor Observation Service–Observations and Measurements (SOS/O&M) standards. Data could be visualized from the NeXOS Sensor Web client (Fig. 4.1.13).

4.1.5 CONCLUSION AND FUTURE WORK

Passive acoustics are an emerging field in ocean environmental monitoring and requires sensors able to measure from existing and future observing systems. Sound differs from other variables though, as it has this in common with images and video that it is very costly to transmit from the open ocean through conventional channels. Solutions are becoming available as technology enables compact and low-power pre-processing. This chapter has presented one example of recent innovations in acoustic sensors. NeXOS A1 is a compact, low-power, low-noise digital hydrophone with embedded processing. The chapter has also highlighted some current challenges and opportunities of passive acoustic monitoring in the ocean, where autonomous systems are emerging as cost-efficient solutions. The Internet of Things also calls for standard solutions and one has been presented, in which sensor metadata and services can be discovered and accessed with relative ease and allow for improved user experience. Future work involves ensuring the new sensors are further validated in terms of usability (ease of installation and operation), bioacoustics detection, and identification processors. Field validation and demonstration with profilers and gliders have taken place in the North Sea and the Central Atlantic, with interesting results on noise statistics measurement for MSFD Descriptor 11. Bioacoustic processing was tested in the lab with encouraging results for future field validation that will take place in 2018–2019 in the framework of the MARCET project). Once the new sensors are made commercially available and reach a level of adoption by the community, arrays of synchronized sensors with standard services may allow for open-ocean near real-time acoustic monitoring, including identification and localization opportunities in the medium term, with the increasing use of autonomous vehicles. Open-source recoding of sensor firmware and fleet synchronization of detected sources will open new horizons, such as in the tracking of long-distance migrating species and ocean-noise mapping.

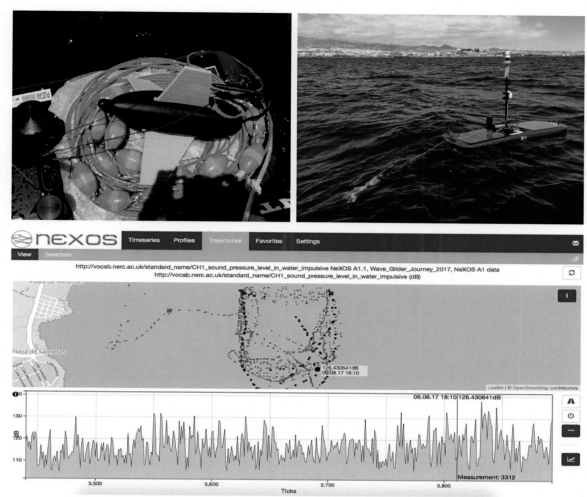

FIGURE 4.1.13 Demonstration of A1 sensor on Wave Glider from tow-body (PLOCAN, off Gran Canaria).

Acknowledgments

NeXOS has received funding from the European Union's Seventh Program for research, technological development and demonstration under grant agreement No 614102 (Ocean of Tomorrow). The authors wish to thank other members of the NeXOS team involved in this work: Enoc Martinez (UPC, Spain), Diego Pinzani, Alberto Figoli, Alessandra Casale (SMID Technology and SITEP, Italy), Ayoze Castro Eduardo Caudet, Ruben Marrero, Rayco Moran, Carlos Barrera, Simone Memè, Gabriel Juanes, Hector Rodriguez (PLOCAN), Françoise Pearlman, Jay Pearlman (IEEE, France), for their active contributions. For the Norway demonstration, in addition: Virginie Del Marro (Alseamar), Céline Bonaldi (Alseamar), Karsten Kvalsund (Runde Center), Claudia Erber (Runde Center), Nils-Roar Hareide (Runde Center), Tom Knudsen (Photographer), Per Eide (Photographer), Roger Kvalsund (Runde Center), Jenny Ullgren (Observer from 'FjUF' – Fjernstyrte Undervassfarkoster, local project), Emily Eide (Photographer), Arild Hareide (Photographer).

References

[1] Directive 2008/56/EC of the European Parliament and of the Council of 17 June 2008 establishing a framework for community action in the field of marine environmental policy (Marine Strategy Framework Directive).

[2] Tasker ML, Amundin M, Andre M, Hawkins A, Lang W, Merck T, et al. Marine strategy framework directive - task Group 11 Report - underwater noise and other forms of energy. 2010.

[3] Köller J, Köppel J, Peters W. Research on marine mammals. Offshore wind energy - research on environmental impacts. Springer; 2006.

[4] Olmstead TJ, Roch MA, Hursky P, Porter MB, Klinck H, Mellinger DK, et al. Autonomous underwater glider based embedded real-time marine mammal detection and classification. J Acoust Soc Am 2010;127(3):1971.

[5] Luczkovich JJ, Mann DA, Rountree RA. Passive acoustics as a tool in fisheries science. Trans Am Fish Soc 2008;137(2):533–41.

[6] Mann DA. Remote sensing of fish using passive acoustic monitoring. Acoustics Today; 2012.

[7] Arangio R. Minimising whale depredation on longline fishing. Australian Government - Fisheries Research and Development Corporation; 2012.

[8] Leighton TG, White PR. Quantification of undersea gas leaks from carbon capture and storage facilities, from pipelines and from methane seeps, by their acoustic emissions. Proc R Soc A Math Phys Eng Sci 2012;468(2138):485–510.

[9] Sagen H. Acoustic technologies for observing the Interior of the Arctic Ocean. Paper presented at the Ocean Obs '09; 2009.

[10] Medwin H. Sounds in the sea. In: Press CU, editor. From ocean acoustics to acoustical oceanography. New York: Cambridge University Press; 2005.

[11] Zedel L, Gordon L, Osterhus S. Ocean ambient sound instrument system: acoustic estimation of wind speed and direction from a subsurface package. J Atmos Ocean Technol 1999;16(8):1118–26.

[12] Nystuen JA. Listening to aaindrops from underwater: an acoustic disdrometer. J Atmos Ocean Technol 2001;18(10):1640–57.

[13] Ma BB, Nystuen JA. Passive acoustic detection and measurement of rainfall at sea. J Atmos Ocean Technol 2005;22(8):1225–48.

[14] Potter JR, Delory E. Noise sources in the sea and the impact for those who live there. Acoustic and Vibration, Asia' 98; 1998; [Singapore].

[15] Delory E. Nouveaux concepts et développements pour l'étude acoustique des cétacés dans deux disciplines: Modélisation, localisation et identification acoustique. Acoustique sensorielle et diagnostic auditif. [Ph.D.]. Barcelona: Universitat Politècnica de Catalunya; 2010.

[16] Potter JR, Delory E, Constantin S, Badiu S. The thinarray; a lightweight, ultra-thin (8 mm OD) towed array for use from small vessels of opportunity. Underwater Technology 2000, Tokyo, Japan. June 2000. 2000.

[17] André M, Houégnigan L, van der Schaar M, Delory E, Zaugg S, Sánchez AM, et al. Localising cetacean sounds for the real-time mitigation and long-term acoustic monitoring of noise. In: Strumillo DP, editor. Advances in sound localization. InTech; 2011.

[18] André M, van der Schaar M, Zaugg S, Houégnigan L, Sánchez AM, Castell JV. Listening to the deep: live monitoring of ocean noise and cetacean acoustic signals. Mar Pollut Bull 2011;63(1–4):18–26.

[19] Baumgartner MF, Fratantoni DM, Hurst TP, Brown MW, Cole TVN, Van Parijs SM, et al. Real-time reporting of baleen whale passive acoustic detections from ocean gliders. J Acoust Soc Am 2013;134(3):1814–23.

[20] Zaugg S, van der Schaar M, Houégnigan L, André M. A framework for the automated real-time detection of short tonal sounds from ocean observatories. Appl Acoust 2012;73(3):281–90.

[21] Delory E, Corradino L, Toma D, Del Rio J, Brault P, Ruiz P, et al. Developing a new generation of passive acoustics sensors for ocean observing systems. Sensor Systems for a Changing Ocean (SSCO), 2014. IEEE; October 13–17, 2014.

[22] Dekeling RPA, Tasker ML, Van der Graaf AJ, Ainslie MA, Andersson MH, André M, et al. Monitoring guidance for underwater noise in European seas, Part II: monitoring guidance specifications. 2014. Luxembourg, EUR 26555 EN.

[23] Dekeling RPA, Tasker ML, Van der Graaf AJ, Ainslie MA, Andersson MH, André M, et al. Monitoring guidance for underwater noise in European seas, Part III: background information and annexes. 2014. Luxembourg, EUR 265556EN.

[24] Dekeling RPA, Tasker ML, Van der Graaf AJ, Ainslie MA, Andersson MH, André M, et al. Monitoring guidance for underwater noise in European seas, Part I: executive summary. 2014. Luxembourg, EUR 26557 EN.

[25] Delory E, Waldmann C, Fredericks J. A proposed architecture for marine mammal tracking from globally distributed ocean acoustic observatories. Passive '08: 2008 new trends for environmental monitoring using passive systems. New York: IEEE; 2008. p. 128–33.

[26] Delory E, Waldmann C. GEOSS, sensor web enablement, and ocean acoustic monitoring. GEOSS sensor web Workshop. Geneva, Switzerland: World Meteorological Organization; 2008.

[27] Pearlman J, Zielinski O. A new generation of optical systems for ocean monitoring: matrix fluorescence for multifunctional ocean sensing. Sea Technol February 2017. p. 30–33.

[28] Rolin J-F, Bompais X, Choqueuse D, Delory E, Hüber R, Waldman C, et al. ESONET Label Definition. 2010. http://www.esonet-noe.org/content/download/42247/574588/file/Deliverable_D68_esonet-label-definition_1.0.pdf.

Glossary

ARGOS Worldwide satellite-based location and data collection system created in 1978 by the French Space Agency (CNES), the National Aeronautics and Space Administration (NASA) and the National Oceanic and Atmospheric Administration (NOAA), originally as a scientific tool for collecting and relaying meteorological and oceanographic data around the world.

CTD Oceanography instrument used to measure the conductivity, temperature, and pressure of seawater (the D stands for "depth," which is closely related to pressure).

GES Good environmental status, assessed by a number of descriptors defined in the marine strategy framework directive.

MARCET Interregional and multidisciplinary macaronesian network for the transfer of knowledge and technologies for the protection, observation and monitoring of cetaceans and the marine environment, as well as the analysis and sustainable exploitation of associated tourism activity. Marcet is co-funded by the European Union INTERREG MAC 2014-2020 programme. http://www.marcet-mac.eu

MCU Microcontroller unit

MSFD The European Union Marine Strategy Framework Directive

NeXOS European collaborative research and development project funded by the European Commission under the Ocean of Tomorrow Program. NeXOS is short for Next-generation, Cost-effective, Compact, Multifunctional Web-Enabled Ocean Sensor Systems Empowering Marine, Maritime, and Fisheries Management. http://www.nexosproject.eu

RMS Root Mean Square, used in the context of this chapter to calculate sound pressure level over a given duration (SPL_{rms}).

SCPI Standard Commands for Programmable Instruments

SMID Italian company recently acquired by SITEP (www.sitepitalia.it)

SPL Sound-pressure level, in decibels is 20 times the logarithm of the ratio of the sound pressure to the reference sound pressure of 1 µPa in water, 20 µPa in air. See also RMS to calculate SPL_{rms}.

USV unmanned surface vehicles (USV) or autonomous surface vehicles (ASV) are vehicles that operate on the surface of the water (watercraft) without a crew.

CHAPTER

4.2

Acoustic Telemetry: An Essential Sensor in Ocean-Observing Systems

Michelle R. Heupel[1,2], *E.J.I. Lédée*[3], *Vinay Udyawer*[4], *Robert Harcourt*[5,6]

[1]Australian Institute of Marine Science, Townsville, QLD, Australia; [2]Centre for Sustainable Tropical Fisheries and Aquaculture and College of Science and Engineering, James Cook University, Townsville, QLD, Australia; [3]Fish Ecology and Conservation Physiology Lab, Carleton University, Ottawa, ON, Canada; [4]Arafura Timor Research Facility, Australian Institute of Marine Science, Darwin, NT, Australia; [5]Sydney Institute of Marine Science, Mosman, NSW, Australia; [6]Department of Biological Sciences, Macquarie University, North Ryde, NSW, Australia

4.2.1 INTRODUCTION

4.2.1.1 What Is Acoustic Tracking?

Acoustic telemetry (from the Greek *akoustikós, tele* = remote, and *metron* = measure) refers to tracking the movements of aquatic animals using active transmission of sound pulses as an encoded signal. This differs from passive acoustics, which monitors the presence or movements of animals using sounds naturally emitted by them [1] (also see Chapter 4.1 in this volume), and from acoustic surveys that use an echo sounder to measure schools of fish and other animals that coalesce in the water [2]. In acoustic telemetry, an acoustic tag is implanted or externally attached to a fish, crustacean, medusa, or indeed any aquatic animal large enough to not be severely impeded by the tag. The acoustic tag transmits a signal made up of acoustic pulses or "pings" that encodes specific information about the tagged animal to one or a network of hydrophone receivers. The encoded information may be as simple as individual identity, or include more complex information such as depth, temperature, behavior such as acceleration, and mortality. The distance over which a signal will propagate is limited by physical features of the ambient environment, such as complex topography, high-intensity ambient sources, waves, and surface noise, greatly limiting successful detection ranges ([3]; see later). Typically, signal propagation for most acoustic telemetry systems usually measures in hundreds of meters at best [4]. One consequence of this is that high-precision tracking requires high densities of receivers. However, networks of receivers strategically located can reveal both large-scale movements and fine-scale behavior [5,6].

4.2.1.2 How Acoustic Tracking Is Applied—What Areas, Species, Questions

Acoustic tracking has been applied in all aquatic environments of the earth, from high-altitude freshwater lakes, in extensive river systems, under the sea ice at the poles, in coastal environments from cold temperate to the tropics, through to transocean basin movements [7]. Each habitat poses its own challenges, from noisy environments degrading signal quality, through loss of receivers due to inclement weather such as floods, storms, cyclones, scouring icebergs, to biological hazards such as sharks, crocodiles, and venomous marine creatures. Later we outline the main technical challenges and how to mitigate them.

Acoustic telemetry has now been successfully implemented on an incredibly diverse range of species. Acoustic telemetry includes both passive acoustic telemetry, in which a network of static receivers detects animals moving through their environment, and active acoustic tracking in which an individual animal is followed around by a small boat with a mobile receiver attached (Fig. 4.2.1). In this paper, we only discuss the former as it comprises the overwhelming majority of acoustic telemetry research outputs. Hussey et al. [7] identified 555 aquatic acoustic telemetry studies published prior to December 2013, and the level of uptake continues at an extraordinary rate. Taxa acoustically tracked range from marine invertebrates (crustacea, cephalopoda, medusa) to all classes of marine vertebrates (elasmobranchs, teleosts, reptiles, aves, mammals). Limitations are principally size and hearing spectrum. Acoustic tags can be tailored to the type of species or the type of environment being studied, with sound parameters such as frequency and modulation method chosen for optimal detectability and signal level. Animals weighing only a few

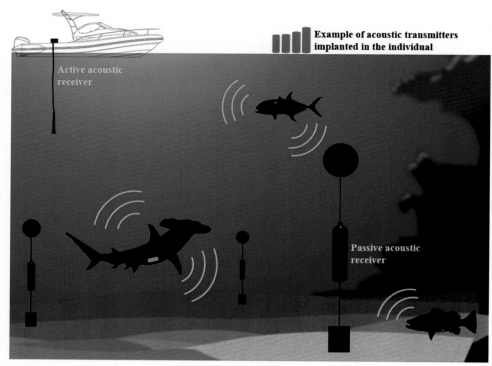

FIGURE 4.2.1 Diagram of acoustic tracking of fish using a receiver monitored on a vessel, or via passive receivers anchored in the study site. The size, transmission rate, and battery life of transmitters is variable, programmable, and designed to suit the research questions and study species.

grams (e.g., salmon smolt) can be tagged, admittedly for limited durations and with limited detectability range. In marine environments, frequencies are usually less than 100 kHz, though new higher frequency tags (e.g., 180 kHz) are becoming more common in the marine environment, whereas in freshwater, as sound propagates more efficiently, frequencies may be well above that (see Ref. [8]). Accordingly, although acoustic tags usually transmit frequencies above the hearing of most marine organisms, many odontocetes can hear well above those frequencies, precluding their use with those taxa.

As acoustic telemetry has matured, the scope of questions it is addressing has broadened significantly. From early descriptive studies on single species movements, acoustic telemetry now covers everything from habitat use and long-range movements [28], abundance estimation [9,10], fisheries management [11], long-term monitoring using decadal tags [12], through range extensions and climate change impacts [7,13] to conservation physiology, energetics, and life history [14].

4.2.2 CHALLENGES

4.2.2.1 The Underwater Environment

Application of sensor technologies in marine environments is difficult as efforts are hampered by a variety of factors. The accessibility of instruments is influenced by weather conditions, water visibility, and depth; whereas processes such as currents and storm surges can displace or damage equipment. Acoustic technologies provide an additional layer of difficulty given they rely on the transmission of sound through water. Transmission of sound is affected by a suite of variables including salinity, turbidity, and temperature (e.g., Refs. [3,4,15,16]). These factors, which dampen or reflect acoustic signals, can alter the performance of acoustic equipment (e.g., Refs. [3,4,17,18]), and thus limit the data collected. Passive acoustic telemetry relies on a two-component system: a transmitter and a receiver. The transmitter signal must be loud enough to travel an adequate distance to reach the receiver, and the receiver must be deployed in an environment conducive to receiving the signal. Although simple in concept, the application of acoustic telemetry in marine environments has numerous challenges.

One of the most common challenges for acoustic telemetry studies is determining the distance transmission signals will travel, and subsequently how many receivers are required to answer relevant research questions. There are several factors that can impact the design and performance of acoustic telemetry studies. The first is conditions that are not favorable for sound transmission. For example, complex habitats (e.g., coral reefs, seagrass beds) can reflect or absorb sound, altering its energy and path [16,19]. Water-column conditions are also a factor with pycnoclines and thermoclines altering sound transmission. Wave action and aeration of the water column are also key factors that affect the transmission of sound [3,17]. Successful detection and decoding of acoustic signals requires the receiver to be able to adequately hear the transmission. Receiver performance can be affected by noise in the environment (e.g., strong current flow, oyster beds, wind, and rain), biofouling growth on the receiver, time of day, positioning of the receiver, and location within the water column (e.g., Refs. [3,17,19,20]). Therefore, researchers are faced with a series of challenges in deploying receivers in locations and configurations suitable to facilitate acoustic telemetry research.

Given the complexities outlined earlier, it is recommended that researchers conduct thorough range testing in a study site prior to developing research plans and deploying acoustic receivers to understand the conditions and potential limitations within the study site [3,4]. Range testing should be conducted over a minimum of one tidal cycle in shallow areas to account for any effects from changes in water depth. Although this testing will be useful in designing a study, it cannot account for seasonal variation or other longer-term events that might affect acoustic telemetry studies. Because conditions can continue to change through time, many studies now employ sentinel transmitters ("sentinel tags") within a study site (e.g., Refs. [4,21]). These transmitters are located in a fixed position relative to an acoustic receiver to help determine whether transmissions are still being detected, at what rate tags are detected, and assist interpretation of data. One of the most compelling studies employing sentinel tags was used to explain anomalies in detection data. Payne et al. [20] used sentinel tags to reveal changes in transmitter detections diurnally, which explained the apparently aberrant behavior observed in tracked individuals. These results indicated that sentinel tags can be crucial in differentiating behavioral patterns from environmentally driven anomalies in detection patterns and can be used to help standardize data detected from tracked individuals. Sentinel tags are also useful for identifying changes in the environment or conditions. For example, if detections decrease through time, it is probable that biofouling may be covering a receiver and affecting its performance. In this way a sentinel tag can act as a form of calibration for conditions through time as well as equipment performance and possible causes for changes in the amount of data collected.

Another significant challenge in acoustic telemetry studies is the need to retrieve receivers to offload archived data. Accessing acoustic receivers can be logistically difficult and labor intensive. Traditionally, most studies have deployed and recovered acoustic receivers via divers which require a large time commitment, good weather conditions, and high confidence in equipment locations. If working in turbid or low-light conditions, it can be difficult to locate receivers for servicing. Likewise, if receivers are deployed in deep areas, dive time to recover or replace the unit is limited. Therefore, the type and style of mooring system, conditions in the area (e.g., currents), accuracy of position recording, and duration between recovery periods all need careful consideration to help facilitate receiver recovery and data offloading. Many acoustic receivers have a restricted battery life (e.g., 12 months) relative to transmitter life. This requires servicing at least once per year. If data are needed more frequently then servicing schedules must be planned to match research needs.

Due to the difficulty in diver deployment of receivers, along with the increasing awareness of hazards involving diving, alternative approaches have been developed. These include use of acoustic release systems to allow the receiver to be freed from the mooring, retrieved at the surface, downloaded, and redeployed. Other systems are cable or satellite linked to provide near real-time data without having to recover the receiver. Some receivers also have the capacity to communicate to a surface-based modem, and groups such as the Ocean Tracking Network (OTN) are employing oceanographic gliders to interrogate receivers to reduce demands and costs of ship time. As battery technologies advance and acoustic telemetry systems become more common, additional options for longer-term deployments or reduced need to service receivers are likely to be developed.

4.2.2.2 Equipment Performance

Receiver detection range is a significant issue in the efficacy of acoustic telemetry studies. Receiver range directly determines how much of a study site can be monitored. Accordingly, if researchers require continuous data from tagged individuals (e.g., 24 h monitoring of a habitat), receivers must be deployed with overlapping detection ranges. Unlike other tracking systems such as GPS and satellite telemetry, acoustic data are limited to the extent of receiver detection range and/or the extent of the deployed receiver network coverage [4,16,22]. This is a constraint on passive acoustic telemetry studies requiring continuous or high-resolution data, limiting them to relatively small study sites or collaborative programs that can cover broader areas. The implications of limitations due to detection range are that researchers must be careful in their study design to ensure adequate data are collected to address their research questions. They must also be prepared to deal with the potential of individuals moving outside of the study site or monitored area. Loss of detection of individuals must be treated with caution. The individual could still be in the study site, but moving just outside detection range of the receiver, or it could have moved a large distance from the area. This means that the interpretation of negative data (i.e., no detections) must be carefully interpreted, and can be somewhat alleviated by data sharing over networks (see later).

Lack of detection of individuals can also be an issue that results from a number of factors including signal collision and environmental noise effects (e.g., wave action, rain, a boat causing temporary noise effects; [19,22,23]). Transmitter signal collisions occur when two transmissions arrive simultaneously at a receiver. Most transmitters have a coding scheme or other mechanisms to reduce the probability of code collision. For example, some transmitters are programmed on a pseudorandom repeat to ensure two transmitters do not continue to transmit at the same time and block detection. As technology and coding schemes have improved, issues with code collision have been reduced, but researchers still need to consider the number of transmitters deployed in a study site and the rate that they will transmit to ensure adequate detection rates [16,19,22,23]. Given the potential for signal collision, noise disruption, or other factors influencing signal reception, the lack of detection of individuals must be treated with caution, because the individual may still be in the area. However, most of these are short-term issues that are alleviated or resolved through the sheer volume of transmissions and detection opportunities. If detections are missed due to a short-term factor, detections of that individual, if still present, will reoccur once the source of disturbance has passed.

Transmitter signal collisions can produce two outcomes in the data: (1) no data are recorded for either individual the signal for which overlapped, or (2) a false detection is recorded. False detections occur when part of one transmitter code overlaps with part of another transmitter code to match the coding scheme of a third (and not detected) code. Effectively, the two signals combine to produce an erroneous detection of a third individual. False detections tend to occur as singular events in time (i.e., no other detections of this code within hours or days). The pseudorandom repeat rate of most coded tags prevents the same two transmitter signals continuing to collide and produce the false detection. The singular nature of these events allow them to be easily detected and discarded or ignored in the dataset.

4.2.2.3 Interpretation of Data

The unit of measure in acoustic telemetry studies is a single detection. However, the applicability, reliability, and use of individual detections must be handled with care. The number of detections received is a function of the tag transmission rate, receiver range, swimming speed of the tagged individual, and the environment in which the study occurs. The capacity to customize acoustic tag transmission rates provides many advantages, but may also prove disadvantageous. For example, the ability to increase transmitter life to multiple years often comes with an increase in the time between transmissions. Increased duration between transmissions limits the temporal resolution of the data, and depending on the swimming speed of the tagged individual and range of the receiver can result in animals swimming in and out of receiver range during a transmission interval. Application of long transmission intervals on highly mobile and/or fast-swimming individuals can potentially reduce detections. This scenario is complicated by the incidence of code collision or noise issues listed previously. Based on these concerns, most studies will not use a single detection as evidence of presence of an individual. This precaution reduces the risk that false detections are being treated as real data and animal presence [23]. However, in some scenarios a single detection may prove to be a highly valuable data point that requires further consideration. In these instances, greater exploration of receiver data is required. The number of detections recorded on the receiver during the day or time of the critical single detection should be considered as a factor. If the receiver recorded a large, continuous number of detections, there is a greater probability that the single detection is the result of signal collision, and thus a false detection. However, if the receiver had limited data and/or the critical detection was isolated in time from other detections, it is more likely to be a real record of animal presence. Another way of trying to determine the veracity of this data point is to place it in spatial and temporal context of other detections of this individual. For example, if an individual was detected in geographic locations and relevant timelines that imply a broader movement pattern (e.g., north and then south of the questionable detection location), there is a case for including the data point as a real detection. Therefore, single-detection data can be used if investigated to rule out other factors, but these data points should be used with caution.

Similar to single-detection data, defining directionality of movement can be complicated by receiver range and array design. Array design is often crucial to exploring directionality of movement by an individual. Grid arrays have a distinct advantage in defining directionality by providing receiver coverage in multiple directions. The movement of

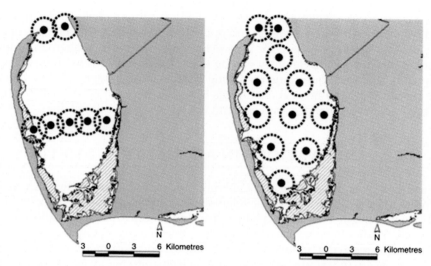

FIGURE 4.2.2 Examples of gate or curtain array design (left) versus grid design (right). *From Heupel MR, Semmens JM, Hobday AJ. Automated acoustic tracking of aquatic animals: scales, design and deployment of listening station arrays. Mar Freshw Res 2006;57:1–13.*

an individual from receiver to receiver is easily traced through this sampling design [16,24]. However, it is not always feasible or necessary to deploy a gridded array. In many cases researchers employ gates or lines to determine when individuals depart a region (Fig. 4.2.2; [16,25]). Although this can be successful, it is not possible to determine which direction an individual was moving when detected at the gate. For example, the individual could: (1) pass through the gate and be detected as it transits, (2) remain on one side of the gate but swim close enough to be detected, or (3) have swum through the gate undetected and then get detected as it passes back through the gate in the opposite direction [16]. If there are additional receivers or other gates in geographic sequence (e.g., north to south) it may be possible to use detection in multiple locations to define directionality (e.g., Refs. [6,15,25,26]). The use of existing knowledge or supplemental data, however, may not always be available. If directionality is an important component of the analysis and study conclusions, a parallel gate design is likely required (Fig. 4.2.2). This configuration allows individuals to be detected on subsequent lines to confirm passage through both as an indication of directional movement.

There are a variety of ways acoustic telemetry data can be analyzed and interpreted. As described previously, data from acoustic gates or lines can be used to define movement past specific areas and/or help document dispersal distance of individuals. This approach is most often applied to species that are known to make distinct movement such as alongshore migrations (e.g., Refs. [25,27,28]). However, most acoustic telemetry studies are designed around defining localized movements of individuals of a single species in a particular habitat or site (e.g., Refs. [29–32]). In these cases, metrics such as residency index can be useful. This metric examines the number of days an individual was detected relative to the number of days it could have been detected (based on transmitter battery life) to provide a comparative metric reflecting duration of residence in a study site (e.g., Refs. [33,34]). This type of metric may be a useful tool for defining the amount of time an individual spends inside a specific region such as a marine-protected area [35–37]. Most other metrics applied to acoustic telemetry data are spatial analyses of movement and activity space. These often take the form of home range estimators (e.g., minimum convex polygon, kernel utilization distribution; e.g., Heupel et al., 2001; [31,38,39]), which may also be analyzed relative to factors such as time of day, season, salinity, etc. (e.g., Refs. [40–42]). Although many of these metrics are standard approaches, they lack standardization in the methods used to calculate estimates (see later for more detail). In addition to horizontal space use, the potential to include a depth (pressure) sensor in transmitters provides the ability to define use of the water column. Analysis of depth-use patterns are simple and readily calculated from acoustic telemetry data (e.g., Refs. [21,43]). Several groups have gone a step further to integrate horizontal and vertical location data to produce three-dimensional estimates of space use ([44–46]; Lee et al., 2016).

The metrics applied and types of analysis completed are dictated in large part by the research questions and array design. The need to match the analysis to the question or the array design has led to a practice of largely customised data analyses. This approach is possible due to the wide versatility in application of acoustic telemetry, the broad array of species that can be monitored simultaneously, and thus the capacity to address increasingly complex research questions. Researchers have investigated social interactions of individuals [47–50], repeatability of movement patterns among cohorts and across years (Heupel et al., 2001; [12,51]), responses to environmental variables [40,41,52,53], efficacy of management practices ([11,54]; Espinoza et al., 2015), and many other aspects of animal movement and behavior. With these options and opportunities come the challenges of continuing to develop and adapt methodologies that are appropriate for answering research and management questions. In addition to research to understand species, acoustic telemetry can be used to answer questions regarding size and efficacy of marine-protected areas and other key questions regarding ecosystem management. However, quantitative analysis and interpretation remain to be the biggest challenges to researchers using acoustic telemetry as technological advances far outstrip our capacity to manage, process, and interpret the associated data.

4.2.3 INNOVATIONS IN ACOUSTIC TELEMETRY

Acoustic telemetry technology is developing and evolving quickly as battery technology and marine sensors continue to improve. With the increased use of acoustic telemetry techniques, novel approaches are being developed to answer research questions that are more complex, which drive advances in the application of the technology as well as outcomes of the research. Studies have begun to use acoustic telemetry to define species interaction by using large-bodied species as platforms for receivers (mobile receivers), through development of transmitters that can determine how close together individuals are (proximity sensors), and via transmitters that can indicate when an individual has been consumed by a predator (predation tags). How individuals move and respond to their environment is now being explored in increasing detail through novel environmental sensors and sensors that quantify movement types, such as burst swimming, which may indicate feeding events or predator avoidance (accelerometer tags). Accelerometer tags are also being used to explore physiological aspects of species such as defining free-swimming metabolic rates based on coupling tracking data with laboratory studies. Receiver advances are also occurring with new frequencies being used to improve efficacy, increased capacity to remotely download data, built-in transmitters to help relocate displaced units and much more. Here we examine some of the more recent advances in transmitter and receiver technology.

4.2.3.1 Mobile Receivers

One major constraint of passive acoustic telemetry is the relatively limited spatial coverage resulting from the static nature of receiver grids and arrays. However, dynamic reception of tag transmissions is feasible and a number of innovative methods have been proposed to facilitate this. Bioprobes, i.e., miniaturized animal-borne transceivers that are attached to large-bodied animals (such as sharks, seals, or turtles) can provide a mobile receiving unit that can potentially detect tagged animals across large spatial scales as the large animal moves through its environment (e.g., Lidgard et al., 2014). An example of this technology is the VEMCO Mobile Transceiver (VMT) attached to seals to study whether seals interact with tagged fish (Fig. 4.2.3). Bioprobes provide the opportunity to simultaneously

FIGURE 4.2.3   A Sable Island gray seal with a head-mounted satellite GPS transmitter, and a back-mounted VEMCO Mobile Transceiver (VMT) for detecting acoustically tagged animals. Acoustic detections are stored in the VMT, although technology exists to relay these detections to the satellite transmitter via Bluetooth, and in near real-time via satellite to the investigators. *Photo Credit: Damian Lidgard.*

identify social interactions of predators and record predator–prey interactions. The addition of small, mobile acoustic receivers to the payload of marine autonomous vehicles (such as underwater and wave gliders; [55]) or on ships of opportunity such as fishing boats provide similar opportunities.

4.2.3.2 Proximity Tags

Studying group behavior of smaller, mobile fish in the absence of mobile predators is more difficult, but new approaches are being constantly refined. This is important because group behavior is common across many fish species and yet there is still comparatively little known about how, when, and where free-ranging individuals interact. Business card tags (Holland et al., 2009) were initially trialed on large elasmobranchs but successful trials of smaller versions of this technology called Proximity Sensors have been developed. These units are animal-borne acoustic proximity receivers, which record close-spatial associations between free-ranging fish by detecting acoustic signals from tags on other individuals, have proven highly effective [48].

4.2.3.3 Accelerometer Tags

The use of triaxial accelerometers loggers has revolutionized our understanding of bioenergetics [56], activity budgets [57], prey capture, handling, and foraging success [58] of many marine vertebrates. Accelerometers have recently been incorporated into acoustic tags (Fig. 4.2.4) to provide similar avenues of exploration combining laboratory studies of oxygen consumption and metabolic rate to calibrate free-ranging behavior to produce estimates of energetics in free-living marine animals and holds immense promise (e.g., Refs. [59–61]).

4.2.3.4 Predation/Mortality Tags

Identifying predation events is an important aim of ecosystem studies. Yet although biologgers have very successfully been able to quantify feeding events by marine predators using accelerometry (see Ref. [58]), the identification of predation via acoustic sensor is limited. The sensor-tag approach has been to identify aberrant changes in temperature, depth, or behavior of an individual, and assume this is due to predation. However, this approach is severely limited as transmitters with sensors are often too large for small fish, or the predator does not differ enough in physiology or behavior to reliably discern predation. One new approach is to deploy a predation tag that does not rely on an active sensor, but instead is physically altered during digestion within the predator's gastrointestinal tract and then transmits a new identification code [62]. Based on the current state of technology, additional advances are required to fully address questions of predation and predator–prey relationships.

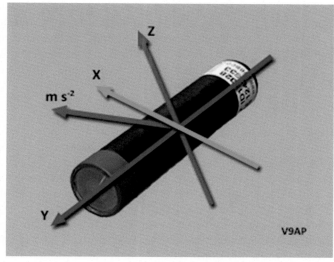

FIGURE 4.2.4   Diagram of a VEMCO pressure and accelerometer tag which transmits acceleration on three axes (X,Y,Z) in m/s².

4.2.3.5 Environmental Sensor Tags

A new avenue for acoustic telemetry, but one now well-proven with biologging, is the concept that animal-mounted tags can provide reliable environmental measurement (see Ref. [63] for an overview). New technical advances mean that miniaturized sensors with sufficient precision to provide acceptable data for ocean observing are becoming available and measures such as salinity and dissolved oxygen can be incorporated into acoustic telemetry tags, albeit still for specialized purposes.

4.2.3.6 Receiver Advances

With advances in processing power and digital processors, along with tag-coding advances, high-resolution positioning using high-frequency transmissions has become available in recent times. New receivers that detect high-frequency tags (e.g., 180 kHz) sacrifice range for the capability to detect larger numbers of smaller tags with submeter precision. New systems are becoming more flexible with positioning coordinated by inbuilt synchronization transmission, and real-time monitoring capabilities available if receivers are hard-wired (cabled) instead of the classic autonomous units. Other remote-downloading capabilities that have become available and facilitate real-time tracking include data transmission via satellite modem or mobile-phone telemetry, providing such capabilities as early warning by text message (Short Message Service - SMS) for tagged dangerous sharks conveyed to the public via mobile-phone applications such as *Sharksmart* [64].

4.2.4 ANALYTICAL APPROACHES

4.2.4.1 Standardizing Methods

Despite broad international use of acoustic telemetry tracking systems, tracking data are commonly collected, processed, and analyzed by individual research groups. The result of this disparate approach, and site- or species-specific analyses means that there is limited consistency in how and what metrics are estimated and reported using acoustic detection data. Although there are some analytical methods that are commonly used such as home-range analyses, there are a variety of models and means of calculating these metrics (e.g., Minimum Convex Polygons; fixed Kernel Utilization Distributions (KUDs); Brownian bridge KUDs). For example, home-range analyses for five species that were tracked in the Ningaloo Reef region applied separate smoothing parameters to home-range estimates [12,65–67]. This difference in parameter selection means the home ranges calculated among these species are not directly comparable (see Ref. [68]). In addition to difficulties in comparing results across studies, acoustic tracking networks produce large volumes of data, which warrant a standardized analysis approach that can be applied consistently across large datasets [69]. To fill this need, researchers from the Australian Integrated Marine Observing System Animal Tracking Facility (IMOS ATF) have developed an analytical framework and tool to provide standardized analysis of acoustic telemetry data [70]. This standardized framework provides capacity to process and analyse large datasets to conduct comparative metaanalyses across species and geographic scales. As acoustic telemetry networks expand and projects become more integrative, it will be increasingly important to be able to compare research outputs among regions, species, or habitats. For this to be effective a standardized approach and common movement metrics must be applied. Ultimately, the application of a standardized framework will advance telemetry study outputs beyond a species- or region-specific focus to tackle ecosystem-level questions in animal movement ecology.

4.2.4.2 Network Analysis and Other Approaches

Network Analysis is an analytical method that offers a potential new framework for analysis of telemetry data. Network Analysis examines the structure of systems that are characterized by connections (or edges) between nodes [71]. A node can symbolize a range of objects, from individuals and species to patches in a landscape [72] whereas an edge signifies interactions between individuals or connectivity between patches. Network Analysis has successfully contributed to the understanding of spatial ecology by examining social interaction [47,73,74], space use [24,75], movement patterns [76–79], and habitat use [80,81] of a variety of marine species. Network analysis can also be used to help develop, guide, and evaluate management measures by examining/modeling habitat loss and fragmentation, climate change, and fisheries exposure [82–84]. For example, Nicol et al. [85] and Rayfield et al. [86] used network analysis metrics to determine the importance of each patch (node) or corridor (edge) in maintaining or contributing to landscape connectivity to help prioritize areas for management and conservation. Application of

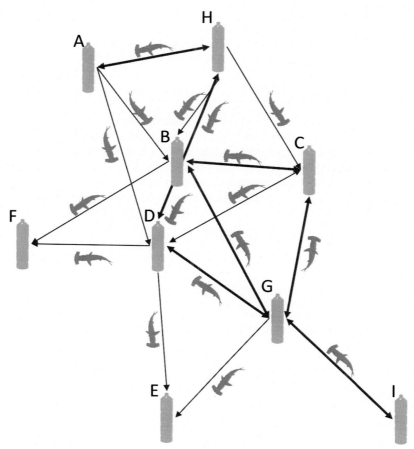

FIGURE 4.2.5 Example of a hammerhead shark movement network. Acoustic receivers, identified using letters from A to I, represent network nodes. Edges (or paths) represent individual movements between acoustic receivers. *Thicker lines* indicate more frequent path use by the individual. *Arrows* represent the direction of movement.

Network Analysis to acoustic telemetry data is based on acoustic receivers functioning as nodes and edges representing the movement of individuals between receivers (Fig. 4.2.5). This method is proving to have the capacity to replicate data produced using traditional home-range analyses such as KUDs, but because Network Analysis is scale independent, it produces data that are more comparable across study sites than traditional approaches. For example, [24] defined the core and general use areas (equivalent to 50 and 95% KUD) of two nearshore shark species and identified the importance of movement corridors within core areas for both species. Results from Network Analysis not only replicated the outputs of KUD analysis, it provided information on the amount of connectivity between receivers that KUDs do not provide. Use of Network Analysis and other quantitative approaches is likely to increase as researchers realize the utility and applicability of this kind of analysis.

4.2.4.3 Networks of Receiver Arrays and Data Sharing

By its very nature acoustic telemetry is almost predestined for large-scale collaboration and data sharing. Animals tagged in one project's array are likely to leave the area covered by those receivers unless it is very extensive or they have extremely small home ranges. With standardized approaches, shared metadata standards, and data quality control, value adding to individual projects through collaborative networks of receiver arrays is a win–win [5,70,87]. The benefits are large and appreciable. For example, by sharing data across a national network, Australian researchers gained the ability to identify and filter false detections generated by coded identification systems and remove false detections [23]. This then led to the development of universal database query tools and quality control methods [87]. Large telemetry networks are now present at regional (e.g., the Florida Atlantic Coast Telemetry network), continental (e.g., Australia's IMOS ATF, ocean or freshwater basin (e.g., the Great Lakes Acoustic Telemetry Observation System [GLATOS]), and global (e.g., the Ocean Tracking Network [OTN]) scales ([7]; Lennox et al., 2017). Early reluctance to share data through these large networks has been further alleviated by recent research demonstrating the immediate and tangible benefits to individual researchers of sharing their data, with greater collaborative success, and faster publication rates by researchers that share data [88].

4.2.5 THE FUTURE OF ACOUSTIC TELEMETRY

Acoustic telemetry has become an important tool in the aquatic ecologists' armory and an increasingly important component of ocean-observing systems. From early descriptive studies, acoustic telemetry is now used to delve into important physiological and ecological questions, monitor environmental parameters, and address important management questions on population dynamics, conservation, and harvest. As acoustic telemetry technology has advanced, so has its capabilities and the requirement for complex analysis tools has also grown. In parallel with the advances in technology, the need for collaboration, data sharing, and open access have also advanced. As broader-scale, acoustic telemetry networks are established in conjunction with other marine-observing technologies the applicability of acoustic telemetry will continue to expand.

Glossary

Accelerometer An instrument for measuring the acceleration of a moving body.
Kernel utilization distribution A distribution based on locations of an individual to determine where an individual is likely to spend a proportion of its time, typically defined as core areas (50% of time) and extent of movement (95% of area used).
Minimum convex polygon A polygon surrounding the outermost position locations of an animal which is used to estimate activity space or home range.
Odontocete A toothed whale.
Pycnocline A layer in the ocean in which water density increases with depth.
Thermocline An abrupt temperature gradient in a body of water consisting of a layer above and below which the water is different temperatures.
Turbidity The cloudiness or haziness of water.

Acknowledgments

We thank the IMOS Director Tim Moltmann for supporting the IMOS Task Team that indirectly led to the four of us collaborating on this chapter. V Udyawer is supported by an Australian Institute of Marine Science post-doctoral fellowship; E Lédée is a post-doctoral fellow funded by an NSERC Discovery Grant awarded to Dr. Steven J. Cooke.

References

[1] Mellinger DK, Stafford KM, Moore SE, Dziak RP, Matsumoto H. An overview of fixed passive acoustic observation methods for cetaceans. Oceanography 2007;20(4):36–45.
[2] MacLennan DN, Simmonds EJ. Fisheries acoustics, vol. 5. Springer Science and Business Media; 2013.
[3] Huveneers C, Simpfendorfer CA, Kim S, Semmens J, Hobday AJ, Pederson H, Stieglitz T, Vallee R, Webber D, Heupel MR, Peddemors V, Harcourt R. The influence of environmental parameters on the performance and detection range of acoustic receivers. Methods Ecol Evol 2016;7:825–35.
[4] Kessel ST, Cooke SJ, Heupel MR, ' NE, Simpfendorfer CA, Vagle S, Fisk AT. A review of detection range testing in aquatic passive acoustic telemetry studies. Rev Fish Biol Fish 2014;24:199–218.
[5] Brodie S, Lédée EJI, Heupel MR, Babcock R, Campbell H, Gledhill D, Hoenner X, Huveneers C, Jaine F, Simpfendorfer C, Taylor MD, Udyawer V, Harcourt R. Realizing the potential of animal telemetry networks: value-adding through a nationally coordinated array. Sci Rep 2018;8(1):3717.
[6] Welch DW, Melnychuk MC, Rechisky ER, Porter AD, Jacobs MC, Ladouceur A, McKinley RS, Jackson GD. Freshwater and marine migration and survival of endangered Cultus Lake sockeye salmon (*Oncorhynchus nerka*) smolts using POST, a large-scale acoustic telemetry array. Can J Fish Aquat Sci 2009;66(5):736–50.
[7] Hussey NE, Kessel ST, Aarestrup K, Cooke SJ, Cowley PD, Fisk AT, Harcourt RG, Holland KN, Iverson SJ, Kocik JF, Mills Flemming JE, Whoriskey FG. Aquatic animal telemetry: a panoramic window into the underwater world. Science 2015:1255642. https://doi.org/10.1126/science.1255642.
[8] Cooke SJ, Midwood JD, Thiem JD, Klimley P, Lucas MC, Thorstad EB, Eiler J, Holbrook C, Ebner BC. Tracking animals in freshwater with electronic tags: past, present and future. Anim Biotelem 2013;1:5.
[9] Dudgeon CL, Pollock KH, Braccini JM, Semmens JM, Barnett A. Integrating acoustic telemetry into mark–recapture models to improve the precision of apparent survival and abundance estimates. Oecologia 2015;178:761–72.
[10] Lee K, Huveneers C, Gimenez O, Peddemors V, Harcourt R. To catch or to sight? A comparison of demographic parameter estimates obtained from mark-recapture and mark-resight models. Biodivers Conserv 2014. https://doi.org/10.1007/s10531-014-0748-9.
[11] Crossin GT, Heupel M, Holbrook CM, Hussey N, Lowerre-Barbieri S, Nguyen VM, Raby GD, Cooke SJ. Acoustic telemetry and fisheries management. Ecol Appl 2017;27:1031–49.

[12] Pillans RD, Bearham D, Boomer A, Downie R, Patterson TA, Thomson DP, Babcock RC. Multi year observations reveal variability in residence of a tropical demersal fish, *Lethrinus nebulosus*: implications for spatial management. PLoS One 2014;9:e105507.

[13] Williams J, Hindell JS, Jenkins GP, Tracey S, Hartmann K, Swearer SE. The influence of freshwater flows on two estuarine resident fish species show differential sensitivity to the impacts of drought, flood and climate change. Environ Biol Fish 2017;100(9):1121–37.

[14] Brownscombe JW, Cooke SJ, Danylchuk AJ. Spatiotemporal drivers of energy expenditure in a coastal marine fish. Oecologia 2017;183(3):689–99.

[15] Finstad B, Okland F, Thorstad EB, Bjorn PA, McKinley RS. Migration of hatchery reared Atlantic salmon and wild anadromous brown trout post-smolts in a Norwegian fjord system. J Fish Biol 2005;66:86–96.

[16] Heupel MR, Semmens JM, Hobday AJ. Automated acoustic tracking of aquatic animals: scales, design and deployment of listening station arrays. Mar Freshw Res 2006;57:1–13.

[17] Clements S, Jepsen D, Karnowski M, Schreck CB. Optimization of an acoustic telemetry array for detecting transmitter-implanted fish. N Am J Fish Manag 2005;25(2):429–36.

[18] Scherrer SR, Rideout BP, Giorli G, Nosa,l E-M, Weng KC. Depth- and range-dependent variation in the performance of aquatic telemetry systems: understanding and predicting the susceptibility of acoustic tag–receiver pairs to close proximity detection interference. PeerJ 2018;6:e4249.

[19] Heupel MR, Reiss KL, Yeiser BG, Simpfendorfer CA. Effects of biofouling on performance of moored data logging acoustic receivers. Limnol Oceanogr Methods 2008;6:327–35.

[20] Payne NL, Gillanders BM, Webber DM, Semmens JM. Interpreting diel activity patterns from acoustic telemetry: the need for controls. Mar Ecol Prog Ser 2010;419:295–301.

[21] Matley JK, Heupel MR, Simpfendorfer CA. Depth and space use of leopard coral grouper (*Plectropomus leopardus*) using passive acoustic tracking. Mar Ecol Prog Ser 2015;521:201–16.

[22] Simpfendorfer CA, Heupel MR, Collins AB. Variation in the performance of acoustic receivers and its implication for positioning algorithms in a riverine setting. Can J Fish Aquat Sci 2008;65:482–92.

[23] Simpfendorfer CA, Huveneers C, Steckenreuter A, Tattersall K, Hoenner X, Harcourt R, Heupel MR. Ghosts in the data: false detections in VEMCO pulse position modulation acoustic telemetry monitoring equipment. Anim Biotelem 2015;3:55. https://doi.org/10.1186/s40317-015-0094-z.

[24] Lédée EJI, Heupel MR, Tobin AJ, Knip DM, Simpfendorfer CA. A comparison between traditional kernel-based methods and network analysis: an example from two nearshore shark species. Anim Behav 2015;103:17–28.

[25] Steckenreuter A, Hoenner X, Huveneers C, Simpfendorfer C, Buscot M, Tattersall K, Babcock R, Heupel M, Meekan M, van der Broek J, McDowall P, Peddemors V, Harcourt R. Optimising the design of large -scale acoustic telemetry curtains. Mar Freshw Res 2016. https://doi.org/10.1071/MF16126.

[26] Cooke SJ, Hinch SG, Farrell AP, Patterson DA, Miller-Saunders K, Welch DW, Donaldson MR, Hanson KC, Crossin GT, Mathes MT, Lotto AG. Developing a mechanistic understanding of fish migrations by linking telemetry with physiology, behavior, genomics and experimental biology: an interdisciplinary case study on adult Fraser River sockeye salmon. Fisheries 2008;33(7):321–39.

[27] Espinoza M, Heupel MR, Tobin AJ, Simpfendorfer CA. Evidence of partial migration in a large coastal predator: opportunistic foraging and reproduction as key drivers? PLoS One 2016:e0147608. https://doi.org/10.1371/journal. pone.0147608.

[28] Heupel MR, Simpfendorfer CA, Espinoza M, Smoothey A, Tobin AJ, Peddemors V. Conservation challenges of sharks with continental scale migrations. Front Mar Sci 2015;2:1–12. https://doi.org/10.3389/fmars.2015.00012.

[29] Heupel MR, Simpfendorfer CA. Long-term movement patterns of a coral reef predator. Coral Reefs 2015;34:679–91.

[30] Huveneers C, Harcourt RG, Otway NM. Observation of localised movements and residence times of the wobbegong shark *Orectolobus halei* at Fish Rock, NSW, Australia. Cybium 2006;30(4):103–11.

[31] Lowe CG, Topping DT, Cartamil DP, Papastamatiou YP. Movement patterns, home range, and habitat utilization of adult kelp bass *Paralabrax clathratus* in a temperate no-take marine reserve. Mar Ecol Prog Ser 2003;256:205–16.

[32] Topping DT, Lowe CG, Caselle JE. Site fidelity and seasonal movement patterns of adult California sheephead *Semicossyphus pulcher* (Labridae): an acoustic monitoring study. Mar Ecol Prog Ser 2006;326:257–67.

[33] Abecasis D, Bentes L, Erzini K. Home range, residency and movements of *Diplodus sargus* and *Diplodus vulgaris* in a coastal lagoon: connectivity between nursery and adult habitats. Estuar Coast Shelf Sci 2009;85(4):525–9.

[34] Reubens JT, Pasotti F, Degraer S, Vincx M. Residency, site fidelity and habitat use of Atlantic cod (*Gadus morhua*) at an offshore wind farm using acoustic telemetry. Mar Environ Res 2013;90:128–35.

[35] Chapman DD, Pikitch EK, Babcock E, Shivji MS. Marine reserve design and evaluation using automated acoustic telemetry: a case-study involving coral reef-associated sharks in the Mesoamerican Caribbean. Mar Technol Soc J 2005;39(1):42–55.

[36] Chateau O, Wantiez L. Human impacts on residency behaviour of spangled emperor, *Lethrinus nebulosus*, in a marine protected area, as determined by acoustic telemetry. J Mar Biol Assoc UK 2008;88(4):825–9.

[37] Knip DM, Heupel MR, Simpfendorfer CA. Evaluating marine protected areas for the conservation of tropical coastal sharks. Biol Conserv 2012;148:200–9.

[38] Marshell A, Mills JS, Rhodes KL, McIlwain J. Passive acoustic telemetry reveals highly variable home range and movement patterns among unicornfish within a marine reserve. Coral Reefs 2011;30(3):631–42.

[39] Meyer CG, Holland KN, Wetherbee BM, Lowe CG. Movement patterns, habitat utilization, home range size and site fidelity of whitesaddle goatfish, *Parupeneus porphyreus*, in a marine reserve. Environ Biol Fish 2000;59(3):235–42.

[40] Heupel MR, Simpfendorfer CA. Influence of salinity on the distribution of young bull sharks in a variable estuarine environment. Aquat Biol 2008;1:277–89.

[41] Heupel MR, Simpfendorfer CA. Importance of environmental and biological drivers in the presence and space use of a reef-associated shark. Mar Ecol Prog Ser 2014;496:47–57.

[42] Schlaff AM, Heupel MR, Simpfendorfer CA. Environmental effects on space use of a common reef shark on an inshore reef. Mar Ecol Prog Ser 2017;571:169–81.

[43] Matley JK, Tobin AJ, Heupel MR, Lédée EJI, Simpfendorfer CA. Contrasting patterns of vertical and horizontal space use of two exploited and sympatric coral reef fish. Mar Biol 2016;163:1–12.

[44] Cooke SJ, Niezgoda GH, Hanson KC, Suski CD, Phelan FJ, Tinline R, Philipp DP. Use of CDMA acoustic telemetry to document 3-D positions of fish: relevance to the design and monitoring of aquatic protected areas. Mar Technol Soc J 2005;39(1):31–41.

[45] Simpfendorfer CA, Olsen EM, Heupel MR, Moland E. Three dimensional kernel utilization improve estimates of space use in aquatic animals. Can J Fish Aquat Sci 2012;69:565–72.

[46] Udyawer V, Simpfendorfer CA, Heupel MR. Diel patterns in the three-dimensional use of space by sea snakes. Anim Biotelem 2015;3:29.

[47] Armansin NC, Lee KA, Huveneers C, Harcourt RG. Integrating social network analysis and fine-scale positioning to characterize the associations of a benthic shark. Anim Behav 2016;115:245–58.

[48] Guttridge TL, Gruber SH, Krause J, Sims DW. Novel acoustic technology for studying free-ranging shark social behaviour by recording individuals' interactions. PLoS One 2010;5(2):e9324.

[49] Heupel MR, Simpfendorfer CA. Quantitative analysis of aggregation behaviour in juvenile blacktip sharks. Mar Biol 2005;147:1239–49.

[50] Mourier J, Vercelloni J, Planes S. Evidence of social communities in a spatially structured network of a free-ranging shark species. Anim Behav 2012;83(2):389–401.

[51] Heupel MR, Simpfendorfer CA, Olsen EM, Moland E. Consistent movement traits indicative of innate behavior in neonate sharks. J Exp Mar Biol Ecol 2012;432–433:131–7.

[52] Heupel MR, Simpfendorfer CA, Hueter RE. Running before the storm: blacktip sharks respond to falling barometric pressure associated with Tropical Storm Gabrielle. J Fish Biol 2003;63:1357–63.

[53] Ortega LA, Heupel MR, van Beynen P, Motta P. Movement patterns and water quality preferences of juvenile bull sharks (*Carcharhinus leucas*) in a Florida estuary. Environ Biol Fish 2009;84:361–73.

[54] Espinoza M, Farrugia TJ, Webber DM, Smith F, Lowe CG. Testing a new acoustic telemetry technique to quantify long-term, fine-scale movements of aquatic animals. Fish Res 2011;108(2):364–71.

[55] Lin Y, Hsiung J, Piersall R, White C, Lowe CG, Clark CM. A multi-autonomous underwater vehicle system for autonomous tracking of marine life. J Field Robot 2017;34:757–74.

[56] Gleiss AC, Wilson RP, Shepard EL. Making overall dynamic body acceleration work: on the theory of acceleration as a proxy for energy expenditure. Methods Ecol Evol 2011;2(1):23–33.

[57] Ladds MA, Thompson AP, Slip DJ, Hocking DP, Harcourt RG. Seeing it all: evaluating supervised machine learning methods for the classification of diverse otariid behaviours. PLoS One 2016;11(12):e0166898.

[58] Carroll G, Slip D, Jonsen I, Harcourt R. Supervised accelerometry analysis can identify prey capture by penguins at sea. J Exp Biol 2014;217:4294–302.

[59] Brodie S, Taylor MD, Smith JA, Suthers IM, Gray CA, Payne NL. Improving consumption rate estimates by incorporating wild activity into a bioenergetics model. Ecol Evol 2016;6(8):2262–74.

[60] Udyawer V, Simpfendorfer CA, Heupel MR, Clark TD. Temporal and spatial activity-associated energy partitioning in free-swimming sea snakes. Funct Ecol 2017;31:1739–49.

[61] Wilson SM, Hinch SG, Eliason EJ, Farrell AP, Cooke SJ. Calibrating acoustic acceleration transmitters for estimating energy use by wild adult Pacific salmon. Comp Biochem Physiol Part A Mol Integr Physiol 2013;164(3):491–8.

[62] Halfyard EA, Webber D, Del Papa J, Leadley T, Kessel ST, Colborne SF, Fisk AT. Evaluation of an acoustic telemetry transmitter designed to identify predation events. Methods Ecol Evol 2017;8:1063–71. https://doi.org/10.1111/2041-210X.12726.

[63] Roquet F, Boehme L, Block B, Charrassin J-B, Costa D, Guinet C, Harcourt RG, Hindell MA, Hückstädt LA, McMahon CR, Woodward B, Fedak MA. Ocean observations using tagged animals. Oceanography 2017;30(2):139. https://doi.org/10.5670/oceanog.2017.235.

[64] Taylor MD, Babcock RC, Simpfendorfer CA, Crook DA. Where technology meets ecology: acoustic telemetry in contemporary Australian aquatic research and management. Mar Freshw Res 2017;68(8):1397–402.

[65] Escalle L, Speed CW, Meekan MG, White WT, Babcock RC, Pillans RD, Huveneers C. Restricted movements and mangrove dependency of the nervous shark *Carcharhinus cautus* in nearshore coastal waters. J Fish Biol 2015;87:323–41.

[66] Pillans RD, Babcock RC, Thomson DP, Haywood MDE, Downie RA, Vanderklift MA, Rochester WA. Habitat effects on home range and schooling behaviour in a herbivorous fish (Kyphosus bigibbus) revealed by acoustic tracking. Marine and Freshwater Research 2017;68:1454–67.

[67] Speed CW, Meekan MG, Field IC, McMahon CR, Harcourt RG, Stevens JD, Babcock RC, Pillans RD, Bradshaw CJA. Reef shark movements relative to a coastal marine protected area. Reg Stud Mar Sci 2016;3:58–66.

[68] Kie J. A rule-based ad hoc method for selecting a bandwidth in kernel home-range analyses. Anim Biotelem 2013;1:1–12.

[69] Hampton SE, Strasser CA, Tewksbury JJ, Gram WK, Budden AE, Batcheller AL, Duke CS, Porter JH. Big data and the future of ecology. Front Ecol Environ 2013;11:156–62.

[70] Udyawer V, Babcock RC, Brodie S, Campbell HA, Jaine F, Harcourt RG, Hoenner X, Huveneers C, Simpfendorfer CA, Taylor MD, Heupel MR. A framework for analysing animal movement patterns and space use from individuals detected by an array of static receivers. 2018. [Submitted for publication].

[71] West DB. Basic definitions and concepts. Introduction to graph theory. London: Prentice Hall; 2001. p. 2.

[72] Schick RS, Lindley ST. Directed connectivity among fish populations in a riverine network. J Appl Ecol 2007;44(6):1116–26.

[73] Mourier J, Bass NC, Guttridge TL, Day J, Brown C. Does detection range matter for inferring social networks in a benthic shark using acoustic telemetry? R Soc Open Sci 2017;4(9).

[74] Wilson ADM, Brownscombe JW, Krause J, Krause S, Gutowsky LFG, Brooks EJ, Cooke SJ. Integrating network analysis, sensor tags, and observation to understand shark ecology and behavior. Behav Ecol 2015;26(6):1577–86.

[75] Stehfest KM, Patterson TA, Barnett A, Semmens JM. Markov models and network analysis reveal sex-specific differences in the space-use of a coastal apex predator. Oikos 2015;124(3):307–18.

[76] Finn JT, Brownscombe JW, Haak CR, Cooke SJ, Cormier R, Gagne T, Danylchuk AJ. Applying network methods to acoustic telemetry data: modeling the movements of tropical marine fishes. Ecol Model 2014;293:139–49.

[77] Fox RJ, Bellwood DR. Herbivores in a small world: network theory highlights vulnerability in the function of herbivory on coral reefs. Funct Ecol 2014;28(3):642–51.

[78] Heupel MR, Lédée EJI, Simpfendorfer CA. Telemetry reveals spatial separation of co-occurring reef sharks. Mar Ecol Prog Ser 2018;589:183–96.

[79] Jacoby DMP, Brooks EJ, Croft DP, Sims DW. Developing a deeper understanding of animal movements and spatial dynamics through novel application of network analyses. Methods Ecol Evol 2012;3(3):574–83.

[80] Lea JS, Humphries NE, von Brandis RG, Clarke CR, Sims DW. Acoustic telemetry and network analysis reveal the space use of multiple reef predators and enhance marine protected area design. Proc R Soc Lond B Biol Sci 2016;283(1834).

[81] Papastamatiou YP, Meyer CG, Kosaki RK, Wallsgrove NJ, Popp BN. Movements and foraging of predators associated with mesophotic coral reefs and their potential for linking ecological habitats. Mar Ecol Prog Ser 2015;521:155–70.

[82] Borrett SR, Moody J, Edelmann A. The rise of Network Ecology: maps of the topic diversity and scientific collaboration. Ecol Model 2014;293(0):111–27.

[83] Cumming GS, Bodin Ö, Ernstson H, Elmqvist T. Network analysis in conservation biogeography: challenges and opportunities. Divers Distrib 2010;16(3):414–25.

[84] Galpern P, Manseau M, Fall A. Patch-based graphs of landscape connectivity: a guide to construction, analysis and application for conservation. Biol Conserv 2011;144(1):44–55.

[85] Nicol S, Wiederholt R, Diffendorfer JE, Mattsson BJ, Thogmartin WE, Semmens DJ, López-Hoffman L, Norris DR. A management-oriented framework for selecting metrics used to assess habitat- and path-specific quality in spatially structured populations. Ecol Indic 2016;69:792–802.

[86] Rayfield B, Pelletier D, Dumitru M, Cardille JA, Gonzalez A. Multipurpose habitat networks for short-range and long-range connectivity: a new method combining graph and circuit connectivity. Methods Ecol Evol 2016;7(2):222–31.

[87] Hoenner X, Huveneers C, Steckenreuter A, Simpfendorfer C, Tattersall K, Jaine F, Atkins N, Babcock R, Brodie S, Burgess J, Campbell H, Heupel M, Pasquer P, Proctor R, Taylor MD, Udyawer V, Harcourt R. Australia's continental-scale acoustic tracking database and its automated quality control process. Australian Ocean Data Network; 2017. https://doi.org/10.4225/69/5979810a7dd6f.

[88] Nguyen VM, Brooks J, Young N, Lennox RJ, Haddaway N, Whoriskey FG, Harcourt R, Cooke SJ. To share or not to share in the emerging era of big data: perspectives from fish telemetry researchers on data sharing. Can J Fish Aquat Sci 2017;74(8):1260–74. https://doi.org/10.1139/cjfas-2016-0261.

Further Reading

[1] Lee KA, Huveneers C, Macdonald T, Harcourt RG. Size isn't everything: movements, home range, and habitat preferences of eastern blue gropers (*Achoerodus viridis*) demonstrate the efficacy of a small marine reserve. Aquat Conserv Mar Freshw Ecosyst 2015;25(2):174–86.

[2] Lee K, Huveneers C, Duong T, Harcourt R. The ocean has depth: two-versus three-dimensional space use estimators in a demersal reef fish. Mar Ecol Prog Ser 2017;572:223–41. https://doi.org/10.3354/meps12097.

CHAPTER

4.3

Increasing Reliability: Smart Biofouling Prevention Systems

Laurent Delauney[1], Emily Ralston[2], Kelli Zargiel Hunsucker[2]

[1]Detection, Sensors and Measurement Laboratory Manager, Research and Technological Development Unit, Ifremer, Plouzane, France; [2]Florida Institute of Technology, Melbourne, FL, United States

4.3.1 INTRODUCTION

4.3.1.1 Why Sensors Should Be Protected From Biofouling

Within the oceanographic community, there is a growing request to acquire more in situ data in various locations and on various platforms to initiate and calibrate environmental models or operate environmental or underwater industrial process supervision. Thus, there is a need for longer-term in situ instrument deployments. However, longer immersion periods make oceanographic instrumentation and associated sensors susceptible to increased rates of biofouling. This is particularly true for observations that are near the ocean surface. The accumulation of biofouling can inhibit operation of the sensors, require increased maintenance of the system, and impact the quality of data generated by the instrument. Biofouling can also diminish water flow to and around sensors, inhibit the mechanical movement (of some sensor devices), and add weight and increased hydrodynamic drag of the instrumentation. To be operational, these systems must be equipped with biofouling protection dedicated to the instrumentation used in situ. Indeed, in less than 2 wks [1], biofouling will modify the transducing interfaces of the sensors and cause unacceptable bias on the measurements provided by in situ monitoring systems. Autonomous monitoring systems should provide, sometimes in real time, reliable measurements without costly and/or frequent maintenance. In deep-sea conditions, this maintenance is nearly impossible to achieve; for coastal applications, it is quite accepted that a 2-mo interval for maintenance is the minimum duration for economically viable in situ monitoring systems [2].

4.3.1.2 Biofouling Mechanism: Microfouling and Macrofouling

Biofilms are a conglomeration of cells, water, and excreted or released extracellular macromolecules attached to a surface. Biofilms forming on artificial surfaces are referred to by several different names including: microfouling, slime, bacterial–algal films, or films. The formation of a biofilm is generally thought of as a several-step process (Fig. 4.3.1). A clean surface will adsorb an organic layer within seconds of immersion, called a conditioning film. This is followed by the adhesion of microbes such as bacteria and algal cells, thus developing a slimy layer. Following biofilm formation is the settlement of larval spores and higher organisms. These stages can be successional, occur in parallel, or occur simultaneously.

Higher organisms, or macrofouling, consist of plants and animals that grow in many different forms. Macrofouling can be divided into soft and hard fouling.

Soft-bodied fouling organisms include seaweed, tunicates, hydroids, and arborescent bryozoans. Hard foulers consist of those organisms which have a hard outer layer, generally composed of calcium carbonate. Examples of hard-fouling organisms are barnacles, calcareous tubeworms, mussels, oysters, encrusting bryozoans, and calcareous algae. There are also organisms, referred to as fouling associates, that have the ability to move around submerged surfaces, often times hiding in cracks and crevices created by fouling. Examples include crabs, snails, and sea stars. Macrofouling is generally of a larger concern on oceanographic instrumentation due to its contribution to weight, hydrodynamic loading, and interference with optical sensors.

For both microfouling and macrofouling, the extent of accumulation will be dependent on the geographical location, time of year, environmental conditions (e.g., salinity, temperature, nutrient levels, depth), substrate, mechanical arrangement (cavities), and the biofouling prevention system. Attachment-point theory states that an organism will select settlement surfaces that maximize the attachment surface, roughness features much smaller or larger than the settling organism will be unattractive, as the ability to attach is limited. A roughness feature slightly larger than the settling organism provides a large attachment area and a refuge from hydrodynamic stress and predation.

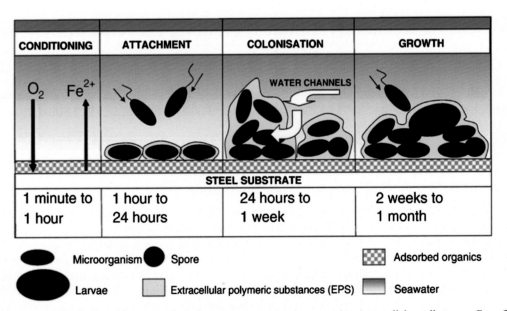

FIGURE 4.3.1   The stages of biofouling formation. Note: the stages may occur in succession, in parallel, or all at once. *From Chambers LD, Stokes KR, Walsh FC, Wood RJK. Modern approaches to marine antifouling coatings. Surf Coat Tech 2006;201:3642–52.*

4.3.1.3 Effect on Sensors Measurements

Biofouling in seawater can grow very rapidly, particularly during productive periods (blooms), and lead to poor data quality in less than a few weeks' time. As shown on Fig. 4.3.2 the biofouling species involved can be very different from one location to another [4].

This biofouling development gives rise to a continuous shift or drift in the measurements. Consequently, the measurements can be out of tolerance and then data become unusable. Video systems such as cameras and lights can also be disrupted by biofouling. Pictures become blurred or noisy, and lights lose efficiency because the light intensity decreases due to the screening effect of biofilm and macrofouling.

As shown on Fig. 4.3.3, after 7 da, a drift can be observed on measurements produced by a fluorescence sensor due to biofouling settled on the sensitive part of the sensor made of an optic system [5]. This type of optical sensor is very sensitive to biofouling, because even a very thin biofilm on the optics can interfere with the measurement process and give rise to incorrect measurements. An increase of the sensor response due to biofouling is quite specific to fluorescence sensors; the reason is that the fluorescent emitter is closer to a dense biomass and the fluorescent signal is stronger. Otherwise, the drift observed due to biofouling is usually a decrease in response. This decrease can be observed with conductivity sensors (electrode-based cells), transmissometers, pH sensors, and oxygen sensors (Clark electrodes and Optodes) [6].

4.3.1.4 Real World Experiences and Special Challenges in High-Fouling Environments

In high-fouling environments, instrumentation that is not equipped with a preventative fouling measure will begin to see problems in mechanics and data quality within a few days of immersion. Even instruments that do have biofouling prevention systems may begin to see drift in data after a few weeks, especially during peak fouling seasons (warmer months; Fig. 4.3.4). Therefore, to prevent degradation in the data, it is strongly recommended that the instrumentation be outfitted with a biofouling-prevention system appropriate for the geographical location of deployment. Often one preventative measure is not enough, and two methods may need to be combined to eliminate or reduce fouling accumulation on oceanographic sensors.

4.3.1.5 Objectives of Biofouling Protection for Sensors

Biofouling protection for oceanographic sensors is a difficult task in which the specifications should be driven by three important characteristics:

- It should not affect the measurement or the environment.
- It should not consume too much energy, to preserve the endurance specifications of the autonomous monitoring system.
- It should be reliable even in aggressive conditions (seawater corrosion, sediments, hydrostatic pressure, etc.).

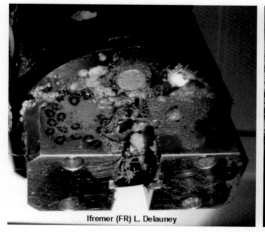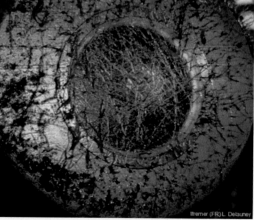

FIGURE 4.3.2   Left: Fluorometer after 30 days in Helgoland (Germany) during summer. Right: Transmissometer after 40 days in Trondheim harbor (Norway) during summer. *Source: L. Delauney (IFREMER).*

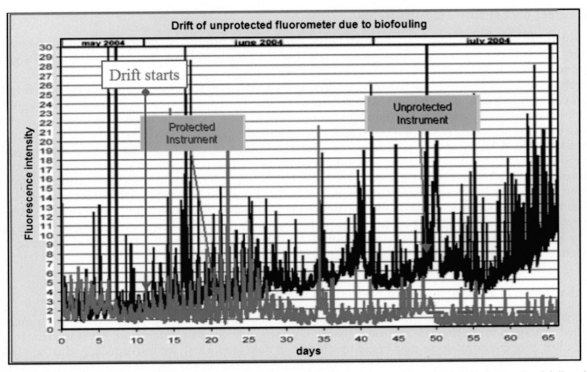

FIGURE 4.3.3 Drift of an unprotected fluorometer due to biofouling development on the optics (vertical axis, % of full scale). *Source: L. Delauney (IFREMER).*

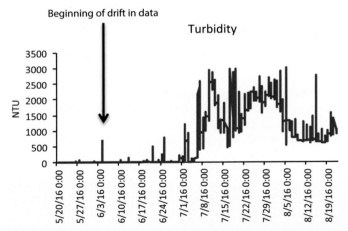

FIGURE 4.3.4 Example of drift occurring on an optical turbidity sensor after several weeks (June 3, 2016) in a high-fouling environment. The units for turbidity are expressed in Nephelometric Turbidity Units (NTUs). *Source: K. Hunsucker (CCBC).*

The protection of oceanographic sensors requires a method that does not interfere with the more sensitive components. In this case, some of the more common biofouling techniques, such as antifouling paints, are not well suited to protect water-quality sensors. For sensors such as optical sensors (fluorometer, turbidimeter, transmissometer, dissolved oxygen, CO_2), membrane sensors (pH, dissolved oxygen, methane, CO_2) or electrochemical sensors (conductivity), the interface between the measurement medium and the sensor's sensitive components must remain intact. Some of the newer and promising biofouling-protection techniques include localized Ultraviolet (UV) emission and surface chlorination on optical transduction windows. Many biofouling-prevention methods can also be purchased in conjunction with oceanographic sensors. Within the following sections, methods for biofouling protection are discussed and separated by mode of operation (passive vs. active). Methods may include antifouling products as well as antifouling regimes or maintenance.

4.3.2 STRATEGIES FOR SENSOR BIOFOULING PROTECTION

4.3.2.1 Passive

4.3.2.1.1 Antifouling Paints, Fouling-Release Paints

In the marine environment, the most common method of preventing fouling is through the application of antifouling coatings. These coatings either use a biocide to kill the fouling organisms (antifouling systems), create a surface to which they find it difficult to attach (silicone-based fouling release systems), or form a hard inert surface that requires cleaning [7].

Historically, the most effective biocide-based coatings were the tributyltin (TBT) self-polishing copolymers. These were biocidal at low levels, which prevented colonization, self-smoothing, and had a long lifetime. Unfortunately, TBT is detrimental to the environment, which led to the ban for commercial use in 2008 [8]. Currently used antifouling coatings are broadly separated into copper systems and copper-free systems. Many marine algal species are copper resistant, and as a result, booster biocides (e.g., zinc pyrithione) are now incorporated within the paint to enhance both the life and the performance of the copper systems. Copper and copper-free antifouling coatings can be effective at preventing settlement of fouling organisms, as long as they consistently leach the biocide(s), which can be achieved as a ship moves through water or through wave and current action. If an oceanographic sensor is deployed in a static environment, this will pose a problem as the coating will not adequately leach the biocide, and result in fouling buildup over time. These coatings would be more effective if deployed in an area with high flow. However, as with TBT, there is a growing movement to eliminate the use of copper-based antifouling paints due to its persistence in sediments, bioaccumulation, and other environmental impacts [3,9].

Currently, the most environmentally friendly coating systems are the nonstick and fouling-release coatings. Fouling-release systems have a low surface energy, low modulus, low microroughness, and may contain additives, which prevent a fouling organism from generating a strong bond to the surface. This weak bond allows for removal of the organism either through their weight or hydrodynamic pressure as a ship or platform moves through water [10,11]. These coatings are considered environmentally friendly, easy to clean, offer a smooth surface when application is correct, and reduce skin-friction drag [7]. However, they are costly, not easy to apply, are not as durable as their biocidal counterparts, and the environmental impacts of many of the additives are unknown [7].

More recently, coatings are being designed that integrate a biocide into a fouling-release matrix. These coatings demonstrate effectiveness in static environments for longer-term immersion. Regardless of the coating system being utilized, care should be taken not to coat optical sensors, conductivity–temperature sensors, or moving parts. If a copper-based paint is used, it should not be applied directly to certain metals, as undesirable galvanic reactions will result.

4.3.2.1.2 Substrate Selection, Hydrophobicity, Copper Mesh Screen

Substrate selection may be another consideration for biofouling prevention, as certain surfaces are more attractive to fouling organisms. Particular components of the sensor can be fabricated out of copper alloy, which is effective at preventing fouling buildup (e.g., 90:10 copper:nickel). Additionally, different tapes and wraps (i.e., copper tape, plastic wrap, or duct tape) can be used to cover components of the instrumentation that comprise plastic or other inert surfaces. Copper tape can be applied directly onto the sensors and can last for several months. An example of copper-tape protection is shown on a fluorometer [12], Fig. 4.3.5. To maximize the effectiveness of the protection, it was necessary to build up a copper cell and to coat the entire sensor head with copper.

Although wrap material (plastic wrap or duct tape) themselves may foul, it can be removed after the immersion period, thus removing the fouling with it and reducing the time needed for cleaning maintenance. These are acceptable biofouling-prevention systems for short-term deployments (months).

Sensor faces can also be designed to be very smooth, especially optical sensors. The lower-surface roughness will act to decrease the fouling settlement and attachment as well as improve mechanical cleaning techniques. Additionally, hydrophobic sprays consisting of nanopolymer formulations can be applied to sensors in a thin coat, which acts to repel water, and allow for weak attachment of fouling organisms. Combined with a cleaning or wiping regiment (either with a device or a hand-clean), the sensors can be maintained free of fouling.

4.3.2.2 Active

4.3.2.2.1 Wipers and Pads, Brushes, Water-Jetting, Electrolysis

Many optical sensors and electrodes are now implemented with a wiper, which will move over the sensor face at a set frequency (i.e., about once an hour or before a measurement) to remove any newly settled fouling organisms.

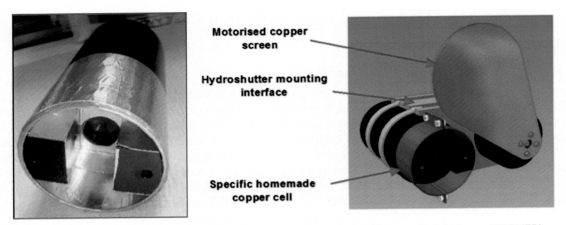

FIGURE 4.3.5   Biofouling protection with a motorized copper shutter. *Photos credit: L. Delauney (IFREMER).*

FIGURE 4.3.6   Example of wipers used to keep optical sensors free of fouling. (A) New deployment. (B) 16 days immersion, probe faces remain free of fouling where wipers clean. Biofilm and encrusting bryozoans have formed along probe edges. (C) One month of growth in a high-fouling environment. *Photo credit: K. Hunsucker (CCBC).*

The more-effective wipers will cover the entire top part of the sensor face not just the optical port. Keeping a swath around the optical port free of fouling will help keep organisms from settling nearby and growing inward toward the optical port. The wiper material is nonabrasive in pad or blade form, and designed not to scratch the sensor face. These devices can be effective at preventing the buildup on a sensor for 3 to 6mo. However, in high-fouling environments the wipers themselves will foul (Fig. 4.3.6). This eventually causes functional problems with the wipers, which in turn cause the sensor face to foul. And, in a highly turbid environment such as an estuary, the wiper can little by little scratch the optical port of the sensors.

4.3.2.3 Adverse Effects

Special care must be taken to avoid adverse effects because surface treatments of an optical window may modify the interfacial properties. Electrodes in the vicinity of a conductivity sensor may perturb measurement. Biocide molecules can interfere with an oxygen sensor membrane or induce local water property modifications. An example of adverse effect on measurements is given in Fig. 4.3.7; in case of high chlorine generation (residual chlorine content: 0.55mg/L) in the vicinity of an oxygen sensor, the dissolved oxygen measurements are no longer comparable with reference measurements performed in the laboratory on seawater sampling using the Winkler[1] titration method.

[1] http://www.quae.com/fr/r449-hydrologie-des-ecosystemes-marins-parametres-et-analyses.html.

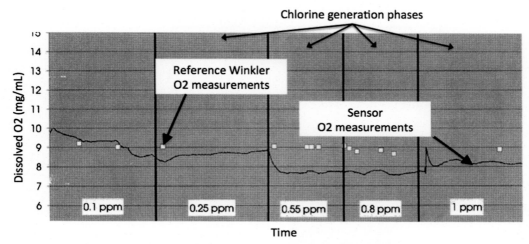

FIGURE 4.3.7 Adverse effects of high chlorine generation on dissolved oxygen measurements [4].

4.3.3 OFF-THE-SHELF SENSOR BIOFOULING PROTECTION

4.3.3.1 Wipers, Pads, Brushes

The use of wipers and brushes is usually incorporated during the design stage of the instrument, and thus usually sold as a unit with the sensors. Numerous companies have equipped their oceanographic instruments with wipers, such as: Yellow Springs Instruments (YSI) multiparameter sondes, Western Environmental Testing Laboratories (WET Labs) Environmental Characterization Optics (ECO) Series, and Forest Technology Systems (FTS) Digital Turbidity Sensor (DTS)-12 Turbidity Sensor. Separate wiper devices are available on the market (e.g.: Zebra-Tech Ltd.). They need to be adapted to the existing sensor very carefully to be efficient. The important thing to remember while using wipers is that they must also remain free of fouling. The buildup of fouling on the wipers can begin to affect the mechanical movement, which will in turn allow for settlement of fouling organisms on the optical sensor. Additionally, as the wipers foul, they may begin to scratch or damage the face of the sensors, affecting data quality.

4.3.3.2 Copper: Screen, Tape, Cu–Ni Antifouling Guard

Components of the sensor can be constructed out of a copper alloy, such as 90:10 copper–nickel. For water-quality instrumentation, the sensor or probe guard can be retrofitted completely out of the copper–nickel. This has been found effective as it has low dissolution rates, can last for long-term exposures, and provides better performance than the common plastic probe guards provided by the manufacturer. Oceanographic sensors such as YSI, WET Labs, and Turner Designs sell copper–alloy probe guards and plates.

Copper tape can be applied to plastic sensors, housings, wipers, and any other inert (nonreading) surface. Like copper-based antifouling paints, the copper tape acts to prevent the settlement of fouling organisms to the surface. Depending on the fouling environment, copper tape may be effective for 2–6 mo. Copper tape can be found as accessories through manufacturers such as Turner Designs[2] and YSI.[3] It can also be purchased through other retail locations; just make sure the copper tape is water resistant. The downside to using copper tape is, once it is removed, a sticky residue may remain on the probe. To avoid this, regular Scotch tape/packing tape can be applied underneath the copper tape directly to the probe.

Certain nonoptical sensors (e.g., conductivity and temperature) can be protected with copper screens. This prevention incorporates the copper antifouling, but openings made on the screen still allow water to pass over the sensor. The screens need to be small enough to prevent the passage of biofouling larvae through to the sensor. Several manufacturers offer copper screens for purchase, such as YSI. The screens can be effective in low-fouling environments.

[2] https://www.turnerdesigns.com/t2/doc/tech-notes/S-0185.pdf.

[3] https://www.ysi.com/Accessory/id-616189/Anti-Fouling-Copper-Tape.

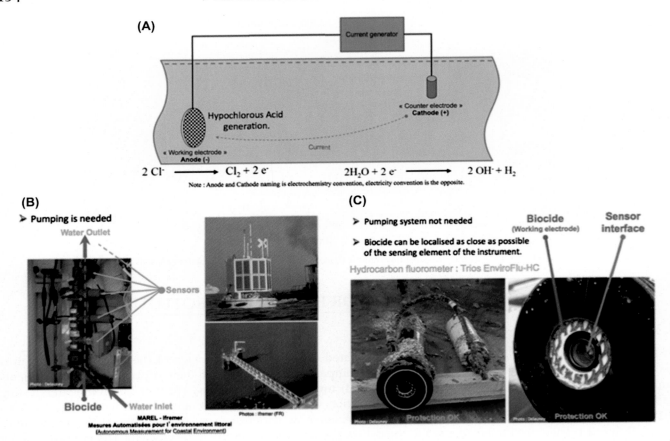

FIGURE 4.3.8   (A) Seawater electrolysis principle. (B) Global protection. (C) Localized protection. *Source: L.Delauney.*

In the high-fouling environments, the copper screen surface area is not large enough to provide a significant antifouling surface. Additionally, the small holes in the screens trap suspended sediment, impeding the flow of water to the sensors and allowing for the settlement of organisms such as amphipods that thrive in sediment.

4.3.3.3 Bleach

Chlorination has long been used in industrial applications to protect systems from biofouling. Recently it has been used for biofouling protection of in situ oceanographic instruments. Bleach-injection methods can be found on freshwater monitoring stations, Ferrybox systems (see Chapter 5.3), and on very few autonomous monitoring instruments such as the Wet Labs/Sea-Bird Water Quality Management (WQM)'s instrument[4] for which we can read: "BLeach Injection System (BLIS) protects Dissolved Oxygen (DO) sensor and CTD flow path." This scheme requires a reservoir for the chlorine solution and a pump.

4.3.3.4 Seawater Electrolysis

The electrolysis chlorination system (Fig. 4.3.8A) can be found on monitoring stations (Fig. 4.3.8B) [13,14] and "Ferry Box" instruments that use pumping circuitry, the protection is known as a "global chlorination" scheme [15]. In this way the whole circuitry is protected at the same time as the sensors.

Another electrolysis chlorination scheme can be found on a few autonomous sensors; it consists of protecting only the sensing area of the sensor (Fig. 4.3.8C). The electrolysis is performed on a very restricted area and, consequently, the energy needed is very low and compatible with autonomous deployment. Very few commercial instruments are equipped with such a scheme [16]. This is because the adaptation to the sensor-transducing interface can be difficult

---

[4]See http://www.wetlabs.com/products/wqm/wqm.htm.

for some applications. Additionally, the energy required for this technique (240 mW) is still too high for use in long-term, low-energy deployments such as deep-sea standalone observatories.

4.3.3.5 UV Irradiation

Recently, the Canadian company AML has proposed a UV irradiation biofouling-protection scheme for sensors. The system, called UV·Xchange,[5] is based on UV Bulbs that irradiate the area to be protected. When energy expenditure is not so much of a concern,[6] and when the area to protect is not too large, this solution may provide sufficient sensor-biofouling prevention.

4.3.3.6 Freshwater

Freshwater has been investigated as a way to prevent colonization of fouling organisms in marine systems. One system stored a conductivity sensor in freshwater between readings, only exposing the cell to seawater during flushing and readings. The authors were able to keep the sensor clean for 19 days in biologically rich natural water during a lab test of system [17]. This method holds promise, but issues with maintaining a clean water source and the possibility of moving parts breaking down may limit this for long-term deployments. Etheridge and colleagues were able to extend time between cleanings without data degradation of a UV Spectrometer deployed in a brackish stream by integrating a high-pressure freshwater rinse and only exposing the device to the natural water during sampling [18].

4.3.3.7 Optically Clear Coating

Optically clear coatings have been designed to prevent biofouling settlement and are formulated for high light transmission. The coatings can be applied to instrument housings and optical sensors. One such coating is the commercially available Clear Signal by Severn Marine Technologies.[7]

4.3.3.8 NanoPolymer Coating (e.g., C-Spray)

There are several nanopolymer coatings on the market. Once applied, these create a hydrophobic slippery surface, making it difficult for fouling organisms to settle. Care should be taken not to apply the coating directly onto optical sensor faces, as the coating is not optically clear and will interfere with readings. An example of nanopolymer sprays includes C-spray sold by YSI.

4.3.3.9 Routine Mechanical Maintenance, Vinegar

If the sensors are unprotected or the antifouling protection fails, cleaning must be performed. In high-fouling environments, this may need to occur as frequently as twice a week to prevent colonization by calcareous macrofouling organisms and the degradation of data. This need for routine maintenance severely limits the ability to collect data remotely for long periods. The body of the probes and the sonde may be cleaned with brushes and scrapers. The probe faces require a gentler approach. If fouling consists of biofilms and soft fouling, a cotton swab may be used. However, if calcareous fouling organisms have colonized the probe faces, a chemical cleaner such as vinegar or other acid may be required to avoid scratching the sensor/membrane. This is especially critical on sensors like the optical dissolved oxygen probes, which have an oxygen-permeable cap with a delicate coating that is easily damaged by wiping/cleaning or etched by organisms. UV spectrometers are sensitive not only to biofouling, but to signal degradation caused by adsorbed chemical fouling. These also respond well to cleaning with a mild acid [18]. Usually, the data can be examined to determine when a cleaning must be performed. This is often seen as a dramatic change in measurements over a short time period with no mitigating factors. For example, a large sudden increase in chlorophyll might be seen without any rain or climactic event. Marrs and colleagues were able to use the transmission spectra from their UV-visible spectrophotometer to not only determine when a biofilm had covered the sensor, but to identify the major components of the biofilm [19].

---

[5] https://amloceanographic.com/biofouling-control/.

[6] UV·Xchange consumption: 100 mA up to 190 mA (12–26 V) depending of the version.

[7] http://www.clearsignalcoating.com.

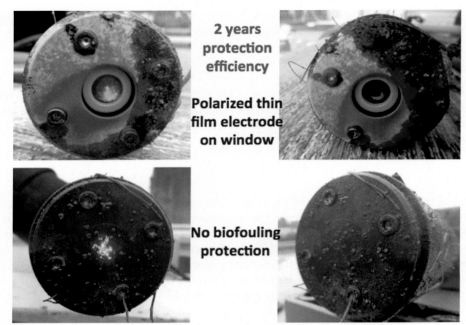

FIGURE 4.3.9   Coating seawater electrolysis biofouling-protection efficiency test on fluorometer sensors. *Source: L. Delauney, IFREMER.*

4.3.4 NOVEL TECHNIQUES FOR BIOFOULING PROTECTION OF SENSORS

4.3.4.1 Electrolysis on Conductive Layer for Optical Windows

Optical sensors, cameras, or lights can be protected using a polarized thin-film electrode that generates biocides through seawater electrolysis. In this technique, a conductive, transparent coating doped with tin dioxide is sprayed on the optical window [20]. This technique offers a high level of robustness (no moving parts), a high level of integration, and uses 20 times less energy than the conventional method based on titanium electrode (localized seawater [SW] electrolysis). It has the advantage of an active technique that can be turned on and off, thus enabling full control of biocide generation.

More recently, in the EU project NeXOS,[8] this technique has been matured up to Technology Readiness Level (TRL)[9] 8. Fluorometers, which are known to be very sensitive to biofilm interference, have been successfully protected for 2 yrs during an open-sea deployment. Fig. 4.3.9 shows the transducing interfaces of nonprotected versus protected sensor. The protected sensor has its window completely free of any biofouling organisms; the one with no protection is covered with biofouling that can be characterized as slime (heavy biofilm) that is sufficient to induce a drift on the fluorescence measurement.

4.3.4.2 Biofilm Sensor-Controlled Loop for Optimized Biofouling Protection

To optimize the use of energy to protect sensors from biofouling, a biofilm sensor can be used to drive the active operation of the biofouling protection. The biofilm sensor can give a signal to the management unit when biofouling activity is reaching a prescribed level. Then, for example, the duty cycle of the active biofouling protection can be increased. An example of such a system is the ALVIM[10] biofilm sensor adapted to marine applications.

During the EU NeXOS project, an ALVIM biofouling sensor worked for 1 month without interruption on the IFREMER test site. A first approach of the gain obtained by this device has been established. In comparison to an average chlorination rate used to protect optical sensors, biofouling protection for a sensor controlled by the ALVIM

[8] NeXOS - Next generation Low-Cost Multifunctional Web Enabled Ocean Sensor Systems Empowering Marine, Maritime and Fisheries Management, is funded by the European Commission's seventh Framework Program - Grant Agreement No 614102.

[9] TRL, Technology Readiness Level, a scale from 1 to 9 used to qualify technological maturity of oceanographic instrumentation development [21].

[10] http://www.alvimcleantech.com/cms/en/products.

biofilm sensor reduces the global duration of chlorination needed to get a fouling-free system by a factor of six. The energy demand is consequently divided by the same factor. This is a very promising first result in terms of overall efficiency improvement. Now, the ALVIM biofilm sensor is not optimized for use on mobile platforms; its size remains too important for such platform. Its power consumption remains similar to a conventional oceanographic sensor. Therefore, the biofilm sensor ALVIM would rather be dedicated to the coastal observatories where several sensors are to be protected simultaneously and where the available space is not too tight.

4.3.5 BIOMIMETICS FOR BIOFOULING CONTROL

Studying the structure and function of biological systems and processes to design engineering solutions, or biomimicry, is one possible way to discover novel antifouling for oceanographic sensor arrays. Biomimicry is attractive because many organisms are able to maintain fouling-free surfaces in a complex, highly competitive environment for long periods in a nontoxic or environmentally friendly way [22]. Additionally, the methods and chemicals used are able to be synthesized in the natural environment under normal temperature, UV, salinity, etc. conditions, with limited chemical resources, even in dirty water [23]. Organisms may use chemical, physical, mechanical, or behavioral methods, either singly or more often in combination to protect their surfaces [24]. Often the difficulty with the practical application of biomimetic solutions lies in identifying the effective mechanism or in transferring the mechanism to an artificial surface [25].

Antifouling through natural chemistry is perhaps the best-studied mechanism for biomimicry, with over 20,000 identified secondary metabolites with antifouling function [23,26]. These include low pH, anesthetics, deterrents, attachment, metamorphosis inhibitors, and toxic chemicals [26]. The difficulty is that natural chemistries frequently have complex structures, are only produced in small amounts by the organism, are difficult to synthesize artificially, and are only effective for a short time because organisms are able to reproduce them as needed [24,25]. Additionally, any novel chemistry must be proven environmentally safe and approved for use before it can be adopted [25,26].

Physical antifouling may be accomplished through deterrent surface texture or tailored wettability. The efficacy of antifouling-surface texture or roughness depends on the size of the roughness element, the size of the settling organism or its sensory apparatus, and the adult growth form. Solitary or limited-attachment area organisms like barnacles and arborescent bryozoans are more likely to seek these refuges than colonial or sheet-forming organisms [23,25,26]. Because of the size specificity, a surface with a complex roughness containing features of different sizes arranged hierarchically would be expected to be more effective than a single texture size class [23].

Hydrophobic surfaces with surface energy in the 20–30 mN/m are said to fall into a biologically minimally adhesive range and should experience self-cleaning. However, this is rarely proven without other mechanisms to enhance the effect of wettability [23,26]. Preference for specific surface energy often changes for different organisms, for example, diatoms adhere more strongly to hydrophobic surfaces like silicone fouling-release coatings from which algae and barnacles are easily removed [26]. For this reason, amphiphilic surfaces that combine hydrophobic and hydrophilic regions have proven more effective than a surface with just one wettability [23].

Mechanical antifouling is achieved through grooming or surface renewal [24,26]. Many oceanographic sensors already incorporate some form of grooming by using wipers or scrapers on the probe faces and through regular maintenance on the probe bodies. The use of disposable tape or plastic wrap mimics surface renewal to remove fouling easily from probe bodies. One emerging area is the use of active polymers that change surface area and topology in response to stimuli such as electrical, pneumatic, or chemical changes [23]. Basically, stimulus is applied between readings causing the polymer, perhaps on the face of a probe, to twitch and vibrate, causing self-cleaning.

Organisms may also avoid fouling by moving between freshwater and salt water, emerging into air, burrowing, moving into areas with low oxygen for a time, and primarily emerging at night [26]. Closed, semienclosed, and shuttered instruments may mimic this behavior by separating the sensor from the fouling environment between readings. Shuttered sensors are less susceptible to algal fouling because light is prevented between readings. Closed and semienclosed sensors may be isolated, rinsed with freshwater, or closed off entirely, and allowed to become anoxic between readings. These systems may also be combined with a slow-release biocide that is released when the sensor is inactive to assist with antifouling [27].

By far the most promising direction for biomimicry lies in combining antifouling mechanisms as most organisms do. Sharks are theorized to use a combination of deterrent-surface topography, a slippery mucus, constantly flexing denticles, an entrapped water layer (hydrophilic surface), and beneficial hydrodynamics to prevent and/or remove fouling [22,25,28]. Dolphins also use a combination of surface texture, deterrent antiadhesive enzyme chemistry, surface compliance, and hydrodynamics to maintain a clean surface [24,26,28]. The combination of micro- or nanoroughness and a hydrophobic or superhydrophobic substance leading to self-cleaning is a common theme in biomimetic research [28]. Promising work has been done to create algal mimics that combine deterrent surface texture with halogenated furanones [29].

Membranes are a rich field to mine. Many biological membranes are selectively permeable and nonfouling, for example diatom frustrules and fish skin [25,28]. Phosphorylcholine-based polymers that mimic cell membranes have been shown to protect oxygen sensors from microfouling [3]. Eyes are kept clean by a combination of wiping and flushing with an optically clear fluid. Hydrogels, or polymers that are highly swollen with water and mimic mucus, offer some possible antifouling solutions. Hydrogels or other tunable matrices can be used to regulate biocide-release rate; for example, a superhydrophobic matrix with silver ions was used to prevent attachment of microbes [30]. Conversely, a hydrophilic poly 2-hydroxyethylmethacrylate (PHEMA) hydrogel combined with benzalkonium chloride was used to keep acrylic surfaces fouling free for up to 5 mos in situ [31]. Hempel utilizes a slippery hydrogel that slowly releases copper pyrithione in their Hempaguard coating which guarantees up to 120 days antifouling efficacy to a ship sitting pier side. Using hydrogels or encapsulation/microvascular techniques, self-healing can be conferred to the antifouling mechanism. Considering how fragile some sensor faces are and the cost replacement and of lost data, this would be immensely useful. Slippery liquid-infused porous surfaces (SLIPS) technology uses slippery liquid-infused microporous structures to prevent fouling. These surfaces are self-healing because the entrapped liquid can flow to an area where damage has occurred [23,30]. If these surfaces could be made optically clear and/or selectively permeable, they could offer an excellent antifouling option for oceanographic sensors.

4.3.6 CONCLUSION AND DISCUSSIONS

With no biofouling prevention, the sensors eventually foul, resulting in a significant decline in data quality. There are currently many methods to prevent the settlement and accumulation of biofouling on sensors. Selecting the appropriate prevention method will be dependent on the type of sensor being utilized as well as the geographical location for deployment. Many of the methods outlined in this chapter are effective for short-term periods or low-fouling environments. However, it is still important to incorporate routine check on the prevention systems, to make sure they do not fail or become fouled themselves. In high-fouling environments, there is no perfect fouling-prevention system. Ideally, the sensor would need to be protected by combining several different systems. For example in a high-fouling environment, a UV diode or an SW electrolysis on conductive coating for optical sensors will be effective in a very localized area of the sensor for preventing growth. The unprotected areas would require another type of fouling prevention (i.e., antifouling paint, copper tape) to prevent fouling accumulation, which can start to impede the UV diode area. Additionally, in high-fouling areas, wipers are not as effective, unless they have some type of biofouling-prevention system such as copper tape, copper alloy, or copper antifouling paint. Sometimes in the high-fouling environments, the alternative fouling-prevention system can be routine mechanical maintenance by the user. A biweekly (twice a week) cleaning will remove any biofilm or newly settled larvae, keeping the surfaces and sensors free from fouling. But such strategy is not acceptable in terms of operating expenditures (OPEX)[11] for routine ocean monitoring that is more involved in the various initiatives of global ocean monitoring. Biofouling protection for marine environmental sensors also needs further evaluation and collaboration between researchers and manufacturers/developers.

References

[1] Delauney L, Compère C, Lehaitre M. Biofouling protection for marine environmental sensors. Ocean Sci 2010;6(2):503–11. ISSN: 1812-0784-2010.

[2] Blain S, Guillou J, Tréguer P, Woerther P, Delauney L. High frequency monitoring of the coastal marine environment using the MAREL buoy. J Environ Monit 2004;6:569–75.

[3] Chambers LD, Stokes KR, Walsh FC, Wood RJK. Modern approaches to marine antifouling coatings. Surf Coat Tech 2006;201:3642–52.

[4] Lehaître M, Delauney L, Compère C. Real-time coastal observing systems for marine ecosystem dynamics and harmful algal blooms, Chap 12, biofouling and underwater measurements. UNESCO; 2008. p. 463.

[5] Delauney L, Cowie P. Biofouling resistant infrastructure for measuring, observing and monitoring BRIMOM Report, Project number EVR1-CT-2002–40023. 2002.

[6] Delauney L, Lepage V. Biofouling resistant infrastructure for measuring, observing and monitoring BRIMOM Report, Project number EVR1-CT-2002–40023. 2002.

[7] Swain G. The importance of ship hull coatings and maintenance as drivers for environmental sustainability. In: Proc ship design operation environmental sustainability. London: RINA; March 10–11, 2010.

---

[11] **Opex**: Operational expenses, i.e., running costs for operating a system, complementing and not including capital expenses (CAPEX).

[8] International Maritime Organization. In: International Convention on the control of harmful antifouling systems on ships. London, 18 October 2001. 2003.

[9] Thomas KV, Brooks S. The environmental fate and effects of antifouling paint biocides. Biofouling 2010;26(1):73–88.

[10] Swain G. Redefining antifouling coatings. J Prot Coat Lin 1999;16(9):26–35.

[11] Omae I. General aspects of tin-free antifouling paints. Chem Rev 2003;103:3431–48.

[12] Delauney L, Compere C. Biofouling protection for marine environmental optical sensors. In: International Conference on recent advances in marine antifouling technology (RAMAT), Chennai (Madras), India. 2006.

[13] Woerther P, Grouhel A. Automated measurement network for the coastal environment. In: OCEAN'S 98 IEEE - Conference proceeding. 1998.

[14] Woerther P. Coastal environment of the Seine bay area monitored by a new French system of automated measurement stations. In: EUROGOOS Second International Conference proceeding. 1999.

[15] Hengelke CJ, et al. The stationary FerryBos Helgoland: supporting the Helgoland roads time-series, European operational oceanography: present and future. In: Proceedings on the Fourth International Conference on EuroGOOS. 2005. p. 174–8.

[16] Delauney L, Compère C, Biofouling protection for marine environmental sensors by local chlorination. In: Springer Series on Biofilms, vol. 4. In: Marine and Industrial Biofouling; 2009. pp. 119–34.

[17] Villagran V, Alarcon G, Pizarro O. Environmentally friendly anti-fouling system for oceanographic equipment. Sea Technol 2014;55:43–7.

[18] Etheridge J, Birgand F, Burchell II M, Smith B. Addressing fouling on *in situ* ultraviolet-visual spectrometers used to continuously monitor water quality in brackish tidal marsh waters. J Environ Qual 2013;42:1896–901.

[19] Marrs S, Head R, Cowling M, Hodgkiess T, Davenport J. Spectrophotometric evaluation of micro-algal fouling on marine optical windows. Estuar Coast Shelf Sci 1999;48:137–41.

[20] Laurent D, et al. Biofouling protection by electro-chlorination on optical windows for oceanographic sensors and imaging devices. In: OCEANS 2015 Genova. 2015.

[21] Prien RD. The future of Chemical in-situ sensor. Mar Chem 2007;107:422–32.

[22] Sullivan T, Regan F. The characterization, replication and testing of dermal denticles of *Scyliorhinus* canicula for physical mechanisms of biofouling prevention. Bioinspir Biomim 2011a;6. https://doi.org/10.1088/1748-3182/6/4/046001.

[23] Zhao N, Wang Z, Cai C, Shen H, Liang F, Wang D, Wang C, Zhu T, Guo J, Wang Y, Liu X, Duan C, Wang H, Mao Y, Jia X, Dong H, Zhang X, Xu J. Bioinspired materials: from low to high dimensional structure. Adv Mater 2014;26:6994–7017.

[24] Ralston E, Swain G. Bioinspiration - the solution for biofouling control? Bioinspir Biomim 2009;4:015007. https://doi.org/10.1088/1748-3182/4/1/015007.

[25] Sullivan T, Regan F. Biomimetic design of novel antifouling materials for application to environmental sensing technologies. J Ocean Tech 2011b;6:41–54.

[26] Ralston E, Swain G. Can biomimicry and bioinspiration provide solutions for fouling control? Mar Tech Soc J 2011;45(4):216–27.

[27] Manov D, Chang G, Dickey T. Methods for reducing biofouling of moored optical sensors. J Atmos Ocean Tech 2004;21:958–68.

[28] Bhushan B. Biomimetics: lessons from nature – an overview. Philos Trans R Soc A 2009;367:1445–86.

[29] Chapman J, Hellio C, Sullivan T, Brown R, Russell S, Kiterringham E, Le Nor L, Regan F. Bioinspired synthetic macroalgae: examples from nature for antifouling applications. Int Biodeterior Biodegrad 2014;86:6–13.

[30] Huang X, Zacharia N. Functional polyelectrolyte multilayer assemblies for surfaces with controlled wetting behavior. J Appl Polym Sci 2015. https://doi.org/10.1002/App.42767.

[31] Cowling M, Hodgkiess T, Parr A, Smith M, Marrs S. An alternative approach to antifouling based on analogues of natural processes. Sci Total Environ 2000;258:129–37.

CHAPTER

4.4

Material Advances for Ocean and Coastal Marine Observations

Hassan Mahfuz, Seyed Morteza Sabet

Department of Ocean and Mechanical Engineering, Florida Atlantic University, Boca Raton, FL, United States

4.4.1 OCEAN AND COASTAL MARINE OBSERVATIONS

Ocean-observance systems are usually network of sensor systems to measure the physical, chemical, geological, and biological variables in the ocean and seafloor [1] for improved detection and forecasting of environmental changes. These variables would include temperature, oxygen, salinity, turbidity, density, etc. Equipment for inshore mooring (Fig. 4.4.1) are designed to examine coastal-scale phenomena and withstand the challenging conditions of shallow coastal waters. In addition, to inshore mooring, there are shelf surface mooring and offshore surface mooring. These autonomous buoy platforms are key to capturing large-scale ocean–atmosphere phenomena. In Europe,

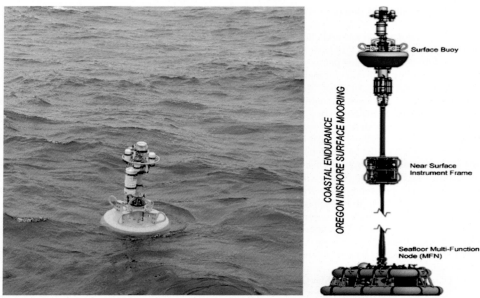

FIGURE 4.4.1   Inshore surface mooring systems. From US National Science Foundation's Ocean Observatories Initiative.

FIGURE 4.4.2   Autonomous underwater vehicles for ocean observations. From US National Science Foundation's Ocean Observatories Initiative.

similar fixed-point open-ocean observations have also been developed for deep subsurface mooring at 700–2200 m depths. There are other fixed-point moorings throughout the oceans and with greater-depth observations to monitor ocean–atmosphere phenomena, e.g., sediment traps, physical and biological phenomena, etc. For such equipment, a long life over a span of 25 year is expected and needs to be incorporated into the design.

Ocean observations also include mobile platforms such as the one shown in Fig. 4.4.2. Mobile platforms also known as autonomous underwater vehicles (AUVs) provide flexibility in ocean observations as they allow for the movement of sensors through the water in three dimensions (3D) [2]. AUVs can house multiple sensors and travel for long distances, communicating with and accepting instructions from operators onshore. These vehicles can provide high-resolution mapping of seafloor, water movement such as eddies and turbulence across the depth, sea animal migration, threat detection for coastal security, etc. AUVs can glide from the sea surface to great ocean depths and back, putting an excessive demand on material properties and structural integrity. For longer life and repeated use, fatigue would also become an issue. Although materials are important for structures, it is equally important for sensors. Due to the presence of salt, dissolved oxygen, carbon dioxide, and microorganisms, seawater is highly corrosive and one should be careful in selection of materials for both sensors as well as structures.

4.4.2 MATERIALS FOR OCEAN OBSERVATION AND SENSORS

For centuries, metals and alloys have reliably served our requirement for ocean-related structures. The properties, constitutive equations, and failure characteristics of metals and alloys are well known and can be readily used for design purposes. Rules for testing and certification of these materials are available in American Bureau of Shipping (ABS) [3], Det Norske Veritas–Germanischer Lloyd (DNV-GL) [4], and American Society for Testing and Materials (ASTM) [5] standards. On the other hand, for composites, although constitutive equations are available, complete failure and fatigue characterization under marine environments are still ongoing. In the case of nanocomposites, interactions between particles and polymers are governed by quantum mechanics, whereas macroscale phenomena are dictated by continuum mechanics. This requires a multiscale approach for complete understanding of the constitutive relationships. Such multiscale modeling is evolving and establishment of design rules are still in the making.

4.4.2.1 Marine Composites

Composites also known as fiber-reinforced composite materials offer a combination of strength and modulus that are either comparable or better than many traditional metallic materials [6]. Due to their low specific gravities, the specific strength and modulus of composites are much higher compared to those of metallic materials. The other advantage of composite materials is that they have directional (anisotropic) properties unlike metals. One can align the fiber in the direction of loading for example to take tensile load. In other words, properties can be engineered into the composites according to specific requirement. As such, the design of fiber-reinforced structure is considerably more difficult than isotropic materials. However, anisotropy provides a unique opportunity to tailor properties. This design flexibility allows selectively reinforcing the material in a particular direction by increasing its stiffness in a preferred direction, and produce parts with curvature without any forming operation [7]. In addition, to possessing light weight and directional properties, composites are also damage tolerant, corrosion resistant, have better vibrational-energy absorption, and low signature for detection. Fibrous composites are usually made of a polymer matrix reinforced with a fiber such as carbon, glass, graphite, boron, or ceramic, etc. The interface between the fiber and the matrix provides a mechanism for load transfer and can be tuned to control the elastic deformation. Multiple but sequential failure criteria such as matrix crack, interface separation, delamination, and fiber failure allow gradual progression of damage in composites instead of a catastrophic failure.

4.4.2.2 Polymer Matrix and Hybrid Composites

Polymer matrix composites (PMC) are by far the most widely used composites in ocean-observation platforms such as AUV, remotely operated vehicle buoys, ocean-instrumentation housing, Acoustic Doppler Current Profiler (ADCP), submersibles, etc. PMCs are made of a polymeric matrix with a compatible fiber reinforcement. Polymer matrices can be epoxy, vinylester, polyester, polyurethane, polyamide, etc. The most commonly used reinforcements are carbon, graphite, glass, and thermoplastic fibers. Reinforcement fibers are combined with polymer matrix in various ways: bag molding process, compression molding, pultrusion, filament winding, resin transfer molding, and injection molding. In these manufacturing processes, one has to ensure wetting of the fibers with resin, maintaining proper fiber volume fraction, minimizing void content, and curing of the resin. In most cases, postcuring is required for complete cure.

Although polymer matrix composites are lightweight and corrosion resistant, they do not exhibit plastic deformation, and this may be a disadvantage for underwater equipment. To overcome this problem, hybrid composites with aluminum or titanium are used. Aluminum–carbon fiber-reinforced plastics (Al-CFRP) have been used in recent years for AUV construction [8]. In hybrid composites, thin sheets of metal are sandwiched in between fabric composite layers and cocured in an autoclave. In other applications, pultruded glass/vinyl ester has been used in the Deep-Ocean Long-term Observatory System (DELOS). Pultruded composite profiles were fitted together in a complex triangular structure to construct the scientific monitoring modules [9]. The module is designed for deployment at seafloor and expected to last 20yrs. Recently, a composite lander has been developed at IFREMER, the Multidisciplinary Autonomous Module (MAP) in the MAST project Autonomous Lander Instrument Packages for Oceanographic Research ("Alipor") [10]. The structure is meant to support various probes and instrumentations on the ocean floor at 6-km depth. It is made of glass/epoxy composites. Pressure-resistant housing of profilers are also made of composites. Glass/epoxy cylinders about 500mm long and 150mm diameter are made through filament winding and used as instrument-protection casings. These instruments descend to a given depth and then make measurements of seawater properties (salinity, temperature, density), while following the ocean currents. They rise to the surface periodically to send data via satellite and then redescend [11]. At great ocean depth, temperature is substantially lower than the surface, which helps polymer composites because viscoelastic effects are minimal at low temperature thus reducing creep effects.

4.4.2.3 Sensors and Nanotechnology

Sensors are devices that produce a response corresponding to a change in physical condition, such as temperature or thermal conductivity, or to a change in chemical concentration [12]. Sensors, therefore, are suitable for making in situ measurements. Sensor technology is likely the area in which nanotechnology has been implemented the most. Nanotechnology is enabling the development of small, inexpensive, and highly efficient sensors, with broad applications [13]. It is envisioned that interaction at the nanoscale would offer significant advantage over conventional sensors. Number of atoms at the surface of a nanoparticle is more than that of micron-size particles allowing more interaction with the surrounding environment. In addition, large specific surface area of nanoparticle promotes further interaction. Improved sensitivity is a major attraction for developing nanotechnology-enabled sensors [13]. With nanotechnology, there exists the potential to detect a single molecule or atom. The small size, lightweight, and high surface-to-volume ratio of nanostructures would allow such detection in chemical and biological species. The speed at which species can be detected is definitely affected by sensor dimensions. Hence, nanoscale modifications provide the opportunity for improving sensor performance. Other advantages of nanotechnology application is in the fabrication of chemical sensors. Recently developed nanostructured thin films, similar to those of polyaniline and TiO_2 thin films, have produced novel classes of sensors. Using these nanoparticles, it is possible to tune and amplify the response of optical sensors for narrow frequency bands, which makes them more accurate and selective [13]. Another advantage of nanotechnology is that nanoparticles can be functionalized with various functional groups to control their interaction with the surroundings. Functional groups attached to nanoparticles can be varied to detect the chemical moiety of interest. As for an example, in a recent study quartz crystal microbalance (QCM) transducers were coated with Nafion film reinforced with COOH-functionalized carbon nanotubes (CNT) for detecting various concentrations of ammonia vapor [14]. Weight loading of nanoparticles was 0.7% of Nafion. Using a frequency analyzer, basic sensor data were collected. Frequency response for neat Nafion and CNT-reinforced films were obtained. It was observed that when Nafion contains nanoparticles, frequency shift between exposed and unexposed sensors was around 4000 Hz. On the other hand, samples without nanoparticle reinforcement showed a frequency shift of only 1000 Hz [14]—demonstrating that inclusion of nanoparticles increased the sensitivity of QCM sensors by a factor of four in detecting ammonia gas. This was possible because CNTs affected the microporous ionic regions of the polymer which were responsible for ion transport properties. For the same reason, coated-gold nanoparticles have become one of the most promising technologies for creating high-performance chemiresistive sensor systems [15]. Chemical sensitivity for most nanoparticle sensors result from swelling when chemical is absorbed, varying the distance between neighboring conductive gold cores. By varying the coating material, degree of selectivity can also be obtained, allowing differentiation between chemical analytes [15].

4.4.2.3.1 Nanotubes as Sensors

CNTs have a 1D cylindrical tube with nanometer-scale diameter and micron-size length [16]. CNTs could be single walled or multiwalled depending on the number of graphene sheets and their chirality. They are formed by rolling a graphene sheet in a specific direction and maintaining the circumference of the cross section. The atomic structure of CNT is described by the tube chirality, or helicity defined by a chiral vector and a chiral angle. Chirality is related to graphene lattice vectors from which CNTs are formed. CNTs are highly conductive both thermally and electrically. Thermal conductivity of individual CNT is around 3000 W/(mK) or even higher at room temperature, which is higher than that of diamond. Electrical conductivity of CNT is in the range of 10^6–10^7 S/m compared to that of 6×10^7 S/m for copper. Young's modulus of CNT is on the order of 270–950 GPa and tensile strength in 11–63 GPa range. Due to these excellent properties, CNTs have huge potential to be used in optical, electrochemical, and biosensors. CNTs need to be purified or functionalized prior to using them as sensors. The design of the sensing interface constitutes the key challenge in sensor development. It needs to take into account both functionalization and transduction steps to enhance the performance of the sensor [17]. CNTs modified with biorecognition elements constitute ideal materials for tailoring nanostructured surfaces for sensor devices. The unique properties of nanotubes and nanoscale materials offer excellent prospects for interfacing biological recognition events with electronic signal transduction [18]. For instance, the surface of nanomaterials can be tailored to a desired use, e.g., to improve the biocompatibility of materials or through the attachment of receptors for targeted-analyte binding or enhanced adhesion to biological structures [19–21]. One-dimensional nanostructures such as CNTs and semiconductors or conducting polymer nanowires are particularly attractive for bioelectronics detection. Because of the high surface-to-volume ratio and novel electron-transport properties, their electronic conductance is strongly influenced by minor surface perturbations. Such 1D material thus offers rapid and real-time detection of chemical species and biological agents [18]. Extreme miniaturization of these nanomaterial-based sensors would allow packaging a huge number of sensors into a small footprint of an array device [18]. As indicated earlier, single-wall carbon nanotubes (SWCNTs) possess a cylindrical nanostructure formed by rolling up a single graphite sheet into a tube. It can thus be viewed as a molecular wire with every atom on the surface [16]. Unique properties of nanoparticles, nanowires, and nanotubes can enhance the performance of electrochemical sensors as well. Nanotechnology-based electrochemical sensors offer several distinct advantages. In particular, such devices provide clear pathways for interfacing at the molecular level, biological recognition, and electronic signal transduction [18]. Such signal transduction pathways involve the binding of signaling molecules to receptors that trigger events inside the cell. The combination of a signaling molecule with a receptor accordingly causes a change in the conformation of the receptor.

4.4.2.3.2 Conducting Polymers

Conducting polymers are organic polymers (for example, polyacetylene) that can conduct electricity and have been studied worldwide for sensors since their discovery in 1977. Conducting polymers are sensitive to environmental conditions and their composites have large potential in sensor applications [21]. Conductive polymers are, in essence, conjugated polymers that can be made to conduct electricity through doping [22,23]. Conjugated polymers are organic macromolecules that are characterized by a backbone chain of alternating double and single bonds. Special conjugation in their chains enables the electrons to delocalize throughout the entire system so that many atoms can share them. The delocalized electrons can move around the whole system and become the charge carriers to make them conductive. A significant amount of research has been carried out in the field of conducting polymers [24]. In a conjugated system, the electrons are loosely bonded and an electron flow is possible. However, as the polymers are covalently bonded, the material needs to be doped for electron flow to occur. Doping is either the addition of electrons or the removal of electrons from the polymer. Once doping has occurred, the electrons in the π-bonds are able to "jump" around the polymer chain and, while electrons are moving along the molecule, an electrical current flows. In addition, as the chemical and physical properties of polymers can be tailored, they gain importance in the construction of sensing devices [25]. These conducting polymers are of great scientific and technological importance because of their unique electrical, electronic, magnetic, and optical properties [26–29]. Nanoscale π-conjugated organic molecules and polymers can be used for biosensors, electrochemical devices, and single-electron transistors, etc. [30–34]. Another advantage of conductive polymers is their processability, mainly through dispersion. Electrical properties can be fine-tuned using the method of organic synthesis and by advanced dispersion techniques. π–electron backbone of conducting polymers is responsible for unusual electronic properties such as high electrical conductivity, low-energy optical transitions, low ionization potential, and high electron affinity. These unique properties are used to enhance speed, sensitivity, and versatility of pH, alcohol, humidity, and biosensors [27].

4.4.2.3.3 Nanosensors for Environment

In recent years, nanosensors have been developed with particular focus in environmental monitoring and analysis. Several nanostructures are used in the development of these sensors: nanoparticles, nanorods, embedded nanostructures, and self-assembled materials [35]. Nanoparticles (NPs) are clusters of a few hundred to a few thousand atoms that are only a few nanometers long. Because of their size, which is of the same order as the de Broglie wavelength associated with valance electrons, nanoparticles behave electronically as zero-dimensional quantum dots (QDs) with discrete energy levels that can be tuned in a controlled way by synthesizing nanoparticles of different diameters. In QDs, the presence or absence of an electron can change the properties in some useful way so that they can be used for information storage or useful transducers in sensors. Nanoparticles also have outstanding size-dependent optical properties that have been used to build optical nanosensors primarily based on noble metal nanoparticles or semiconductor QDs. These QDs, which are nanocrystals of inorganic semiconductors with diameters of 2–8 nm, have been used to develop optical sensors based on fluorescence measurements [36]. The band gap of these semiconductor nanocrystals depends on the size of the nanocrystals. The smaller the nanocrystal, the larger the difference between the energy levels, and therefore, the wider the energy gap and the shorter the wavelength of the fluorescence. QDs have size-tunable fluorescence emission and are highly resistant to photobleaching, thus making them useful for continuously monitoring fluorescence and sensing [36]. Goldman et al. [37] used QDs functionalized with antibodies to perform multiplexed fluoroimmunoassays for simultaneously detecting four toxins. This type of sensors could be used for environmental purposes for simultaneously identifying pathogens in water. Chemical sensing of gases is also critical from an environmental perspective. Using nanoparticle films increases the sensitivity of gas sensors because the surface area of the sensor increases [38]. For example, Baraton et al. [39] used SnO_2 nanoparticles to monitor air quality. The gases were detected through variations in electrical conductivity when reducing or oxidizing gases were absorbed on the semiconductor surface. The gas-detection thresholds of these sensors were 3 ppm for CO, 15 ppb for NO_2 and O_3, and 50 ppb for NO. In other attempts, photoluminescence of cadmium selenide (CdSe) nanocrystals were incorporated into polymer thin films to detect gases such as benzylamine and trimethylamine. The responses were extremely sensitive—only tens of nanocrystals were sufficient for detection [40]. Magnetic nanoparticles have also been used in sensor applications. They can be prepared in the form of superparamagnetic magnetite (Fe_3O_4), greigite (Fe_3S_4), maghemite (γ-Fe_2O_3), and various types of ferrites. Bound to biorecognitive molecules, magnetic nanoparticles can be used to enrich the analyte to be detected. In other words, the sensitivity of the sensors can be substantially improved by using magnetic nanoparticles [41]. Magnetic NP sensors can also be used to detect toxins by functionalizing NPs with antibodies.

Another use of nanotechnology in sensors is in the area of subsea structures. Especially, temperature measurement is a critical parameter for a range of oil and gas production processes. Knowledge of temperature is important for oil and gas pipelines as the parameter can be used to monitor the formation of hydrates [42]. Hydrates are crystalline water-based solids that resemble ice, which can have detrimental effects within an oil flow line, as hydrates can restrict or completely block the flow. New and improved temperature-sensing schemes are therefore vital for subsea oil and gas productions. Recently, it has been demonstrated that temperature has an effect on QD luminescence, opening the field for thermometry applications. QDs have been shown to shift in wavelength with the increase in temperature, ranging from room temperature to 100 and 200°C [43,44]. The highest-reported temperature range using QDs was carried out with QDs encased in a sol–gel environment ranging from 22 to 252°C resulting in a linear red wavelength shift and a linear decrease in luminescence intensity [45]. A reversible wavelength shift was observed due to heat expanding the size of the QD, decreasing the band gap energy, hence less energy is required to excite the QD.

4.4.3 BIOFOULING PROTECTION OF SENSORS

A specific chapter (Chapter 4.3) is dedicated to a novel active sensor antifouling technique for optical windows. In this subsection we only provide a brief review of biofouling and antifouling techniques used in ocean sensors. In ocean observations, data need to be sampled at relatively high frequencies and accuracies for long periods, with minimal drift from predeployment calibrations [46]. Many of the in situ platforms should also be capable of unmanned or autonomous sampling and real-time data telemetry to provide continuous observations. These dedicated observational systems are based on moorings, AUVs, gliders, drifters, profiling floats, and fiber-optic and copper-cabled platforms. The goals of these sensors and platforms are to acquire data for high-frequency, episodic, seasonal, interannual, decadal, and climate-scale phenomena [47]. Study regions also vary widely; it encompasses both open-ocean and coastal settings ranging from equatorial to high-latitude areas. Despite these major accomplishments, biofouling (marine growth) of sensors still limits data quality and effective deployment periods [46]. Biofouling is the accumulation of microorganisms, plants, algae, or animals on wetted surfaces, as seen in Fig. 4.4.3.

The variety among biofouling organisms is highly diverse, and extends far beyond attachment of barnacles and seaweeds. According to some estimates, over 1700 species (a species is often defined as the largest group of organisms comprising over 4000 organisms) are responsible for biofouling [48]. Biofouling is divided into micro- and macrofouling. Microfouling is biofilm formation and bacterial adhesion, whereas macrofouling is the attachment of larger organisms. Biofouling could also be hard or soft depending on the chemistry and type of settling. Hard calcareous fouling includes barnacles, encrusting bryozoans, mollusks, polychetes, and zebra mussels, Examples of soft-fouling organisms are seaweed, hydroids, algae, and biofilm "slime" [49]. Whether micro, macro, soft, or hard, biofouling has long been considered as a limiting factor in ocean monitoring requiring the placement of any materials underwater [50]. Many potential solutions to this problem have been proposed [46]. In the last 20 years or so, innumerable surface buoys or subsurface moorings have been deployed and equipped with sophisticated sensing equipment. Sensors, housing, and support structures—all are undoubtedly subject to fouling problems. Protection of these systems needs adequate knowledge of the chemistry (growth and adhesion) of biofouling and the development of appropriate strategy for protection. Two aspects need to be considered for prevention; one is the protection of the sensor housing, and the other is the protection of the sensing interface. Techniques to protect sensor housing are mainly based on antifouling paints generally used for ship hull protection. Antifouling paints with active biocides such as copper compounds, copper oxides, and cobiocide chemicals are often used for sensor containers [50].

FIGURE 4.4.3   Biofouling in ocean structures.

Another category is self-polishing paints that are effectively used for sensor containers only when water flow is present. Self-polishing paints also contain biocides that can disrupt the environment to be monitored by the sensor [50]. Membrane (bio) fouling is a dynamic process of microbial colonization and growth at the membrane surface [51–53]. Once permanently attached, these organisms start to produce extracellular polymeric secretions (EPS) comprising proteins, glycoproteins, lipoproteins, polysaccharides, and other biomacromolecules. The accumulation of EPS and reproduction of bacteria would lead to the formation of mature biofilm at the surface [51]. To protect membrane surfaces, several approaches can be undertaken such as antiadhesion and antimicrobial approaches. In antimicrobial approaches, antimicrobial polymers, and incorporation of antimicrobial NPs are gaining momentum [51]. Incorporation of biocidal nanoparticles is a promising approach to enhance antimicrobial activity and hence promoting biofouling resistance. It is because inactivated microorganisms would not be able to grow and create an irreversible surface. For instance, silver (Ag) nanoparticles possess antimicrobial properties through several mechanisms. First, released Ag^+ could interact with disulfide or thiol groups of enzymes of DNA and disrupt metabolic processes that generate reactive oxygen species or interrupt replication of DNA leading to the damage or death of bacterial cell. Second, Ag nanoparticles can also be attached to the surface of the cells and disturb their proper function. Finally, nanoparticles with dimension between 1 and 10 nm can penetrate inside the bacteria and cause further damage by interacting with sulfur- and phosphorus-containing compounds [54,55]. One disadvantage with these nanoparticles is that over time the antimicrobial activity declines. Therefore, carbon-based nanomaterials such as CNTs and graphene are introduced as alternatives [56]. They have been demonstrated to inactivate bacteria upon contact with bacterial cells. Graphene, comprising single-atom-thick sheets of sp^2-bonded carbon, has received much recent attention. Graphene oxide–functionalized membranes have shown remarkable performance—by reducing the number of viable *E. coli* cells up to 64.5% after direct contact with graphene oxide [51]. This is possible by inducing membrane damage, physical disruption, charge transfer, formation of reactive oxygen species, and extraction of lipid from the cell membrane [47]. As opposed to sensor housing or platforms, strategies for protection of sensors are somewhat different. Techniques employed for protection can fall into three categories [52]; (1) volumetric, (2) active, and (3) passive. Volumetric action involves a small volume surrounding the sensor area. Active biofouling protection is dependent on energy and can be turned on and off (see Ref. [50] and Chapter 4.3), whereas passive protection is builtin and consequently cannot be turned off [57,58].

4.4.4 CORROSION IN OCEAN OBSERVATIONS AND SENSORS

Exposure to seawater containing high levels of salts, dissolved oxygen, carbon dioxide, and microorganisms usually lead to enhanced corrosion. From a chemical viewpoint, seawater is an aqueous solution of salts. On average, ocean water salinity equals approximately 3.5% [59,60] of water mass. Differences in local salinity are due to water evaporation, which increases salinity levels [61]. A limited water exchange in a given sea area and an inflow of freshwater may also result in a decrease in salinity [59]. Ocean observations and sensor platforms are susceptible to constant corrosive influence of seawater and need protection throughout their deployment. Corrosion rate in seawater is controlled by an oxygen-reduction reaction. In essence, all corrosion reactions are electrochemical in nature. At anodic sites on the surface, the iron goes into solution as ferrous ions and constitutes the anodic reaction. As iron atoms undergo oxidation to ions, they release electrons the negative charge for which would quickly build up in the metal surface. This dissolution will only continue if the electrons released can pass to a site on the metal surface where a cathodic reaction is possible. Accordingly, cathodic protection is used to control corrosion. A simple method of protection connects the metal to be protected to a more easily corroded "sacrificial metal" to act as the anode. Increase in oxygen concentration usually gives an increase in corrosion rate. Also, increased water flow increases oxygen access to the surface and removes protective surface films, enhancing corrosion. The oxygen inflow to the cathode is a function of oxygen concentrations and water flow. Thus, higher corrosion rates are observed in splash zones and in the locations where waving is greater due to increased oxygen migration to the structure's surface [62]. Sensors, including pressure and linear-position sensors are widely used at various depths in offshore platforms, desalination systems, mooring cables, seafloor wellheads, and oil and gas gathering systems that need protection from corrosion [63]. During operation, sensors can be submerged in seawater at varying depths. To operate in these environments, sensors must be constructed from corrosion-resistant materials so that units can provide continuous information under hostile conditions. As demand for sensors in contact with seawater and sea fog increases, specifying sensor materials with proper selection of alloys becomes more and more important. As the depth of seawater changes, oxygen content, temperature, and salinity also changes, and accordingly affects the corrosion rate. Whether corrosion comes from varying seawater depth levels, galvanic effects, or biological attack, matching the proper materials for service application is the top priority for good sensor performance over a long period [63]. Corrosion rates also depend on the combination of location, temperature, and microorganism activities. Stagnant and polluted waters can also trigger corrosion often by sulfate-reducing bacteria [63]. Localized corrosion of stainless steel and other active–passive materials has been observed because of dissolved chlorides and other salts. This type of corrosion occurs in the form of pitting, crevice, or intergranular fracture. High electrical conductivity of seawater promotes macrocell corrosion and increases galvanic corrosion, which accelerates rise in surface temperature further promoting corrosion [63].

4.4.4.1 Corrosion Control

Anticorrosion coating has been successfully used to restrict corrosion damage to metal surfaces. Self-healing coating based on pH-sensitive polyelectrolyte/inhibitor sandwich-like nanostructures have been employed to dramatically reduce the level of corrosion [64]. Standard anticorrosion coatings developed so far passively prevent the interaction of corrosive species with the metal. However, in the next generation of protective coatings, it is important to provide several functionalities with the essential possibility to react on external impact such as pH, humidity changes, or distortion of the coating integrity, so that a self-healing capability is enabled. The multilayer structure of a coating, in which the components are integrated and mutually reactive, is a main point in high corrosion protection. A novel smart multilayer anticorrosion system has been developed that consists of polyelectrolyte and inhibitor layers deposited on aluminum alloy surfaces pretreated by sonication [64]. A very effective solution for the preparation of self-healing anticorrosion coatings is the layer-by-layer (LbL) deposition procedure, which involves the stepwise electrostatic assembly of oppositely charged species (e.g., polyelectrolytes and inhibitors or others: proteins, nanoparticles) on a substrate surface with nanometer-scale precision, and allows the formation of a coating with multiple functionality [65,66]. It has been observed that even the nanometer-thick polyelectrolyte/inhibitor coating can provide good corrosion protection for the aluminum alloy (Fig. 4.4.4). When LbL films consisting of polyelectrolytes and inhibitor are employed, they can provide three mechanisms of corrosion protection. The schematic representation of the corrosion protection of the polyelectrolyte/inhibitor coating is shown in Fig. 4.4.5. The approach to prevention of corrosion propagation on metal surfaces achieving the self-healing effect is based on suppression of the accompanying physicochemical reactions. The corrosion processes are followed by changes of the pH value in the corrosive area and metal degradation [67]. Self-healing or self-curing of the areas damaged by corrosion is performed by three mechanisms: pH neutralization, passivation of the damaged metal surface by inhibitors entrapped between polyelectrolyte layers, and repair of the coating [64]. The polyelectrolyte layers promote pH-buffering activity and can stabilize the pH between 5 and 7.5 at the metal surface in corrosive media. The inhibitors are usually released from polyelectrolyte layers only after the initiation of the corrosion process, directly preventing the corrosion propagation in the rusted area. Polyelectrolyte itself that form the coating is relatively mobile and have a tendency to

FIGURE 4.4.4   Corrosion prevention by polyelectrolyte/inhibitor coating.

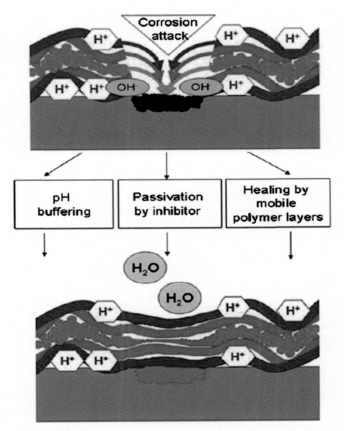

FIGURE 4.4.5   Schematic mechanism of corrosion protection.

seal and eliminate the cracks in the coating. In particular, corrosion processes develop fast as protective barriers fail and a number of reactions cause a change in the composition and properties of both the metal surface and the local environment (e.g., formation of oxides, diffusion of metal cations into the coating matrix, local changes of pH, and electrochemical potential) [64].

In other studies, influence of nanoclay particle modification by polyester–amide hyperbranched polymer on the corrosion protective performance of the epoxy nanocomposites has been investigated [68]. Surface modification of nanoclay particles was carried out by various amounts of polyester-amide hyperbranched polymer (HBP). Nanoclays are nanoparticles of layered mineral silicates. Most common nanoclays used in materials application are montmorillonite. It consists of approximately 1 nm thick aluminosilicate layers surface substituted with metal cations and stacked in about 10μm-sized multilayer stacks. Depending on surface modification of clay layers, montmorillonite can be dispersed in a polymer matrix to form polymer–clay nanocomposites. Within the nanocomposite, individual nm-thick clay layers become fully exfoliated to form platelike nanoparticles with very high aspect ratio. It was observed that by addition of 1.0 wt% of modified clay nanoparticles, corrosion resistance increased significantly as evaluated by electrochemical impedance spectroscopy (EIS). EIS measurements revealed that addition of modified clay nanoparticles to the epoxy coating resulted in less water permeation compared to unmodified clay. To improve compatibility with epoxy coating, it is necessary to reduce the polarity of the clay. The HBP chains adsorption on the surface of clay may restrict the access of water molecules to its surface, which can be attributed to the amphiphilic nature of the HPB chains. This is responsible for the uniform dispersion of platelike shaped clay particles in the epoxy coating matrix. X-ray diffraction (XRD) analyses have shown that surface modification of clay particles with HBP caused an increase of interlayer distance (Fig. 4.4.6). The clay particles dispersion in the coating matrix is both perpendicular and parallel to the substrate. However, for the modified clay particles, a more uniform distribution of the particles in the coating body mostly parallel to the substrate was observed [69]. This effectively increased the length of diffusion pathways for corrosive electrolyte. In addition, modified clay particles can occupy the free volumes, i.e., voids and defects, which are preferential diffusion paths for water molecules, ions, and oxygen—thus essentially reducing the corrosion [68].

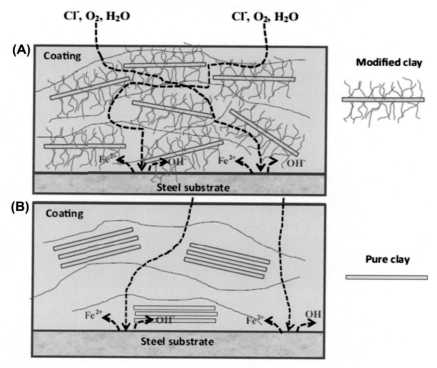

FIGURE 4.4.6 Schematic illustration of (A) exfoliated, and (B) intercalated clay nanoparticles in epoxy coating.

4.4.5 CONCLUSION

Material advances for ocean observation and sensors follow the pattern observed in other industries. Composites and nanostructured materials have taken a giant leap forward in the new millennium. The reason being their light-weight, functionality, and design flexibility at their nascent stage. Fibrous composites are leading the charge in the case of large platforms, housing, and coastal structures. Fundamental advantage of polymeric fibrous composites is their resistance to corrosion and biofouling in marine environments. Although constitutive relationships are formidable due to high degree of anisotropy, design approaches using finite-element codes are becoming more and more familiar. Sufficient amount of data for composites under marine environment and under cyclic loading is very limited. But efforts are underway to produce them. Life prediction under hostile ocean environments is also an issue. Such prediction requires accurate assessment of load. Hydrodynamic loading and environmental conditions in the ocean are affected by wind, wind-driven currents, tides, eddies, turbulence, and other seasonal variations. Most often, probabilistic approaches are taken to simulate those loads that are not quite simple. In other words, material development is one thing but deploying new materials into structures or sensors is another. Sufficient materials testing and characterization data must be available before these novel materials can be used.

Compared to fibrous composites, nanostructured materials and nanocomposites are more recent and their development is following a somewhat different path. As far as sensors are concerned, nanotechnology has had its biggest impact. Both miniaturization and sensing capability have been significantly enhanced by nanostructured materials. A large number of atoms at the surface of a nanoparticle, and ease of attaching chemical moiety at the nanoparticle surface, are the main advantages of nanotechnology-based sensors. They are more precise and dramatically shorten the response time. In sum, both composites and nanocomposites will continue to dominate material advances to satisfy challenging applications in ocean structures and sensors.

References

[1] http://ceoas.oregonstate.edu/ooi/.
[2] http://oceanobservatories.org/marine-technologies/robotic-auvs/.
[3] Houston (Texas): American Bureau of Shipping (ABS).
[4] Baerum, Akershus (Norway): Det Norske Veritas (Norway) and Germanischer Lloyd (Germany).
[5] West Conshohocken (PA): American Society for Testing Materials (ASTM).
[6] Marine composites. 2nd ed. Annapolis (Maryland): Published by Eric Greene Associates, Inc. ISBN: 0-9673692-0-7.
[7] Mallick PK. Fiber-reinforced composites. 2nd ed. New York: Marcel Dekkar, Inc.
[8] Kim YH, Jo YD, Sin SJ. Material design of Al/CFRP hybrid composites for the hull of autonomous underwater vehicle. In: Oceans 2010, IEEE – Sydney. October 2010. https://doi.org/10.1109/OCEANSSYD.2010.5603587.
[9] http://www.compositesworld.com/articles/composites-the-right-choice-for-subsea-ocean-observatory.
[10] Davies P, Chauchot P, Composites for Marine Applications, part 2 Underwater Structures. In: Mechanics of Composite Materials and Structures. In: NATO Science Series, vol. 361, pp. 249–60.
[11] Davies P. Behavior of marine composites under deep submergence. In: Marine applications of advanced fiber-reinforced composites. https://doi.org/10.1016/B978-1-7842-250-1.
[12] Mills G, Fones G. A review of in situ methods and sensors for monitoring the marine environment. Sens Rev 2012;32(1):17–28. https://doi.org/10.1108/02602281211197116.
[13] Kalantar-Zadeh K, Fry B. Nanotechnology-enabled sensors. Springer Science; 2008. ISBN: 978-0-387-32473-9.
[14] Davis CM. Studies of nanoparticle reinforced polymer coating for trace gas detection. [MS thesis]. Boca Raton (FL): Florida Atlantic University; December 2013.
[15] Lazarus N, Jin R, Fedder GK. Nanosensors for chemical and biological applications. 2014. p. 231–53. https://doi.org/10.1533/9780857096722.2.2.231.
[16] Boujitita M. Chemical and biological sensing with carbon nanotubes (CNT). In: Honeychurch KC, editor. Nanosensosrs for chemical and biological applications. 1st ed. Woodhead Publishing; 2014. ISBN: 9780857096722.
[17] Wang J. Nanomaterials-based electrochemical biosensors. Analyst 2005;130:421–6.
[18] Rauch J, Kolch W, Laurent S, Mahmoudi M. Big signals from small particles: regulation of cell signaling pathways by nanoparticles. Chem Rev 2013;113:3391–406.
[19] rao CNR, Satishkumar BC, Givindaraj A, Manashi Nath. Nanotubes. ChemPhysChem 2001;2:78–105.
[20] Deshpande M, Patil P, Conducting polymer nanocomposites for sensor applications. In: Conducting Polymer Hybrids. https://doi.org/10.1007/978-3-319-46458-9_8.
[21] Rajesh, Ahuja T, Kumar D. Recent progress in the development of nano-structured conducting polymers/nanocomposites for sensor applications. Sens Actuators B 2009;136:275–86.
[22] Shirikawa H, Louis EJ, MacDiarmid AG, Chiang CK, Heeger AJ. Synthesis of electrically conducting organic polymers: halogen derivatives of polyacetylene (CH)x. J Chem Soc Chem Commun 1977:578–80.
[23] Skotheim TJ, Elsenbaumer RL, Reynolds JR. Handbook of conducting polymer. 2nd ed. New York: Marcel Dekker; 1998.
[24] Adhikari B, Majumdar S. Polymers in sensor applications. Prog Polym Sci 2004;29:699–766.

[25] Retama RJ. Synthesis and characterization of semi conducting polypyrrole/polyacrylamide micro particles with GOx for biosensor application. Colloids Surf A Phys Chem Eng Asp 2005;270–271:239–44.

[26] Lange U, Roznyatovskaya NV, Mirsky VM. Conducting polymers in chemical sensors and arrays. Anal Chim Acta 2008;614:1–26.

[27] Gupta N, Sharma S, Mir IA, Kumar D. Advances in sensors based on conducting polymers. J Sci Ind Res 2006;65:549–57.

[28] Cosnier S. Affinity biosensors. Electroanalysis 2005;17:1701–15.

[29] Alsana S, Gabrielli C, Perrot H. Influence of antibody insertion on the electrochemical behavior of polypyrrole films by using fast QCM measurements. J Electrochem Soc 2003;150:E444–9.

[30] Kim BH, Kim MS, Kang KT, Lee JK, Park DH, Joo J, Yu SG, Lee SH. Characteristics and field emission of conducting poly (3,4-ethylenedioxythiophene) nanowires. Appl Phys Lett 2003;83:539–41.

[31] Huang J. Syntheses and applications of conducting polymer polyaniline nanofibres. Pure Appl Chem 2006;78:15–27.

[32] Hatchett DW, Josowicz M. Composites of intrinsically conducting polymers as sensing nanomaterials. Chem Rev 2008;108:746–69.

[33] Malinauskas A, Malinauskiene J, Ramanavicius A. Conducting polymerbased nanostructurized materials: electrochemical aspects. Nanotechnology 2005;16:R51–62.

[34] Naarmann H. Polymers, electrically conducting. In: Ullmann's encyclopedia of industrial chemistry. 2000. ISBN: 3527306730. https://doi.org/10.1002/14356007.a21_429.

[35] Riu J, Alicia Maroto, Xavier Rius F. Nanosensors in environmental analysis. Talanta 2006;69:288–301.

[36] Smith AM, Nie S. Chemical analysis and cellular imaging with quantum dots. Analyst 2004;129:672.

[37] Goldman ER, Clapp AR, Anderson GP, Uyeda HT, Mauro JM, Medintz IL, Mattoussi H. Multiplexed toxin analysis using four colors of quantum dot fluororeagents. Anal Chem 2004;76:684.

[38] Hoel A, Reyes LF, Heszler P, Lantto V, Granqvist CG. Nanomaterials for environmental applications Wo$_3$-based gas sensors made by advanced gas deposition. Curr Appl Phys 2004;4:547.

[39] Baraton MI, Merhari L. J Nanopart Res 2004;6:107.

[40] Nazzal AY, Qu L, Peng X, Xiao M. Photoactivated CdSe nanocrystals as nanosensors for gases. Nano Lett 2003;3:819.

[41] Jianrong C, Yuqing M, Nongyue H, Xiaohua W, Sijiao L. Biotechnol Adv 2004;22:505–18.

[42] McDowell GR, Holmes-Smith AS, Uttamlal M, Wallace PA, Faichnie DM, Graham A, McStay D. The use of nanotechnology in the development of a distributed fiber-optic temperature sensor for subsea applications. J Phys Conf Ser 2013;450:012008. https://doi.org/10.1088/1742-6596/450/1/012008. Sensors & their Applications XVII.

[43] De Bastida G, Arregui FJ, Goicoechea J, Matias IR. Quantum dots-based optical fiber temperature sensors fabricated by layer-by-layer. IEEE Sens J 2006;6:1378–9.

[44] Chen KJ, Chen HC, Shih MH, Wang CH, Kuo MY, Yang YC, Lin CC, Kuo HC. The influence of the thermal effect on CdSe/ZnS quantum dots in light-emitting diodes. J Lightwave Tech 2012;30:2256–61.

[45] Pugh-Thomas D, Walsh BM, Gupta MC. CdSe(ZnS) nanocomposite luminescent high temperature sensor. Nanotechnology 2011;22:1–7.

[46] Manov Derek V, Chang Grace C, Dickey TD. Methods for reducing biofouling of moored optical sensors. J Atmos Ocean Technol 2003Vol. 21:958–68.

[47] Dickey T. The emergence of concurrent high-resolution physical and bio-optical measurements in the upper ocean and their applications. Rev Geophys 1991;29:383–413.

[48] Almeida E, Diamentino TC, De Sousa O. Marine Paints: the particular case of antifouling paints. Prog Org Coat 2007;59(1):2–20.

[49] Marianne S. Biofouling: it's not just barnacles anymore. ProQuest; May 21, 2012.

[50] Delauney L, Compere C, Lehaitre M. Biofouling protection for marine environmental sensors. Ocean Sci 2010;6:503–11.

[51] Misdan N, Ismael AF, Hilal N. Recent advances in the development of (bio) fouling resistant thin film composite membranes for deslination. Desalination 2016;380:105–11.

[52] Baker JS, Dudley LY. Biofouling in membrane systems: a review. Desalination 1998;118:81–9.

[53] Kochkodan VM, Sharma VK. Graft polymerization and plasma treatment of polymer membranes for fouling reduction: a review. J Environ Sci Health Part A 2012;47:1713–27.

[54] Rahaman MS, Therien-Aubin H, Ben-Sasson M, Nielsen M, Elimelech M. Control of Biofouling on reverse osmosis polyamide membranes modified with biocidal nanoparticles and antifouling polymer brushes. J Mater Chem B 2014;2:1724–32.

[55] Yin J, Yang Y, Hu Z, Deng B. Attachment of silver nanoparticles (AgNPs) onto thin film composite (TFC) membranes through covalent bonding to reduce membrane biofouling. J Membr Sci 2013;441:73–82.

[56] Perreault F, Tousley ME, Elimelech M. "Thin-film composite polyamide membranes functionalized with biocidal graphene oxide nanosheets. Environ Sci Technol Lett 2013;1:71–6.

[57] Mejias Carpio IE, Santos CM, Wei X, Rodrigues DF. Toxicity of a polymer–graphene oxide composite against bacterial planktonic cells, biofilms, and mammalian cells. Nanoscale 2012;4:4746–56.

[58] Lehaitre M, Delauney L, Compere C. "Real-time coastal observing systems for marine ecosystem dynamics and harmful algal blooms," Chapter 12, biofouling and underwater measurements. UNESCO; 2008. p. 463.

[59] Zakowski K, Naroznv M, Szocinski M, Darowicki K. Influence of water salinity on corrosion risk – the case of the southern Baltic sea coast. Environ Monit Assess 2014;186(8):4871–9.

[60] Williams PD, et al. The role of mean ocean salinity in climate. Dyn Atmos Oceans 2010;49:108–23. https://doi.org/10.1016/j.dynatmoce.2009.02.001.

[61] Da-Allada CY, et al. Seasonal mixed-layer salinity balance in the tropical Atlantic Ocean: mean state and seasonal cycle. J Geophys Res Oceans 2013;118(1):332–45.

[62] Al-Fozan SA, Malik AU. Effect of seawater level on corrosion behavior of different alloys. Desalination 2008;228(1–3):61–7. https://doi.org/10.1016/j.desal.2007.08.007.

[63] https://www.sea-technology.com/features/2013/0313/2_material_selection.php.

[64] Daria V, Andreeva, Fix D, Mohwald H, Shchukin DG. Self-healing anticorrosion coating based on pH-sensitive polyelectrolyte/inhibitor sandwichlike nanostructures. Adv Mater 2008;20:2789–94.

[65] Schlenoff JB. Retrospective on the future of polyelectrolyte multilayers. Langmuir 2009;25(24):14007–10.

[66] Shiratori SS, Rubner MF. Ph-dependent thickness behavior of sequentially adsorbed layers of weak polyelectrolytes. Macromolecules 2000;33(11):4213–9.

[67] Zheludkevich MI, Shchukin DG, Yasakau KA, Mohwald H, Ferreira MGS. Anticorrosion coatings with self-healing effect based on nanocontainers impregnated with corrosion inhibitor. Chem Mater 2007;19(3):402–11.

[68] Potvin E, Brossard L, Larochelle G. Corrosion protective performances of commercial low-VOC epoxy/urethane coatings on hot-rolled 1010 mild steel. Prog Org Coat 1997;31:363–73.

[69] Dong Y, Ma L, Zhou Q. Effect of the incorporation of montmorillonite-layered double hydroxide nanoclays on the corrosion protection of epoxy coatings. J Coat Technol Res 2013;10(6):909–21.

CHAPTER

5

Innovative Sensor Carriers for Cost-Effective Global Ocean Sampling

CHAPTER

5.1

Maturing Glider Technology Providing a Modular Platform Capable of Mapping Ecosystems in the Ocean

Oscar Schofield[1], David Aragon[1], Clayton Jones[2], Josh Kohut[1], Hugh Roarty[1], Grace Saba[1], Xu Yi[1,3], Scott Glenn[1]

[1]Center of Ocean Observing Leadership, Department of Marine and Coastal Sciences, School of Environmental and Biological Sciences, Rutgers University, New Brunswick, NJ, United States; [2]Teledyne Webb Research, Falmouth, MA, United States; [3]IMBER Regional Project Office, State Key Laboratory of Estuarine and Coastal Research, East China Normal University, Shanghai, China

OUTLINE

5.1.1 THE NEED

Understanding of the oceans' physical–chemical–biological variability, their respective feedbacks, and the responses to anthropogenic forcing (climate change, resource extraction and utilization, waste production, and nutrient pollution), remains a fundamental challenge for oceanography. Our lack of understanding limits our ability to predict future trajectories of the ocean in the face of accelerating change. To meet this challenge, oceanographers need to maintain a sustained presence in the ocean to measure not only the mean state of the system but also the high-frequency changes that can have disproportionately large effects on physics, biology, and chemistry. This requires a range of new approaches for platforms that are capable of carrying a range of sensors as ships cannot provide the needed cost–effective sustained presence that is required.

Fortunately, oceanographic technologies have been undergoing a technical revolution as autonomous underwater vehicles (AUVs) have matured and are becoming reliable tools to collect data for sustained periods of time. The range of AUVs is maturing rapidly and all of them are tuned to sample specific time and space domains [1]. One particular useful technology is the underwater buoyancy-driven glider. These systems are optimized to collect data over regional scales (100s–1000s of kilometers) and maintain a long-term mobile adaptive sampling presence in the subsurface ocean over ecologically relevant scales. A range of buoyancy gliders [2–5] currently exists and have strong track records for studying a wide range of science topics [6]. All the major gliders used today (Seaglider, Spray, Slocum, and Sea Explorer) have demonstrated great potential; however, for this chapter we will focus on the Slocum glider, which is the system that we have the greatest experience operating [5,7]. This chapter will provide an overview of glider technology, our experience using the platform, and the current range of sensors they have been demonstrated to successfully carry.

5.1.2 GLIDER TECHNOLOGY

Buoyancy gliders are increasingly filling mesoscale sampling needs for ocean science. Slocum gliders (Fig. 5.1.1) maneuver through the ocean at a forward speed of 20–30 cm/s in a vertically undulating sawtooth motion, deriving their forward propulsion by means of a buoyancy change. The change in buoyancy is driven by sucking in a small amount of seawater in the glider's front end via an onboard piston pump. Pitch is regulated by shifting batteries back and forth within the glider. For sinking, the glider's front-end batteries are pitched forward as the piston pulls in the seawater making the nose heavy and the platform begins to sink. When the glider moves into floating mode, it pushes the water out of its front end and pulls the batteries back from the front end. Steering is by means of a tail fin rudder. An altimeter and depth sensor enable preprogrammed sampling of the full water column. Depth ranges of the systems are from greater than 5 m to typically 1500 m. The primary vehicle navigation system uses an

FIGURE 5.1.1 A picture of a Slocum glider at the surface. The tail is raised above the water line to initiate an Iridium call with a shore-based laboratory. The *dark circles* on the top middle section of the glider are sensors that measure optical backscatter and chlorophyll fluorescence.

onboard Global Positioning System receiver coupled with an attitude sensor, depth sensor, and altimeter to provide dead–reckoned navigation, with backup positioning and communications provided by an Argos transmitter. Two-way communication with the vehicle is maintained by radiofrequency modem or the global satellite phone service Iridium. Iridium is the primary mode of communication given its global range. Currently, Iridium maximum capacity is on the order of 2400 bits/s; however, often communication speed is less for gliders (1200 bits/s). This is likely to improve in the coming years with the launch of Iridium NEXT, which is going to provide transmission speeds as much as 50 times faster than the current systems. This would decrease a glider's surfacing time, typically now ~15 min, to about 20 s. Our team typically assumes monthly costs of Iridium during a mission at about $1000. It should be noted that this can be adjusted as an operator can decide to limit data transmission from the glider to the shore if desired. Communication allows for shore–based operators to control and navigate the gliders by remotely adjusting the mission as needed based on either recently collected subsurface ocean data from the glider or satellite data. Ocean model forecasts are increasingly being used by glider operators as tools to optimize flight planning. Generally, the glider operator during a mission continuously updates glider "waypoints" of where the glider should transit over the next few days. Global communications allow for this technology to be scalable with a single shore team able to coordinate activities of many gliders simultaneously.

Gliders have transitioned from experimental platforms to becoming critical science tools since the early 2000s. While initially the number of groups flying gliders was small, the number of labs now flying gliders has grown dramatically over the last decade. The statistics of the success of individual groups have been documented [8,9], which show a range of successes. While some [8] suggest a mission success rate of ~50%, others [9] show much higher success rates. Given this range, we highlight our personal experience in operating gliders. Our team began to operate gliders in 2003 for science missions without the support of large engineering teams. Since that time, we have conducted 406 missions, which have mapped ocean properties over 170,000 km during 9000 days at sea around the world when this chapter was written (Fig. 5.1.2). During that time, as numerous new glider groups have formed, we have been frequently asked to assist them in glider preparation, piloting a glider during the mission, and/or providing a glider data-mirroring role (Fig. 5.1.3A). Combining these various activities, our group has seen missions conducted per year increase from five in 2003 to about 30 over the last 5 years. These annual efforts represent a significant number of days at sea (Fig. 5.1.3B). A major advance over the decade of operations was the improvement in lithium batteries that now allow gliders to be operated continuously for up to a year and allow for long transocean missions (Fig. 5.1.1). This is in contrast to past alkaline batteries that can last typically for 1.5 months depending on the sensors that were carried by the glider. Generally, the lithium batteries are purchased through the glider vendor. As battery life has allowed for extended missions, other issues become larger factors for determining mission success. Biofouling can significantly affect glider mobility and it becomes critical to apply antibiofouling coatings, as well as utilize antifouling flight behaviors, which include minimizing time at the surface where the concentration of organisms is highest. It also can have significant negative impacts on sensor performance. These issues combined with designing the appropriate duty cycling of sensors are now major challenges for conducting long missions.

We are often asked how big does a glider team have to be? This is really a function of how many missions the group is expecting to conduct in a given year. Our current team consists of 2.5 glider technicians, which is then augmented by less experienced faculty, graduate, and undergraduate students to maintain our pace of ~30 missions/year. Having one to two experienced glider specialists directing less experienced workers is a major means to cut the overall costs of the operations. For nonuniversity operators, the team will need at minimum one experienced glider technician who could potentially support upward of five to seven missions simultaneously. This again is influenced by the type of mission being conducted. For example, a nearshore shallow mission often requires more oversite, which results in a more frequent call-in schedule (typically every few hours) compared to an open ocean glider that

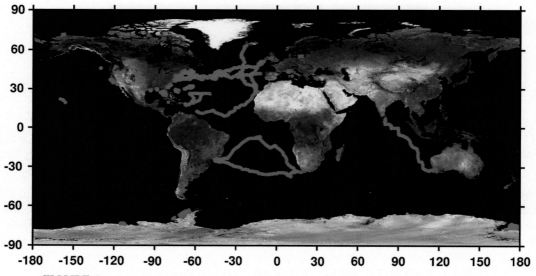

FIGURE 5.1.2   The *red marks* indicate glider missions conducted by the Rutgers glider team.

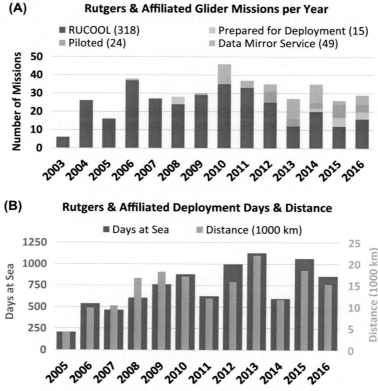

FIGURE 5.1.3   The time series of Rutgers glider operations. (A) The number of missions conducted annually. (B) Distance and number of days at sea by Rutgers gliders.

might only call the shore side lab twice a day. Table 5.1.1 is the overall success rate of our team, using the metrics specified by [9], where we were in control of aspects of the mission (preparation, piloting, recovery). Overall, our success rate was 84% (Table 5.1.1). This success rate mirrors those reported by Ref. [9]. These statistics demonstrate that gliders are a robust technology for supporting ocean science. In fact, we find that glider technology is evolving (currently, the third generation Webb glider is available) more rapidly than the turnover time of the gliders and our team still operates our first glider purchased almost 15 years ago. Over the hundreds of missions, our team has

TABLE 5.1.1 Glider Statistics by the Rutgers Glider Team

Standardized[a] Rutgers Glider Statistics		
Total Rutgers-only missions	330	Success rate **84%**
Short missions excluded	58	Loss rate **2.7%**
Total missions considered	272	
Significant problems	43	
Total successful missions	229	
Number of losses	9	

[a]*as per Ref. [9], Spray Underwater Glider Operations, American Meteorological Society.*

had nine glider losses since 2003. The majority of those losses were associated with external factors (storms, sharks, ships) and not associated with the glider itself. For those losses, only one can be related to a mechanical failure of the system. Additionally, four of the losses were during the early years of the gliders representing the transition from an engineering experiment to a science tool. Currently, losses are largely associated external factors. Storms can be problematic as they induce currents faster than the glider flight speed and this can result in pushing gliders into the shore resulting in damage. Gliders are occasionally attacked by sharks almost always during ascent, which is not surprising given the underneath profiles with the wings and slow ascent speeds. Luckily, the majority of the attacks are not catastrophic and while the hull is nicked, the glider often lives to fly another day.

Data flow for gliders is rapidly maturing. Gliders, especially during coastal missions, can collect thousands of profiles (>5000/month) depending on the mission profile, therefore one of the next community needs will be developing community tools for efficient standardized processing and visualization tools. There is currently a transition occurring from where data only reside in an individual database to a shared data environment. Increasingly, data are beginning to flow to a centralized glider US Data Assembly Center (DAC) that is continuously growing through National Oceanic and Atmospheric Administration support. The marine glider DAC page is at https://gliders.ioos.us/data/. Another useful page allows for data searches and can be accessed at data.ioos.us/gliders/erddap/search/index.html?page=1&itemsPerPage=1000. The DAC provides a repository where glider data collected by the operator can be transmitted allowing it to be visible to the wider community. The DAC is available to accept national and international data streams. Beyond collating many glider missions, it allows glider data to flow directly to the Global Telecommunication System supporting global weather and ocean models.

5.1.3 WHAT SENSORS CAN GLIDERS CARRY?

The utility of glider technology is increasing as the number of sensors integrated into these platforms has increased. This in the past has been limited by the size and energy consumption of the sensor as gliders are optimized for long duration with slow horizontal speeds (~20 km/day). Fortunately, there are accelerating improvements in instrument miniaturization and more low-power sensors. The increased availability of high-density energy batteries has increased the power availability over traditional alkaline batteries. It will be critical for any researcher to decide the length of the mission to be balanced against the science needs, given variable sensor energy requirements. This will require groups to focus mission strategies on what sensors should be incorporated into any glider experiment. Later we outline the current sensor capabilities presently deployed on gliders and map these capabilities onto a simplified food web perspective that is typical for many regional and global models (Fig. 5.1.4). We focus discussion on sensors now routinely carried by gliders given that development efforts under way are beyond the scope of this chapter. We refer the reader to Sections 2, 3, 4, and 6 of this book for examples of recent sensor innovations for biology and biogeochemistry, or new integration schemes, some of which have already been tested on power- and size-limited autonomous platforms.

The most complete routine set of measurements are for physical properties (temperature, salinity, and currents). Beyond mapping the hydrography, it is possible to estimate physical transport through geostrophic currents from a density section and/or directly measuring with a mounted acoustic Doppler current profiler. Turbulence, as estimated through shear probes, has also been measured on a glider. Optical measurements available today can be decomposed into the inherent (absorption, attenuation, and backscatter) and the apparent (downwelling irradiance,

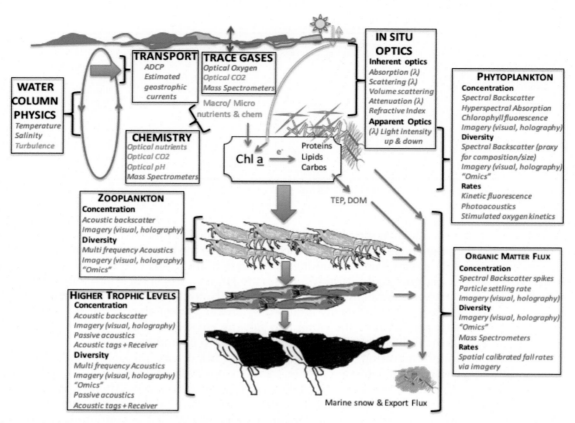

FIGURE 5.1.4   A simplified marine food web with energy flowing from the sun through the primary producers to secondary and higher trophic levels. Breakout boxes listing the relevant measurements to specific parts of the food web are indicated. The level of readiness is indicated by the color of the text within the break--out box with green font indicating is currently a routine measurement made by gliders today, orange indicating a measurement that has been made on higher powered Remotely Operated Vehicles (*ROV*) or propellered Autonomous Underwater Vehicles (*AUV*) or has demonstrated on glider rarely, and red indicates a parameter has not been measured on a glider. *Figure modified from Schofield O, Jones C, Kohut J, Kremer U, Miles T, Saba G, Webb D, Glenn S. Building a coordinated community fleet of autonomous gliders for sampling coastal systems. Mar Technol Soc 2015;49(3):9–16.*

upwelling radiance) optical measurements. An advantage of the inherent optical properties is that they can be measured at night and at depth when there is no light present. Radiative transfer models are then used to link the inherent and apparent optical properties to derive optical parameters of interest that are not measured [10]. These measurements are often available as either single wavelength, multispectral, or hyperspectral sensors.

For in situ gas and chemistry a range of sensors is rapidly evolving. Optically based estimates of oxygen are now common and provide a stable means for collecting data over the long duration of a glider's deployment. Approaches for nitrate, carbon dioxide, and pH also often employ optically based approaches and groups are currently beginning to incorporate these systems into gliders. As some of these sensors are now becoming standard on Bio–Argo profiling floats [11], they are very amenable for glider technology with a small spatial footprint on the glider and low power usage.

Phytoplankton are the base of the marine food web and for most of the ocean are the dominant optical constituent, excluding water molecules, influencing in situ optical spectral properties. Not surprisingly, optical measurements (spectral backscatter and/or absorption) can be used as a proxy for phytoplankton biomass. Another standard approach is by measuring chlorophyll-*a* fluorescence to estimate phytoplankton biomass. However, an easy measurement can be difficult to interpret as the chlorophyll*a* fluorescence is sensitive to physiological status of the phytoplankton (cf. [12]). Deriving estimates of the composition of the phytoplankton is more problematic with currently the most effective demonstrated means being the use of spectral absorption signatures associated with specific phytoplankton ([13], see also Chapter 3.2 in this book). Ultimately, better estimates of phytoplankton diversity will require new sensors. Beyond making measurements of the biomass and community composition, it will be critical to develop the means to assess the physiological status of phytoplankton populations as this influences the growth rates of the phytoplankton and higher elements of the food chain (see Use Case 1 in Chapter 8.2).

Measurements capable of characterizing secondary and higher trophic-level activity are beginning to be incorporated into gliders. For zooplankton, acoustic backscatter from echosounders is being used and can provide a general proxy for biomass; however, the identification of specific zooplankton is difficult. Multifrequency acoustics is closer to reality and could provide some broad characterization of the secondary and higher trophic levels. Recent demonstrations of deployments show this technology is now available to gliders and should become a standard package in the near future. Higher trophic levels (fish, seals, birds, and whales) offer the possibility of carrying acoustic tags, which allow the specific individuals to be identified by a glider outfitted with a listening device [14]. Finally, passive acoustical recording of the in situ soundscape is offering a new means to characterize ocean activity by identifying species or groups of species depending on the contextual information available. Chapter 4.1 reports on a newly developed digital hydrophone of small form factor with open embedded processing capability, currently under test on gliders in the European project NeXOS.

5.1.4 CONCLUSIONS AND FUTURE DIRECTIONS

Gliders have been demonstrated as a mature platform for conducting oceanographic research. As the platform is now robust, efforts are now rightfully focused on increasing the number of sensors capable of being carried by a glider.

Moving forward there is a critical need to continue the development of new sensors to allow the platform to address a wider array of questions. Improving the ability to accelerate the incorporation of the new sensors into the glider platform would have great benefit for the community. Incorporation of any sensor into a glider results in an immediate new "sink" of power, which ultimately limits the lifetime of a mission. Therefore increasing energy efficiency is absolutely critical. Tools to allow for increased efficiency and assist in mission planning are becoming critical as the newer instrument suites tend to have high-power requirements.

Finally, as the number of glider operators grows, the distributed communities should coordinate activity whenever possible. Increasing the utility of the gliders will be dependent on growing this community to provide coherent sampling for national and international ocean-observing systems. As these distributed networks of operators and diverse sensors increasingly become available, tools for optimizing mission planning will increasingly become important, therefore we recommend that efforts should also be focused on developing mission-planning tools. The ability to maintain a cost-effective sustained presence in the ocean will enable critical and exciting science.

Acknowledgments

This work was not possible without the dedicated members of COOL. We thank the large team of technicians, collaborators, and graduate and undergraduate students who have been critical for this work. Glider missions over the last two decades have been due to funding provided by the ONR, Gordon and Betty Moore Foundation, EPA, NASA, NJ DEP, NSF, Vetelsen Foundation, and NOAA.

References

[1] Moline MA, Schofield O. Remote video assisted docking of untended underwater vehicles. J Atmos Ocean Technol 2009. https://doi.org/10.1175/2009JTECHO666.1.

[2] Davis RE, Eriksen CC, Jones CP. In: Griffiths G, editor. Autonomous buoyancy-driven underwater gliders. Technology and applications of autonomous underwater vehicles. Taylor and Francis; 2003. p. 37–58.

[3] Eriksen CC, Osse TJ, Light RD, Wen T, Lehman TW, Sabin PL, Ballard JW, Chiodi AM. Seaglider: a long-range autonomous underwater vehicle for oceanographic research. IEEE J Ocean Eng 2001;26:424–36. https://doi.org/10.1109/48.972073.

[4] Sherman J, Davis RE, Owens WB, Valdes J. The autonomous underwater glider "Spray". IEEE J Ocean Eng 2001;26:437–46. https://doi.org/10.1109/48.972076.

[5] Webb DC, Simonetti PJ, Jones CP. SLOCUM: an underwater glider propelled by environmental energy. IEEE J Ocean Eng 2001;26:447–52. https://doi.org/10.1109/48.972077.

[6] Rudnick DL. Ocean research enabled by underwater gliders. Ann Rev Mar Sci 2015;8:1–23.

[7] Schofield O, Kohut J, Aragon D, Creed E, Graver J, Haldeman C, Kerfoot J, Roarty H, Jones C, Webb D, Glenn SM. Slocum gliders: robust and ready. J Field Robot 2007;24(6):473–85. https://doi.org/10.1009/rob.20200.

[8] Brito M, Smeed D, Griffiths G. Underwater glider reliability and implications for survey design. J Atmos Ocean Technol 2014;31:2858–70. https://doi.org/10.1175/JTECH-13-00138.1.

[9] Rudnick DL, Davis RE, Sherman JT. Spray underwater glider operations. J Atmos Ocean Technol 2016;33:1113–22. https://doi.org/10.1175/JTECH-D-15-0252.1.

[10] Mobley CD. Light and water: radiative transfer in natural waters. Academic Press; 1994. 442 pp.

[11] Johnson KS, Berelson WM, Boss ES, Chase Z, Claustre H, Emerson SR, Gruber N, örtzinger A, Perry MJ, Riser SC. Observing biogeochemical cycles at global scales with profiling floats and gliders: prospects for a global array. Oceanography 2009;22(3):216–25.

[12] Dubinsky Z, Schofield O. Photosynthesis under extreme low and high light in the world's oceans. Hydrobiologia 2009. https://doi.org/10.1007/s10750-009-0026-0.

[13] Kirkpatrick G, Millie DF, Moline MA, Schofield O. Absorption-based discrimination of phytoplankton species in naturally mixed populations. Limnol Oceanogr 2000;42:467–71.

[14] Oliver MJ, Breece MW, Fox DA, Haulsee DE, Kohut JT, Manderson J, Savoym T. Shrinking the haystack: using an AUV in an integrated ocean observatory to map Atlantic sturgeon in the coastal ocean. Fisheries 2013;38(5):210–6.

[15] Schofield O, Jones C, Kohut J, Kremer U, Miles T, Saba G, Webb D, Glenn S. Building a coordinated community fleet of autonomous gliders for sampling coastal systems. Mar Technol Soc 2015;49(3):9–16.

C H A P T E R

5.2

Sensor Systems for an Ecosystem Approach to Fisheries

Emilie Leblond, Patrice Woerther, Mathieu Woillez, Loïc Quemener

Ifremer - Centre de Bretagne, Plouzané, France

5.2.1 INTRODUCTION

Despite several countries' efforts devoted to the implementation of Fisheries Information Systems in recent years, especially in relation to the EU Data Collection Framework (https://datacollection.jrc.ec.europa.eu/), the lack of reliable data to accurately assess catches and fishing effort remains undeniable. The evaluation of both variables, along with their spatial distribution, is fundamental to assess the states of exploited resources and to make a diagnosis of fisheries. Data currently available for French fisheries come mainly from fishermen's declarations (log books), at the scale of the International Council for the Exploration of the Sea statistical rectangles (30 min latitude, 1 degree longitude). This scale is inadequate for most research studies and a fine analysis of the fishing sector. Moreover, the coverage of these data is sometimes partial and their reliability sometimes hard to appreciate.

In addition, the local environmental conditions and their variability, especially on the continental shelf, are often undersampled, especially because of specific conditions: low depth, significant current (especially tidal current), and concurrent human activities (professional and recreational) making measurement devices vulnerable. Thus even for basic parameters such as temperature or salinity, most of the available measurements are limited to oceanographic campaigns.

Faced with this lack of data, especially in areas that are precisely fishing sectors, Ifremer has been implementing since 2005 a new network called RECOPESCA, consisting in fitting out a sample of voluntary fishing vessels with sensors recording data on fishing effort and physical parameters such as temperature or salinity. RECOPESCA aims at setting up a network of sensors for scientific purposes to collect data to improve resources assessment and diagnostics on fisheries and environmental data required for an EAF or to feed oceanographic models, e.g., for circulation of water masses. Specific sensors are implemented on the fishing gears and aboard a sample of the vessels.

RECOPESCA is a project of national scale, including the outermost regions (such as Martinique, Balearic Islands, etc.), and is a concrete achievement of a participative approach: scientists and fishermen team up to give to voluntary fishermen the role of scientific observer. RECOPESCA provides an innovative tool to collect data, especially through integrated multidisciplinarity. The collected data can be used by both fisheries scientists and physicists, who will have information on areas that are nonaccessible or poorly accessible until now (Fig. 5.2.1).

5.2.2 SENSORS FITTED TO CONDITIONS AND CONSTRAINTS ABOARD FISHING VESSELS

The RECOPESCA project involved the development of sensors and measurement devices. The challenge was to develop devices robust enough to be fitted on fishing gears, and be self-powered, autonomous, affordable, and able to run without any intervention from the fisherman or interference from fishing activity. Because they had to be

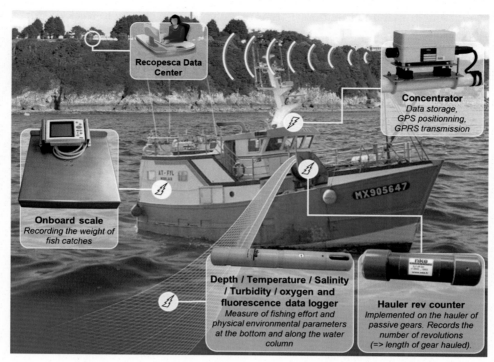

FIGURE 5.2.1   RECOPESCA infography, example of a netter. *GPRS*, General Packet Radio Service; *GPS*, Global Positioning System.

implemented on different kinds of fishing activities and fleets, the sensors are modular and scalable to collect new data. Different sensors have been developed and implemented on board.

A specific sensor allows the recording of physical parameters at the bottom and along the water column for pressure (and thus depth) and duration of immersion. A first version included only a temperature sensor; a second version has been developed to measure salinity. Autonomous and compact, it is robust enough to be fixed up on all types of fishing gears, active (trawls, dredges) or passive (nets, long lines, pots). The sensor records the parameters along each stage of the fishing operation (descent, fishing action, and raising of the gear) with a frequency configurable according to the gears and their operation. The device builds temperature or salinity series and profiles. The maximum immersion depth varies from 300 to 1200 m, depending on the version. In addition to the environmental parameters, the measurement of duration of immersion is a good indicator for the fishing time of the gear, active or passive.

Another specific sensor, called the "turns-counter," has been designed to equip the hauler of passive gear (gill nets, pots, or lines). Fixed on the rotation axis, the sensor records the number of turns from which the number or length of passive gears hauled at each fishing operation is deduced. Like the other sensors, this device is autonomous and compact.

To know the position of the physical measurements and follow the course and areas of fishing activity of the vessels, a Global Positioning System (GPS) is implemented on board and tracks the location of the vessel at a configurable and regular frequency (most of the time, a quarter of an hour). The knowledge of its speed characterizes the different actions of a fishing trip (fishing, route).

Finally, an "antirolling" weigh scale onboard, compatible with the RECOPESCA devices, has been developed. Recording the catches per species and fishing operations, and in association with the other RECOPESCA sensors, this device is expected to link fishing effort and catches at the finest scale.

Since the beginning of the project, the developments and tests carried out analyzed mechanical tolerance of the sensors, improving their resistance, validating their autonomy and their maintenance needs, and optimizing their placement onboard and on the gears. The sensors have considerably evolved, especially to improve the quality and reliability of data, taking into account the autonomy constraints and giving more security and durability to their use. Furthermore, each sensor has been equipped with a radio device transferring the data to a receiver on board, called a "concentrator" (containing the GPS device), that sends the data to Ifremer central databases. The automatic transmission of the data to land is done by General Packet Radio Service (GPRS) once the vessel is within range of a

GPRS network, without any human intervention. This approach has been chosen to track dysfunction, interruption, or loss of sensors quickly.

The main characteristics of the different sensors are as follows:

SP2T: Measurements of Temperature and Depth (Pressure)

Designation	SP2T10	SP2T100	SP2T300	SP2T 600	SP2T 1200	SP2T 4000
Depth:						
Range	10 m	60 m	300 m	600 m	1200 m	4000 m
Resolution	3 mm	3 cm	9 cm	18 cm	36 cm	1.2 m
Accuracy	0.03 m	0.30 m	0.90 m	1.80 m	3.60 m	12 m
Max. depth (sensor breaking)	30 m	250 m	1200 m	1200 m	4300 m	5000 m
Temperature:						
Range	−5°C to +35°C					
Resolution	11 m°C at 0°C, 13 m°C at 10°C, 20 m°C at 20°C					
Accuracy	0.05°C in range 0 + 20°C/0.1°C out of this range					
Time	Internal clock with calendar (±1.3 min/month)					
Automatic starting	Programmable level trigger pressure					
Sample rate	Programmable from 1 s to 99 h					
Mechanical:						
Size	120 mm × 25 mm				120 mm × 25 mm	
Weight	75 g in air, 9 g in water				80 g in air	

SP2T; RECOPESCA sensor for measurements of temperature and depth.

STPS: Measurements of Temperature, Depth, and Salinity

Designation	STPS 100	STPS 300
Depth:		
Range	100 m	300 m
Resolution	30 cm	90 cm
Accuracy	2.4 cm	7.2 cm
Temperature:		
Range	−5 to +35°C	
Resolution	<0.05°C	
Accuracy	0.001°C at 10°C	
Conductivity:		
Range	0–70 mS/cm	0–2000 μS/cm
Resolution	0.0012 mS/cm	0.04 μS/cm
Accuracy	<0.05 mS/cm in the range 0–60 mS/cm	0.04 μS/cm
Salinity:		
Range	2–42°/°° at 20°C	0°–1.137°/°° at 20°C
Resolution	0.0011°/°°	0.0001°/°°
Accuracy	±0.1°/°°	<0.002°/°°
Time	Internal clock with calendar (±1 min/month)	
Automatic starting	Programmable level trigger pressure	
Sample rate	Programmable from 1 s to 99 h	
Mechanical:		
Size	217 mm × 30 mm	
Weight in air	180 g	

STPS; RECOPESCA sensor for measurements of temperature, depth, and salinity; °/°°, per thousand.

STBD: Measurements of Temperature, Depth, and Turbidity

Designation	STBD 300	STBD 1200
Depth:		
Range	300 m	1200 m
Resolution	0.09 m	0.35 m
Accuracy	0.9 m	3.6 m
Temperature:		
Range	−5 to +35°C	
Resolution	<0.02°C	
Accuracy	±50 m°C between 0 and 20°C	
Turbidity:		
Range	0–2000 NTU	
Resolution	<0.007–0.5 NTU (0–750 NTU). Nonlinear from 750 to 2000 NTU	
Time	Internal clock with calendar (±1 min/month)	
Automatic starting	Programmable level trigger pressure	
Sample rate	Programmable from 1 s to 99 h	
Mechanical:		
Size	290 mm × 40 mm	
Weight in air	445 g	

NTU, Nephelometric turbidity unit; STBD; RECOPESCA sensor for measurements of temperature, depth, and turbidity.

Regarding the database infrastructure, RECOPESCA relies on existing operational data centers:

- Coriolis, for operational oceanography.
- The Fisheries Information System of Ifremer and its data center *Harmonie*.

Once Ifremer receives the data emitted by the "concentrator" on board the vessel, the physical data (i.e., temperature and salinity series and profiles) are stored in the *Coriolis* database. The fisheries data (fishing trips and operations) are stored in the *Harmonie* database. This ensures quality control and dissemination of data to the users. Confidentiality of individual data sets (especially fishing data) is also guaranteed.

5.2.3 A NEW SOURCE OF OBJECTIVE FISHERIES DATA

The fisheries data resulting from direct measurements, and not from fishermen's declarations or estimations, supply the Fisheries Information System at Ifremer. However, information from sensors and GPS cannot be used directly and has to be processed to become usable. Two generic algorithms have been developed:

- The first algorithm aims to rebuild the fishing trip of a given vessel on the basis of GPS positions (generally with a frequency of 15 min) or vessel monitoring system (VMS) positions (1 h). The beginning and end of the fishing trip (date and port) are identified on the basis of the distance from the nearest port and the speed of the vessel. The algorithm also identifies the fishing and steaming period of the vessel on the basis of the speed between two positions. Depending on the type of fishing gear used, a threshold of vessel speed determines whether or not it is fishing.
- The second algorithm processes the data of the physical sensors implemented on fishing gears, and especially the time, depth, and duration of immersion, and rebuilds the different fishing operations of a trip. The depth profiles of the sensor are analyzed to reconstruct the key stages of each fishing operation: launching of the gear, arrival at the bottom, beginning of the rise, and end of the operation. Thanks to the position of the vessel and the recording of the date of the profile, the exact point of the physical data is calculated.

RECOPESCA can provide an objective measure of activity and fishing effort for all kinds of vessels, including those of less than 12 m overall length or coastal vessels, and allows:

- A comprehensive and detailed view of all the fishing trips by the vessels.
- A map of the path traveled by the ship during the trip and the precise location of the fishing sector exploited by the vessel.
- For each trip, the identification and description of all the fishing operations: duration, location, depth, and environmental conditions such as temperature and salinity.
- A fine estimation of the fishing effort by the fishing sector.

5.2.4 GEOLOCATION BACKGROUND, ISSUES, AND CHALLENGES

Understanding the dynamics of fishing vessels is essential to characterize the spatial distribution of the fishing effort on a fine spatial scale, and thus to estimate the impact of fishing pressure on the marine ecosystem [1]. It is also a prerequisite for understanding fishermen's reactions to management measures [2] or to the dynamics of exploited fish resources [3,4] (Fig. 5.2.2).

Recently, the mandatory satellite-based VMS, implemented for legal controls and safety [5], has led to massive acquisition of fishing vessel movement data. VMS data, as well as GPS data collected with volunteer fishermen by national research projects (e.g., French project RECOPESCA; [6]), offered new means of studying the spatiotemporal dynamics of fishermen.

The analysis of GPS positions from VMS data has received growing attention in the last decade [7–10]. Various approaches, from simple statistical speed filters to more elaborate state-space models, have been developed to detect fishing activity from VMS data. All relied on the idea that trajectory metrics (e.g., step length, average speed between successive positions, or instantaneous speed) provides critical information to infer fishing activity. Basically, during a fishing trip, fishing vessel activities can be mainly split into two regimes: traveling to fishing grounds (or back from fishing grounds), which occurs at high speed, and fishing, which occurs at low speed.

To segment a fishing trip into activities, a practical and operational method considers that fishing occurs when vessel speed is below a certain threshold value [11]. A more mechanistic approach is possible using hidden Markov models (HMM). Mostly inspired from animal movement ecology [12], models that have been implemented are aimed at estimating jointly fishing vessel movement and activities along trips. HMM have proven to be well suited for this purpose. Numerous models have been developed in fishery sciences [13–18].

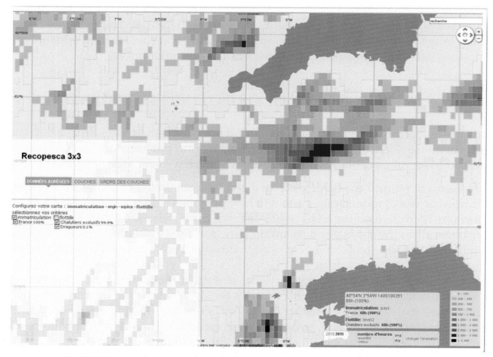

FIGURE 5.2.2   Spatial distribution of RECOPESCA vessels on a 3′ × 3′ grid.

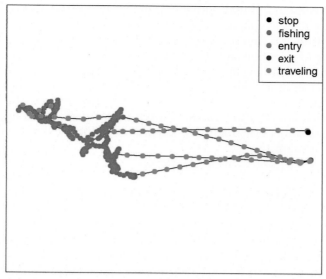

FIGURE 5.2.3 Segmented trajectories with a five-states (stop, fishing, entry harbor, exit harbor, traveling) hidden Markov model very similar to the one developed by Ref. [16].

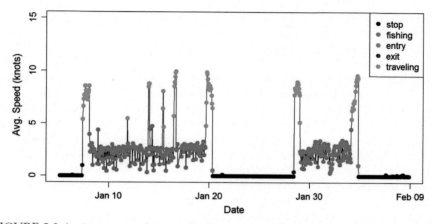

FIGURE 5.2.4 Average speed process for these segmented trajectories with estimated states.

An HMM is a stochastic time series involving two layers: an observable state-dependent process and an unobservable state process [19]. In the context of fishing vessel dynamics, an HMM assumes that an observation at a particular time step (e.g., average speed) results from a distribution associated with an activity state (e.g., fishing, traveling, stopping, and searching—Figs. 5.2.3 and 5.2.4). The movement is modeled conditionally to hidden activity states, usually either by a random walk [17,18] or by an autoregressive process on vectorial speeds [14]. The time series of these hidden activity states is ruled by a first-order Markov or semi-Markov chain [15]. Along that chain, the probabilities of switching from one state to the others are determined by a transition matrix. The probability of a state at current time only depends on the probabilities of the different states at a previous time and of their transition probabilities to this state at the current time.

However, several key issues related to geolocation remain a challenge in fisheries science. Improving the fishing effort estimation is one of the first challenges. The relevance of fishing vessel dynamic models in comparison with simple statistical filters should be evaluated with respect to validation data to derive accurately and efficiently fishing effort from movement data [20]. This improved definition of fishing effort should also provide an improved typology of fishermen's spatial behavior according to fleets, gears, target species, fishing communities, or seasons. Although these models seem appealing, they do not consider a fleet as a whole. New research should also assess the individual variability and the interactions among individuals to provide fleet dynamics models and investigate the link between changes in fleet dynamics and changes in ecosystem conditions.

In addition, spatial fishery data offer a unique opportunity to better understand fish spatial distribution. Indeed, as the spatial behavior of fishing vessels can be partially driven by the spatiotemporal distribution of their target species, once integrated over several individuals, modeling their tracks opens the possibility to infer the spatiotemporal dynamics of their target species at a large scale. These novel approaches could provide a new type of indices of abundance. To address this issue, it is first necessary to estimate the fishing grounds' position and area, and their link with the spatial distribution of scientific and commercial catches.

Finally, ecosystem monitoring with an in situ sensor system on board volunteer geolocalized fishing vessels (e.g., the RECOPESCA project) also raises new challenges in integrating the biotic and abiotic environment within fishing vessel movement modeling. A significant part of fishing vessel displacement may be driven by environmental cues (e.g., specific water masses, bathymetry features, sediment types), the presence of target species, and social interactions (e.g., collaborative fishing or not). Formalizing the integration of those external covariates to movement models is clearly an interesting step to develop.

5.2.5 BENEFIT FOR THE STUDIES OF SEAWATER PHYSICAL PHENOMENA

In addition to fish assessment, measurements with net- and line-based sensors can also be used for seawater and ecological assessments, particularly along the coast. The amount of needed observations close to the coast increases due to the wide range of spatial and temporal scales to monitor (e.g., tide impact, river plumes, exchanges between open and coastal ocean) their direct impacts on human activities. Coastal ocean monitoring is becoming a major goal for public authorities in an effort to predict and mitigate various types of events. The irregular and steep bathymetry in a region or a continental shelf combined with seasonal wind regimes drives a wide range of processes to study. The RECOPESCA network is a solution to collecting low-cost data but with a good level of quality.

The RECOPESCA network is able to detect a number of seawater physical phenomena, for example, temperature dynamics in the sea. The measurements can be compared on qualitative and quantitative observations from other sources. The low-cost data help and guide improvements of physical mathematical models and reduce modeling errors. Analyzing the results for various seasons, the RECOPESCA networks are efficient for most of the year, except in winter when there are too few observations. Indeed, the added value of a sustainable network, based on fishing vessels of opportunity, is confirmed by the efficiency of the collection of profiles at detecting the main modes of variability in a region. The need to keep the profiles has been confirmed. For salinity and turbidity, multiparameter assessment is helpful in regions driven by river dynamics. For example, the French network High-frequency Observation network for the environment in coastal SEAs, which is supporting a sustainable RECOPESCA network in the Bay of Biscay and the English Channel as a key component of the continental shelf, illustrates the benefits of measurements made from opportunity vessels. A scientific publication from Lamouroux and Charria, physicists at Noveltis and Ifremer, with results of the measurements made with RECOPESCA sensors of temperature and salinity, give a demonstration of the benefit of the network. This work was published 8 years after the operational deployment of numerous sensors on several fishing vessels in the Bay of Biscay and English Channel [22].

5.2.6 THE NEW ECOSYSTEM APPROACH TO FISHERIES: RECOPESCA SENSORS OF DISSOLVED OXYGEN AND CHLOROPHYLL

The existing RECOPESCA instrumentation was developed in collaboration between the private company nke and Ifremer. Several devices are available such as the concentrator (hub), the sensor of depth/temperature, the sensor of depth/temperature/salinity, the sensor of depth/temperature/turbidity, and the sensor that measures the length of nets or the low-cost marine scale.

There are other variables reflecting the water conditions that are important to fisheries. Two of these are the oxygen and nourishment available in the water for fish. The opportunity to use new miniaturized probes to measure new variables such as dissolved oxygen and fluorescence through development of new sensors was provided through the NeXOS project [21], building on the experience and structure of RECOPESCA. The two new sensors were called RECOPESCA sensor for measurements of temperature, depth, and oxygen (STPO$_2$) (for oxygen) and RECOPESCA sensor for measurements of temperature, depth, and fluorescence (STPFlu) (for fluorescence).

The NeXOS project extends this approach to a European scale, involving volunteer fishermen and adding new features. The experience of use of RECOPESCA temperature and salinity probes has been demonstrated and satisfies the needs of coastal oceanography and 3D monitoring [22]. We anticipate that a similar service can be provided with oxygen and fluorescence probes. These two essential variables are still poorly sampled. The need for measurements of oxygen and chlorophyll for coastal oceanographic research was identified. This would enable the study of eutrophication in the near-coast areas where data are sparse. Developing these devices presented a technical challenge since the probe must be small, robust, stable, and inexpensive. They must operate on board the fishing vessels for at least a 6-month period, so energy consumption and time drift must be minimized.

Experience with RECOPESCA led to design specifications for the sensors. Each sensor keeps the original functionalities from RECOPESCA, e.g., they are equipped with a radio device to transfer the data to the onboard "concentrator," which sends the data to central databases. The challenge was to develop cost-efficient, trouble-free sensors for the fishermen, robust enough to be attached to fishing gears, self-powered, and autonomous, i.e., similar in mechanical characteristics to the initial set of sensors (temperature, salinity, turbidity, and depth). Insofar as the selection of targeted vessels intends to be representative of all the trades and fleets, the sensors must be modular and scalable to collect new data. The sensors record the parameters along each stage of the fishing operation (descent, fishing action, and raising of the gear) with a frequency configurable according to the gears and their operation. Each sensor cannot be used on all the fishing gears. Sometimes the gear has an impact on the measurement, for example, the trawl modifies the turbidity, oxygen, or chlorophyll near the bottom. Other times the gear can cause damage to the sensors. The time response of the sensor is also to be considered for profile measurements. Both probes and concentrator can withstand 5 years' marine operation without corrosion. Reliability is limited by the tough operating conditions. Five years' "mean time between failure" is expected for the handling hazard during fishing.

Dissolved oxygen is used to trace water masses, to assess mixing processes, and to understand the biogeochemical conditions of their formation regions. Currently on the market are different types of sensors for measuring dissolved oxygen. Oxygen sensors can be electrochemical or optical. The electrochemical method for the determination of dissolved oxygen in water uses an electrochemical cell, which is isolated from the sample by a gas permeable membrane. The optical method for the determination of dissolved oxygen in water uses a sensor working on the basis of fluorescence quenching. For the EAF $STPO_2$ sensor we chose to use an optical sensor. The advantages of optical sensors are their excellent long-term stability and high precision. They also appear to be accurate provided that they have sufficient time to come into equilibrium with the surrounding temperature and oxygen concentration. The low maintenance of optical sensors makes this type of sensor more useful for the project. The $STPO_2$ for NeXOS uses a cost-effective Original Equipment Manufacturer (OEM) sensor, manufactured by the company PyroScience, the Pico2-OEM.

EAF.1 $STPO_2$ Probe Performance

| Sensor | Resolution (Minimum Detection Limit) | Range | Accuracy | Response Time | | Maintenance Periodicity | Memory Capacity |
				Stationary	Profiler		1 Measurement/Min
Depth	10 cm	0–300 m	50–100 cm	<3 s at 63%	<0.5 s at 63%	6 months–1 year	6 months
Temperature	0.01°C	−2 to 35°C	<0.05°C	<3 s at 63%	<0.5 s at 63%	6 months–1 year	6 months
Oxygen (OEM)	0.02% (0.01 mg/L)	0%–100% O_2 (from 0 to 10 mg/L)	To be determined after tests	~10 s		3 months	3 months

EAF, Ecosystem Approach to Fisheries; *OEM*, Original Equipment Manufacturer; *$STPO_2$*, RECOPESCA sensor for measurements of temperature, depth, and oxygen.

Monitoring the distribution of phytoplankton (microscopic algae) in the water column is crucial to understand many large-scale physical and biological processes. Chlorophyll-*a* fluorescence serves as a valuable indicator of active phytoplankton biomass and chlorophyll concentrations in waters. This measurement is used for tracking biological variability and abundance in the water column. The fluorescence technique exposes in situ, living phytoplankton cells to a flash of artificial light at one wavelength and measures the chlorophyll-*a* fluorescence response at a lower wavelength [23]. The intensity of the response is approximately proportional to the chlorophyll concentration. For this sensor we used an optical Turner Designs fluorescence cell.

EAF.1 STPFlu Probe Performance

| Sensor | Resolution (Minimum Detection Limit) | Range | Accuracy | Response Time | | Maintenance Periodicity | Memory Capacity |
				Stationary	Profiler		1 Measurement/Min
Depth	10 cm	0–300 m	50–100 cm	<3 s at 63%	<0.5 s at 63%	6 months–1 year	6 months
Temperature	0.01°C	−2 to 35°C	<0.05°C	<3 s at 63%	<0.5 s at 63%	6 months–1 year	6 months
Chlorophyll-*a* OEM (fluorescence)	0.025 μg/L	0–500 μg/L	To be estimated after tests			3 months	3 months

EAF, Ecosystem Approach to Fisheries; *OEM*, Original Equipment Manufacturer; *STPFlu*, RECOPESCA sensor for measurements of temperature, depth, and fluorescence.

Taking into account that being "low cost" from an integrated system perspective is one of the main objectives in the development of the two new probes; these sensors were chosen because they are also relatively easy to integrate. The accuracy, depth range, and dynamic range have an impact on the final price of the product and have been revealed to be crucial. Robustness, accuracy, uncertainty, and response time have been considered in relation to the use on fishing vessels and fishing gears. In addition, the operational costs (configuration, maintenance of the sensor—calibration and battery replacement) have been studied in compliance with other RECOPESCA probes. A good compromise has been reached with the choice of the new sensors (Figs. 5.2.5 and 5.2.6).

As usual we carried out a series of tests on these sensors to validate our choices. The prototypes of the $STPO_2$ and STPFlu probes have been qualified by laboratory tests. Prior to testing, the $STPO_2$ and STPFlu probes were calibrated for temperature and pressure at the nke Instrumentation laboratory (i.e., OEM calibration). The $STPO_2$ sensor was also calibrated for oxygen saturation at nke Instrumentation facilities. The STPFlu sensor was calibrated at the Turner Designs factory. They were first tested to validate their behavior in the marine environment, according to the standards AFNOR NF X10-812 for mechanical strength, pressure, climate exposure, and salt spray on the simulators of Ifremer. Ifremer assessed the accuracy, linearity, and hysteresis for pressure, temperature, and dissolved oxygen.

FIGURE 5.2.5   The RECOPESCA Ecosystem Approach to Fisheries fluorescence probe with a temperature probe on a cod end.

FIGURE 5.2.6   RECOPESCA sensor mounted on an otter door.

Finally, with respect to deployment, here is the course of action. In preparation for deployment, the sensors are calibrated and adjusted in a metrological laboratory. The state of the battery is checked to allow acquisition for at least 6 months. The sensors are protected against shocks by a polyurethane housing. A technician sets up the probe with the appropriate measuring cycle. The probe is placed on the fishing gear, for example, inside the trap, on top of the cod end, or on the anchoring rope of the net. The sensor is then ready for recording as soon as the measured pressure increases.

The sensors are not protected against fouling. This functionality is difficult to implement on the RECOPESCA sensors due to cost, power consumption, and size. Maintenance periodicity must also be reduced.

An example of deployment in the English Channel during the first tests on a fishing vessel can be seen in Figs. 5.2.7–5.2.9.

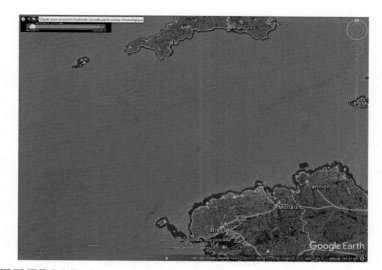

FIGURE 5.2.7   Vessel trajectory during a trip from March 16 to March 23, 2017.

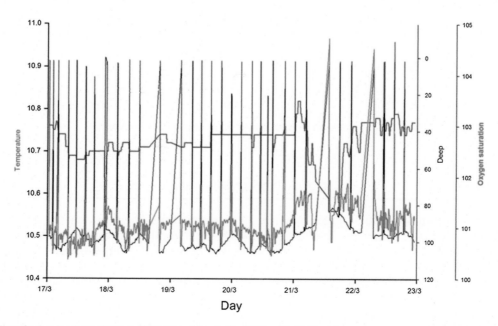

FIGURE 5.2.8   Data from March 17, 2017 at 00:38:30 to March 22, 2017 at 23:00:39—Ecosystem Approach to Fisheries oxygen sensor. Each return to zero of the depth corresponds to a new profile.

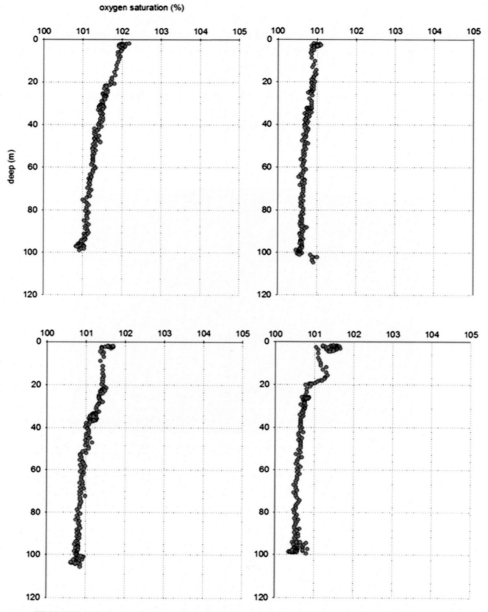

FIGURE 5.2.9   Examples of oxygen (saturation in %) profiles according to the depth.

5.2.7 CONCLUSION

The process of developing a reliable operational sensor requires time. The new sensors developed during the NeXOS project follow this rule. However, the NeXOS project was a good opportunity to design two new sensors in compliance with the RECOPESCA system. A little time will be needed after the project to achieve all of our goals of a robust and routinely operational system. Thanks to the demonstrations and experiments made during the last project phase, as well as our reflection on future plans (market, manufacturing, safety) coming from these experiences, we will soon reach Technology Readiness Level 9, (i.e., operational technology). In the near future, we could study the possibility of adding to the $STPO_2$ sensor a reference salinity measurement to automatically make a correction to the firmware of the oxygen measurement. However, the RECOPESCA sensors must remain cheap, small, and robust with very low power consumption. The measure of fluorescence is complicated in seawater.

We could progress in the calibration procedure to have a better link between the measurement of fluorescence and chlorophyll concentration.

A wide range of equipment is now available to deploy new RECOPESCA networks. A number of tools were developed to process the data, for example, the analysis of the speed of the ship. It allows the generation of speed profiles (average and instantaneous) per trip of the ship over the period. With these profiles, we can deduce when the ship is at port, en route, or fishing. We made a plotting on a map of the route and the position of the ship. A tool creates profiles from the measurements made with the various sensors. The depth profiles analyze fishing operations and temperature profiles allowing the visualization of temperatures during operations. Another piece of software handled the data from the hauler rev counter and generated data profiles in jpeg format. For each ship, automatic processing validated the quality of the data and established a periodic report of fishing activity and effort from the processing of RECOPESCA data.

Glossary

AFNOR Association française de normalisation
CORIOLIS Database for operational oceanography
EAF Ecosystem Approach to Fisheries
EUROCEAN European Centre for information on Marine Science and Technology
FOS/FOOS Fishery and Oceanography Observing System
GPRS General Packet Radio Service
GPS Global Positioning System
Harmonie Database for fisheries information
HMM Hidden Markov models
HOSEA High frequency Observation network for the environmental in coastal SEAs
ICES International Council for the Exploration of the Sea
INRA Institut national de la recherche agronomique (France)
IRD Institut de recherche pour le développement (France)
nke French company for marine instrumentation
NOVELTIS Société innovante au service du spatial, de l'environnement et du développement durable (France)
NTU Nephelometric Turbidity Unit
OEM Original Equipment Manufacturer
RECOPESCA Réseau de mesure de l'activité de pêche spatialisé et de données environnementales, à usage scientifique, par la mise en œuvre de capteurs sur un panel de navires volontaires
SP2T RECOPESCA sensor for measurements of temperature and depth
STBD RECOPESCA sensor for measurements of temperature, depth, and turbidity
STPFlu RECOPESCA sensor for measurements of temperature, depth, and fluorescence
STPO$_2$ RECOPESCA sensor for measurements of temperature, depth, and oxygen
STPS RECOPESCA sensor for measurements of temperature, depth and salinity
TRL Technology Readiness Level
VMS Vessel Monitoring System

Acknowledgments

We thank the French multidisciplinary ETAJERRE consortium (EUROCEAN/INRA/IRD/IFREMER/AgroParitech/Agrocampus Ouest) of marine and fisheries ecologists and statistical methodologists, all focusing on movement analysis in marine ecosystems. The reflection on geolocation was based partly on several workshops that produced a synthesis of the commonly used models, their strength, their weaknesses, and their future methodological developments.

References

[1] ICES. Interim report of the working group on spatial fisheries data (WGSFD). May 2016. p. 17–20. Brest, France. ICES CM 2016/SSGEPI:18. 244 pp.

[2] Vermard Y, Marchal P, Mahévas S, Thébaud O. A dynamic model of the Bay of Biscay pelagic fleet simulating fishing trip choice: the response to the closure of the European anchovy (*Engraulis encrasicolus*) fishery in 2005. Can J Fish Aquat Sci 2008;65(11):2444–53.

[3] Bertrand S, Díaz E, Ñiquen M. Interactions between fish and Fisher's spatial distribution and behaviour: an empirical study of the anchovy (*Engraulis ringens*) fishery of Peru. ICES J Mar Sci 2004;61(7):1127–36.

[4] Walker E, Rivoirard J, Gaspar P. From forager tracks to prey distributions: an application to tuna vessel monitoring systems (VMS). Ecol Appl 2015;25(3):826–33.

[5] Kourti N, Shepherd I, Greidanus H, Alvarez M, Aresu E, Bauna T, Chesworth J, Lemoine G, Schwartz G. Integrating remote sensing in fisheries control. Fish Manage Ecol 2005;12(5):295–307.

[6] Leblond E, Lazure P, Laurans M, Rioual C, Woerther P, Quemener L, Berthou P. The Recopesca project: a new example of participative approach to collect fisheries and in situ environmental data. CORIOLIS Q Newsl 2010;37:40–8.

[7] Hintzen NT, Bastardie F, Beare D, Piet GJ, Ulrich C, Deporte N, Egekvist J, Degel H. VMStools: open-source software for the processing, analysis and visualisation of fisheries logbook and VMS data. Fish Res 2012;115–116:31–43.

[8] Lee J, South AB, Jennings S. Developing reliable, repeatable, and accessible methods to provide high-resolution estimates of fishing effort distributions from vessel monitoring system (VMS) data. ICES J Mar Sci 2010;67:1260–71.

[9] Mills CM, Townsend SE, Jennings S, Eastwood PD, Houghton CA. Estimating high resolution trawl fishing effort from satellite-based vessel monitoring system data. ICES J Mar Sci 2007;64:248–55.

[10] Russo T, Parisi A, Prorgi M, Boccoli F, Cignini I, Tordoni M, Cataudella S. When behaviour reveals activity: assigning fishing effort to métiers based on VMS data using artificial neural networks. Fish Res 2011;111:53–64.

[11] Berthou P, Bégot E, Laurans M, Campéas A, Leblond E, Habasque J. Présentation de la suite logicielle AlgoPesca. Rapport interne Ifremer. 2013.

[12] Patterson TA, Thomas L, Wilcox C, Ovaskainen O, Matthiopoulos J. State space models of individual animal movement. Trends Ecol Evol 2008;23(2):87–94.

[13] Charles C, Gillis D, Wade E. Using hidden Markov models to infer vessel activities in the snow crab (*Chionoecetes opilio*) fixed gear fishery and their application to catch standardization. Can J Fish Aquat Sci 2014;71(12):1817–29.

[14] Gloaguen P, Mahévas S, Rivot R, Woillez M, Guitton J, Vermard Y, Etienne MP. An autoregressive model to describe fishing vessel movement and activity. Environmetrics 2015;26(1):17–28.

[15] Joo R, Bertrand S, Tam J, Fablet R. Hidden Markov models: the best models for forager movements? PLoS One 2013;8:e71246.

[16] Peel D, Good NM. A hidden Markov model approach for determining vessel activity from vessel monitoring system data. Can J Fish Aquat Sci 2011;68:1252–64.

[17] Vermard Y, Rivot E, Mahévas S, Marchal P, Gascuel D. Identifying fishing trip behaviour and estimating fishing effort from VMS data using Bayesian hidden Markov models. Ecol Model 2010;221:1757–69.

[18] Walker E, Bez N. A pioneer validation of a state-space model of vessel trajectories (VMS) with observers' data. Ecol Model 2010;221:2008–17.

[19] Rabiner LR. A tutorial on hidden Markov models and selected applications in speech recognition. Proc IEEE 1989;77(2):257–86.

[20] Mahévas S, Bertrand S, Bez N, Delattre M, de Pontual H, Etienne MP, Fablet R, Gloaguen P, Joo R, Monestiez P, Nerini D, Rivot E, Vermard Y, Walker E, Woillez M. Validation data: keystone to move state-space models for movements to operational models for fisheries and marine ecology. 2014.

[21] Delory E, Castro A, Zielinski O, Waldmann C, Golmen L, Rolin JF, Woerther P, Garello R. Objectives of the NeXOS project in developing next generation ocean sensor systems for a more cost-efficient assessment of ocean waters and ecosystems, and fisheries management. In: Paper presented at the IEEE OCEANS 2014-TAIPEI. April 7–10, 2014. https://doi.org/10.1109/OCEANS-TAIPEI.2014.6964574.

[22] Lamouroux J, Charria G, De Mey P, Raynaud S, Heyraud C, Craneguy P, Dumas F, Le Hénaff M. Objective assessment of the contribution of the RECOPESCA network to themonitoring of 3D coastal ocean variables in the Bay of Biscay and the English Channel. Ocean Dyn 2016;66(4):567–88.

[23] Hersh D, Leo WS. A new calibration method for in situ fluorescence. Boston: Massachusetts Water Resources Authority. Report 2012-06. 2012:11.

Further Reading

[1] Gloaguen P, Woillez M, Mahévas S, Vermard Y, Rivot E. Is speed through water a better proxy for fishing activities than speed over ground?. Aquat Living Res 2016;29(2). 8 pp.

CHAPTER

5.3

Platforms of Opportunity in Action: The FerryBox System

Wilhelm Petersen

Helmholtz-Zentrum Geesthacht, Institute of Coastal Research, Geesthacht, Germany

5.3.1 INTRODUCTION

The idea of using "ships of opportunity" (SOO) for science has been around for a long time. Centuries ago, water temperature and some meteorological parameters were recorded in log books. For example, Wheeler and Wilkinson generated a daily climatological record for the period from 1750 to 1850 by studying log books from different European countries [1]. The 1853 Brussels Conference set the first international standard in systemizing the observation practices over sea initiated by marine meteorology.

In the 1930s, Sir Alistair Hardy [2] started to regularly collect data of the distribution of zooplankton and fish larvae in the North Sea with a newly developed Continuous Plankton Recorder towed behind research vessels and voluntary ships. At the same time, the Norwegians used the "Hurtigruten" along the Norwegian coast to collect salinity and temperature data on a regular basis.

In the 1980s the United States initiated an international Ship of Opportunity Program (SOOP) as a component of the Global Ocean Observing System (GOOS) with a focus on the operation of eXpendable BathyThermographs and thermosalinographs; the program was later extended to pCO_2 measurements. Over the last two decades other international observing programs started using ships of opportunity: several projects at the Atlantic Oceanographic and Meteorological Laboratory GOOS Center (http://www.aoml.noaa.gov/phod/goos.php), as well as ongoing CO_2 measurements within the International Ocean Carbon Coordination Project (http://www.ioccp.org/underway-co2-measurements) [3].

Operational monitoring of coastal areas and shelf seas is mainly carried out by manual sampling and analysis during ship cruises. In addition, automatic measurement systems on buoys allow routine measurement of standard oceanographic parameters (temperature, salinity, currents), and in some cases turbidity, oxygen, and chlorophyll fluorescence. There are increasing demands for reliable and cost-effective ocean observations with high density in space and time. Most systems are heavily affected by biofouling, and the maintenance/operation costs of ship cruises are quite high. Compared to these platforms, the FerryBox systems are very cost-effective. They have fewer limitations related to space, power consumption, or harsh environmental conditions, and they even allow for the operation of experimental and less robust sensors. Furthermore, the FerryBox research platforms always come back to the home port so that ship costs are negligible.

In 1993, the Finnish Institute of Marine Research (now SYKE) initiated regular ferry-based observations on the distribution of algal blooms and nutrients within the Alg@line project [4] in the Baltic Sea. These ongoing measurement systems are rather more sophisticated because they can take both chemical and biological measurements. Meanwhile, the name FerryBox has become virtually synonymous with the use of ships of opportunity for biogeochemical measurements [5].

5.3.2 FERRYBOX SYSTEM

In general, all FerryBox systems employ a similar design. There are differences in the design of the flow-through system, the degree of automation and biofouling prevention, as well as in the possibilities of supervision and remote control. The FerryBox is a modular system that can be easily extended with additional sensors. Fig. 5.3.1 shows a schematic diagram of a FerryBox set-up.

FerryBoxes have a water inlet, from where the water is pumped into the measuring circuit of multiple sensors. This inlet may be located at the sea chest or at a special valve through the hull of the ship specifically designed for the purposes of the FerryBox. The FerryBox is typically positioned as close as possible to the inlet, so that the seawater is not influenced by the long residence times in the sea chest. For correct seawater temperature measurements an extra temperature sensor should be installed close to the inlet or on the hull of the vessel. An optional debubbling unit removes air bubbles, which could enter the system during rough seas. At the same time, coarse sand particles, which may be introduced in shallow harbors and which settle and tend to block the tubing of the FerryBox, are removed as well. A basic FerryBox system [6] has sensors for temperature, salinity, turbidity, and chlorophyll-a fluorescence, as well as a Global Positioning System (GPS) receiver for position control if the GPS signal does not come from the ship system. Many FerryBoxes also include an inline water sampler and additional sensors for dissolved oxygen, pH, pCO_2, or algal groups, as well as meteorological instruments (air pressure, air temperature, and wind). Whenever nutrients are measured, a small part of the water is filtered by a hollow-fiber cross-flow filter module for automatic nutrient analysis. Further sensors suitable for FerryBox applications are under development.

Within the EU initiative "The Ocean of Tomorrow," projects were funded for the development of new sensors for more biologically relevant measurements. For instance, in the European project NeXOS (http://www.nexosproject.eu/), a flow-through sensor for hyperspectral absorption measurements has been developed addressing phytoplankton biomass and taxonomic composition, as well as total suspended matter and dissolved organic matter [7,8]. Other projects in this program deal with an algae detection unit for toxic marine microalgae species (EnviGuard, http://www.enviguard.net/) [9] or sensors for organic pollutants (MariaBox, http://www.mariabox.net). Due to the modular design of the FerryBox, such new instruments are easy to implement, and can give better insights into biogeochemical processes in the surface ocean.

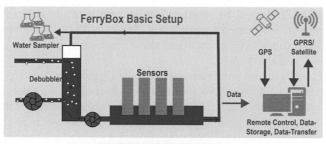

FIGURE 5.3.1 Schematic diagram of a FerryBox system (*top panel*) and installed system on board the cruise liner *Mein Schiff 3* (*bottom panel*). A basic system is equipped with sensors for temperature, salinity, turbidity, and chlorophyll-*a* fluorescence. In many cases, sensors for dissolved oxygen, pCO$_2$, and pH are added. *GPRS*, General Packet Radio Service; *GPS*, Global Positioning System.

For a reliable, unattended operation, the FerryBox system is controlled by a computer that also records the data. Data are transmitted to the shore via mobile phone connection or satellite communication, if available. Normally, support of the ship's crew is not necessary. A major issue of long-term unattended operation of monitoring systems is biofouling, especially for optical sensors (e.g., for turbidity or for chlorophyll-*a* fluorescence). In several FerryBoxes, biofouling is prevented by automatic cleaning of the sensors with tap water, and rinsing with acidified water after each cruise (controlled according to the vessel position). Other systems use pressurized air for removing biofilm. Using such measures against biofouling, FerryBox systems can be operated several months without any maintenance even in coastal areas, where biological activity is particularly high. More details about the FerryBox systems can be found in Ref. [10]. A typical FerryBox profile along one transect is shown in Fig. 5.3.2.

A unique feature of the FerryBox data is the high frequency of measurements along a seawater transect, as most lines are operated on a daily, or at least on a weekly, interval. This is shown in Fig. 5.3.3, where salinity data along one transect are pooled over several months. This allows the observation of particular events, such as the detection of an extraordinary inflow of freshwater at a certain position and time [11].

One of the biggest advantages of the FerryBox data is that they can be used together with other platforms, such as satellites, to study environmental processes within a larger geographic context. Fig. 5.3.4 shows that large-scale remotely sensed observations by the instrument medium-spectral resolution, imaging spectrometer (MERIS) installed on the satellite Environmental Satellite (ESA) over the North Sea are consistent with measurements made on board various FerryBox lines. The European Copernicus programs, which involved new environmental satellites such as the Sentinel satellites (primarily Sentinel 2a launched June 2015, Sentinel 2b launched March 2017, and Sentinel 3 launched February 2016), will benefit clearly from strengthening the connection between in situ data from FerryBoxes and satellite data for direct validation and downstream service developments. On the one hand, in combination with hydrodynamic models, as well as ecosystem models, FerryBoxes can fill the gaps of the satellites due to cloud coverage and, on the other hand, satellites can provide the spatial distribution of algae blooms detected by FerryBoxes. The challenges to combine these data are still the different principles used for chlorophyll-*a* determination. From satellite images, chlorophyll-*a* is derived from reflectance at different wavelengths. Especially in estuarine

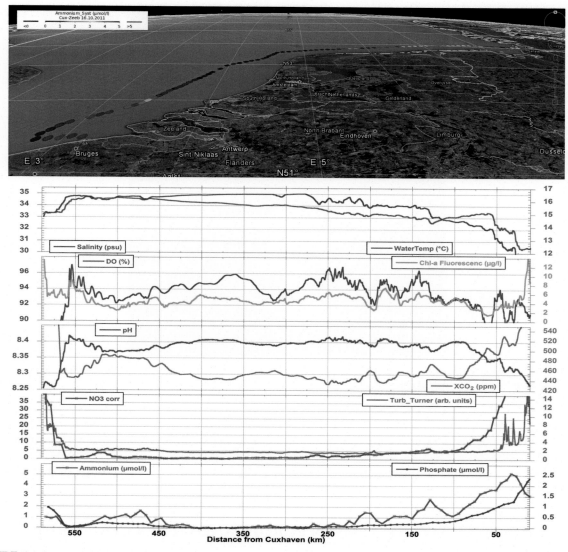

FIGURE 5.3.2   Example of FerryBox data sampled on one transect from Cuxhaven (DE) to Zeebrugge (BE). *DO*, Dissolved oxegen.

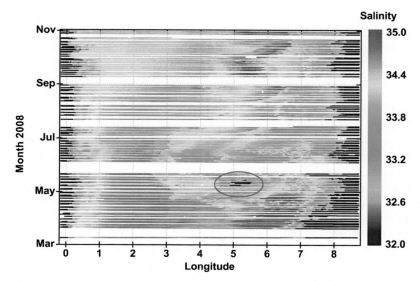

FIGURE 5.3.3   Pooled salinity data along a route in the North Sea (Immingham, UK, to Cuxhaven, DE) in 2008. The *gray circle* indicates the occurrence of a freshwater inflow.

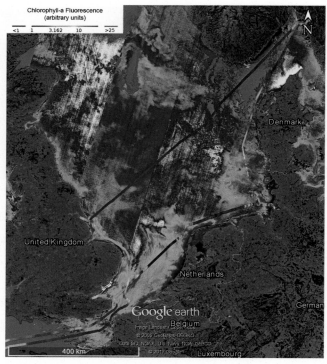

FIGURE 5.3.4   Comparison between large-scale remote sensing of chlorophyll-*a* (MERIS onboard satellite ENVISAT) in the North Sea with approximately simultaneous observations of chlorophyll-*a* fluorescence made with FerryBox systems along a few routes (May 2011).

and coastal waters (Case-2 waters) it is sometimes difficult to determine chlorophyll-*a* correctly because reflectance is often affected by site-specific factors. In FerryBox systems, only chlorophyll-*a* fluorescence can be measured, which is influenced by light conditions and physiology of the living cells. Newly developed sensors for measuring chlorophyll-*a* absorption instead of fluorescence [8] may overcome this problem.

5.3.3   FERRYBOX SYSTEMS IN EUROPE AND OTHER COUNTRIES

FerryBoxes or similar systems are used worldwide [6]. Codiga et al. [12] describe the usage of ferry-based monitoring systems in the United States. For example, the FerryMon (www.ferrymon.org) project monitors a large estuary in North Carolina to measure water quality indicators, human and climatic drivers of water quality, as well as patterns of water quality variability [13]. In Canada, observations of chlorophyll-*a* biomass from an instrumented ferry have been used to determine how the Fraser River plume affects phytoplankton biomass in the Strait of Georgia [14]. A special partnership has been established between the University of Miami and a cruise line operator (Royal Caribbean Cruises Ltd.) creating over 8 years a large data set of comprehensive atmospheric and oceanographic measurements along the cruise ship's itinerary. These data have been applied to investigate the ocean acidification of the greater Caribbean region [15]. In the Oleander Project (http://www.po.gso.uri.edu/rafos/research/ole/), continuous shipboard acoustic Doppler current profiler (ADCP) measurements have been used to investigate the variability of the Gulf Stream over 17 years [16]. The project is an inter-institutional effort to collect oceanographic data such as ocean currents, sea-surface temperatures (SST), sea-surface salinity (SSS), and surface carbon dioxide (pCO_2) in the highly dynamic region between New Jersey and Bermuda.

From 2003 to 2005, a European-funded multinational collaborative project FerryBox was initiated to develop and optimize the use of so-called FerryBox systems for automated measurements and water sampling using ships of opportunity, e.g., merchant vessels and ferries. The consortium in this project focused on four core parameters (temperature, salinity, turbidity, and chlorophyll-*a* fluorescence) to be measured along each route. In addition, nonstandard sensors have been tested for observation of currents and sediment transport (hull-mounted ADCP, [17]), as well as pH, oxygen, nutrients, and algal species [18]. After completion of this project in 2005, most of the FerryBox operators continued their monitoring activities through institutional and other sources of funding. This reflects the attractiveness of ferry-based measurements for research, as well as for monitoring programs. Furthermore, FerryBox systems have been installed on research vessels as well. Through the years, some 30 ships are involved in FerryBox monitoring of sea-surface parameters. A map of the routes currently operating FerryBoxes in Europe is shown in Fig. 5.3.5.

Under the umbrella of EuroGOOS (www.eurogoos.eu) the European FerryBox community is organized in the FerryBox task team (www.ferrybox.org) to promote the FerryBox idea, to organize regular international workshops, to build up a common European FerryBox database, and to deliver quality-controlled (QC) real-time data to the European marine monitoring service (Copernicus Marine Environment Monitoring Service [CMEMS; http:// marine.copernicus.eu]), which can be used, for instance, for data assimilation in model forecasts of temperature and salinity [19,20] to improve the estimates of numerical models.

In the **North Atlantic**, the Norwegian institute Norwegian Institute for Water Research (NIVA) operates FerryBox systems close to the Norwegian coast (www.ferrybox.no). Main routes are the "Hurtigruten" from Bergen to Kirkenes and two other routes in the Skagerrak and Baltic Sea, respectively (Bergen–Hirtshals and Oslo–Kiel). Another route operates in the arctic region between Tromsø and Longyearbyen (Svalbard). The focus is on monitoring Norwegian coastal waters, the detection of algal blooms, and validation of satellite data (chlorophyll-*a*), by combination of information from the FerryBoxes with data from environmental satellites and collected water samples [21–23]. In addition to data from the water phase, as a special feature, shipborne sea-surface reflectance from above-surface (ir)radiance measurements are recorded aboard some ferries for comparison and validation of remotely sensed data from satellites.

In the **Baltic Sea**, several institutions are involved in ferry-based monitoring programs. The first ferry-based program to conduct routine water quality monitoring was the Finnish Marine Institute (Marine Research Centre

FIGURE 5.3.5   FerryBox routes from different institutions in Europe in 2016.

at the Finnish Environment Institute [SYKE]), which started with a ship-based system on a ferry between Helsinki (FI) and Tallin (EE) in the early 1990s. In 1993, this effort was expanded to the entire Baltic Sea, with a continuously operated shipborne monitoring system on the route between Helsinki (FI) and Travemünde (DE). Especially on this route, additional measurements have been carried out for monitoring carbon dioxide [24] and methane fluxes [25]. More recently, additional lines were added, including one between Helsinki and Stockholm, one between Tallinn and Helsinki (with special focus on upwelling processes [e.g., Ref. [26]], operated by the Marine Systems Institute, Tallinn University of Technology, Estonia), and one between Tallinn and Stockholm (SW) (maintained by the Estonian Marine Institute, University of Tartu). In the meantime, even more lines are operated in the Baltic Sea by SYKE covering the southwest Finnish coastal area and the Archipelago Sea. Other lines in the Baltic Sea are operated by the Swedish Meteorological and Hydrological Institute cruising along the route Gothenburg (SE)–Kemi (FI)–Oulu (FI)–Lübeck (DE)–Gothenburg (SE), on a weekly basis. The main activity is on monitoring the phytoplankton dynamics in the Baltic Sea [27,28] with special focus on the cyanobacterial blooms [29]. The results are continuously reported in the Baltic Sea portal Alg@Line webpage (https://www.finmari-infrastructure.fi/ferrybox/).

In the **North Sea**, the Helmholtz-Zentrum Geesthacht (HZG), Germany, operates FerryBox systems mainly aboard cargo ships cruising on fixed routes on a regular basis. Only one short line between Büsum and the island Helgoland uses a small passenger ferry. One main route crosses the entire southern North Sea from Immingham (UK) to Cuxhaven (DE) about five times per week. Among others, chlorophyll-*a* fluorescence data were combined with chlorophyll-*a* data from the instrument MERIS on board the satellite ENVISAT [30]. Another vessel cruises from Moss/Halden (NO) to Ghent/Zeebrügge (BE) to Immingham (UK) and then back to Norway every week. In addition to the core parameters (temperature, salinity, chlorophyll-*a*, and turbidity), the FerryBox systems from HZG are equipped with sensors for oxygen, pH, algal groups, and pCO_2. In former times, there was also a FerryBox line between Amsterdam (NL) and Bergen (NO), operated by Rijkswaterstaat (NL). This line will be probably reactivated in the near future operating along the route Amsterdam (NE)–Esbjerg (DK)–Bergen (NO). A very short transect with special focus on monitoring the current and sediment transport is operated by NIOZ (NL) aboard a ferry between Den Helder and the island Texel [31]. In this special case, the ferry is equipped with a moonpool for operating an ADCP, and it often deploys additional scientific instruments. It should be noted that the installation of the moonpool (used only for scientific purposes) was financially supported by the ferry company. In return, NIOZ presents the results of the FerryBox data onboard the vessel, in real-time with access for the public. Another ferry, operated by the University of Rhode Island and Stony Brook University (USA) together with NIVA (NO), makes weekly runs between Denmark and Iceland, and is equipped with an ADCP mounted on the hull of the ship. The ADCP measurements are focused on monitoring the northward flow of the North Atlantic waters through the Faeroes–Shetland Channel and into the Greenland and Norwegian seas. Furthermore, there are FerryBoxes installed aboard research vessels operated by the water authorities CEFAS (UK), Rijkswaterstaat (NL), as well as the Alfred Wegener Institute (DE), which monitor water quality parameters in the entire North Sea along random tracks.

In the **Atlantic Ocean**, in the Western Channel and the Bay of Biscay, two FerryBox lines have been operated aboard car ferries by the French institutions Ifremer (Brest) and CNRS (Station Biologique Roscoff) since 2010 and 2011, respectively. These two vessels travel between Plymouth (UK) and Roscoff (FR), and between Cork (IR), Portsmouth (UK), and Santander (ES). In addition to monitoring the standard parameters (temperature, salinity, turbidity, chlorophyll-*a* fluorescence, oxygen), these FerryBox systems are used to observe the variability of the carbonate system by a pCO_2 sensor [32], and the distribution of algal species by the temporary operation of a flow cytometer [33].

In the entire **Mediterranean Sea**, one FerryBox line is operated between Piraeus and Heraklion (Crete) by the Hellenic Centre for Marine Research in Greece. This system is an integrated part of the monitoring system for the Greek seas (http://www.poseidon.hcmr.gr). The focus of this FerryBox is on its integration with other monitoring platforms, such as buoys, and using the FerryBox data for data assimilation to improve model estimates (e.g., Ref. [20]). In the western part of the Mediterranean Sea, a program monitoring SST and SSS (TransMED) has been set up by the Mediterranean Science Commission (www.ciesm.org) using ships of opportunity. In the pilot phase, many problems were encountered during the operating of SeaKeeper modules due to biofouling, etc. Since 2010, a simplified system only measuring temperature and salinity has been installed on a container ship, which operates between Genoa (IT), Malta, and Libyan harbors. In recent years, two other ferries in the Mediterranean Sea were equipped with FerryBox systems. One FerryBox is installed by the University of Cagliari on a ferry traveling between Toulon (FR) and Ajaccio (IT) and, during the summer, also between Livorno (IT) and Golfo (IT). The other FerryBox is installed on a Tunisian ferry cruising between Tunis (TN) and Marseille (FR). This is a joint project between the French University of Marseille (Mediterranean Institute of Oceanography) and the Tunisian National Marine Institute of Science. Besides the core parameters, this FerryBox is additionally equipped with sensors for pCO_2 and colored dissolved organic material, and occasionally with a flow cytometer for detection of algae species. Furthermore, a FerryBox operated by HZG on board a cruise liner (*Mein Schiff 3*) is cruising in the entire Mediterranean Sea during the summer period. From a scientific point of view, data recorded with a FerryBox operating aboard a cruise liner on random routes is of less interest than measurements on fixed routes repeated on a daily or weekly basis. In particular, the high density of recorded data along a certain transect over a long time period is an outstanding feature of the FerryBox concept.

5.3.4 FERRYBOX DATA MANAGEMENT

Continuous operation of FerryBox lines creates a large volume of data and requires appropriate data management. Despite the recent increase in the number of operational FerryBoxes in Europe, there is still a lack in visibility and accessibility of the FerryBox data. Data sets are often only available upon request to the data originator and mostly based on files (ASCII or NetCDF). Only subsets of real-time FerryBox data (mainly temperature and salinity and partly oxygen and chlorophyll-*a* fluorescence) are delivered to the regional data centers (ROOSs), through CMEMS. In addition, within CMEMS, data are aggregated per day and per month to fulfill the need of operational oceanography users and the original very specific structure of FerryBox data (one data set for each performed transect) is being lost. Therefore it was established that there is a need to enhance the access of all FerryBox data both in real-time and in delayed mode. For specific regions, the sampled data should be easily accessible and visible for a variety of users in a timely manner.

As an example, the FerryBox data management at HZG is described in more detail in Fig. 5.3.6, which shows the data flow from the vessel to the database on shore.

The data are transmitted to the shore either via mobile phone connection (GSM or UMTS) or satellite communication, after completing a transect and the ship has arrived at the harbor. Before starting the transfer to the database, the FerryBox data are quality checked and flagged in real-time, according to recommendations for near-real-time quality control of the EuroGOOS working group "Data Management, Exchange and Quality" (DATA-MEQ, http://www.eurogoos.org). Among other issues, these quality checks deal with frozen values, spikes, as well as flow rates within the flow-through system or the speed of the ship. The data are stored in a relational database (http://ferrydata.hzg.de), which is embedded in the data portal of the coastal observatory Coastal Observing System for Northern and Arctic Seas (COSYNA) (www.cosyna.de), where FerryBox data can be additionally combined and compared with data from other sources [34]. The database is structured in such a way that FerryBox data can be stored as a single transect with high performance for data access and different types of database queries. The web interface to the database has a lot of tools for graphical presentation of the FerryBox data online (e.g., to track the activity and the status of a certain FerryBox system), and this can also be used for quality assessment and further data processing (e.g., changing quality flags, correcting, deleting faulty data, etc.).

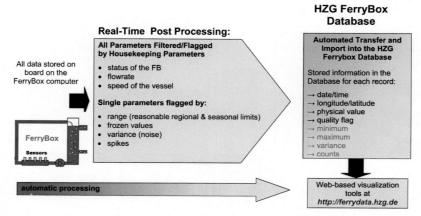

FIGURE 5.3.6   Flow diagram of FerryBox real-time data processing. *HZG*, Helmholtz-Zentrum Geesthacht.

During the last FerryBox workshop in 2016 it was decided that a separate European FerryBox portal will be developed, housing a FerryBox database. This portal will provide free access to the highest quality of European FerryBox data, fed directly by FerryBox users in near real-time or in delayed mode, once the data have been processed. Furthermore, this common European FerryBox database of all FerryBox operators will also serve as a showcase for the joint FerryBox activities in Europe and will increase the visibility of the FerryBox community. Fig. 5.3.7 shows the proposed design of such a FerryBox data portal and database.

The data will be freely available to the operational services and research communities. Moreover, web-based tools can be used by each FerryBox operator in a similar way to track their FerryBox data. Password-controlled access can be offered, so that users can have individual access to their own data sets, and they can control which data will be shown to the public (e.g., experimental test data of a newly developed sensor may remain hidden).

As a first step in the development of such a common database, a selected number of parameters (temperature, salinity, turbidity, oxygen, chlorophyll-*a*, etc.) of all FerryBox routes in Europe will be delivered in real-time, or near real-time, to a central relational database. Data can be uploaded to the database either file structured (e.g., via ftp server) in near real-time (at the end of one cruise) or in real-time by direct communication (e.g., via satellite communication) of a certain FerryBox computer with the database (machine-to-machine communication). All data in the database should have undergone a real-time quality control with agreed standards and flagging scheme of all operators, according to the recommendations of the EuroGOOS DATA-MEQ group (http://eurogoos.eu/download). The internal structure of the database will retain the structure of single FerryBox transects. In combination with a web-based pan-European data portal, this ensures that all FerryBox data are easily accessible and visible. Also, data that have been corrected in delayed mode should be made available to provide an integrated access to FerryBox transects.

Furthermore, this European database will act as a central and reliable provider for selectable subsets of FerryBox data that can be further delivered to, or downloaded from, other portals such as the portals of the ROOSs, CMEMS in situ TAC, and EMODnet. For this purpose, a few download options will be provided: (1) a direct data download from the database keeping the transect structure; (2) data for each observed parameter between two users using the Open Geospatial Consortium (OGC) Sensor Observation Service (SOS) web service (http://www.opengeospatial.org) for a certain time period; (3) export of a subset of regularly observed parameters in OceanSite

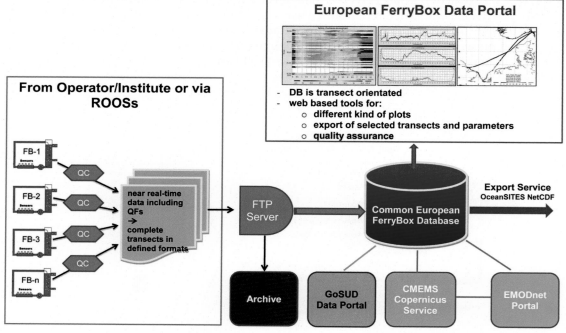

FIGURE 5.3.7 Schematic diagram of a proposed European FerryBox database and the connected data portal. *CMEMS*, Copernicus Marine Environment Monitoring Service; *DB*, database; *QC*, quality controlled; *QF*, quality flags; *ROOSs*, regional data centers.

NetCDF format, on regular time intervals, and retaining the transect structure. These data can be downloaded using the OpenDAP web service (https://www.opendap.org/). The Sensor Web Enablement framework of OGC was also the technological basis of the NeXOS project (see Section 6 of this book). Together with SOS functionality of the database the developments in this project could be used for an improved visualization of FerryBox data via an SOS web server.

5.3.5 CONCLUSION

Long-term operation of FerryBox systems for more than one decade have demonstrated the reliability and cost-effectiveness of such systems for observing the state of the ocean over a wide range of temporal and spatial scales. FerryBox systems have meanwhile reached a state of maturity. Within Europe, numerous lines are in operation, particularly in the marginal seas of northern Europe. The FerryBox program possesses numerous strengths: no ship operation costs, no energy restrictions, regular maintenance is easily possible as the platform always comes back to port, transects are sampled repeatedly, and biofouling can be controlled. In the United States, Ref. [12] suggested that the FerryBox system should become a part of the US Integrated Ocean Observing System. They claim that in the United States, FerryBox-based sampling is currently underutilized to a great extent. Apart from Europe, currently several international activities have been started. Thus the applicability of FerryBox systems has been successfully demonstrated in the Chilean fjords [35], as well as in the Yellow Sea in China [36]. Furthermore, the FerryBox community is associated with other international programs using voluntary ships or ships of opportunity such as the JCOMM Ship Observation Team (http:/www.jcommops.org/sot/) or the IODE project GOSUD (www.gosud.org) collecting mainly salinity and temperature data of observations on the ocean surface.

Different applications reveal the usefulness of FerryBox measurements not only for monitoring purposes but for scientific questions as well. For instance, there is still a particular lack of robust biogeochemical observations in the oceans and especially in the coastal regions with their high biological activity. Parameters such as dissolved oxygen, chlorophyll-a, pH, nutrients, and turbidity are not observed to the same extent as parameters such as salinity. This situation makes it difficult or even impossible to validate and provide boundary data for ecosystem models, or to understand and parameterize the underlying biological processes better. However, numerous promising technologies for better automated measurements of different biologically relevant parameters are under development (e.g., EU projects within "The Ocean of Tomorrow"), or are even at a mature stage. FerryBoxes are an ideal platform to integrate such sensor systems, even if they are in the development stage, as they offer a protected environment and easy access for maintenance and other issues. One example is the development of novel carbon sensors (e.g., high-precision pH and alkalinity sensors based on spectrophotometric detection developed within the NeXOS project), which can be cost effectively operated by FerryBox systems to investigate the not so well-known role of coastal areas as a sink or source of atmospheric carbon dioxide. FerryBoxes provide the opportunity to obtain continuous measurements in these areas for better knowledge of the upper ocean carbon cycle.

Most FerryBox systems are equipped with a cooled water sampler for subsequent lab analysis of bottle samples. Recently, it was shown that a range of contaminants, including several pharmaceuticals and antibiotics of human and veterinary use, as well as synthetic food additives such as artificial sweeteners, could be detected in FerryBox-originated samples even at subpart-per-trillion levels starting from samples of a few liters of marine waters [37].

The weaknesses of FerryBoxes are that sampling is limited to fixed transects and surface water only, and that the operation relies on voluntary vessels that may be affected if the ship operator switches routes often at short notice. This could partly be overcome if newly built ships could already be prepared for the installation of ongoing systems such as a FerryBox; this would not create significant additional costs in the construction phase. The working group "OceanScope" within the Scientific Committee on Oceanic Research proposed a partnership between the ocean-observing community and the maritime industry [38] to facilitate such possibilities.

Due to large amounts of data, suitable data management must be established, which includes sophisticated quality control in both real-time and delayed modes. It should be mentioned that even from automated systems such as a FerryBox, the quality of the data strongly depends on sufficient system maintenance and reliable quality data control on a regular basis. Quality assessment, in particular, must be harmonized and standardized according to internationally accepted standards to make the data comparable and exchangeable. In Europe, the FerryBox operators are organized within a task team supported by EuroGOOS (http://eurogoos.eu/ferrybox-task-team/). To increase the visibility and availability of FerryBox data the operators started to deliver QC real-time data from different FerryBoxes to a common European FerryBox database and data portal. Such a database enables the better use of FerryBox data, including the combination of overlapping routes for cross-calibration, etc.

Even ongoing systems such as FerryBoxes can contribute significantly to a GOOS, but these systems must still be complemented by conventional monitoring strategies, such as fixed platforms (e.g., buoys), autonomous moving platforms (e.g., gliders, autonomous underwater vehicles), remotely sensed data, and research surveys to obtain an integrated picture of the oceans.

Acknowledgments

This work was partly supported by the EU projects JERICO (grant agreement no. 262584), NeXOS (grant agreement no. 614102), and JERICO-NEXT (grant agreement no. 654410).

References

[1] Wheeler D, Wilkinson C. The determination of logbook wind force and weather terms: the English case. Clim Change 2005;73:57–77. https://doi.org/10.1007/s10584-005-6949-1.

[2] Reid P, Colebrook J, Matthews J, Aiken J, C.P.R. Team. The continuous plankton recorder concepts and history, from plankton indicator to undulating recorders. Prog Oceanogr 2003;58:117–73.

[3] Sabine CL, Ducklow H, Hood M. International carbon coordination: Roger Revelle's legacy in the intergovernmental oceanographic commission. Oceanography 2010;23:48–61. https://doi.org/10.5670/oceanog.2010.23.

[4] Rantajärvi E, Olsonen R, Hällfors S, Leppänen J-M, Raateoja M. Effect of sampling frequency on detection of natural variability in phytoplankton: unattended high-frequency measurements on board ferries in the Baltic Sea. ICES J Mar Sci 1998;55:697–704. https://doi.org/10.1006/jmsc.1998.0384.

[5] Hydes D, et al. The way forward in developing and integrating FerryBox technologies (21–25 September 2009). In: Hall J, Harrison DE, Stammer D, editors. Proceedings of OceanObs'09: sustained ocean observations and information for society, vol. 2. Venice, Italy: ESA Publication WPP-306; 2010. https://doi.org/10.5270/OceanObs09.cwp.46.

[6] Petersen W. FerryBox systems: state-of-the-art in Europe and future development. J Mar Syst 2014;140:4–12. https://doi.org/10.1016/j.jmarsys.2014.07.003. ISSN 0924-7963.

[7] Wollschläger J, Grunwald M, Röttgers R, Petersen W. Flow-through PSICAM: a new approach for determining water constituents absorption continuously. Ocean Dyn 2013;63:761–75. https://doi.org/10.1007/s10236-013-0629-x.

[8] Wollschläger J, Voß D, Zielinski O, Petersen W. In situ observations of biological and environmental parameters by means of optics-development of next-generation ocean sensors with special focus on an integrating cavity approach. IEEE J Ocean Eng PP 2016;99:1–10. https://doi.org/10.1109/JOE.2016.2557466.

[9] Metfies K, Schroeder F, Hessel J, Wollschläger J, Micheller S, Wolf C, Kilias E, Sprong P, Neuhaus S, Frickenhaus S, Petersen W. High-resolution monitoring of marine protists based on an observation strategy integrating automated on-board filtration and molecular analyses. Ocean Sci 2016;12:1237–47. https://doi.org/10.5194/os-12-1237-1247.

[10] Petersen W, Petschatnikov M, Schroeder F. FerryBox systems for monitoring coastal waters. In: Dahlin H, Flemming NC, Nittis K, editors. Third international conference on EuroGOOS. Amsterdam (The Netherlands): Elsevier Oceanography Series Publication; 2003. p. 325–33.

[11] Petersen W, Schroeder F, Bockelmann F-D. FerryBox — application of continuous water quality observations along transects in the North Sea. Ocean Dyn 2011;61:1541–54.

[12] Codiga DL, Balch WM, Gallager SM, Holthus PM, Paerl HW, Sharp JH, Wilson RE. Ferry-based sampling for cost-effective, long-term, repeat transect multidisciplinary observation products in coastal and estuarine ecosystems. In: Community white paper, IOOS summit, Herndon, VA. November 2012.

[13] Paerl HW, Rossignol KL, Guajardo R, Hall NS, Joyner AR, Peierls BL, Ramus JS. FerryMon: ferry-based monitoring and assessment of human and climatically driven environmental change in the Albemarle–Pamlico sound system. Environ Sci Technol 2009;43:7609–13.

[14] Halverson MJ, Pawlowicz R. High-resolution observations of chlorophyll-a biomass from an instrumented ferry: influence of the Fraser River plume from 2003–2006. Cont Shelf Res 2013. https://doi.org/10.1016/j.csr.2013.04.010 59.

[15] Gledhill DK, Wanninkhof R, Millero FJ, Eakin M. ocean acidification of the greater Caribbean region 1996–2006. J Geophys Res 2008;113. https://doi.org/10.1029/2007JC004629 (C10031).

[16] Rossby T, Flagg C, Donohue K. On the variability of gulf stream transport from seasonal to decadal timescales. J Mar Res 2010;68:503–22.

[17] Buijsman MC, Ridderinkhof H. Long-term ferry-ADCP observations of tidal currents in the Marsdiep inlet. J Sea Res 2007;57:237–56.

[18] Petersen W, et al. In: Petersen W, Colijn F, Hydes D, Schroeder F, editors. FerryBox: from online oceanographic observations to environmental information. 2007. EuroGOOS Publication No. 25. EuroGOOS Office, SHMI, 601 76 Norkoepping, Sweden (ISBN 978-91097828-4-4).

[19] Grayek S, Staneva J, Schulz-Stellenfleth J, Petersen W, Stanev E. Use of FerryBox surface temperature and salinity measurements to improve model based state estimates for the German Bight. J Mar Syst 2011. https://doi.org/10.1016/j.jmarsys.2011.02.020.

[20] Korres G, Nittis K, Hoteit I, Triantafyllou G. A high resolution data assimilation system for the Aegean Sea hydrodynamics. J Mar Syst 2009;77:325–40.

[21] Folkestad A, Pettersson LH, Durand DD. Inter-comparison of ocean colour data products during algal blooms in the Skagerrak. Int J Remote Sens 2007;28(3–4):569–92.

[22] Kratzer S, Ebert K, Sørensen K. Monitoring the bio-optical state of the Baltic Sea ecosystem with remote sensing and autonomous in situ techniques. Baltic Sea Basin 2011:407–35. Springer Berlin Heidelberg.

[23] Sørensen K, Grung M, Röttgers R. An intercomparison of in vitro chlorophyll a determinations for MERIS level 2 data validation. Int J Remote Sens 2007;28:537–54. https://doi.org/10.1080/01431160600815533.

[24] Schneider B, Gülzow W, Sadkowiak B, Rehder G. Detecting sinks and sources of CO2 and CH4 by ferrybox-based measurements in the Baltic Sea: three case studies. J Mar Syst December 2014;140(Part A):13–25. https://doi.org/10.1016/j.jmarsys.2014.03.014.

[25] Gülzow W, Rehder G, Schneider V, Deimling J, Seifert T, Tóth Z. One year of continuousmeasurements constraining methane emissions from the Baltic Sea to the atmosphere using a ship of opportunity. Biogeosciences 2013;10(1):81–99.

[26] Kikas V, Lips U. Upwelling characteristics in the gulf of Finland (Baltic Sea) as revealed by ferrybox measurements in 2007–2013. Ocean Sci 2016;12:843–59. https://doi.org/10.5194/os-12-843-2016.

[27] Fleming V, Kaitala S. Phytoplankton spring bloom intensity index for the Baltic Sea estimated for the years 1992 to 2004. Hydrobiologia 2006;554(1):57–65.

[28] Lips I, Lips U. Abiotic factors influencing cyanobacterial in the gulf of Finland (Baltic Sea). Hydrobiologia 2008;614:133–40.

[29] Groetsch PMM, Simis SGH, Eleveld MA, Peters SWM. Cyanobacterial bloom detection based on coherence between ferrybox observations. J Mar Syst December 2014;140(Part A):50–8. https://doi.org/10.1016/j.jmarsys.2014.05.015. ISSN 0924-7963.

[30] Petersen W, Wehde H, Krasemann H, Colijn F, Schroeder F. FerryBox and MERIS — assessment of coastal and shelf sea ecosystems by combining in situ and remotely sensed data. Estuar Coast Shelf Sci 2008;77(2):296–307.

[31] Merckelbach LM, Ridderinkhof H. Estimating suspended sediment concentration using backscatterance from an acoustic Doppler profiling current meter at a site with strong tidal currents. Ocean Dyn 2006;56:153–68. https://doi.org/10.1007/s10236-005-0036-z.

[32] Marrec P, Cariou T, Collin E, Durand A, Latimier M, Macé E, Bozec Y. Seasonal and latitudinal variability of the CO_2 system in the western English Channel based on Voluntary Observing Ship (VOS) measurements. Mar Chem 2013;155:29–41.

[33] Bonato S, Christaki U, Lefebvre A, Lizon F, Thyssen M, Artigas LF. High spatial variability of phytoplankton assessed by flow cytometry, in a dynamic productive coastal area, in spring: the eastern English Channel. Estuar Coast Shelf Sci 2015;154(0):214–23.

[34] Breitbach G, Krasemann H, Behr D, Beringer S, Lange U, Vo N, Schroeder F. Accessing diverse data comprehensively – CODM, the COSYNA data portal. Ocean Sci 2016;12:909–23. https://doi.org/10.5194/os-12-909-2016.

[35] Aiken CM, Petersen W, Schroeder F, Gehrung M, Ramírez von Holle PA. Ship-of-Opportunity monitoring of the Chilean fjords using the pocket FerryBox. J Atmos Ocean Technol 2011;28:1338–50.

[36] Liu X, Tang C, Hou C, Zhang H, Petersen W. Seasonal variation of surface water quality in the Chinese Bohai Strait indicated by FerryBox monitoring data. Environmental monitoring and assessment, submitted January 2018.

[37] Brumovsky M, Becanová J, Kohoutek J, Thomas H, Petersen W, Sørensen K, Sáňka O, Nizzetto L. Exploring the occurrence and distribution of contaminants of emerging concern through unmanned sampling from ships of opportunity in the North Sea. J Mar Syst 2016. https://doi.org/10.1016/j.jmarsys.2016.03.004.

[38] Rossby T. Partnership proposed for ocean observation. Eos Trans. Am. Geophys. Union 2012;93(14):144.

CHAPTER

5.4

Quantum Leap in Platforms of Opportunity: Smart Telecommunication Cables

Christopher R. Barnes

School of Earth and Ocean Sciences, University of Victoria, Victoria, BC, Canada

5.4.1 INTRODUCTION

5.4.1.1 Quantum Leap in Platforms of Opportunity: Smart Telecommunication Cables

The last two decades have witnessed some remarkable innovations in in situ sensors for measuring ocean processes and for better understanding of the health of the oceans in the current digital age. These developments have not been without significant technological and commercial challenges, along with those for successful deployment and operation of observing systems and networks, especially those in the deep ocean. However, these have emerged at a most critical time in human history when the fate of social and planetary systems is in jeopardy.

5.4.1.2 Challenges Posed by Climate and Sea Level Change and Tsunami Hazards

Over the first decade and a half of the 21st century, there has been a rapid, if not yet uniform, acceptance of the severity and cost of climate and sea-level change facing societies and a marked increase in the number of major tsunamis and their consequent profound damage on human life and coastal infrastructure. The year 2016 was the warmest on record and followed the previous 15 years in each setting records and being progressively warmer than the year before. It also recorded the date when the global CO_2 level exceeded 400 ppm, one of the key greenhouse gas benchmarks for global warming (Fig. 5.4.1). In the Arctic, scientists working both on land and offshore recorded a multitude of large craters, interpreted as explosive outgassing features of methane release from subsurface permafrost. The Arctic Ocean has experienced a marked reduction in the area of sea ice (Fig. 5.4.2), as have the ice caps and glaciers of Greenland and West Antarctica, which impact ocean and atmospheric circulation patterns and polar communities.

The overall realization of the scale and magnitude of current and future climate change was aided by the series of 5-year international climate assessments by the International Panel of Climate Change (IPCC 5th Assessment,

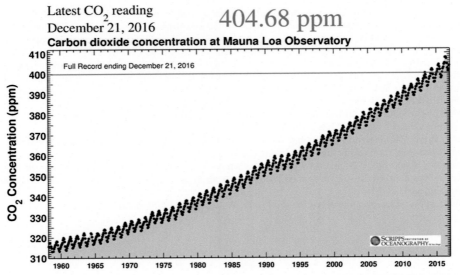

FIGURE 5.4.1   Carbon dioxide record, "Keeling Curve," at Mauna Lao Observatory, Hawaii (1958–2016). *Source: Scripps Institution of Oceanography at UC San Diego:* https://scripps.ucsd.edu/programs/keelingcurve/wp-content/plugins/sio-bluemoon/graphs/mlo_full_record.png.

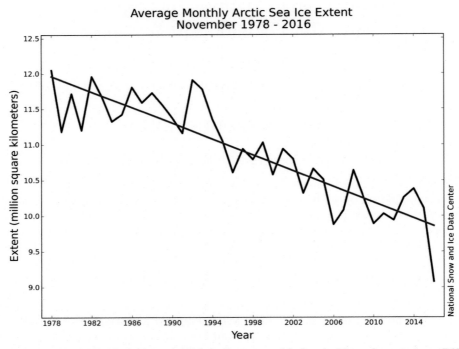

FIGURE 5.4.2   Loss of Arctic sea ice (1978–2016). Through 2016, the linear rate of decline for November extent was 5.0% per decade. *Source: US National Snow and Ice Data Centre:* http://nsidc.org/arcticseaicenews/.

Ref. [1]; 6th Assessment currently being initiated). The last one helped trigger international action by 197 parties that culminated in the Paris Agreement (United Nations Framework Convention on Climate Change [UNFCCC], COP21, 2015 [http://unfccc.int/paris_agreement/items/9485.php]), currently ratified by 153 parties (December 2016).

5.4.1.3 Societal and Economic Costs

Most nations, both developed and less developed, have incurred progressive and large to alarming amounts of national debt. Irrespective of the causes and challenges at reversing debt loads, one consequence is a financial incapacity to deal with the impending scale of environmentally induced catastrophes. Recent events include the 2004 Banda Aceh Sumatra earthquake and tsunami (227,898 dead or missing and damages of $10 billion), the 2005 Hurricane Katrina (over 1800 deaths with the National Oceanic and Atmospheric Administration [NOAA] estimating damage at $198 billion to New Orleans and adjoining regions), and the 2011 Tohuko-Oki earthquake and tsunami, offshore northeast Japan (230,000 homes lost and total damages of $300 billion). For climate change, the intermediate effects are seen as issues of desertification in many parts of Africa, and major droughts in the eastern Mediterranean (2016 was the worst in the last 900 years), India, Indonesia, Cambodia, and California. Over longer timescales, sea-level rise is impacting low-lying areas of many Pacific and Indian Ocean islands, countries (Bangladesh, Vietnam, the Netherlands), and coastal regions/cities (Florida/Miami, New Jersey/New York, Venice). This will inevitably result in mass migration and resettlement of peoples on a far greater scale than the war/economic refugees currently streaming into western Europe.

The key issue is likely to be the incapacity of nations to have the financial resources, political will, global empathy, and organizational support structures to cope with these natural environmental forcing agents. New ways of societal cooperation, global scientific investigations, and long-term planning to allow meaningful management of the planet need to be developed, with at least some ideas being formulated by the new Future Earth Program (http://www.futureearth.org/). Whereas these issues and their potential resolution are well beyond the scope of this chapter, they serve to illustrate the global scale and urgency that are pertinent to the contribution that can be facilitated by progressively installing cost-effective ocean in situ observing infrastructures, such as SMART (Science Monitoring and Reliable Telecommunications) telecommunication cable networks, throughout the world's oceans.

5.4.1.4 New Technological Solutions Offered by SMART Telecommunication Cables

The oceans are the dominant controlling factor in the Earth's climate and by extension in future climate change. They contain much of the Earth's surface heat in contrast to the atmosphere, and absorb about 30% or more of the CO_2 being added to the atmosphere through anthropogenic processes. This in turn is leading to a significant increase in ocean acidification, reef destruction, and expansion of ocean dead zones (hypoxia) [2]. Repeated scientific measurement of ocean conditions is a paramount requirement to properly understanding ocean processes and in assessments of the health of the world's oceans. In recent decades this has been only partially achieved by shipborne observations limited by weather (sea state) conditions and commonly by an inability to make repeated observation from specified sites or transects. Newer programs and technologies have helped to improve these deficiencies with, for example, moored buoys (e.g., OceanSITES program with some moorings down to 5000 m [http://www.oceansites.org/about.html]) and drifting floats (e.g., the Argo program with many floats extending down to 2000 m [http://www.argo.net], see also Chapter 5.7). Some international programs have provided decadal measurements along a few longitudinal transects (pole to pole in the Atlantic and Pacific oceans; [3]) to ascertain deep-ocean bottom temperatures, and through other studies, including passive acoustic thermometry [4]. However, the scientific community has to this stage not captured many long-time series of data from the deeper parts of the world's oceans (i.e., between about 2000 and 7000 m) to fully understand ocean circulation changes and conveyor belt systems. This emphasizes the necessity of securing real-time data from many oceanographic sources to track such potential rapid changes as possible preludes to climate tipping points [5].

One such approach is the introduction of SMART subsea telecommunication cables. A prime advocate is the Joint Task Force (JTF) established in 2012 by the International Telecommunication Union (ITU), World Meteorological Organization (WMO), and UNESCO's Intergovernmental Oceanographic Commission (UNESCO IOC). Details of annual JTF workshops in Rome, Paris, Madrid, Singapore, and Dubai, and many presentations and publications, are available on the JTF website (http://www.itu.int/en/ITU-T/climatechange/task-force-sc/Pages/default.aspx).

5.4.2 CHALLENGES POSED BY CLIMATE CHANGE AND GLOBAL HAZARDS

Key aspects of the health of the oceans, possible future changes, and impacts on the environment and human societies are climate change, sea-level change, and hazards from tsunamis and submarine slope failures. These are considered separately next as the basis for advocating the installation of sensors on subsea telecommunication cables to obtain real-time environmental data over decades to respond to these urgent societal threats.

5.4.2.1 Climate Change

The decadal time series of atmospheric CO_2 at the Mauna Loa Observatory in Hawaii (https://scripps.ucsd.edu/programs/keelingcurve/wp-content/plugins/sio-bluemoon/graphs/mlo_full_record.png) clearly demonstrated the progressive anthropogenic input of greenhouse gases since the industrial revolution (Fig. 5.4.1). The direct correlation to increasing mean global temperatures is now evident (e.g., Ref. [1]). Polar regions are warming at a rate more than most other areas, and satellite data show the remarkable loss of sea ice, especially in the Arctic Ocean (about 5%/decade in the Arctic Ocean and accelerating; Fig. 5.4.2), where the loss directly follows anthropogenic CO_2 emissions [6]. The last decade has witnessed an almost exponential increase in these temperatures, with each year setting new records. With the scale of future climate change noted earlier and the consequences for changing ocean circulation and deep-ocean temperature increases, it is now imperative that new approaches be implemented to secure ocean bottom environmental conditions in real-time over decades to properly understand and assist in predictions of the changing Earth's climate, especially the potential for short-term tipping points [5]. Ref. [7] cautioned that the scale and duration of anthropogenic atmospheric greenhouse gas input has no parallel within the last 66 million years and hence no comparable analogs can be derived from the geologic record. The climate issue has been recognized more formally through the Paris Agreement (http://unfccc.int/paris_agreement/items/9485.php) involving 197 nations, with world leaders, bankers, and environmentalists urging immediate action to curb emissions and implement firm targets for reductions. The time is ripe for consideration to adopt other technologies such as SMART telecommunication cables.

5.4.2.2 Sea-Level Change

The progressive increase in global temperature, noted earlier, has resulted in an increase in sea-level rise. A rise of about 1 m by 2100 is predicted by the modeling used in the IPCC 5th Assessment ([1]; Fig. 5.4.3). This is attributed mainly to three factors: polar ice cap melting, glacier melting, and ocean thermal expansion. However, sea ice loss and melting of the Arctic and West Antarctica ice caps seem to be occurring faster than the earlier models predicted [6,8]. NOAA's 2016 Arctic Report Card noted that average air temperatures were unprecedented, being the highest on observational record, and Arctic temperatures continue to increase at double that of global temperature increase. The possibility of faster ice shelf calving suggested a sea-level rise of over 2 m [9,10]. Ref. [11] argued that these estimates of sea-level rise may be substantially low and "that continued high fossil fuel emissions leading to CO_2 ~600–900 ppm will cause exponential ice mass loss up to several meters of sea level." Globally, sea-level increases are not uniform due to different gravitational forces, ice cap gravitational pull, isostatic rebound and differential loading, and thermal expansion. Sea-level rise will have enormous impact on populations and cities occupying coastal and low-lying areas (e.g., some Pacific and Indian ocean island states, Bangladesh, Vietnam, Netherlands, Florida, Louisiana). Many populations will have to migrate and be resettled, causing political and economic difficulties. The result is an urgent need to generate accurate sea-level measurements and track changes with precise time series. Some data are provided by GRACE satellites, but these need to be ground-truthed by several fixed observatories. SMART subsea telecommunication cables with pressure sensors every 50–100 km would provide complementary precise sea-level data.

5.4.2.3 Tsunamis and Submarine Slope Failures

Tsunamis are generated by differential vertical displacement of the seafloor by earthquakes and faulting, and also by rapid seafloor displacement through large submarine slope failures. Large earthquakes can trigger such failures, so these factors can commonly be interrelated and potentially magnified. Regions that are tectonically active generate the most tsunamis such as around the Pacific Ocean and within the Mediterranean Sea; evidence of large slope failures is seen along continental margins, off submarine canyons, and off large islands such as Hawaii and Taiwan with many resulting in severed telecommunication cables. Several major tsunamis occurred worldwide in the last decade and a half, notably associated with earthquakes between M_w 7.7 and 9.1 as in Sumatra (2004), Java (2006),

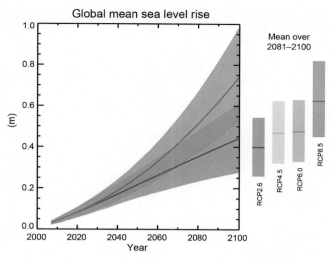

FIGURE 5.4.3   Sea-level change 2000–2100. A new set of scenarios, the Representative Concentration Pathways (RCPs), was used for the new climate model simulations carried out under the framework of the Coupled Model Intercomparison Project Phase 5 (CMIP5) of the World Climate Research Programme. Time series of projections and a measure of uncertainty (shading) are shown for scenarios RCP2.6 (blue) and RCP8.5 (red). *Source: IPCC WG1 Technical Summary (2013); Figure TS.3 from Stocker TF, Qin D, Plattner G-K, Alexander LV, Allen SK, Bindoff NL, Bréon F-M, Church JA, Cubasch U, Emori S, Forster P, Friedlingstein P, Gillett N, Gregory JM, Hartmann DL, Jansen E, Kirtman B, Knutti R, Krishna Kumar K, Lemke P, Marotzke J, Masson-Delmotte V, Meehl GA, Mokhov II, Piao S, Ramaswamy V, Randall D, Rhein M, Rojas M, Sabine C, Shindell D, Talley LD, Vaughan DG, Xie S-P. Technical summary. In: Stocker TF, Qin D, Plattner G-K, Tignor M, Allen SK, Boschung J, Nauels A, Xia Y, Bex V, Midgley PM editors. Climate change 2013: the physical science basis. Contribution of working group I to the fifth assessment report of the intergovernmental panel on climate change. Cambridge (United Kingdom) and New York (NY, USA): Cambridge University Press; 2013:* https://www.ipcc.ch/report/graphics/index.php?t=Assessment%20Reports&r=AR5%20-%20WG1&f=Technical%20Summary.

US Samoa (2009), Mantawai (2010), Chile (2010), and Japan (2011). While tsunami waves may spread across oceans affecting many coastal communities, those initiated near dense populations have the most dramatic impact and cost to populations and societal infrastructure. Such destruction over the last decade has resulted in billions of dollars' worth of damage and killed hundreds of thousands of people. Reducing such losses and mitigating damage is a key factor in developing tsunami warning systems [12,13]. A network of SMART subsea telecommunication cables with pressure sensors and accelerometers every 50–100 km across the oceans would ultimately produce a remarkable seismic and tsunami monitoring network. Current tsunami warning systems such as DART buoys are less widely distributed, prone to vandalism, and have high maintenance costs.

5.4.3 A NEW ERA OF SEAFLOOR OBSERVATIONS, CABLED OBSERVATORIES, AND POTENTIAL MINI-OBSERVATORIES

The past century of oceanographic scientific observations has been primarily from research vessels, limited by funding, logistics, and by inclement weather and sea state in the higher latitudes and open ocean. In 2001, Walter Munk (Scripps Institution of Oceanography) observed that the last century of physical oceanography was marked most by the degree of undersampling that allowed poorly substantiated hypotheses. The past few decades have seen major developments in ocean instrument technology—some with remote deployments through remotely operated vehicles, autonomous underwater vehicles, gliders, drones, and more sophisticated buoys. Sensor data are commonly transmitted back via satellite communication (expensive; some latency issues). Recently, scientific cabled ocean observatories were successfully deployed, hosting hundreds of sensors with real-time instrument or robotic control and data return accomplished by using state-of-the-art telecommunication cables to shore stations. These have proven that sensors can be deployed for extended periods with high reliability and connected in a variety of ways to the backbone or secondary cables, as needed for SMART cable systems.

5.4.3.1 Current Distribution of Subsea Telecommunication Cables

Over a million kilometers of subsea optical fiber telecommunication cables have been deployed across the world's oceans (Fig. 5.4.4). They provide global high-speed interconnectivity (currently up to 80 Tb/s) for telephone, digital,

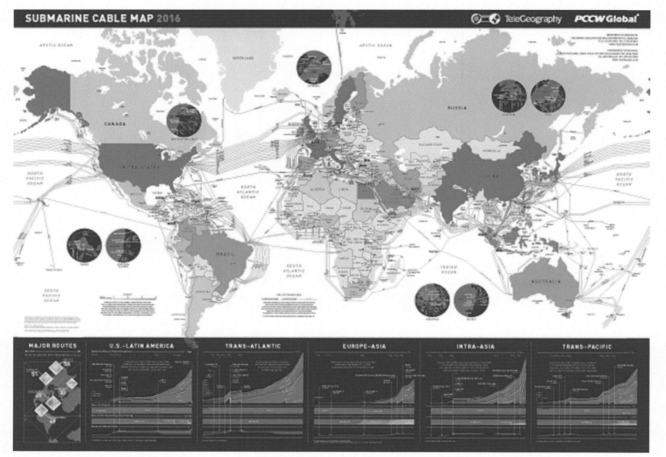

FIGURE 5.4.4   TeleGeography's Submarine Cable Map for 2016 depicts 321 cable systems that are currently active, under construction, or expected to be fully funded by the end of 2016. *Source: TeleGeography:* https://www.telegeography.com/telecom-maps/submarine-cable-map/.

video, high-frequency stock market trading, and petabytes of scientific data. Optical repeaters are regularly spaced at 50–100 km intervals to reamplify the signal carried by the cables' fibers. The International Cable Protection Committee (ICPC) is a non-profit corporation that helps to protect submarine cables from manmade and natural hazards through a variety of services and is a source of data on all deployed cables. JTF advocates for SMART cable systems to progressively replace the traditional systems over the next few decades.

5.4.3.2 Scientific Ocean Observatories, Arrays, and Sensor Developments

Notable examples of scientific cabled ocean observatories with an operational history include: the NEPTUNE and VENUS observatories within Ocean Networks Canada, DONET in Japan, the US Ocean Observatories Initiative's Cabled Array, and the Aloha Ocean Observatory in Hawaii [14–19]. These use commercial cable systems and deployment methods and generate free real-time scientific data from hundreds of networked sensors at selected locations in, on, and above the seafloor. These observatories have collaborated with sensor companies to foster and test the development of more advanced sensors with greater precision, reliability, and depth range.

A wider, government-funded application has been off the Japanese east coast where the Japan Trench Observation and Tsunami Warning System has been deployed recently (following the 2011 moment magnitude [M_w] 9.0 Tohoku-Oki earthquake and tsunami) with over 5200 km of cable supporting 154 instrument stations at depths to over 4000 m. An example of new sensor package technology is the Geodesy + Ocean Disasters Sensors system developed by Paroscientific Inc. and Quartz Seismic Sensors Inc. in Seattle, USA. The package includes deep-sea absolute pressure gauges (APGs), a triaxial accelerometer, and nanoresolution processing electronics, with an in situ calibration system that eliminates pressure sensor drift (http://www.paroscientific.com/pdf/Nanoresolution_Sensors_for_Disaster_Warning_Systems.pdf). RBR Ltd., Ottawa, Canada, produces a combination of RBR*duo* two-channel logger and Paroscientific Digiquartz pressure and temperature transducer, with a depth range to 7000 m. The role of arrays of APGs, which measure changes in the mass of overlying ocean, from which seafloor uplift or subsidence can be determined, has been discussed by Ref. [20] in regard to seafloor deformation observations made during a slow-slip event in the 2014 Hikurangi subduction zone offshore earthquake, New Zealand.

5.4.4 SENSOR PACKAGES APPLIED TO REPEATER SYSTEMS ALONG SMART TELECOMMUNICATION CABLES

The concept advocated by JTF is to add sensor packages to each, or a variable number of, repeaters along commercial telecommunication cables, particularly those spanning major oceans. These would be applied only to new and refurbished cables. For simplicity and to minimize costs, sensor packages would host three basic types to measure temperature, pressure, and acceleration. Currently available sensors are relatively small, reliable, inexpensive, can be clustered as a package without interference, and have been deployed on research cabled observatories.

5.4.4.1 Temperature, Pressure, and Acceleration Sensors—Current Availability and Applications/Limitations

The JTF Engineering team produced a White Paper (June 2016) on "General Requirements for Sensor Enabled and Reliable Telecommunications (SMART) Cable Systems" with specific performance parameters for the sensor package (Table 5.4.1–5.4.3; http://www.itu.int/en/ITU-T/climatechange/task-force-sc/Documents/General-Requirements-of-a-SMART-Cable-Issue-1.0.pdf).

5.4.4.1.1 *Temperature Sensor*

Temperature sensors shall have the performance parameters given in Table 5.4.1.

5.4.4.1.2 *Absolute Pressure Gauge*

Absolute pressure sensors shall have the performance parameters given in Table 5.4.2.

5.4.4.1.3 *Acceleration Sensor*

Acceleration sensors shall have the performance parameters given in Table 5.4.3.

5.4.4.2 Connection Options to the Repeaters, Data Transmission, Maintenance, Reliability, and Wet Demonstrator Facility

Initial consideration of how to attach the sensors to the repeaters examined the option of having them within the repeaters. However, internal heat generation could impact the temperature sensor and the pressure sensor sensitivity could be compromised by the casing. Having the sensor package at a few meters' distance either within a blister on a secondary cable or simply attached by a secondary cable have been the most favored options to date. Importantly, the connection into the repeater must not affect the fibers carrying commercial traffic, so real-time data transmission from the sensors could be via the supervisory (back) channel that monitors overall cable system conditions.

The amount of data from the three basic sensors would be minor. For the data to alert for tsunamis and earthquakes, it would have to be free and immediately available to send to preexisting systems/agencies such as the Incorporated Research Institutions for Seismology and the US Geological Survey for global electronic distribution. The sensor package would be relatively small and with its secondary cable would need to be smoothly deployed

TABLE 5.4.1   Temperature Sensor Parameters

Range	−5.0 to +35°C
Initial accuracy	±0.001°C
Stability	0.002°C/year
Sampling rate	0.1 Hz
Sample resolution	24 bits

TABLE 5.4.2   Pressure Sensor Parameters

Range	0–73 MPa (0–7000 m)
Overpressure tolerance	84 MPa (8000 m)
Accuracy	±1 mm relative to recent measurements 0.01% of full range absolute
Maximum allowable drift after a settling-in period	0.2 dbar/year
Accuracy after drift correction	For further study
Hysteresis	≤±0.005% of full scale
Repeatability	≤±0.005% of full scale
Sampling rate	20 Hz
Noise floor	0.14 Pa/Hz
Sample resolution	32 bits
Temperature sensor sampling rate	20 Hz
Temperature sensor resolution	24 bits

TABLE 5.4.3   Acceleration Sensor Parameters

Configuration	3-Axis
Response	0.1–200 Hz
Resonance frequency	>2000 Hz
Full-scale range	±1.5 g where g is 9.806 m/s
Noise	≤2 ng/√Hz
Amplitude response	±1% across frequency range
Linearity	±1% of full scale
Cross-axis sensitivity	<1%
Sampling rate	200 Hz
Sample resolution	24 bit

from the cable ship along with the repeaters during cable lay. No maintenance is envisaged after deployment. Any failure of one of more sensors in a package would have relatively minor consequences given the hundreds that would be deployed on a transocean system. Sensors are anticipated to last for a decade or more, with good reliability and minimal drift. During that period, additional transocean cables would be laid and supplement or effectively replace ailing sensors on older cables.

The JTF is planning to establish a Demonstrator Project to show the effectiveness of deployment and operation of a SMART cable system by using a cable ship, commercial cables, and existing commercial sensors attached to repeaters on commercial cables, at depths of about 2000 m, with real-time data collected and distributed over a year or two (Fig. 5.4.5). The likely arrangement would be to collaborate with an existing cabled ocean observatory that has systems, staff, proximity to a cable ship, and a data management group. Three such observatories (Ocean Networks Canada, Aloha Ocean Observatory, and the European Multidisciplinary Seafloor and water-column Observatory) have each expressed an interest in such a partnership with JTF with in-kind support; a recent (2017) Request for Expression of Interest in the project received over 30 responses and a Request for Proposals is expected to be issued in late 2018. JTF issued a document in 2015 on "Scope document and budgetary cost estimate for a wet test to demonstrate the feasibility of installing sensors external to the repeater and to provide data from such sensors for evaluation" (http://www.itu.int/en/ITU-T/climatechange/task-force-sc/Documents/Wet-demonstrator-requirements-2015-05.pdf). This was amplified by a more detailed document in which more specific technical details and

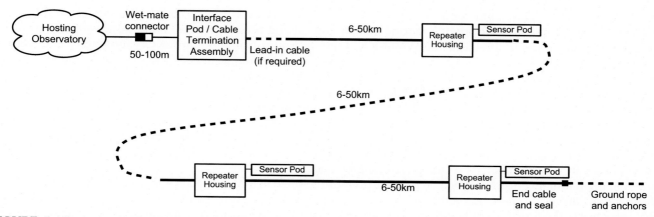

FIGURE 5.4.5   Sensor-enabled Science Monitoring and Reliable Telecommunications (SMART) cable systems: Wet Demonstrator Project Description, Figure 1 in JTF Engineering Team White Paper, Issue 1.0, July 2016. *Source: International Telecommunications Union:* http://www.itu.int/en/ITU-T/climatechange/task-force-sc/Documents/Wet-Demonstrator-Design-Issue-1.0.pdf.

requirements of the Wet Demonstrator Project were provided for "Sensor Enabled Scientific Monitoring And Reliable Telecommunications (SMART) Cable Systems" (July 2016) (http://www.itu.int/en/ITU-T/climatechange/task-force-sc/Documents/Wet-Demonstrator-Design-Issue-1.0.pdf). Other details were addressed at the 5th JTF Workshop, April 2016, Dubai (http://www.itu.int/en/ITU-T/Workshops-and-Seminars/5-ws-smart-cable-systems/Pages/default.aspx). It is anticipated that this demonstration would (1) assist industry to move more rapidly to install SMART cables and (2) show agencies or governments that could/should use such environmental data (e.g., Pacific Tsunami Warning Center, United States Geological Survey, and NOAA/US; Japan Meteorological Agency; Central Weather Bureau, Taiwan; Intergovernmental Coordination Group for the Tsunami Early Warning and Mitigation System in the North-Eastern Atlantic, the Mediterranean and connected seas; and Global Earth Observation System of Systems [GEOSS]) the value of funding partnerships to gather such data to better deal with the global and regional environmental crises and threats.

5.4.5 CHALLENGES AND OPTIONS FOR ACHIEVING PROGRESSIVE INSTALLATIONS ON SENSORS/CABLE NETWORKS

Why have not SMART subsea cable systems been deployed at this stage and what will it take to realize such deployments in the (near) future? Although the concept has been known for some time, highlighted in a short note by Ref. [21] and then advocated at industry and science workshops and conferences over the past 5 years by JTF, there is clearly some inertia or impediments to overcome. Industry appears to acknowledge that the technical issues are relatively minor and many have internal capacity or have already developed SMART systems. Some telecommunication companies have individually noted that this new concept would add a cost (estimated at about 5%–8% of the total cost of a transocean cable system). Suppliers would look to owners of the systems or nontraditional users to cover the extra costs. The latter, such as environmental agencies, commonly do not yet fully realize the potential to have and utilize these environmental data. There is currently a gap in understanding between owners, suppliers, and nontraditional end-users, yet there is a distinct business opportunity for such dual-use cables [22]. What new communications can be used to bridge this gap and by whom, and how can environmental urgency elaborated earlier be better appreciated and accommodated?

5.4.5.1 Roles of JTF, ITU, WMO, UNESCO IOC, Paris Climate Agreement, IPCC, International Agencies (GOOS, GEOSS), National Agencies (NOAA, ESA), Commercial Sector, and ICPC

There is a continuing role for advocacy, planning, funding, and actions by the JTF, with an increasing involvement through the energy and funding of the three sponsoring UN agencies. ITU provides the secretariat for the work of the JTF; ITU's Telecommunication Standardization Sector (ITU-T) assembles experts from around the world to develop international standards known as ITU-T Recommendations, which act as defining elements in the global infrastructure of information and communication technologies. WMO has responsibilities for climate and IOC for oceans and tsunami hazards. Worldwide marine meteorological and oceanographic communities are also working in partnership under the umbrella of the WMO-IOC Joint Technical Commission for Oceanography and Marine Meteorology to respond to interdisciplinary requirements for meteorological/ocean observations, data management, and service products. One group that should become more involved is the Global Ocean Observing System (GOOS), a program executed by the IOC of UNESCO. These UN agencies and programs clearly have an opportunity and responsibility to ensure that the JTF's SMART subsea cable initiative is advanced expeditiously. Another group is the Group on Earth Observations that is creating a GEOSS to better integrate observing systems and share data by connecting existing infrastructures using common standards. There are over 200 million open data resources in GEOSS from more than 150 national and regional providers such as NASA and ESA, international organizations such as WMO, and the commercial sector such as Digital Globe.

International climate concern and related political pressure resulted in the UNFCCC COP21 Paris Climate Agreement in 2015 (http://unfccc.int/paris_agreement/items/9485.php). "The Paris Agreement's central aim is to strengthen the global response to the threat of climate change by keeping a global temperature rise this century well below 2 degrees Celsius above pre-industrial levels and to pursue efforts to limit the temperature increase even further to 1.5 degrees Celsius. Additionally, the agreement aims to strengthen the ability of countries to deal with the impacts of climate change. To reach these ambitious goals, appropriate financial flows, a new technology framework and an enhanced capacity building framework will be put in place, thus supporting action by developing countries and the most vulnerable countries, in line with their own national objectives." A year after the initial agreement, 153 Parties (nations) have ratified of the 197 Parties to the Convention. The time is ripe to fully promote the concept of SMART subsea cable systems to deliver critical new data for better understanding global climate and sea-level change and tsunami hazards.

Industry should also seize the commercial opportunities of a dual cable system, marketing the environmental data component directly through other partners. This is likely to emerge on a new individual system where environmental agencies in the participating countries would see the clear benefits. An example would be between Japan and the United States, where societal concern and mature agencies exist. There may be a potential to bring new attitudes with the new IT players in the telecommunication cable business such as Google, Microsoft, and Apple that have financial muscle and can promote concern for the environment through company vision. These new players may influence changes to the work of the ICPC that could broaden its vision statement and activities beyond "To be the international submarine cable authority providing leadership and guidance on issues related to submarine cable security and reliability" (https://www.iscpc.org/).

5.4.5.2 The Argo Float Program as a Comparison for Developing a New Ocean-Observing Program

The advocacy of SMART ocean-observing systems by JTF and other agencies to the nontraditional user community to build greater inclusion and adoption may require additional funding and organization. A comparative example for the upper ocean is the Argo floats program of JCOMMOPS (http://www.argo.net; http://www.jcommops.org/board) within the GOOS (within UNESCO IOC). This has seen over the past 18 years the progressive increase in the number of floats with key sensors throughout the world's oceans as the user community increased and more nations participated in the organization. The Argo website offers the following pertinent description of a comparative program to the SMART systems primarily in the deep ocean: "Argo is a global array of 3800 free-drifting profiling floats that measures the temperature and salinity of the upper 2000 m of the ocean. This allows, for the first time, continuous monitoring of the temperature, salinity, and velocity of the upper ocean, with all data being relayed and made publicly available within hours after collection. Deployments began in 2000 and continue today at the rate of about 800 per year. The array is made up of 30 different countries' contributions that range from a single float, to the U.S. contribution, which is roughly 50% of the global array. Funding mechanisms differ widely between countries and involve over 50 research and operational agencies. Each national program has its own priorities but all nations subscribe to the goal of building the global array and to Argo's open data policy. The project is overseen by an International Argo Steering and a Data Management Team that are comprised of representatives of float-providing countries. The array's growth is monitored by the Technical Coordinator at the Argo Information Center (AIC) that is located in Toulouse as part of the JCOMMOPS monitoring and coordinating system for operational ocean observations." These Argo observations have revealed in detail the extent of ocean warming over the last decade [23] and resulting thermosteric sea-level rise [24].

With the successful development of the Argo program over almost 20 years [25], it continues to evolve, specifically by adding floats to penetrate below 2000 m (deep Argo) and by adding biogeochemical sensors (Biogeochemical-Argo) to extend initially to 1000 m. Ref. [26] noted that the biochemical sensors would provide real-time data for pH, oxygen, nitrate, chlorophyll, suspended particles, and irradiance to address issues of ocean ecology, metabolism, and carbon uptake. This initiative has recently been endorsed by the G7 nations; a global system would cost about US$25M annually.

5.4.6 SUMMARY

As detailed earlier, there is an urgent global societal imperative to address the issues of climate and sea-level change and tsunami and submarine slope failure hazards. The costs of inaction will be enormous. A key gap in scientific knowledge and monitoring capability is in understanding the deep ocean. Installing SMART telecommunication cable systems would contribute profoundly to the precise measurement in real-time of temperature, pressure, and acceleration. There is no significant technical impediment. Suitable sensors are already commercially available and three-sensor packages (T, P, Accel) can be attached with some additional engineering design to repeaters along new or refurbished transocean and regional cable systems. The JTF is planning a deep-sea Demonstrator Project in partnership with an operational scientific cabled ocean observatory and interested industry companies, using industry deployment methods, technical components, and operational procedures as proof-of-concept.

Advocacy will be expanded to the nontraditional user community interested in acquiring real-time environmental data. It is anticipated that the latter agencies and governments will become engaged in one or more future transocean cable systems ahead of the commercial financing and deployment decisions to ensure adequate funding to cover the additional modest costs of a dual-purpose SMART cable system. This would apply to new and/or refurbished cable systems for transocean or shorter routes such as between western Pacific island states. Given the need and potential commercial market potential, it is envisaged that once one or two SMART cable systems were in service, then a majority of future cables would be of this type and thus progressively establish a global network over a decade or two.

The real-time environmental data could readily be received by one or more national/international agencies for rapid dissemination, especially in response to short-term events such as tsunamis, submarine slope failures, and for monitoring longer-term trends and short-term tipping points in climate and sea-level change.

Acknowledgments

The Executive and Plenary members of the Joint Task Force and the parent agencies of ITU, WMO, and UNESCO IOC are acknowledged for their contributions and interest in developing the concept of SMART ocean cabled systems. This chapter attempts to summarize much of the work of JTF to date (2016) and the prospects for the future progressive commercial implementation of such systems, which represent a quantum leap in platforms of opportunity that address the challenges and innovations in ocean in situ sensors.

References

[1] Stocker TF, Qin D, Plattner G-K, Alexander LV, Allen SK, Bindoff NL, et al. (and 28 other co-authors). Technical summary, in climate change. 2013. In: Stocker TF, Qin D, Plattner G-K, Tignor M, Allen SK, Boschung J, et al., editors. The physical science basis. Contribution of working group I to the fifth assessment report of the intergovernmental panel on climate change. Cambridge (United Kingdom) and New York (NY, USA): Cambridge Univ. Press; 2013. p. 33–115.

[2] Levin LA, Le Bris N. The deep ocean under climate change. Science 2015;350:766–8.

[3] Purkey SG, Johnson GC. Warming of global abyssal and deep Southern Ocean waters between the 1990s and 2000s: contributions to global heat and sea level rise budgets. J Clim 2010;23:6336–51.

[4] Sabra KG, Cornuelle B, Kuperman WA. Sensing deep-ocean temperatures. Phys Today 2016;69:32–8.

[5] Turney CSM, Fogwill CJ, Lenton TM, Jones RT, von Gunten L, editors. Tipping points. Full issue of past global changes (PAGES) magazine, vol. 24. 2016 (1), August, PAGES-Future Earth, 51 p.

[6] Notz D, Stroeve J. Observed Arctic sea-ice loss directly follows anthropogenic CO_2 emission. Science 2016;354(6313):747–50.

[7] Zeebe RE, Zachos JC. Long-term legacy of massive carbon input to the Earth system: anthropocene versus eocene. Philos Trans R Soc A 2013;371:20120006.

[8] Paolo FS, Fricker HA, Padman L. Volume loss from Antarctic ice shelves is accelerating. Science 2015;348(6232):327–31.

[9] DeConto RM, Pollard D. Contribution of Antarctica to past and future sea-level rise. Nat Geosci 2016;531:591–7.

[10] Oppenheimer O, Alley RB. How high will the seas rise? Science 2016;354(6318):1375–7.

[11] Hansen J, Sato M, Hearty P, Ruedy R, Kelley RM, Masson-Delmotte V, et al. Ice melt, sea level rise and superstorms: evidence from paleoclimate data, climate modeling, and modern observations that 2°C global warming could be dangerous. Atmos Chem Phys 2016;16:3761–812.

[12] Bernard EN, Robinson AR. Introduction: emergent findings and new directions in tsunami science. In: Bernard EN, Robinson AR, editors. The Sea, vol. 15. Cambridge (MA) and London (England): Harvard University Press; 2009. p. 1–22.

[13] Whitmore PM. Tsunami warning systems. In: Bernard EN, Robinson AR, editors. Tsunamis. The sea, vol. 15. Cambridge (MA) and London (England): Harvard University Press; 2009. p. 401–42.

[14] Barnes CR, Best MMR, Johnson FR, Pautet L, Pirenne B. Challenges, benefits and opportunities in installing and operating cabled ocean observatories: perspectives from NEPTUNE Canada. IEEE J Ocean Eng 2013;28:144–57.

[15] Barnes CR, Best MMR, Johnson FJ, Pirenne B. The NEPTUNE Canada Project: installing the world's first regional cabled ocean observatory. In: Favali P, Santis A, Beranzoli L, editors. Ocean observatories: a new vision of the Earth from the Abyss. Springer Praxis Books, Springer-Verlag Berlin Heidelberg; 2015. p. 415–38. [Chapter 16].

[16] Howe B, Duennebier F, Lukas R. The Aloha cabled observatory. In: Favali P, Santis A, Beranzoli L, editors. Ocean observatories: a new vision of the Earth from the Abyss. Springer Praxis Books, Springer-Verlag Berlin Heidelberg; 2015. p. 439–63. [Chapter 17].

[17] Kaneda Y, Kawaguchi K, Araki E, Matsumoto H, Nakamura T, Kamiya S, et al. Development and application of an advanced ocean floor network system for megathrust earthquakes and tsunamis. In: Favali P, Santis A, Beranzoli L, editors. Ocean observatories: a new vision of the Earth from the Abyss. Springer Praxis Books, Springer-Verlag Berlin Heidelberg; 2015. p. 643–62. [Chapter 25].

[18] Kelley DS, Delaney JR, Juniper SK. Establishing a new era of submarine volcanic observatories: Cabling Axial Seamount and the Endeavour Segment of the Juan de Fuca Ridge. Mar Geol 2014;352:426–50.

[19] Kawaguchi K, Kaneko S, Nishida T, Komine T. Construction of the DONET real-time seafloor observatory for earthquakes and tsunami monitoring. In: Favali P, Santis A, Beranzoli L, editors. Ocean observatories: a new vision of the Earth from the Abyss. Springer Praxis Books, Springer-Verlag Berlin Heidelberg; 2015. p. 211–28. [Chapter 10].

[20] Wallace LM, Webb SC, Ito Y, Mochizuki K, Hino R, Henrys S, et al. Slow slip near the trench at the Hikurangi subduction zone, New Zealand. Science 2016;352(6286):701–4.

[21] You Y. Harnessing telecoms cables for science. Nature 2010;466:690–1.

[22] Barnes CR, Meldrum DT, Ota H. Emerging subsea networks: new market opportunities for, and societal contributions from, SMART cable systems. In: Proceedings of SubOptic 2016, Dubai. April 2016. p. 18–21. 11 p.

[23] Roemmich D, Church J, Gilson J, Monselesan D, Sutton P, Wijffels S. Unabated planetary warming and its ocean structure since 2006. Nat Clim Change 2015;5:240–5.

[24] Levitus S, Antonov JI, Boyer TP, Baranova OK, Garcia HE, Locarnini RA, et al. World ocean heat content and thermosteric sea level change (0–2000 m), 1955–2010. Geophys Res Lett 2012;39. L10603 (5 pp.).

[25] Riser S, et al. Fifteen years of ocean observations with the global Argo array. Nat Clim Change 2016;6:145–53.

[26] Johnson KS, Claustre H. Bringing biogeochemistry into the Argo age. EOS November 08, 2016;97. https://doi.org/10.1029/2016EO062427.

Further Reading

[1] Abraham JP, Baringer M, Bindoff NL, Boyer T, Cheng LJ, Church JA. A review of global ocean temperature observations: implications for ocean heat content estimates and climate change. Rev Geophys 2013;51:450–83.

C H A P T E R

5.5

Innovations in Cabled Observatories

S. Kim Juniper, Benoît Pirenne, Adrian Round, Scott McLean

Ocean Networks Canada, University of Victoria, Victoria, BC, Canada

5.5.1 INTRODUCTION

Connecting ocean sensors by cable to shore facilities extends the Internet under the ocean. Cabled observatories enable continuous, high-frequency, real-time ocean observing by providing power and communications to sensors monitoring the seabed, the water column, and the sea surface, generating time series of ocean data over design lives of 25 years or more. These networked ocean-observing systems can also include shore-based components, such as high-frequency oceanographic radars (e.g., CODAR, WERA, WaMoS),[1] coastal weather stations, and automatic

---

[1] CODAR—Coastal Ocean Dynamics Applications Radar. A high-frequency (3–50 MHz) oceanographic radar designed for near-surface monitoring of ocean waves and currents.
WaMoS—Wave Monitoring System—X-band or microwave frequency (7–12 GHz) marine radar that allows for very high-resolution imagery of objects and ocean surface features.
WERA—WavE Radar—A high-frequency (4–50 MHz) over-the-horizon radar that is deployed with an array of receivers to produce ocean surface current and wave maps.

identification system (AIS) receivers that monitor vessel traffic. The primary focus of this chapter will be technological. Following a brief overview of the scientific and operational needs for continuous, real-time, high-bandwidth ocean observing (Section 5.5.2), the chapter will examine the basic hardware components of cabled observatories (Section 5.5.3), the logistics of deploying and maintaining cabled observing systems (Section 5.5.4), and shore-based observatory control and data acquisition, archiving, and distribution (Section 5.5.5). Section 5.5.6 will touch briefly on technologies and policies related to access to observatory data and the development of data products for different user sectors, from ocean scientists to the general public. Sections 5.5.3–5.5.6 will draw on our knowledge of Ocean Networks Canada's (ONCs) cabled observatories, with which the authors are most familiar, and include examples from other cabled ocean-observing systems worldwide. Forward-looking Section 5.5.7 will consider current and future technological solutions for continuous, real-time ocean observing in remote locations.

5.5.2 WHY DO WE NEED CABLED OBSERVATORIES?

Cabling is obviously not the only solution for powering oceanographic instruments or handling their data output. Autonomous and shipboard deployments of sensors are much more common ways of collecting ocean data, and long-term autonomous deployments on stationary and mobile platforms are benefiting from technological advances that are reducing power requirements (thereby lengthening battery life) and increasing onboard data storage capacity. Nevertheless, energivorous (power hungry) devices such as acoustic Doppler current profilers (ADCPs), video lights, and sampling pumps are important components of observing systems, as are devices producing high data volumes such as scanning sonars, hydrophones, and high-definition video cameras. As well, the lengthy, continuous operation of large numbers of sensors and sensor networks often requires external power and data storage solutions. The primary research driver for the development of cabled ocean observatories has been the need for uninterrupted, long-term, high-resolution time series of ocean and Earth variables, unfettered by power requirements and data storage constraints, to address priority questions about process connections in the ocean, ocean change, and geohazards. Additional drivers have arisen from increasing societal need for observational data, especially real-time, to support maritime safety, ocean health monitoring, and hazard mitigation. Cabled observatory technology can represent the best solution for these research and operational requirements, where there is sufficient scientific and/or societal interest to support the capital cost of observatory construction and subsequent operation.

5.5.3 CABLED OBSERVATORY TECHNOLOGIES—BRIEF HISTORY AND COMPONENT DESCRIPTION

The first major cabled ocean observatory initiative was the NEPTUNE project [1], a United States/Canada consortium that sought to build a cabled ocean observatory at the scale of the Juan de Fuca tectonic plate in the northeast Pacific Ocean. NEPTUNE emerged from increasing recognition of the limitations of ship-based expeditions for directly observing high-impact submarine volcanic events on the Juan de Fuca Ridge, and the need for long-term, continuous monitoring to study these and other dynamic ocean systems [2–4]. NEPTUNE was conceived in the late 1990s during a period of exponential growth of the Internet[2] and its adoption by the research community. Much of the technology that made cabled ocean observatories possible emerged from Internet-driven advances in transoceanic telecommunications, such as optical amplification systems, cable manufacturing facilities, and cable deployment and repair vessels. These technological advances enabled the construction of low-cost, high-bandwidth commercial data links between continents. According to a recent estimate, 99% of today's intercontinental information traffic is routed through submarine fiber-optic cables [5], with only a small fraction using costlier satellite connections. Similarly, while satellite data links permit widespread collection of ocean data by autonomous and often mobile sensor packages, cabled systems can represent the most cost-effective solution for high-volume data transmission from fixed-point observing systems where a cabled connection to land is feasible.

High-power and high-bandwidth connections from land-based control centers to ocean sensors are the essential underlying feature of cabled ocean observatories. They enable the continuous, real-time acquisition of ocean data and real-time control of sensors and other devices, with virtually none of the data storage or power constraints that limit ocean observing by autonomous (fixed or mobile) instrument platforms. Subsea observatory cable connections to shore stations can range in length from tens of meters for coastal platforms to hundreds of kilometers for offshore

---

[2] Between 1990 and 2016, global Internet traffic grew by seven orders of magnitude, from 0.001 PB/month to >95,000 PB/month [30].

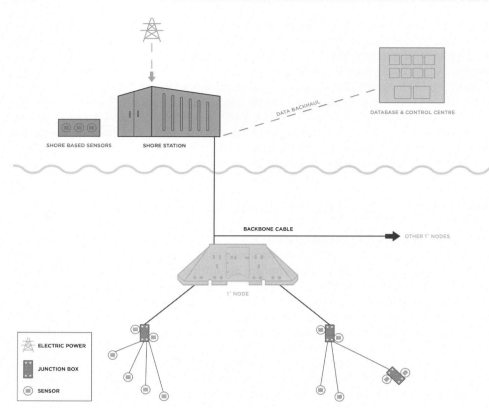

FIGURE 5.5.1 Schematic representation of major components of the primary and secondary subsea infrastructure for cabled ocean observatories. Primary infrastructure consists of a backbone cable that provides power and communications from shore station to single or multiple, networked primary nodes. Secondary subsea infrastructure set-ups vary between systems but most consist of extension cables to instrument platforms supporting junction boxes or similar local hubs that provide required voltage and data links to platform-mounted or off-platform sensor(s).

networks. Cables with length exceeding a few hundred meters have optical fiber cores for data transmission, accompanied by copper power conductors. Shorter cables can make use of twisted-pair Ethernet or serial data links.

Backbone cables branch into or terminate at seafloor power and data conversion nodes, which in turn link to secondary power distribution and data hubs that support individual instruments and their embedded sensors (Fig. 5.5.1). Little of this secondary infrastructure has been borrowed from subsea telecom technology but rather represents innovations driven by scientific requirements. Shore-based network control and data acquisition systems are also technologies that have been developed specifically for ocean observatories.

5.5.4 DEPLOYMENT AND MAINTENANCE OF CABLED OBSERVATORIES

The shift from short-term autonomous instrument deployments to multiyear deployments on cabled observatories requires different approaches in instrument design, testing, and maintenance. The location of the observatory and limitations on operations and maintenance budgets may restrict maintenance to once a year. In addition, the costs associated with deploying and recovering an instrument from a deep-ocean cabled observatory are often an order of magnitude more than the instrument capital cost. The cost of failure is therefore high, both in lost research opportunities and funds required to support the logistics of sensor replacement.

It is common practice to deploy a number of instruments on a single platform to simplify deployment and servicing logistics. Wherever possible, using a standardized platform design has a number of benefits, including reduced engineering costs, simplified planning for testing, transport, and ship loading, and the ability to develop standard procedures for subsea operations (Fig. 5.5.2). The location of the observatory drives the design requirements of the instruments and supporting platforms, and the need (or not, in the case of deep-water deployments) for antifouling technologies[3] and maintenance schedules for cleaning. Shallow coastal observatories need to take into account the limitation of divers and the deployment vessels when determining the layout of the platform. As the depths increase,

---

[3] The UV antifouling technology now installed on ONC's coastal observing systems is briefly described in Section 5.5.8.5 under "Multiuse, multisector connections." For a more detailed discussion of biofouling, see Chapter 4.3.

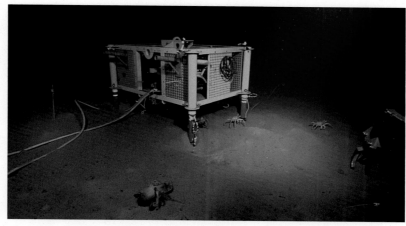

FIGURE 5.5.2 Standard Ocean Networks Canada deep-ocean instrument platform supporting junction box and onboard sensors. Visible cables connect the platform to external sensors up to 70 m away, avoiding interference from onboard devices. Cable connection to primary node is not visible.

FIGURE 5.5.3 Standard Ocean Networks Canada shallow water instrument platform during deployment from a research vessel. Once the platform is released from the ship's wire, it is repositioned and connected to the seafloor network by a remotely operated vehicle.

platform design must take into account the limitations of the remotely operated vehicle (ROV) systems that may be used to deploy/recover or maintain them (Fig. 5.5.3). Weight of the instruments and platforms in water, the location of connector panels, and the mounting of the instruments must all be taken into consideration to ensure that the ROV can successfully manipulate the platform on the seafloor.

The colocation of sensors on a platform can give rise to interference issues between sensor types. For example, active acoustic instruments can interfere with other active sources and may inject unacceptable signals into passive acoustic instruments. Instruments with significant power requirements, such as cameras with lights, may introduce transients into the secondary infrastructure power systems that may result in interference with other sensors. Identifying potential sources of interference early in the planning process and rigorous predeployment testing of the entire platform will help mitigate interference issues.

FIGURE 5.5.4   Cable ship deployment of an armored extension cable for Ocean Networks Canada's NEPTUNE cable network. Cable ships of this type routinely deploy and repair subsea telecommunications cables that can be thousands of kilometers in length. Armoring is the preferred design solution for longer extension cables and for deployments in difficult terrain and high current areas, despite higher build and deployment costs compared to lighter cables.

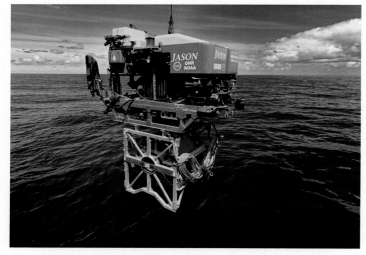

FIGURE 5.5.5   Remotely operated vehicle (ROV) extension cable-laying system. Spool suspended beneath ROV (*Jason* in this case) deploys lightweight cable to build local observatory networks around node sites.

The science goals for the observatory may require that the secondary infrastructure has a large spatial footprint on the seafloor. In these situations, extension cables will be required to lengthen the reach of the platforms. As with the instrument platforms, a number of factors will drive the design of the extension cables. The design must take into account the deployment environment, power, and communication requirements of the supported platforms and the planned life of the cable. Hazardous environmental conditions, such as those occurring in hydrothermal vent fields, on rocky seafloors, or in areas of high currents, may drive the need for armored extension cables. A range of deployment methods exist, including ship lay from traditional cable ships (Fig. 5.5.4) and ROV support vessels or ROV lay using specialized ROV tooling (Fig. 5.5.5).

Instrument workflows are a critical tool for maximizing the chance of success for each instrument deployment. For example, ONC has developed a documentation process that tracks each instrument from arrival at the observatory engineering facility until it is deployed at sea. An instrument workflow typically consists of the following steps:

1. Instrument receipt: The instrument is received from the manufacturer either new or after recalibration. All of the instrument documentation and calibration sheets are collated to support data capture and data quality assurance/quality control (QA/QC). For some sensors, additional calibrations are performed in-house.

2. Bench testing: The basic instrument parameters are confirmed (voltage, current, in–rush current, communication links, firmware version) and recorded. Instrument function in a known environment is tested and sample data streams are collected for analysis.

3. System integration: The instrument is connected to the power and communication infrastructure that will support it when deployed. The performance of the instrument is tested to ensure that there are no mechanical/electrical interference issues that can affect the instrument or the network. The ability of the observatory to control and capture data from the instrument is confirmed. Additional sample data streams are collected for analysis.

4. Test tank: The complete instrument, platform, and supporting power and communication infrastructure is deployed for 1–2 weeks in a test tank to prove long-term stability. Additional sample data streams are collected for analysis.

5. Final rigging for deployment: The instrument, platform, and supporting power and communication infrastructure is prepared for deployment at sea and annotated as ready for deployment.

A robust instrument workflow process ensures that every instrument is configured correctly and tested and documented prior to deployment. The data streams and instrument parameters captured during the process are a valuable tool for troubleshooting should postdeployment issues arise. Workflows should be reviewed and amended annually to reflect new lessons learned.

The maintenance of the cabled observatory will often be one of the largest items in the observatory budget. ONC's maintenance activities, including salaries for marine operations personnel, account for >60% of the annual operation budget. The specialized support vessels, divers, or ROV systems needed to conduct the maintenance are costly and must be employed to maximum effectiveness to ensure that the maintenance can be conducted within the funds available. Detailed planning of maintenance operations is essential, and a structured approach is required that captures any new science goals, known maintenance requirements, and planned infrastructure changes. Once the tasks are clearly understood, the resources necessary to conduct the work can be specified. Some observatories may have the resources within their organizations while others will contract for ship, diver, and/or ROV services. The final stage is the detailed planning of each operation to ensure that the maintenance team deploys with all the correct spares, support equipment, and staff to conduct the work. These plans will vary depending on the observatory and the complexity of the tasks. For ROV operations, detailed dive plans are used to capture:

1. Dive aim;
2. ROV set-up and additional tooling requirements;
3. Navigation information, including seafloor layouts;
4. Required predive notifications for ship-based and shore-based personnel;
5. Specific descent/ascent requirements, such as camera positions or vertical speeds; and
6. Detailed sequence of operations for the dive, including interactions with ROV staff, support ship staff, and observatory operations shore staff.

The use of telepresence technology (high-bandwidth satellite communications) can be a valuable tool during observatory operations. The live streaming of dive video and audio can enable science and maintenance staff to participate in the operations from shore. Tasks that were traditionally done at sea (e.g., real-time logging of operations) can be moved ashore, reducing the size of the team that must be accommodated on the vessel and the cost. This can provide the operations team with flexibility in vessel choice and at-sea staffing. There are also significant outreach advantages of having a shore interface, such as the ability to engage funders, other scientists, and the general public in the operations.

5.5.5 DATA ACQUISITION AND ARCHIVING

5.5.5.1 Data Streams and Diversity of Sensors

A major advantage of a cabled system is that it can serve the needs of many science disciplines relying on different types of instruments to achieve their goals, specifically physical oceanographers and chemical oceanographers will have sensors directly measuring phenomena of interest while biologists will usually rely on proxies to derive populations, species, and abundances (e.g., using cameras). This is reflected in the instrumentation that is hosted on the system.

Typical instruments will therefore usually fall into one of three categories from a data management point of view: scalar, complex, or streaming. Scalar sensors produce single measurements from time to time (e.g., a temperature value every second); complex instruments return multidimensional matrices of numbers every so often (e.g., ADCP); and streaming devices, such as camera and hydrophones, continuously stream their high-resolution observations.

Capturing all data from connected devices includes not only data from science instruments and their sensors, but also data from all other elements of the system even if they do not participate in the collection of the science data. Nonscience elements are important for control of the infrastructure and for monitoring the health and safety of the entire observatory. In systems as complex as those considered here, elements at the shore station, in the nodes, and in junction boxes can produce large volumes of important data to help detect, understand, or predict system failures. The use of engineering sensors for observatory control and failure detection/forecasting is discussed in a following section.

5.5.5.2 Stages of Data Stream Processing

With cabled observatories, data coming from different types of instruments converge to shore-based computers with very low latency. The arrival of new data triggers a sequence of processing steps, from record parsing and QA/QC checks, to event detection and archival. The use of a publish and subscribe model is useful to deal with the mostly asynchronous nature of the data stream. This is the model selected for the Data Acquisition Framework of Oceans 2.0 [6] and other systems.

One of the goals of an underwater observatory is to ensure cross-correlation between data from various sensors. This necessitates that precise time be shared between all observations. Another goal is to foster discoveries of process connections and oceanographic events. Such discoveries can be facilitated by automating event detection and enabling the combination of data from multiple sensors. Both require that data are not only accompanied by appropriate descriptions (metadata) but also are organized in a way that will permit comparisons, matches, and other correlations. Moreover, applying standard operators on the data (averages, decimation, identification of minima and maxima within a period of time) will be easier if data are organized in a systematic, standard way within a database or file system. Hence it is essential to standardize data organization at the earliest possible stage.

The first stage in the stream/event model illustrated in Fig. 5.5.6 is access to instrument data. Regardless of the selected design of the data management system, such an interface is necessary and has to be adapted to each

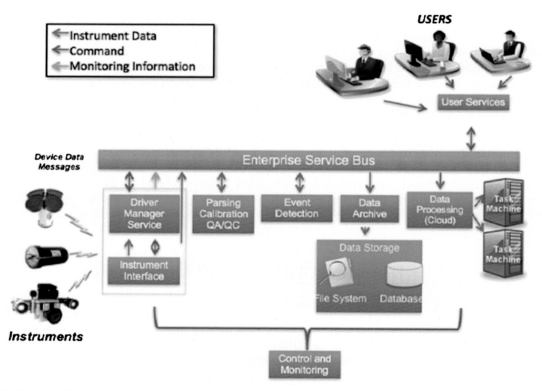

FIGURE 5.5.6   Example of an end-to-end cabled observatory framework for command/control of subsea infrastructure, and data acquisition, archiving, and distribution. The diagram is based on the Oceans 2.0 system developed by Ocean Networks Canada for its cabled observatory systems. *QA/QC*, Quality assurance/quality control.

type of instrument, each vendor, each model, and each firmware revision. Therefore many resources have to be devoted to understand, code, test, debug, and maintain access to a plethora of instruments. Approaches such as the Programmable Underwater Connector with Knowledge system allow smart interfaces between the platform and the sensors with similar characteristics from sensor to sensor (see Chapter 6.1). This reduces the number of custom sensor interfaces that need to be addressed.

5.5.5.3 Interoperability

Recognizing the challenges and the cost of supporting a wide variety of instruments and efforts to propose data interface standards to be shared between manufacturers and implemented throughout their palette of models have been ongoing for many years. One of the standards considered a few years ago was Sensor Model Language (SensorML) [7]. It attempts to accompany sensor readings with all relevant metadata. All values returned are surrounded by descriptive tags predefined in agreed-upon and published dictionaries or ontologies.

Reaching agreement at this level and having instruments return their measurement and accompanying metadata following SensorML conventions promises to achieve interoperability at the sensor level. However, a standard for data representation is not sufficient. Two additional elements have to be considered for sensor-level interoperability to truly take place: a transport protocol and a discovery mechanism. Sensor Web Enablement is a proposed solution for both. The examples described in this section are being proposed and supported by the Open Geospatial Consortium (OGC; www.ogcnetwork.net/) and represent one of the more dynamic initiatives in the area of interoperability at sensor level. OGC is described in Chapter 6.2.

Other interoperability mechanisms have been considered. The combination of NetCDF (data container; https://www.unidata.ucar.edu/software/netcdf/docs/) and ERDDAP (access and transport protocol; http://dap.onc.uvic.ca/erddap/information.html) allows for transparent data access at the level of data centers. ERDDAP (and its variants) addresses the ability to submit queries for archived data at any organization supporting the standard, and allows users to forgo prior knowledge of how data are stored. NetCDF obviously imposes an extra processing step to convert the initial raw instrument output into the NetCDF container with all the necessary accompanying metadata.

5.5.5.4 Engineering Versus Science Sensors

While scientific data collection is the primary goal of any cabled ocean observatory, the actual operation of the science sensors and instruments that are attached to an infrastructure represents a major effort. The oversight of these often large systems is usually a 24/7 task that involves the monitoring of engineering data (power distribution, electrical faults, data transmission) from a significant number of subsystems and their sensors. The engineering data have to be acquired, converted, verified, and checked against ceilings and thresholds on a continuous basis. Any value beyond preset bounds will generate alerts for observatory personnel or trigger preprogrammed reactions.

Such engineering sensors typically return data at the rate of 1 Hz. Nodes and junction boxes connected to ONC's NEPTUNE offshore observatory together have more health and safety sensors than the science sensors they enable.

Engineering data are essential to help predict trends and foresee instrument failures, and offer the ability to conduct forensic analysis to understand why an element has failed or to simply detect sporadic issues. An example where trending will help observatory managers extend the lifetime of the infrastructure and establish a priority list for maintenance and recovery is the analysis of the stability of the various ground leak current sensors (Fig. 5.5.7). A slowly increasing leak current (or reduced resistivity to ground) is an indication that something is amiss somewhere and could lead to accelerated corrosion or failure of subsystems. Switching them off early will increase the lifetime of the rest of the system.

Tools have to be provided to engineers and "wet plant" system managers to access, examine, and react to events occurring underwater. Monitoring such large numbers of individual engineering sensors requires systems to automatically and constantly verify that all variables remain within their preset boundaries. A network management system collects all alerts that come from any subsystem (power or communication) and draws the attention of system operators when they occur. Automating such tasks is essential for constraining infrastructure operating costs by avoiding the need for 24/7 staffing of system operations and limiting the service requirement for having personnel on call.

5.5.5.5 Observatory Assets Management and Operation Support

In addition to the many physical components, a large underwater observatory represents a facility that has to have a long lifetime and host several generations of caretakers. The complexity is so great that it is impossible for a

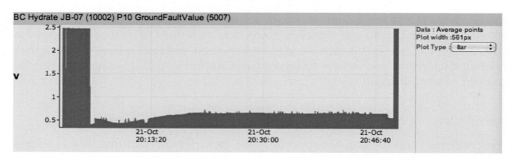

FIGURE 5.5.7   Example of the changes of ground fault (GF) value measured at a junction box port between the <off> and <on> status of the faulty instrument connected to it. The diagram shows about half an hour of data during a test run following the deployment of an instrument platform at the Barkley Canyon node of Ocean Networks Canada's (ONCs) NEPTUNE network. These measures were monitored remotely at ONC system's operations center. At the beginning of the time series the GF value drops from 2.5 to around 0.5 when the port is switched on providing power to the faulty instrument. The value later returns to 2.5 when the port is switched off. Ideally, the GF value should have remained at a value of around 2.5 V, corresponding to an almost infinite resistance between the power lines and the seawater.

single person or small group of people to remember everything about the system. Examples of essential information are: installation dates and locations; date of recalibration of individual instruments and the formulae that have to be used for each of its sensors; and an audit trail of instrument power up and down, etc. This information is absolutely critical to understanding the data that any instrument produces. Moreover, when dealing with a multiyear archive of data from instruments with a complicated history, understanding that history is necessary for data users to build trust in the data quality.

These considerations imply that the volume of information to be recorded, maintained, and presented to users about any component of the observatory (metadata) will be considerable. Fig. 5.5.8 illustrates some the information associated with a single instrument, one of the hundreds that ONC operates. In this example, the "Site" tab indicates the successive locations of that particular instrument throughout its lifetime at the observatory. Such information is essential for understanding and interpreting the data produced from its sensors through the recorded time series.

5.5.6 DATA DELIVERY TECHNOLOGIES FOR CABLED OBSERVATORIES AND DATA ACCESS

A number of components must be in place before cabled observatory data can be successfully delivered from the instrument to the user. Most accesses to data require searching on associated metadata first (i.e., being able to search for data using source location, date/time range, type of variable/parameters measured, and data quality level, etc.). Moreover, and in particular for complex data types for which expressing metadata search constraints is not sufficient, having the ability to preview the actual data is important. In the latter case, it is often simpler to create summary plots of scalar data (Fig. 5.5.9) or low-resolution versions of complex data from current profilers, cameras, hydrophones, etc., and make them available in a simple visual form for quick perusing/browsing. Indeed, the human eye is in many cases the fastest tool for pinpointing anticipated or unexpected patterns in the data. The alternative is to have the user express precisely what they are looking for to code these filter constraints in a data-mining application, perhaps relying on artificial intelligence techniques. For example, ONC's video imagery archive can be explored using metadata and text annotations as search parameters (Fig. 5.5.10).

Not all data access requires human interaction. More and more data mining applications are being developed to address the challenge of data volume. Such applications, together with modeling systems, require direct access to historical or current data without human intervention. For that purpose, an application programming interface to discover and download data needs to be available (e.g., http://searchmicroservices.techtarget.com/definition/RESTful-API).

Both access methods (web/interactive and programmatic) deliver data in specific formats (Fig. 5.5.11) following international, sensor-independent formats such as NetCDF, or data in formats specific to the instrument or instrument type (e.g., .jpg for a still camera or .wav for a hydrophone).

Irrespective of data access technologies and data formats, the promise of open access to observatory data to anyone with an Internet connection is one of the potentially transformative features of cabled observatories. Since these observatories tend to be large, publically funded facilities, such as OOI, ONC, and EMSO-ERIC, they have open access as a basic operating principle.

Ocean Networks Canada Device Details
Oceans 2.0

| Data Preview | Data Search | Plotting Utility | SeaTube | Digital Fishers | Cameras | More | Admin |

Device Id: 19 Device Name: **Sea-Bird SeaCAT SBE16plus 4996**

| General | Sensor | Ip | Electrical Rating | Data Rating | Nameplate | Port | Physical Characteristics | Device Action | Event | Site | Pr |

Site Device Id	Site	Location	Region	Date From	Offset Latitude	Offset
15	VIP-03	Saanich Inlet Central Node	Saanich Inlet	05-Aug-2006 00:00:00		
30018	On Ship or Shore	Expedition - Field Maintenance	ONC Headquarters	07-Nov-2006 00:00:00		
75	VIP-04	Saanich Inlet Central Node	Saanich Inlet	08-Nov-2006 00:00:00		
30019	On Ship or Shore	Expedition - Field Maintenance	ONC Headquarters	01-Feb-2007 00:00:00		
1123	VIP-05	Saanich Inlet Central Node	Saanich Inlet	03-Feb-2007 00:00:00		
22607	VENUS office/integration Lab	Ocean Networks Canada Offices	ONC Headquarters	11-Sep-2007 00:00:00		
1552	VIP-08	Saanich Inlet Central Node	Saanich Inlet	27-Sep-2008 00:00:00		
30126	MTC VENUS Integration Lab	Marine Technology Centre	ONC Headquarters	12-Feb-2009 02:00:00		
30154	VIP-09	Saanich Inlet Central Node	Saanich Inlet	15-Feb-2009 17:00:00		
117461	On Ship or Shore	Expedition - Field Maintenance	ONC Headquarters	26-Sep-2009 00:00:00		
117489	VENUS office/integration Lab	Ocean Networks Canada Offices	ONC Headquarters	30-Sep-2009 23:42:36		
117541	VIP-03	Strait of Georgia Central Node	Strait of Georgia	19-Feb-2010 21:00:00		
117811	VENUS office/integration Lab	Ocean Networks Canada Offices	ONC Headquarters	28-Aug-2010 00:00:00		
117887	DCF-11	Saanich Inlet Central Node	Saanich Inlet	12-Dec-2010 17:30:00		
118016	On Ship or Shore	Expedition - Field Maintenance	ONC Headquarters	09-May-2011 03:30:00		
118067	MTC VENUS Integration Lab	Marine Technology Centre	ONC Headquarters	12-May-2011 20:00:01		
118543	VIP-14	Saanich Inlet Central Node	Saanich Inlet	01-Jul-2011 15:30:01		
118563	MTC VENUS Integration Lab	Marine Technology Centre	ONC Headquarters	01-Oct-2011 23:59:59		
118553	MTC VENUS Integration Lab	Marine Technology Centre	ONC Headquarters	02-Oct-2011 00:00:00		
118599	DCF2-01 - 96m	Saanich Inlet Central Node	Saanich Inlet	10-Dec-2011 19:00:00		
119102	On Ship or Shore	Expedition - Field Maintenance	ONC Headquarters	09-Aug-2012 01:30:00		
119122	DCF2-02 - 96m	Saanich Inlet Central Node	Saanich Inlet	11-Aug-2012 23:30:14		
1189040	MTC VENUS Integration Lab	Marine Technology Centre	ONC Headquarters	08-May-2013 17:30:00		
1189776	VIP-19	Saanich Inlet Central Node	Saanich Inlet	23-Oct-2013 01:12:00		
1190168	On Ship or Shore	Expedition - Field Maintenance	ONC Headquarters	05-Mar-2014 23:37:34		
1190182	VIP-20	Saanich Inlet Central Node	Saanich Inlet	10-Mar-2014 16:52:00		
1191054	MTC VENUS Integration Lab	Marine Technology Centre	ONC Headquarters	17-Sep-2014 17:35:06		
1192370	ONC Integration Testing	Marine Technology Centre	ONC Headquarters	20-Aug-2015 21:59:37		
1192376	VENUS office/integration Lab	Ocean Networks Canada Offices	ONC Headquarters	24-Aug-2015 16:42:05		
1192394	3D_Camera-02	Saanich Inlet Central Node	Saanich Inlet	26-Aug-2015 04:43:56		
1194993	MTC VENUS Integration Lab	Marine Technology Centre	ONC Headquarters	06-Oct-2016 00:02:11		
1196163	ONC Decommissioned	Marine Technology Centre	ONC Headquarters	13-Apr-2017 00:00:00		

FIGURE 5.5.8 Records of successive deployments of a single instrument on the ONC network. Sensor deployment information is available to all users. *ONC*, Ocean Networks Canada.

5.5.7 CABLED SOLUTIONS FOR REMOTE LOCATIONS

In ocean locations far from major population centers, cost and logistic considerations can preclude installation of a dedicated observatory cable connection to shore, or a high-bandwidth Internet connection from a shore station to an observatory control center and data archive. This section considers some recent examples of cabled solutions for extending continuous, real-time ocean observing to remote ocean and coastal locations.

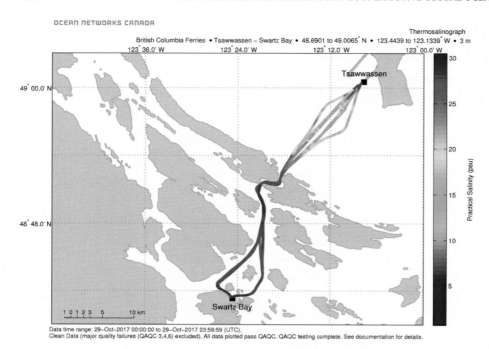

FIGURE 5.5.9 Surface salinity values in the southern Strait of Georgia along the transit path of a British Columbia ferry traveling between Vancouver Island and mainland Canada several times per day. The plot was generated by Ocean Networks Canada's Oceans 2.0 system. *QA/QC*, Quality assurance/quality control.

FIGURE 5.5.10   Example of Ocean Networks Canada's video search/preview interface that allows a search based on metadata (including camera, time, location) but also on annotations describing the image content.

FIGURE 5.5.11 Example data selection interface with cart system for data download on Ocean Networks Canada's Ocean 2.0 data download tool web interface. Tools allow users to select multiple variables from multiple locations and specify a desired data format for the server to convert data prior to download. In this example, 10 different formats are offered.

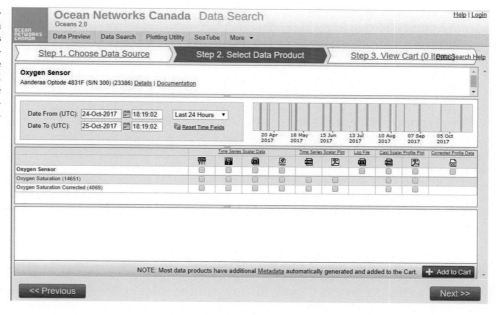

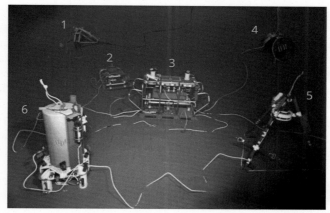

FIGURE 5.5.12 ALOHA cabled observatory seafloor installation. (1) Cable termination assembly; (2) junction box that converts fiber-optic signals to electrical Ethernet; (3) observatory module that provides power and data ports to sensors; (4) thermistor array acoustic modem communicates with moored thermistor string; (5) camera tripod; (6) secondary node (provides additional power and communications ports). *Background figure courtesy of ALOHA cabled observatory, supported by the US National Science Foundation; photo by ROV Jason.*

5.5.7.1 Cable Reuse

An extensive network of commercial submarine telecommunications cables crosses the major ocean basins, primarily to support the Internet. These cables are typically abandoned in place at the end of their commercial life and are replaced by new cable technologies. The ALOHA Cabled Observatory (ACO; http://aco-ssds.soest.hawaii.edu/index.html) is a deep-ocean observatory system operated by the University of Hawaii since June 2011 that has repurposed a retired first-generation fiber-optic telecommunications cable to connect a deep-sea observatory at a location 100 km north of the island of Oahu (22° 45′N, 158°W) in the North Pacific Ocean. ACO is currently the world's deepest operating cabled ocean observatory that allows underwater connection and disconnection of sensors, providing real-time oceanographic observations from a depth of 4728 m. The cable comes ashore at Makaha, on Oahu. In addition to ocean sights and sounds, continuous observations of temperature, salinity, oxygen, turbidity, chlorophyll fluorescence, and ocean currents are acquired and shared with the oceanographic community and general public. The ACO architecture uses a donated section of the AT&T HAW-4 transoceanic cable to provide power and communications bandwidth to a local seafloor network of sensors (Fig. 5.5.12). Since the cable was already in place and was designed to operate for well beyond its commercial lifetime, costs of conversion to scientific use were substantially lower than for new systems. As more interest is now being focused on obtaining observations in the deep ocean, the science community should consider obtaining access to out-of-service cables that traverse areas of interest and may be suitable for ocean observing nodes similar to ACO.

5.5.7.2 Hybrid Local Cable and Satellite—EMSO Açores

At the remote Lucky Strike hydrothermal vent field on the Mid-Atlantic Ridge near the Azores, cost and logistic considerations preclude the laying of a cable connection to shore. Here, France and Portugal operate EMSO-Açores, a hybrid cabled and satellite observatory that is now a component of the European Multidisciplinary Seafloor and water-column Observatory (EMSO). The system uses centralized battery power and data storage nodes on the sea-floor to support locally cabled networks of sensors. The SEAMon nodes (Fig. 5.5.13) provide more power and data storage than would normally be available for individual oceanographic sensors operating autonomously. The sea-floor nodes are linked by acoustic modem to a surface relay buoy that uses a satellite link to enable real-time monitoring of observatory operations and low-bandwidth data transmission. The SEAMon nodes are recovered to the surface during annual maintenance that includes battery replacement and downloading of the data archive.

5.5.7.3 Community-Based Cabled Observatories

There is a growing need in Canada for long-term monitoring of coastal areas where significant increases in marine traffic and/or large-scale marine projects are proposed, most related to natural resource development. Remote coastal communities need the capacity to establish environmental baselines and detect environmental change to support evidenced-based decision-making. ONC is piloting a community-based, cabled ocean observatory program in six locations along the coast of British Columbia with each system either located within or close to First Nation (Indigenous) communities. Typical sensor suites include standard CTD systems, dissolved oxygen, chlorophyll fluorescence, turbidity, weather stations, and high-frequency radar, but many other sensors can be included for ocean acidification, nutrients, ADCPs, fish tag receivers, etc. These observatories also incorporate underwater video and hydrophone systems that are key elements for developing lesson plans for K-12 and postsecondary education. ONC's Ocean Sense program aims to build local capacity for knowledge- and evidence-based decision-making related to coastal environments. ONC has engaged over 60 coastal communities and First Nations in British Columbia and in the Canadian Arctic in this highly successful program, which includes educational materials in Indigenous languages.

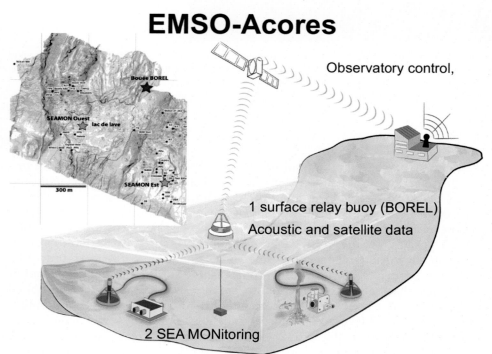

FIGURE 5.5.13 Schematic representation of the EMSO-Açores observatory infrastructure. Year-to-year configurations may vary but basic design of centralized, autonomous power, and data modules that support deployed instruments and communicate with surface mooring via acoustic modem is maintained. *Image—Capsule Graphik, copyright LEP/Ifremer.*

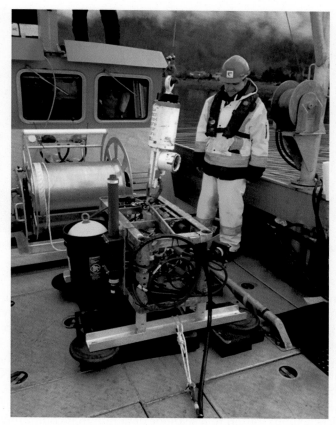

FIGURE 5.5.14 Instrument platform community-based cabled observatory prior to deployment from a fishing vessel.

To ensure the continuity and longevity of data acquisition by these often remote observing systems, careful consideration must be given to system design, and to the integration and selection of sensors. A compact subsea platform designed for installation at 5–50 m depth facilitates deployment and recovery using a small vessel (Fig. 5.5.14), eliminating the need for diver or ROV intervention. Containerized shore stations use locally available electrical power and communication links (Fig. 5.5.15). Depending on instrumentation, the community observatories require 1–8 kW of power and 50–800 Gb/month in bandwidth. Installations are designed for a 1-year service interval, and must be resilient to vandalism, frequent power outages, and communications failures. Systems are remotely controllable, able to self-start after a failure, can operate for a month without communications, and have redundant communications links that include fiber-optic, wireless point-to-point, and cellular data options.

5.5.7.4 Hybrid Local Cable and Satellite—Canadian Arctic

Predictions and observations of accelerated warming of the Arctic have increased the need for continuous year-round observing to support understanding of future sea ice conditions and marine productivity and biogeochemistry. The high-frequency, multisensor observational time series that can be obtained from cabled observatories can provide critical data on the dynamics of sea ice formation and the response of marine ecosystems to changing ice cover and light conditions. Unfortunately, the remoteness of the Arctic Ocean means that there are few reliable local sources of electrical power to support local cabled observatories. More importantly, cable connections to the World Wide Web are essentially nonexistent, and high-bandwidth satellite data transmission is extremely costly. Also, disturbance of the shoreline and seabed by shifting ice presents special challenges for cables that connect shore stations to underwater sensors. There are currently three small hybrid cabled ocean-observing systems operating in the Arctic, all within the Canadian Arctic Archipelago. They provide examples of different hybrid designs that enable continuous, long-term ocean observations in near real-time in waters that are ice covered for most of the year. All three transmit ocean data by low-bandwidth satellite connections from a cable shore station to a distant archive for distribution over the Internet, as illustrated in Fig. 5.5.16.

FIGURE 5.5.15   Shore station for community observatory on Digby Island near Prince Rupert, British Columbia, Canada.

ONC has been operating a real-time cabled observing system in Cambridge Bay, Nunavut, since 2012, the location of the new Canadian High Arctic Research Station. A shore station in the village provides power and communications to a subtidal platform (Fig. 5.5.16A) that supports a suite of oceanographic sensors, a hydrophone, an acoustic ice profiler, and a fish tag receiver. The shore station also supports a weather station, a sea-surface camera, and an AIS receiver. A wireless connection relays data from the shore station to local servers that log all sensor, imagery, and hydrophone data, and transmit scalar data and selected imagery via satellite to the ONC data archive.

The longest operating cabled observing system in the Arctic is the Real-Time Arctic Ocean Observatory installed by Fisheries and Oceans Canada in 2009 in Barrow Strait, at the eastern entrance to the Northwest Passage. There, an 8-km long subsea cable connects a shore station to a seafloor data hub that communicates acoustically with instrument moorings (Fig. 5.5.16B). This observatory provides upper ocean salinity and temperature data for freeze-up forecasting [8]. A two-way Iridium satellite link transmits data bihourly from the shore station to the data archive at the Bedford Institute of Ocean Sciences.

The third cabled arctic observatory is a pilot system installed in 2017 by ONC at 22 m depth in Gascoyne Inlet on the remote and uninhabited Devon Island, near the location of the deeper water Barrow Strait observatory described earlier. The goal of this deployment is to test the performance of several system components that could enable the building of a wider network of low-cost[4] cabled ocean observatories in the coastal arctic. Prototype components include a compact underwater data hub, modular ice protection for seabed cables, and a containerized, self-powered shore station with an Iridium transmitter (Fig. 5.5.16C).

5.5.7.5 SMART Subsea Cable Initiative

The idea of incorporating ocean sensors into future transoceanic telecom cables was articulated in a *Nature* opinion article by Ref. [9] and led the International Telecommunication Union to organize a Joint Task Force (JTF) to explore the idea [10]. Given the current wide spatial distribution of commercial subsea cables (approximately 300 operational systems totaling 1 million kilometers; B. Howe, Univ. Hawaii; pers. comm.), the addition of even a limited number of sensors on future cables would provide a significant window into deep-ocean processes. The JTF is examining the feasibility of incorporating scientific sensors into telecommunications cables in an initiative referred to as SMART (Science Monitoring and Reliable Telecommunications). Integrating ocean sensors into commercial telecommunication cables imposes design and durability constraints. The SMART cable design must not compromise the reliability of commercial data transmission through the cables or require costly modifications to cable manufacturing or cable deployment processes. Sensor durability requirements will need to be much more stringent for SMART cables than for scientific/operational cabled observing systems. For the latter, annual maintenance expeditions permit frequent

---

[4] Capital and installation costs for a network of compact observatories linked to shore by individual km-length cables are expected to be 1–2 orders of magnitude less than the cost of building a continuous cabled system, such as ONC's VENUS network, in the coastal Arctic Ocean.

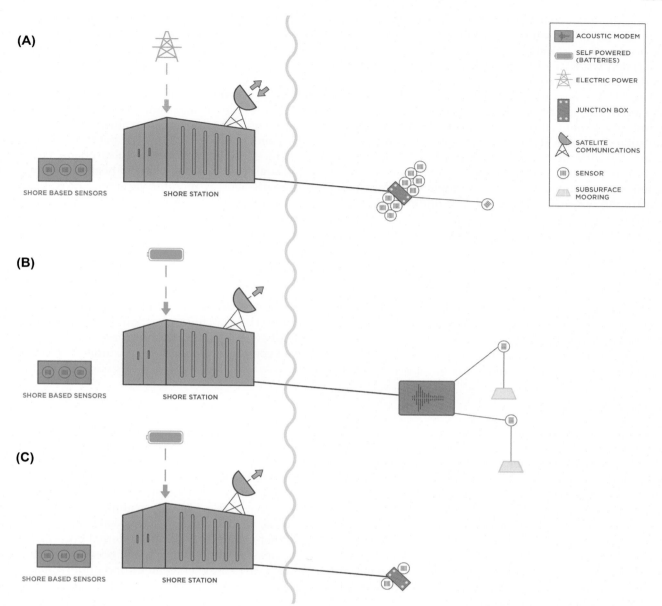

FIGURE 5.5.16 Schematic representation of three types of hybrid cabled coastal observatory currently operating in the Canadian Arctic. (A) Community observatory in Cambridge Bay, Nunavut, receives electrical power from a local grid. Subtidal instrument platform supports junction box and multiple onboard sensor(s). Extension cables support sensors up to 70 m away to avoid interference from onboard devices. Low-bandwidth, two-way data link to commercial telecommunications satellite connects shore station and underwater infrastructure with control center and data archive at the University of Victoria. (B) Self-powered shore station on Devon Island, Nunavut, supports a subsea acoustic modem that uses an acoustic link to acquire data from self-powered sensors(s) on subsurface moorings. Data packets transmitted bihourly via satellite to a data archive at Bedford Institute of Oceanography. (C) Self-powered shore station on Devon Island, Nunavut, supports subsea sensor platform in Gascoyne Inlet. Low-bandwidth data link to commercial telecommunications satellite links shore station to data archive at the University of Victoria.

replacement, calibration, or in situ servicing of sensors. Commercial telecommunications cables are designed to have minimal maintenance during their operational lifetime; thus any sensors deployed as part of a SMART cable would need to survive for up to 20 + years without calibration or repair.

To date, the work of the JTF has identified cable repeater assemblies as the most promising mounting location for oceanographic sensors. These repeaters are spaced at regular intervals along subsea telecommunication cables (from 50 to several 100 km), and house optical amplifiers that maintain signal propagation. A likely solution would have the sensor power and communication interface in the repeater connected with an external sensor module

~20 m from the repeater (similar to sea electrode modules used with branching units). Designs that route data from scientific sensors through engineering communication channels (vs. commercial data channels) and that have connections within individual repeaters would avoid adding risk to the commercial viability of those SMART cables associated with scientific data transmission. Commercial data pathways would not need to be opened to accommodate scientific data.

The JTF has also addressed the question of identifying which sensors would be suitable for deployment in cable repeaters (size, long-term stability, power requirements) and would provide the maximum research and operational monitoring benefits from deployment on the seabed over the long spans of the transoceanic cables. The proposed initial suite of sensors [11] would measure temperature, bottom pressure, and acceleration. Temperature sensors, provided that they are mounted to avoid contamination by heat produced in the repeater modules, enable basin-scale monitoring of deep-ocean temperature variability (seasonal to decadal scales), and tracking of heat flow through the ocean along oceanographic and tectonic boundaries. Bottom pressure sensors monitor variability of waves, tides, barotropic currents and sea level, and improve early warning of tsunamis. Accelerometers improve global coverage of tsunami propagation and seismic events and, depending on sensitivity, could also be used for passive seismic imaging of the solid earth. See Chapter 5.4 for a more detailed discussion.

5.5.8 FINAL THOUGHTS

Providing ocean sensors with cabled power and bandwidth connections to shore facilities permits continuous, high-frequency, real-time ocean observing over long periods of time. This concluding section revisits these technical features of cabled observatories for a brief critical review of their advantages and disadvantages, and then considers some of the less anticipated features that have emerged from the experience of observatory operators and from the results of research that has made use of cabled observatory data.

5.5.8.1 Unlimited Power

Continuous observation with many types of oceanographic sensors, even for up to a year at a time, does not require a cabled connection to shore power. Decades of development of low-power oceanographic sensors for mooring deployment—together with recent improvements in battery technology and data storage—enable long-term, autonomous deployment of instruments like CTDs and individual sensors for dissolved oxygen and fluorescence, etc. In contrast, there are many types of power-hungry devices that could not be operated continuously for months to years without an external power supply. These include active acoustics (e.g., ADCPs, scanning sonars, ice profilers), video camera lights, sampling pumps, and mobile, motorized sensor platforms, such as water column profilers and tethered vehicles. Shore-based oceanographic radars are another example. Providing external power for heavy power-use devices enables cabled observatories to continuously measure a broader range of ocean properties than would be possible from battery power alone. To date, with the exception of tethered vehicles and underwater camera systems, there has not been significant development of new sensors, instruments, or devices that take full advantage of the "unlimited" power supply available on cabled observatories.

5.5.8.2 High Bandwidth

Advances in compact data storage have rendered moot most arguments for the necessity of high bandwidth connections for data logging from scalar sensors. Even requirements for months of continuous data collection can be met with onboard storage available in devices such as CTDs. In contrast, data storage requirements for the uninterrupted, long-term operation of devices that produce complex data and imagery can greatly exceed what can be provided by compact, low-power storage solutions included in instrument pressure housings. High-bandwidth connections to shore-based data storage permit cabled ocean observatories to support continuous measurements that generate high volumes of complex data from devices such as ADCPs, hydrophones, and acoustic plankton sonars, and record imagery from multiple, high-resolution video cameras operating for several hours each day or even continuously. Although challenging to analyze, large-volume time series from cabled devices are yielding important insights into process connectivity in the ocean. There are numerous examples of how analyses of long-term recordings from cabled seafloor video cameras can reveal seasonal and higher-frequency dynamics of marine faunal communities and relationships to environmental drivers (e.g., Refs. [12–19]).

A high-bandwidth data pipe permits cabled observatories to routinely acquire ocean sensor data at frequencies of 1 Hz or greater, allowing ocean scientists to observe high-frequency processes, such as water column and boundary layer turbulence and seismic events, and to use active acoustics and high-frequency radar to measure current speeds, surface currents, and wave height. Several recent examples from ONC's NEPTUNE observatory demonstrate how high-frequency observations from multiple sensors, combined with network time synchronization, can reveal unanticipated coupling of physical processes over considerable distances. For example, Ref. [20] demonstrated the coupling of hydrothermal vent faunal behavior to habitat variability (temperature, dissolved oxygen) that itself was physically coupled to tidal and atmospheric forcing of bottom current velocity and direction in the deep ocean (2400 m). In another example, Ref. [21] was able to combine high-frequency seismometer and bottom pressure data from ONC's NEPTUNE network to reveal tidally induced deformation of the seabed that is in-phase with tidal loading, from the continental shelf to abyssal depths. Finally, Ref. [22] correlated a full year of data from an imaging sonar, CTDs, bottom pressure recorders, a current meter, and a broadband seismometer to identify decreasing tidal pressure as the main triggering mechanisms for bubble emissions from seabed gas seeps.

5.5.8.3 Real-Time Observing

To date, the real-time data delivery capability of cabled observatories has mostly been utilized for operational purposes or research related to hazard mitigation. Tsunami and earthquake detection by seabed sensors is an obvious case in point. Japan's DONET cabled observatory networks were exclusively built to monitor the Nankai and Tonakai seismogenic zones [23] to permit rapid detection of earthquakes and tsunamis. These data are also being used for research to develop earthquake prediction models, and earthquake and tsunami simulations. While earthquake prediction is still an active area of research rather than a practical reality, the Mexican Seismic Alert System has demonstrated how networking seismometers (terrestrial in this case) and public alert systems can provide tens of seconds of advance warning of the arrival of damaging ground shaking, reducing the potential for property damage and loss of life. Such systems rely on the detection of the faster traveling and less damaging "P" waves to predict the arrival time and amplitude of the secondary "S" waves that can cause destructive and life-threatening ground shaking. ONC is currently completing the installation of an integrated network of seabed and terrestrial seismic sensors and the development of earthquake alert software with the goal of creating an earthquake early warning system for southern British Columbia where proximity to the Cascadia subduction zone and risk of a megathrust earthquake in the coming century are of notable concern for emergency management agencies.

Other research use of real-time data streams from cabled observatories is less well developed. Most of the non-geohazard publication output from these facilities has been based on analyses of high-frequency data time series. One exception is the combined use of real-time data streams and two-way communications for optimal data collection or custom configuration of sensors for specific experiments. Examples include the triggering of physical sample collection by observatory instruments in response to events (http://mclanelabs.com/ras-pps-perform-vent-sampling-ooi-cable-array/); the operation of tethered vehicles for video-guided in situ sensing [15,24]; and the project-specific configuration of seafloor camera sweep angles, magnification, and observing cycles (e.g., Refs. [20,25–27]). Expanding the use of this interactive capability is not without challenges for observatory operators and observatory research communities. How do we accommodate custom configurations and user-directed observations while also maintaining long-term time series? How do we balance network security requirements with user requests for access to command and control features of sensors and instrument platforms? One pathway to resolving some of these issues could be the development of event response algorithms and other artificial intelligence features to optimize observations by observatory assets. Such features could be developed in consultation with user communities to maximize their scientific value, and implemented and controlled by observatory facilities operators to minimize risks to network security and to ensure continuity of long-term observing programs.

5.5.8.4 Integrating Management With Cabled Observatory Infrastructure

The complexity of cabled observatories was explained earlier. The many components—from nodes to junction boxes to instruments, their cables and connectors, and the thousands of health and safety sensors—require a comprehensive and meticulously maintained metadata database, tools for its management, as well as a fully integrated monitoring and control system for live operation of the entire system. Over the past decade, ONC has addressed the long-term challenge of managing a vast and growing infrastructure by fully integrating an exhaustive palette of features within its Oceans 2.0 data management system (http://www.oceannetworks.ca/oceans-20-internet-things-ocean). The system helps staff manage the infrastructure on a day-to-day basis, facilitates the sharing of information across many functional teams, maintains corporate knowledge over many years, and supports the planning of maintenance expeditions, the addition of new equipment to the facility, and ultimately enables the end user to better understand the data.

The system is centered on connected devices operated by the organization. Devices can be of an engineering type, such as junction boxes, or they can be scientific instruments. Each device is not only described exhaustively (see https://data.oceannetworks.ca/DeviceSearch and drill down from there), but its entire history is recorded in detail and available to all (see, for example, Fig. 5.5.8). A complete workflow application helps technical personnel and managers across the organization prepare instruments for deployment and assess their readiness throughout the complex lifecycle process. The application ensures that all personnel are part of the process to deliver reliable data from an instrument. Marine engineers, procurement staff, metadata specialists, GIS staff, data quality specialists, software engineers, systems and operations managers, and staff scientists all use permissions-based web forms to view, add, and modify information about individual components of the infrastructure.

5.5.8.5 It's the Network, Not the Cable

This chapter has had a technological focus that reviewed what and how cabled observatories are contributing to ocean sensing. Before closing, it is important to consider some of the less tangible features of cabled observatories that have emerged over the past decade as facilities have been constructed, as data have been acquired and analyzed by researchers, and as access to cabled observatories and observatory data has been extended beyond the oceanographic research community.

5.5.8.5.1 Focus for Transdisciplinarity

With the exception of systems that have primary public safety or operational missions, most cabled observatories serve a broad range of scientific disciplines to justify their capital and operating costs. Early cabled observatory publications described potential opportunities for interdisciplinary research that would arise from the "high resolution, coincident, coordinated sampling of hundreds of data streams across disciplines" and the "analysis of relationships among the disparate data" [28]. A perusal of the scientific literature and the publications cited here would suggest that this potential is being realized and that cabled observatories have indeed become a focus for interdisciplinary research. This is particularly evident in research that uses seafloor cameras and high-frequency, time series data from environmental sensors to study behavioral responses of benthic organisms to habitat variability at time scales of hours to months. For example, multidisciplinary investigations at ONC's deep-water sites have found surprising correlations between the composition of the visible benthic fauna and water column turbulence and bottom current changes attributable to winter-time surface storms [18,20]. Simply stated, this means that what we see in video surveys of deep-water bottom fauna can depend on local habitat conditions at the time of the survey. Since video surveys are becoming an important tool for monitoring deep-sea ecosystems, knowledge of such hours-to-days-scale baseline variability in visible faunal abundances, and the underlying behavioral mechanisms, is important for predicting and interpreting effects of long-term ocean change and natural and anthropogenic disturbances on marine ecosystems.

5.5.8.5.2 Access

It is the practice or the intent of most major cabled observatory operators to provide free, open, and immediate access to data and metadata that are produced by their facilities. Such policies are relatively new in oceanography with the exception of data collected by Earth-observing satellites, international programs such as Argo (http://www.argo.net/), and some government agencies. In theory, open data policies have the potential to expand the community of practicing oceanographers beyond those who have had access to data collected by research vessels and satellites. Anyone who has access to the Internet can locate and download sensor data from cabled ocean observatories. Open data policies have been adopted by cabled observatories in Canada, the United States, and Europe. Free access to ocean data also opens the door to other uses of the data, from the development of derived data products that provide ocean information, to regulatory agencies and interested citizens, to the use of large data sets by computer scientists for research in machine learning to support event detection and pattern recognition. The authors are not aware of any systematic studies of the impact of open data policies on oceanographic research output or on the size and global distribution of the community of researchers using ocean data. However, our experience at ONC with data product development and user tracking provides some insight into the uptake of online ocean data and derived data products by researchers, educators, government officials, and the general public. More than 200 countries represent the ONC community of users. Researchers are using the facility for unanticipated advances in disciplines such as forensics and computer vision. Requests for the translation of research results and ocean data into educational and outreach materials and operational data products, implemented through ONC's Smart Ocean Systems program, are increasing, and there is encouraging support for these activities from traditional and nontraditional funding sources.

5.5.8.5.3 Multiuse, Multisector Connections

The long-term sustainability of large research facilities can be aided by extending the use of the infrastructure to sectors outside basic research. Three obvious operational uses of ocean data are in the areas of maritime safety (sea state nowcasting and forecasting), marine geohazard resilience (earthquake, tsunami, and storm surge alerts), and monitoring of marine mammals for their protection (whale-ship deconfliction alerts). These applications are of considerable importance to the public and to responsible government agencies. Another area of application, particularly for cabled observatories in coastal zones, is the long-term monitoring of ocean health indicators such as temperature, dissolved oxygen, and pH.

Meeting the ocean information needs of sectors outside of science to create value-added data products for society usually requires more than simple tweaking of sensor output from research platforms. For example, ONC's journey toward becoming a multiuse ocean-observing facility has required the installation of observing assets in locations that were not initially identified as priorities for basic research, and investments in software development to meet the requirements of a wide range of data end-users. At the same time, ONC's experience in providing ocean data to support leading-edge research brings a degree of scientific credibility to the data provided, and establishes the observatory as a neutral, third-party provider of ocean information. Although not a cabled facility, Australia's Integrated Marine Observing System (http://imos.org.au/) offers examples of how data from multiple, science-based ocean-observing systems can be integrated to provide national and regional operational summaries of ocean conditions for many stakeholder sectors (https://portal.aodn.org.au).

Cabled observatories can also serve as a test bench for ocean technology development. Prototype sensors and other devices can be mounted on subsea observatory platforms, provided with power and communications links, and monitored in real-time by their developers. The Monterey Accelerated Research System-cabled observatory has a specific mission to support the development of new research instruments by providing opportunities for underwater testing on their facility (see https://www.mbari.org/at-sea/cabled-observatory/). The innovation division of ONC has had a similar mandate, using ONC's observatory infrastructure to support field trials of new sensors by Canada's ocean technology industry. Data from prototype sensors are password protected during this development phase. One success story is in the area of antibiofouling systems, one of the key challenges for subsea sensors in the coastal zone. While most of ONC's larger facilities are well below the photic zone where biofouling is minimal, cabled platforms for the growing network of coastal community-based systems are nominally installed in 20 m of water in high-fouling regions. Working with AML Oceanographic, ONC field tested a UV LED antifouling system on the Folger-shallow instrument platform. Monitored daily with a video camera for over a year, the UV LEDs kept sensors and camera domes clear compared with adjacent sensors that were not exposed to UV light (Fig. 5.5.17). These UV protection systems have been added to all of ONC's coastal and Arctic-observing platforms.

5.5.8.5.4 Focus on Developments in Big Data and Machine Learning

The discovery potential of the large volumes of high-resolution, signal-rich, raw data collected by cabled observatories can be greatly enhanced by automated tools that aggregate, analyze, and distill information to yield domain-specific knowledge. The field of data science has much to offer in the way of computational methods that enable the extraction of salient data features and events of interest in large, heterogeneous data sets. Recent advances in computer vision, pattern recognition, and data mining have the potential to offer significant benefits for oceanographic research and operational use of observatory data. At the same time, there is a surfeit of large data sets that are readily available for development projects in computer science [29]. ONC has enabled several recent collaborations between computer scientists and oceanographers that involve graduate student-scale projects in data exploration, and pattern and object recognition. A recent publication describes one of these collaborations that included the development of a machine vision algorithm and a game interface to enable citizen science contributions to a research project [19]. Data science–oceanography collaborations will be an important and essential area of development for cabled observatories, from the pairing of individual sensors and cameras with artificial intelligences, to predicting future states of complex ocean systems from archived and real-time observations.

FIGURE 5.5.17   Sensor UV-LED antifouling system during test deployment at Ocean Networks Canada's 24-m deep Folger Shallow site on the NEPTUNE network. Three identical CTD units are visible in these video frames from three time intervals over a 280-day test period. The center unit was exposed to UV light at fixed intervals each day as an antifouling measure. The unit on the left received some "overflow" UV light from the LED bank.

References

[1] Delaney JR, Heath BR, Howe B, Chave AD, Kirkham H. NEPTUNE: real-time ocean and earth sciences at the scale of a tectonic plate. Oceanography 2000;13:71–83.

[2] Delaney JR, et al. Scientific rationale for establishing long-term ocean bottom observatory/laboratory systems. In: Teleki PG, Dobson MR, Moore JR, von Stackelberg U, editors. Marine minerals resource assessment strategies. 1987. p. 389–411.

[3] Delaney J, et al. Scientific rationale for establishing long-term ocean bottom observatory/laboratory systems. In: The mid-oceanic ridge — a dynamic global system, proceedings of a workshop. Washington (DC): National Academy Press; 1988. p. 234–57.

[4] Kelley D, Delaney JR, Juniper SK. Establishing a new era of submarine volcanic observatories: Cabling Axial Seamount and the Endeavour Segment of the Juan de Fuca Ridge. Mar Geol 2014;353:426–50. https://doi.org/10.1016/j.margeo.2014.03.010.

[5] Starosielski, N. 2015. https://www.scientificamerican.com/article/undersea-cable-network-operates-in-a-state-of-alarm-excerpt/.

[6] Pirenne B, Guillemot R. Oceans 2.0: a data management infrastructure as a platform. In: EGU general assembly 2012, held 22-27 April, 2012 in Vienna, Austria. 2012. p. 13346. http://adsabs.harvard.edu/abs/2012EGUGA.1413346P.

[7] SensorML description example for a CTD: http://vast.uah.edu/downloads/sensorML/v1.0/examples/sensors/MBARI_CTD.xml. See also: OGC White Paper: OGC® Sensor Web Enablement: Overview And High Level Architecture, Downloadable at http://www.sensorml.net/.

[8] Hamilton JM, Pittman MD. Sea-ice freeze-up forecasts with an operational ocean observatory. Atmos Ocean 2015;53:595–601. https://doi.org/10.1080/07055900.2014.1002447.

[9] You Y. Harnessing telecoms cables for science. Nature 2010;466:690–1. https://doi.org/10.1038/466690a.

[10] Extance A. Can ocean cables go green? Fibre systems, Winter 2015. 2015. https://www.fibre-systems.com/feature/can-ocean-cables-go-green.

[11] Butler R, et al. The scientific and societal case for the integration of environmental sensors into new submarine telecommunication cables. ITU/WMO/UNESCO IOC Joint Task Force; 2014. http://www.itu.int/dms_pub/itu-t/opb/tut/T-TUT-ICT-2014-03-PDF-E.pdf.

[12] Aguzzi J, et al. Challenges to assessments of benthic populations and biodiversity as a result of rhythmic behavior: video solutions from cabled observatories. Oceanogr Mari Biol Ann Rev 2012;50:235–86.

[13] Aguzzi J, Costa C, Robert K, Matabos M, Antonucci F, Juniper SK, Menasatti P. Automated video-imaging for the detection of benthic crustaceans and bacterial mat coverage at VENUS cabled network. Sensors 2012;11(11):10534–56. https://doi.org/10.3390/s111110534.

[14] Delaney JR, Kelley DS, Marburg A, et al. Axial Seamount – wired and restless: a cabled submarine network enables real-time tracking of a mid-ocean ridge eruption and live video of an active hydrothermal system, Juan de Fuca Ridge, NE Pacific. In: Proceedings paper, MTS/IEEE oceans conference, Monterey, CA, USA. 2016. Accession No. WOS:000399939002161. ISBN: 978-1-5090-1537-5.

[15] Doya C, Chatzievangelou D, Bahamon N, Purser A, De Leo F, Juniper SK, Thomsen L, Aguzzi J. Seasonal monitoring of deep-sea megabenthos in Barkley canyon by internet operated vehicle (IOV). PLoS One 2017;12(5):e0176917. https://doi.org/10/1371/journal.pone.0176917.

[16] Matabos M, Aguzzi J, Robert K, Costa C, Menesatti P, Company JB, Juniper SK. Multi-parametric study of behavioural modulation in demersal decapods at the VENUS cabled observatory in Saanich Inlet, British Columbia, Canada. J Exp Mar Biol Ecol 2011;401:89–96.

[17] Matabos M, Tunnicliffe V, Juniper SK, Dean C. A year in hypoxia: epibenthic community responses to severe oxygen deficit at a subsea observatory in a coastal inlet. PLoS One 2012;7(9):e4526. https://doi.org/10.1371/journal.pone.0045626.

[18] Matabos M, Bui AOV, Mihaly S, Aguzzi J, Juniper SK, Ajayamohan RS. High-frequency study of epibenthic megafaunal community dynamics in Barkley Canyon: a multi-disciplinary approach using the NEPTUNE Canada network. J Mar Syst 2014. https://doi.org/10.1016/j.jmarsys.2013.05.002.

[19] Matabos M, Hoeberechts M, Doya C, Aguzzi J, Nephin J, Reimchen T, Leaver S, Marx R, Albu A, Fier R, Fernandez-Arcaya U, Juniper SK. Expert, Crowd, Students or Computer: who holds the key to deep-sea imagery 'big data' processing. Methods Ecol Evol 2017. https://doi.org/10.1111/2041-210X.12746.

[20] Lelièvre Y, Legendre P, Matabos M, Mihály S, Lee R, Sarradin P, Arango C, Sarrazin J. Astronomical and atmospheric impacts on deep-sea hydrothermal vent invertebrates. Proc R Soc B Biol Sci 2017;284(1852). https://doi.org/10.1098/rspb.2016.2123.

[21] Davis EE, Heesemann M, Lambert A, Jianheng H. Seafloor tilt induced by ocean tidal loading inferred from broadband seismometer data from the Cascadia subduction zone and Juan de Fuca Ridge. Earth Planet Sci Lett 2017;463:243–52. https://doi.org/10.1016/j.epsl.2017.01.042.

[22] Römer M, Riedel M, Scherwath M, Heesemann M, Spence GD. Tidally controlled gas bubble emissions: a comprehensive study using long-term monitoring data from the NEPTUNE cabled observatory offshore Vancouver Island. Geochem Geophys Geosyst 2016;17:3797–814. https://doi.org/10.1002/2016GC006528.

[23] Kaneda Y. DONET: a real-time monitoring system for megathrust earthquakes and tsunamis around southwestern Japan. Oceanography 2014;27(2):103. https://doi.org/10.5670/oceanog.2014.45.

[24] Purser A, Thomsen L, Barners C, Best M, Chapman R, Hofbarer M, Menzel M, Wagner H. Temporal and spatial benthic data collection via an internet operated Deep Sea Crawler. Methods Oceanogr 2013;5:1–18. https://doi.org/10.1016/j.mio.2013.07.001.

[25] Cuvelier D, Legendre P, Laes A, Sarradin P-M, Sarrazin J. Rhythms and community dynamics of a hydrothermal tubeworm assemblage at Main Endeavour Field – a multidisciplinary deep-sea observatory approach. PLoS One 2014;9(5):e96924. https://doi.org/10.1371/journal.pone.0096924.

[26] Cuvelier D, Legendre P, Laes-Huon A, Sarradin P-M, Sarrazin J. Biological and environmental rhythms in (dark) deep-sea hydrothermal ecosystems. Biogeosciences 2017;14:2955–77. https://doi.org/10.5194/bg-14-2955-2017.

[27] Robert K, Juniper SK. Surface-sediment bioturbation quantified with cameras on the NEPTUNE Canada cabled observatory. Mar Ecol Prog Series 2012;453:137–49. https://doi.org/10.3354/meps09623.

[28] Barnes CR, Best MMR, Pautet L, Pirenne B. Understanding Earth – ocean process using real-time data from NEPTUNE, Canada's widely distributed sensor networks, Northeast Pacific. Geosci Can 2011. ISSN 1911–4850 https://journals.lib.unb.ca/index.php/GC/article/view/18588/20203.

[29] Gilbert C. Large data sets experience is needed by computer science graduates. Informatica Blog; 2014. https://blogs.informatica.com/2014/06/25/large-data-sets-experience-is-needed-by-computer-science-graduates/#fbid=9ES8Djx1b6m.

[30] Cisco Systems. https://www.cisco.com/c/en/us/solutions/service-provider/visual-networking-index-vni/index.html?CAMPAIGN=VNI+2017&COUNTRY_SITE= us&POSITION=Cisco+link&REFERRING_SITE=Cisco+page&CREATIVE=Go+URL+to+VNI+page; 2018.

CHAPTER

5.6

Drifting Buoys

Pierre Blouch[1], Paul Poli[2]

[1]Météo-France, Brest, France; [2]Centre de Météorologie Marine, Météo-France, Brest, France

5.6.1 INTRODUCTION

For four decades, surface drifting buoys, also referred to as drifters, have been among the main global measuring systems, observing in situ essential parameters over the oceans. Today, drifters measuring surface pressures, surface currents, sea-surface temperature, and sea-surface salinity are essential to enhance satellite measurements and/or numerical weather prediction through data assimilation. This chapter explores the different types of drifting buoys used in international programs for meteorological and ocean data collection.

Historically, drifting buoys were first devised to infer surface currents. It seems the earliest recorded drifting buoy measurements were made by Leonardo da Vinci (1452–1519). A simple float was released in the water flow and its downstream travel was measured after a given period of time [1].

They are among the few platform types to make Lagrangian measurements. The main challenge was to track them. In the mid-1700s, floats fitted with a sea anchor were tracked near the coast thanks to visual triangulation ([2]; [28]). Satellite systems developed in the 1970s, the Eole satellite, followed by Nimbus-6, demonstrated successful drifting buoy experiments [3–5]. From then, drifting buoys could be tracked in all of the world's oceans, and subsequently the first major international metocean observation programs began to organize the deployment of this fabulous new tool for understanding oceanic conditions. We first discuss the two main types of drifting buoys.

5.6.2 FIRST GLOBAL ATMOSPHERIC RESEARCH PROGRAM GLOBAL EXPERIMENT BUOYS

In 1978–79 the First Global Atmospheric Research Program (GARP) Global Experiment (FGGE)[1] recognized the need to fill oceanic data-void areas using free-drifting buoys. During the FGGE year, eight different countries[2] deployed over 300 drifting buoys in the Southern Hemisphere between 20°S and 65°S. Deployments were by ships of opportunity, research vessels, and aircraft. There were six major designs of buoys. In spite of differences, they were all made of a 2.5/3.0-m cylindrical spar fitted with a flotation collar of 0.7–0.9 m in diameter (Fig. 5.6.1), weighing around 100 kg. Some were made of aluminum while others were made of fiber glass and polyester. This experiment was successful in demonstrating the viability of the concept of autonomous platforms remaining in function for several months, able to withstand harsh weather and sea conditions. Although the buoys and network were designed to primarily meet the needs of meteorology (GARP was chiefly initiated by the World Meteorological Organization [WMO] [6]), studies also showed the excellent ability to serve oceanography.

FIGURE 5.6.1   Norwegian First GARP Global Experiment buoy by Bergen Ocean Data.

[1] Also known as Global Weather Experiment.

[2] Australia, Canada, France, New Zealand, Norway, South Africa, United Kingdom, and USA.

FGGE buoys measured air pressure and sea-surface temperature, and more rarely surface currents. Other parameters such as air temperature, humidity, and wind were judged too expensive to be cost-effectively and accurately measured so close to the sea surface.

5.6.2.1 Air Pressure Measurements

On FGGE buoys, the surface pressure accuracy specifications were 1 hPa, with a resolution of 0.1 hPa. The main challenge with such buoys was to prevent water entering the air intake and the pipe connected to the barometer located within the hull, near the waterline level. Meteorological studies demonstrated that the measurements in the Southern Hemisphere had a significant impact on weather analysis and forecasts (e.g., Refs. [7,8]).

5.6.2.2 Sea-Surface Temperature Measurements

The sea-surface temperature accuracy specifications were 1°C, with a resolution of 0.2°C. The measurements proved to be of great value as ground truth to validate satellite products, and to complement coverage in areas where infrared satellite sensors were unable to operate due to the cloud coverage [9].

5.6.2.3 Surface Current Estimates

While most of the FGGE buoys deployed in the southern oceans were free of floating anchors (or drogues), their tracks were in agreement with established concepts of the general ocean circulation, and presented significant correlations with surface currents computed from dynamic ocean heights. Some of the Southern Hemisphere buoys used drogues centered at depths of between 5 and 35 m. Analysis of the mean drift FGGE buoy speeds showed no obvious differences between drogued and undrogued buoys [56]. The large-scale distribution of kinetic energy derived from both types also appeared similar [10–12]. Overall, the use of buoys to estimate surface currents was starting to become a reality.

5.6.2.4 Additional Atmospheric Measurements

After FGGE, some buoys were fitted with wind measurement capabilities thanks to a profiled mast topped with a cup anemometer (Fig. 5.6.2). The mast aligned the buoy with the prevailing wind, and a flux-gate compass estimated the direction. French Marisonde buoys [13,14] and US CASID drifting buoys used this technique. Experiences showed that Ekman's theory could be verified using such buoys [15].

Air temperature was also generally measured on wind FGGE buoys. The sensor was located at the top of the profiled mast, about 2 m above the line of flotation or water line.

5.6.2.5 Additional Oceanic Measurements

After FGGE, some buoys attempted measuring subsurface temperatures on models such as CASID, Marisondes, and XAN buoys [16,17] at depths up to 300 m.

During the Tropical Ocean-Global Atmosphere (TOGA) program in 1989, the National Oceanic and Atmospheric Administration (NOAA) deployed 19 such buoys with thermistor chains in the western equatorial Pacific Ocean [18]. The temperatures were measured at 13 levels down to 300 m depth.

In 1993, Météo-France wind Marisonde buoys were equipped with thermistor chains. The SEMAPHORE experiment in 1993 deployed 29 buoys of this type south of the Azores. The data served to validate a three-dimensional ocean mesoscale model [19]. These buoys were recovered and used in several measurement campaigns until 2013, measuring depths at the bottom of the string and at three other intervals along the string. These were essentially the last buoys of the FGGE type to be used. Since the beginning of the 2000s, no FGGE buoys have been used for operational purposes.

5.6.2.6 Tracking and Data Transmission

Following earlier satellite demonstrations, the budding Argos satellite tracking system, launched in 1978, enabled the locating of FGGE buoys and collecting of their data [20].

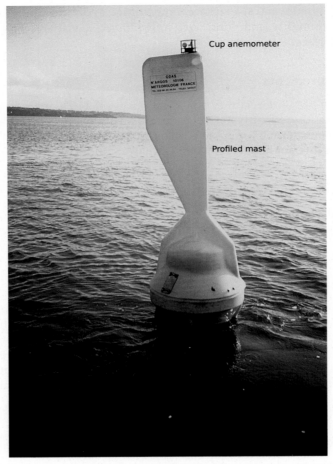

FIGURE 5.6.2   Marisonde wind First GARP Global Experiment buoy by ELTA, France.

Once received, the observations were processed by Collecte Localisation Satellite (CLS), the exclusive operator of the Argos system. Drifting buoy operators were indeed among the main Argos system users during three decades. Once received, the data were sent to meteorological users via the WMO Global Telecommunication System (GTS).

In the 1980s, a unique data format, inherited from FGGE, was supported by CLS for transmitting data onto the GTS. Data transmission required an Argos satellite in view of the buoy. Transmission attempts were renewed every 10–15 min, and fitted with the transmission time. Measured parameters and locations were hence not regularly sampled during a day due to the weakness of the satellite constellation (generally, two nominal satellites plus a few odd ones). In addition, measured parameters and locations were asynchronous.

For GTS purposes, a data format for drifting buoys was created by WMO. Called FM18 DRIBU, it included two timestamps: one for the measured parameters and one for the location. There were also technical parameters and quality flags linked to the specificities of the Argos system. This WMO code was later renamed FM18 DRIFTER, then FM18 BUOY (WMO Manual on Codes).

For nearly 30 years the Argos system remained the only satellite system capable of locating globally and reporting drifting buoy data.

5.6.2.7 Beyond First Global Atmospheric Research Program Global Experiment: Toward Operational Use

After FGGE, drifting buoys continued to be used for experiments in various regions of the globe. For instance, about 100 buoys were deployed in the North Atlantic in 1982 by the University of Kiel [21]. Most were drogued at 100 m depth. However, the lifetime of drogue attachments proved shorter than the buoys'. The investigators found that undrogued FGGE buoys were strongly influenced by wind and waves, while drogued buoys could be used to perform quasi-Lagrangian measurements of the surface current.

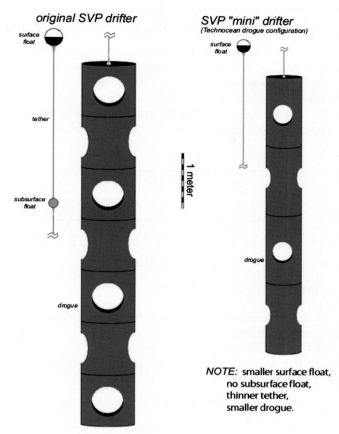

FIGURE 5.6.3   Surface Velocity Program-type buoys.

While these German buoys were primarily used for research, FGGE-type buoys started to be operationally used for weather analysis and forecasting purposes by NOAA/National Data Buoy Center (NDBC) in the United States and the COST-43 action group in Europe. This group became the European Group on Ocean Stations (EGOS) in 1989 before being folded into the operational surface marine component of the EUMETNET[3] observation program (E-SURFMAR) in 2005.

5.6.3 THE SURFACE VELOCITY PROGRAM BUOYS

Following FGGE, international research experiments shifted the focus of drifting buoys toward current estimation. The World Climate Research Programme (WCRP) initiated the TOGA experiment (1985–94) to study the tropical oceans and their relationship with the atmosphere. One important legacy of TOGA was the Tropical Atmosphere Ocean array, to support research and forecasting of ENSO warm cycles. This experiment was followed by the World Ocean Circulation Experiment (WOCE), another WCRP initiative, between 1990 and 1998 [25].

5.6.3.1 The Birth of a New Drifter

TOGA marked a turning point in the area of drifting buoys. While FGGE buoys were still used for atmospheric studies, a new type of drifting buoy appeared.

During TOGA, the NDBC of NOAA was assigned the responsibility of procuring, deploying, and maintaining a grid pattern of 50 FGGE drifting buoys throughout the oceans of the Southern Hemisphere [26]. However, to sample the tropical oceans, and in preparation for WOCE, FGGE buoys were deemed unsuitable to follow ocean currents

[3] Economic Interest Group of European national meteorological services.

BOX 5.6.1 INTERNATIONAL COLLABORATION

As operational use developed, so did international coordination. The Drifting Buoy Cooperation Panel was formed in 1985 under the auspices of the World Meteorological Organization (WMO) and the Intergovernmental Oceanographic Commission (IOC) of UNESCO [22,23]. It was later renamed the Data Buoy Cooperation Panel (DBCP) to cover moored buoys as well. Besides coordination for deployments, the DBCP also maintains best practices. For example, in 1988 the (then) state-of-the-art First GARP Global Experiment (FGGE) drifting buoy technology was described in a guide published by IOC and WMO [1]. Today, Surface Velocity Program (SVP)-B specifications are published in a regularly revised document [24].

BOX 5.6.2 THE SURFACE VELOCITY PROGRAM DRIFTER DESIGN

The Surface Velocity Program (SVP) drifter (Fig. 5.6.3) was designed:

- as a spherical surface float of diameter between 30.5 and 40 cm,
- weighing under 40 kg,
- to be made of fiber glass, or (later) less expensive acrylonitrile-butadiene-styrene,
- to be activated by removing a magnet attached to the hull,
- to be fitted with a "holey-sock" drogue to measure mixed layer currents,
- to contain only: batteries (typically alkaline D-cell batteries), a satellite transmitter, a sensor to identify the presence or absence of drogue, and instrument(s).

The only instrument in the SVP buoy was initially a thermistor to measure sea-surface temperature.

---

due to too large exposure to wind. In addition, FGGE buoys posed major challenges, such as high procurement costs and excessive sizes and weights, which made them difficult to deploy [27].

Consequently, the WCRP expressed the need for a low-cost, lightweight, and easy-to-deploy surface drifter, with a semirigid drogue to maintain shape in high-shear flows [28]. By 1991, after several years of testing, most especially on different drogue types, a clear design of the new drifter emerged [29].

The SVP of TOGA and WOCE was set up in 1988, giving its name to the drifter. Overall objectives were to globally describe the mixed layer circulation, and provide mixed layer velocity and sea-surface temperature observations for testing models.

The design of the SVP drifter is indicated in Boxes 5.6.1 and 5.6.2. The global array of SVP drifters was renamed the Global Drifter Program (GDP), a component of the Global Ocean Observing System, the Global Climate Observing System, and the DBCP of JCOMM. In 2002, to ease deployment and reduce buoy costs, a smaller SVP drifter was proposed in DBCP specification Rev. 1.2 [28]. The size of the "mini" SVP was 30.5 cm in diameter and the drogue was reduced in size accordingly. These drifters proved their efficiency to measure temperature and currents, but they were judged unsuitable to measure the atmospheric pressure. Their smallness made them more often submerged, avoiding the air intake to breathe, and did not carry batteries large enough to measure this parameter for a sufficient period (18 months nominal).

5.6.3.2 Tracking and Data Transmission

The SVP drifters employed Argos almost exclusively until recently, with the main issue of asynchronous positions and measurements. In the 1990s, CLS developed a new GTS processing system that made it possible to report hourly measurements, regularly sampled and stored on board the buoy. This flexibility led to the invention of many different formats, even for operational buoys measuring the same parameters. A few years later, two standard formats were adopted: DBCP-O3 for SVP drifters (without pressure measurement) and DBCP-M2 for barometer drifters.

FIGURE 5.6.4 Satellite communication types for drifting buoys operating during December 2016. Map generated by JCOMMOPS from Global Telecommunication System data received at Météo-France.

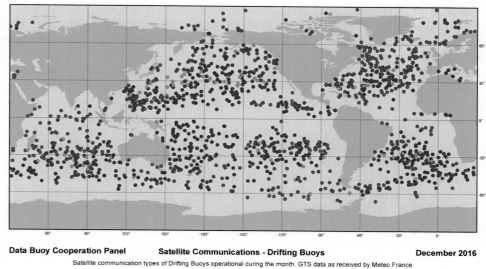

The main drawback of the Argos system for operational use was long transmission delays in some areas. These delays were due to the poor satellite constellation and land receiving stations. Step by step, improvements occurred in land receiving station coverage. However, even in the 2000s, it could still be hours before observations were received.

In 2008, at Météo-France's request, a buoy manufacturer, Metocean, designed a drifter reporting its data through the Iridium Short Burst Data service. The first drifters of that kind were not equipped with a Global Positioning System (GPS). Locations were provided by the Iridium system, though less accurately than Argos. Iridium locations are fitted with a circular error probability radius, helping to screen out large errors. The accuracy of Iridium positions was sufficient for locating meteorological measurements but insufficient to estimate surface currents.

Quickly, new Iridium drifting buoys were fitted with GPS chips, providing a better location than Argos, with hourly reporting. There were many advantages as compared to the old system: more reliable and cheaper communications (Iridium being used for many other applications), shorter transmission time delays, efficient two-way communication, and regular hourly reports. Eventually, this also benefited surface currents, with hourly estimates.

In November 2016, 41% of the drifting buoys reporting their data onto the GTS were Iridium buoys (Fig. 5.6.4). In October 2016, the GDP announced its migration plan to Iridium [30].

In line with other operational observation systems of WMO, drifting buoy data on the GTS have migrated from Traditional Alphanumerical Code to Table Driven Code Form FM94 BUFR.

5.6.3.3 Surface Currents

The SVP drifter was designed to estimate surface currents. These data serve marine service applications and testing and validation of ocean models [31]. Globally, drifting buoys have remained the main source of in situ data for this parameter [32].

Since the start of the SVP program and then the GDP, most of the drifting buoys have been fitted with holey sock drogues centered at 15-m depth. Some users have chosen other depths for their applications. This drogue is made of a cylinder of tissue maintained open thanks to rings. The lowest ring is slightly heavier so the drogue dives at the right depth. The drogue size should yield a drag area ratio of 40:1 between the drag area (cross-section area) of the drogue and the drag area of all other buoy components (Fig. 5.6.3).

As of 2017, the lifetime of the drogue was far shorter than the drifter lifetime. Drogues became quickly damaged or separated from the buoys months before these stop operating. The Lagrangian current can only be inferred from the buoy drift when the drogue is attached, so it is important to detect the presence or absence of drogues.

Several methods have been used. Some use the fact that drogues reduce the buoyancy and cause greater submersion by wave crests. The first SVP drifter design used a submergence detector made of two electrodes fitted on the upper hemisphere of the buoy hull. When the buoy was submerged, the current passed through the salt water and a signal was measured. The signal was sampled over 2–3 min and the output was processed to estimate statistics. The more frequent the submergence, the greater the chance of a drogue still attached.

In 2006, recognizing some deficiencies with this method, the DBCP asked manufacturers to install a strain gauge in the tether of the cable that links the buoy to its drogue [33]. The higher the strain, the greater the chance of a drogue still attached.

Both methods are not handled equally well by all manufacturers. In 2015, it was noted that the time taken by the GPS to obtain a fix was related to the sea state (when the drogue is attached). This relationship may be used in the future to assess the presence of the drogue. However, it must be noticed that all these methods only work well in the presence of waves. In calm waters, doubt remains. In addition, it is difficult to automate the analysis of the signals (submergence, strain, or time). Manual expertise is required to validate the time of loss of the drogue.

Indirect methods have also been used, such as analyzing the relationship between the surface wind (from colocated model outputs) and buoy movement, a decorrelation suggesting the drogue is still attached. Other methods consist in analyzing the accelerations of the buoy due to wind variations, or in considering the effect of wind slippage [34].

The vector is computed between two locations. Until recent years, buoy positions were most often provided by the Argos system. Being unavailable at regular intervals, it was necessary to interpolate them in time. For many years, the interpolations were done every 6 h by the Atlantic Oceanographic and Meteorological Laboratory thanks to a method proposed by Hansen and Poulain in 1995 [35]. Météo-France adopted this method on 3-h intervals for its near-real-time products. A new interpolation method has been proposed by Ref. [36], allowing to estimate errors as well.

5.6.3.4 Sea-Surface Temperature Measurements

On the SVP drifter, a thermistor probe for subskin sea-surface temperature is located at the base of the float to avoid direct radiative heating.

In general, the probe is made of an analog PT100 unit fitted in a metal pin fixed on the buoy hull. The probes are generally calibrated in batches before integration. On SVP-type drifters, the accuracy was estimated to be around 0.5°C.

Recent studies have shown the importance of calibration [37]. In 2009, the Group for High Resolution Sea-Surface Temperature proposed to enhance the quality of sea-surface temperature measurements [38]. The standard Iridium SVP, fitted with a GPS, was qualified as an HRSST-1 drifter. Since higher accuracy (0.02°C) was requested by the group, experimental drifters were fitted with digital sea-surface temperature probes, calibrated individually. Such drifters, called HRSST-2, are still under evaluation.

5.6.3.5 Air Pressure Measurements

In 1993, recognizing that meteorologists and oceanographers were operating drifting buoys with distinct characteristics, because of differing requirements the Global Drifter Design Center at Scripps Institution of Oceanography (SIO) proposed to fit the SVP with a barometer [39]. The initiative, addressing both communities, was supported by the DBCP and became a success. The SVP with barometer (SVP-B) was born. It was gradually adopted by meteorologists. Today, it is considered as the standard for GDP.

While FGGE buoys posed no problem because of their height, designing a waterproof air intake device was essential for SVP-B drifters, because they spend part of their time under the wave crests when the drogue is present. In particular, the "mini" SVP proved incompatible to support a barometer, because the air intake did not emerge enough in rough seas.

For the SVP-B, an air intake designed by the UK Met Office was adopted. It consists of a 20-cm plastic tube put on top of the drifter hull, and leaving only a tight air passage secured by a Goretex filter that only lets air through and no water [24]. When the buoy surfaces between two waves, the water that may have reached the filter falls down outside by gravity. This air intake (Fig. 5.6.5) also reduces wind-induced Bernouilli effect pressure offsets in the pressure signal, ensuring that static air pressure is measured.

The SVP-B Drifter Construction Manual [24] recommends sampling pressure at 1 Hz over 160 s to average over several wave cycles. To smooth data measured only in air, a median is applied to the 10 lowest measurements (assuming the buoy surfaced between waves). The measurement accuracy is estimated at around 0.5 hPa.

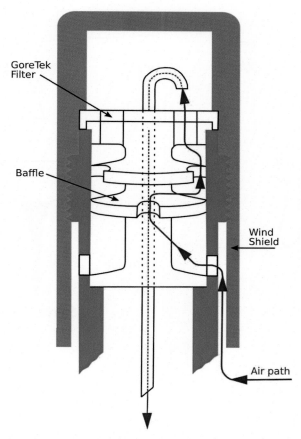

FIGURE 5.6.5 Air intake for Surface Velocity Program with barometer drifters.

Known issues include ice-covered caps on buoys operating in cold and humid regions (boundary between sea ice and open ocean). Also, some designs let water enter the pipe in rough seas, degrading the quality of air pressure measurements.

Recent studies conducted by European Centre for Medium-range Weather Forecasts (ECMWF) [40] and other meteorological agencies confirm the lessons learnt during FGGE: pressure observations reported by drifting buoys are essential to weather analysis and forecasts. This parameter, which cannot be reasonably measured by satellites yet, has a significant impact on numerical weather prediction. Drifting buoy observations contribute to decreasing the total 24-h global forecast error by approximately 3%.

5.6.3.6 Additional Atmospheric Parameters

Air temperature and humidity cannot be accurately measured on SVP buoys because they spend a part of their time below the wave crests.

Measuring wind as done in FGGE was tried on SVP drifters. However, the SVP proved unsuitable to measure the wind velocity through cup anemometers. One buoy manufacturer (Metocean) proposed to apply the Wind Observation Through Ambient Noise (WOTAN) method to SVP drifters to estimate the wind speed [41]. The method was previously experienced on another type of buoy (Compact Meteorological and Oceanographic Drifters [CMOD]). The WOTAN technique was also investigated for classification and quantification of precipitation [17].

A hydrophone put at 10-m depth monitors the noise produced by the wind (cracking air bubbles) at the sea surface. Between 2 and 8 kHz, the noise energy may be empirically linked to the wind force (Figs. 5.6.6 and 5.6.7).

The first wind SVP drifter was deployed by Météo-France in 1997 [42]. A few years later, thanks to the use of several buoys deployed in the surroundings of moored buoys and reporting the noise energy for several frequencies, it was possible to improve the algorithm that estimated the wind speed from the ambient noise spectra [43].

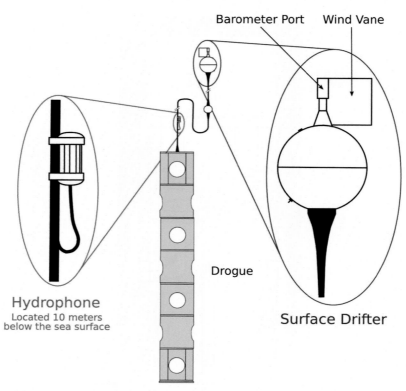

FIGURE 5.6.6   Wind Observation Through Ambient Noise drifter.

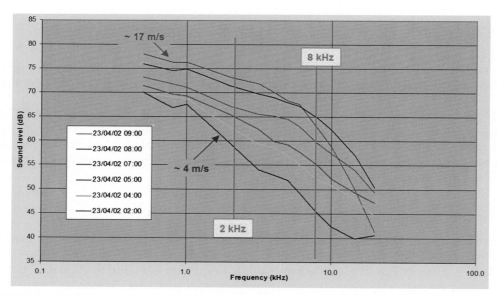

FIGURE 5.6.7   Wind Observation Through Ambient Noise drifter sound spectra in front of a storm on April 23, 2002.

More than 400 drifters of that kind, also built by Pacific Gyre [44], were deployed over all the oceans. However, the method showed its limits. Weak winds and wind above 17 m/s were not correctly estimated. When waves are breaking, the physics changes and the slope of the relationship between noise energy and wind speed is smaller. Errors are greater. Because these drifters did not provide more accurate measurements than satellite scatterometers, the technique was abandoned. Sonic anemometers were also tested on SVP drifters, in particular in tropical cyclones [45].

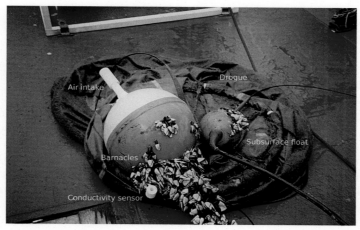

FIGURE 5.6.8   Salinity Surface Velocity Program drifter recovered for expertise after several months at sea.

5.6.3.7 Additional Oceanic Parameters

High-resolution and high-quality sea-surface salinity observations are required for ocean forecasting systems (in assimilation and validation of ocean models). As for sea-surface temperature and wind, satellites such as European Space Agency (ESA) Soil Moisture and Ocean Salinity and NASA Aquarius/SAC-D are able to provide good geographical coverage, but require ground truth for validation [46].

Measuring sea-surface salinity is challenged by biofouling (Fig. 5.6.8). The limited power available on buoys prevents pumping the water.

Inductive conductivity sensors were first tested on SVP drifters in the 1990s though without success. Salinity measurements started to become accurate when the buoys were equipped with Sea-Bird conductivity sensors at Pacific Gyre and Metocean. However, these sensors are generally too expensive for expendable systems like drifters of which the mean lifetime is typically about 12 months.

Subsurface temperature is required for seasonal and interannual forecasts, testing, and validation of ocean forecasting models and marine services. These measurements cannot really be obtained from satellites that only measure electromagnetic waves (which do not penetrate far in water). There are new prospects with gravity missions to infer the displacements of subsurface water masses, though these are not yet providing vertically resolved profiles of temperature. Drifting buoy measurements on these parameters complete those carried out by other networks such as tropical moored buoy arrays, Argo floats, and ship eXpendable BathyThermograph profiles.

Drifting buoys are suitable for air–sea interaction studies. In contrast with Argo floats, they may sample the ocean boundary layer with a higher frequency. These latter perform a temperature profile every 10 days only.

As for FGGE buoys, SVP drifters have been equipped with thermistor chains, some until 80-m depth [47,48,55].

The main challenge for all these buoys is to know the actual depth of each temperature probe. The thermistor string does not remain straight when the buoy moves. The bottom of the string may go several meters up if the buoy is driven by strong winds at the surface or in the presence of current shear.

It would be too costly on expendable buoys to have one pressure sensor per temperature probe. For this reason, the strings are often equipped with a single pressure sensor at its end and an algorithm (based on a catenary model) is used to estimate the depth of each probe according to its distance from the buoy and the pressure measured by this sensor. Recent studies suggest additional depth sensors would be required [49].

5.6.3.8 Power Supply and Lifetimes

SVP drifters are generally fitted with alkaline batteries that confer on them a nominal lifetime of 18 months. For environmental reasons—many buoys run ashore prematurely—lithium batteries are avoided. In practice, true buoy lifetimes depend on many different factors. Some buoys may operate at sea over 4–5 years while others may fail within days after deployment. The use of solar panels, combined with a lower power consumption of the buoys, could be a solution to increase their lifetimes.

5.6.4 OTHER BUOYS

5.6.4.1 Arctic Buoys

Arctic buoys drift on sea ice during part of their life. While some are designed to exclusively operate on sea ice, others may drift in the open ocean after being released by the ice. These latter should also survive through icing periods if possible. This is the most challenging task for them.

Buoy activities in the Arctic started during FGGE. Twenty ice buoys called Ice Experiment (ICEX), measuring air pressure, were air dropped in an approximate 400 km grid in the Arctic Basin [2]. They measured 62 cm in diameter, weighed 38 kg, and were parachuted to the ice below. Later, the experiment was extended for 5 more years (new deployments) to obtain a valuable time series of the factors that affect the ice flow. These ice buoys tracked the ice motion in the Arctic for the first time. Studies could be carried out to assess the parts of the drift due to weather conditions and ocean circulation at various timescales. It was also proven that their pressure measurements were essential for weather analysis and forecasting in this region.

The buoys participated in the Arctic Ocean Buoy Program led by the Polar Science Center of the University of Washington. In 1991, the program became the International Arctic Buoy Programme, an Action Group of the DBCP.

Drifting buoys play an important role in the Arctic since dramatic changes in climate are observed there. Numerous types of buoys have been used:

- ICEX buoys (FGGE type).
- SVP-type drifters.
- Ice mass balance buoys, capable of measuring ice thickness by acoustic sounding.
- Autonomous flux buoys, fitted with an acoustic Doppler current profiler and precision temperature and conductivity sensors.
- UpTempO drifters from Marlin-Yug, fitted with thermistor strings to measure sea temperature up to 60-m depth.

5.6.4.2 Other Types

Drifting buoy outlines would not be complete if we missed a few other types.

CODE drifters were designed and first used in the Coastal Dynamics Experiment in the early 1980s [50]. Their purpose was to measure coastal current in the first meter below the ocean surface, which is important for modeling and for providing better knowledge for several applications: tidal energy, navigation, coastal erosion, for instance. It consists of a slender, vertical, 1-m long negatively buoyant tube with four drag-producing vanes extending radially from the tube over its entire length (Fig. 5.6.9). CODE drifters continued to be used in the 2010s, for instance in the Mediterranean Sea [51] or in the Gulf of Mexico [32].

eXpandable Ambient Noise (XAN) drifters (Fig. 5.6.10) are a series of buoys developed by Metocean, Dartmouth, Nova Scotia, Canada, at the request of the US Navy [17]. All are based on a standard A-size military sonobuoy. The basic one, XAN-1—also called CMOD—measures air pressure and sea-surface and air temperatures. Others measure the ambient noise and/or subsurface temperatures at 10 levels, or directional wave spectra.

These buoys were designed to operate for 90 days at least at sea. More than 1500 buoys of the XAN-3 type (measuring subsurface temperatures) were deployed by the US Navy in the 1990s [16].

XAN buoys are embedded in a 914-mm long tube (123 mm in diameter). This latter can be easily air deployed from any standard sonobuoy launch system, as well as from any surface vessels, stationary or moving at speeds of up to 45 knots. A parachute opens after launch thanks to a wind flap. The buoyancy is ensured by an air bag that inflates when the buoy hits the water surface. The parachute then moves off and the thermistor string, if present, is deployed.

Another type of drifting buoy is devoted to track oil spills. They are simple Lagrangian systems fitted with telemetry. According to their goals, the design of their hull shape is essential.

5.6.5 RECENT EVOLUTIONS AND LOOKING AHEAD

As this short chapter illustrates, general evolutions for drifting buoys have been to improve reliability, increase lifetime, reduce transmission timeliness (from days or hours to a few minutes), reduce costs, and standardize size (from unique laboratory equipment to industry-scale units). This approach, leaning toward "cheap and plentiful," has recently been joined with another one: improving accuracy. Indeed, by their very positions, right at the sea surface, drifters are well placed to serve as fiducial reference measurements for satellite missions.

FIGURE 5.6.9   Code drifter.

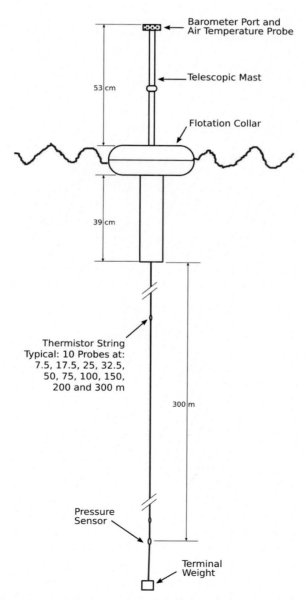

FIGURE 5.6.10   XAN-6 drifter (measuring 10 subsurface temperatures up to 300 m deep).

New emerging trends are to place more accurate sensors on drifters, yet for an affordable price. For example, the AtlantOS project is developing an affordable conductivity sensor to measure sea-surface salinity, reporting also sea-surface temperature with a resolution of 0.001 K and an estimated accuracy of 0.01 K. There are also thoughts to place miniature hydrophones on drifters to track acoustic tags on fish compatible with standards developed by the Ocean Tracking Network.

There are also new efforts to estimate waves from drifters. Knowledge of the sea state is important for research and marine services, for instance. Forecast models and satellite measurements are the most common sources for wave data, but they need in situ observations for calibration and validation. Such in situ data are mainly available in coastal waters from moored buoys. These systems (e.g., Datawell Waveriders and Fugro-Oceanor products) are too expensive to adapt onto expendable drifting buoys. In 2009, a low-cost GPS technology for measuring ocean waves on drifting buoys was tested in California [52]. Most recently, works funded by the US Office of Naval Research were conducted by SIO. Recent results indicate that directional wave spectra can be estimated [32].

Other measurements being considered from drifters include partial pressure of dissolved carbon dioxide (pCO_2). Monitoring pCO_2 at the ocean–air interface is essential to assess the capacity of the ocean to absorb part of the atmospheric CO_2, with a risk of increased acidity affecting marine wildlife. Drifting buoys able to measure surface pCO_2 were developed by LOCEAN. The sensors using chromatography or infrared analysis of the gas phase being too large for drifting buoys, FGGE-type buoy Carbon Interface Ocean Atmosphere (CARIOCA) measurements of pCO_2 use the absorbance variation of a pH-sensitive dye-thymol blue [53].

Another parameter of interest is ocean color. This concerns research in primary productivity and regional or global carbon cycles. Satellites offer excellent spatial and time coverage but, as for sea-surface temperature, they are limited to clear sky conditions and require in situ observations. Optical sensors were experimentally installed on CMOD then SVP drifters [54].

With clear benefits demonstrated from their measurements, drifting buoys occupy today a small niche in global observing. Further evolutions may entail improving their sustainability, both in terms of long-term funding (e.g., serving operational weather prediction and becoming an integral part of multiyear Earth observation space missions) and mitigating environmental impact (e.g., through systematic recovery policy and/or biodegradable materials).

Initiatives in this area are not limited to student projects and there are existing industrial solutions, albeit with limited lifetime (a few months). The buoys typically rely on wood or fruits for the float, bamboo sticks and corn-based (biodegradable) plastic for the structure, and clipboards for assembly; under 20% of nondegradable parts remain, such as electronics for data collection, positioning, and transmission (Fig. 5.6.11).

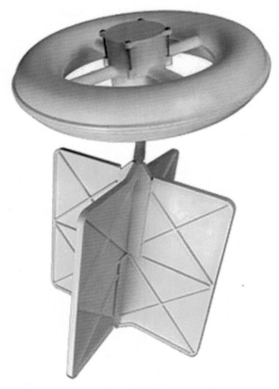

FIGURE 5.6.11   Biodegradable and sacrificial Consortium for Advanced Research on Transport of Hydrocarbon in the Environment (CARTHE) drifter.

References

[1] Hamilton G. Guide to drifting data buoys. IOC WMO Man Guides 1988;20.

[2] Fleming RJ. Buoy systems during the FGGE, proceedings of the first national workshop on the global weather experiment. Curr Achiev Future Dir 1984;2(Part 1):37–50.

[3] Cresswell GR, Richardson GT, Wood JE, Watts R. The CSIRO satellite-tracked Torpedo buoy. 1978. CISRO Rep. 82 available online from: https://publications.csiro.au/rpr/download?pid=procite:bd92498e-ce4c-4d92-a368-9a21f9b0d76c&dsid=DS1.

[4] Kirwan AD, Mc Nally G, Chang M-S, Molinari R. The effect of wind and surface currents on drifters. J Phys Oceanogr 1975;5:361–8.

[5] NASA. Nimbus 6 random access measurement system applications. NASA-SP 457 Goddard Space Flight Center; 1982. Available online from: https://ntrs.nasa.gov/archive/nasa/casi.ntrs.nasa.gov/19830006564.pdf.

[6] WMO. Manual on codes. 2016.

[7] Guymer LB, Le Marshall JF. Impact of FGGE buoy data on southern hemisphere analyses. Aust Meteor Mag 1980;28:19–42.

[8] Puri K, Bourke W, Seaman R. The impact of the FGGE observing systems in the southern hemisphere. In: Proceedings of the first national workshop on the global weather experiment, current achievements and future directions, vol. 2. 1984. p. 146–60. Part 1.

[9] Strong AE. Use of drifting buoys to improve accuracy of satellite sea surface temperature measurements. Trop Ocean Atm Newsl 1984;25:16–8.

[10] Daniault N, Ménard Y. Eddy kinetic energy distribution in the Southern Ocean from altimetry and FGGE drifting buoys. J Geophys Res 1985;90:877–99.

[11] Krauss W, Käse RH. Mean circulation and eddy kinetic energy in the Eastern North Atlantic. J Geophys Res 1984;88:3407–15.

[12] Richardson PL. Eddy kinetic energy in the North Atlantic from surface drifters. J Geophys Res 1983;88:4355–67.

[13] Blouch P, Fusey F-X. Drifting buoy developments at the French meteorological office. In: Proceedings 1983 symposium on buoy technology, New Orleans. 1983. p. 124–30.

[14] Tournadre J, Queffeulou P, Abdellaoui R. Analysis of wind satellite measurements during the SEMAPHORE experiment. In: Conference: OCEANS '94. 'oceans engineering for today's technology and tomorrow's preservation.' Proceedings, vol. 3. 1994.

[15] Daniault N, Blouch P, Fusey F-X. The use of free-drifting meteorological buoys to study winds and surface currents. Deep Sea Res 1985;32:107–13.

[16] Mariette V, Verbeque V, Mouge P, Deveaux M. A new concept for rapid assessment of oceanic environment. In: Proceedings of the second international conference on EuroGOOS, vol. 66. 2002. p. 289–97. Elsevier Ocean. Series.

[17] Selsor HD. Data from the sea: Navy drift buoy program. Sea Technol 1993;34(12):53–8.

[18] McPhaden MJ, Shepherd AJ, Large WG, Niiler PP. A TOGA array of drifting thermistor chains in the western equatorial Pacific Ocean: October 1898-January 1990. 1991. NOAA Data Rep. ERL PMEL-34.

[19] Caniaux G, Planton S. A three-dimensional ocean mesoscale simulation using data from the SEMAPHORE experiment: mixed layer heat budget. J Geophys Res 1998;103(C11):25,081–99.

[20] Garrett J. FGGE buoys: 5 years later. EOS Trans 1983;64:962–3.

[21] Krauss W, Dengg J, Hinrichsen H-H. The response of drifting buoys to currents and wind. J Geophys Res 1989;94C3:3201–10.

[22] DBCP. Drifting buoy Co-operation panel. In: First session, toulouse, final report, JCOMM. 1985. 30 p.

[23] Wallace A. A retospective. 2015. DBCP Technical Document No 56, 32 p.

[24] DBCP. WOCE surface velocity programme barometer drifter construction manual – revision 2.2. 2009.

[25] WCRP/WOCE. Surface velocity programme planning committee. In: Report of the first meeting, Miami. 1988. WMO/TD-No. 323.

[26] Kozak R, Partridge R. The role of drifting buoys in the Tropival Ocean Global Atmosphere (TOGA) program. In: Ocean '85 conference proceedings. (à récupérer). 1985.

[27] Niiler P. The world ocean surface circulation. Ocean circulation and climate. In: Siedler G, Church J, Gould J, editors. International geophysics series, vol. 77. Academic Press; 2001. p. 193–204.

[28] Lumpkin R, Pazos M. Measuring surface currents with Surface Velocity Program drifters: the instrument, its data, and some recent results. In: Griffa A, Kirwan AD, Mariano AJ, Ozgokmen T, Rossby T, editors. Lagrangian analysis and prediction of coastal and ocean dynamics. Cambridge Univ. Press; 2006.

[29] Sybrandy AL, Niiler PP. WOCE/TOGA Lagrangian drifter construction manual. 1992. WOCE Rep. 63, SIO Ref. 91/6, 58 pp., Scripps Inst. of Oceanogr., La Jolla, Calif.

[30] DBCP. Drifting buoy co-operation panel. In: 32th session, La Jolla, final report, JCOMM. 2016. 56 p.

[31] Blockley EW, Martin MJ, Hyder P. Validation of FOAM near-surface ocean current forecasts using Lagrangian drifting buoys. Ocean Sci 2012;8:551–65.

[32] Lumpkin R, Özgökmen T, Centurioni L. Advances in the application of surface drifters. Ann Rev Mar Sci 2017;9:59–81. https://doi.org/10.1146/annurev-marine-010816-060641.

[33] DBCP. Drifting buoy co-operation panel. In: 22th session, La Jolla, final report, JCOMM. 2006. 149 p.

[34] Rio M-H. Use of altimeter and wind data to detect the anomalous loss of SVP-type Drifter's drogue. J Atmos Ocean Tech 2012;29:1663–74. https://doi.org/10.1175/JTECH-D-12-00008.1.

[35] Hansen VD, Poulain P-M. Quality control and interpolation of WOCE-TOGA drifter data. JAOT 1996;13:900–9.

[36] Elipot S, Lumpkin R, Perez RC, Lilly JM, Early JJ, Sykulski AM. A global surface drifter data set at hourly resolution. J Geophys Res Oceans 2016;121. https://doi.org/10.1002/2016JC011716.

[37] Hausfather Z, Cowtan K, Clarke DC, Jacobs P, Richardson M, Tohde R. Assessing recent warming using instrumentally homogeneous sea surface temperature records. Sci Adv 2017;3. https://doi.org/10.1126/sciadv.1601207.

[38] GHRSST. In: Proceedings of the GHRSST XI science team meeting, Lima. 2010. 60 p.

[39] DBCP. Drifting buoy co-operation panel. In: 9th session, Athens, final report, JCOMM. 1993. 108 p.

[40] Horányi A, Cardinali C, Centurioni L. The global numerical weather prediction impact of mean sea level pressure observations from drifting buoys. QJR Meteorol Soc 2016. https://doi.org/10.1002/qj.2981. Accepted Author Manuscript.

[41] Vagle S, Large WG, Farmer DM. An evaluation of the Wotan technique of infering oceanic winds from underwater ambient sound. J Atmos Ocean Tech 1990;4:576–95.

[42] Blouch P, Rolland J. Promising results of the WOTAN technique to provide wind measurements on SBP-BW drifters. In: DBCP technical document n 12, 75–80. Genève, Suisse: Organisation météorologique mondiale; 1998.

[43] Blouch P, Rolland J. Evaluation of SVP-BW drifters thanks to deployments near moored buoys. In: DBCP-18 workshop – martinique. 2002. http://www.jcommops.org/dbcp/doc/DBCP-22/DOCS_DBCP22/03_PierreBlouch.ppt.

[44] Morzel J, Niiler PP. Surface wind observations in tropical cyclones from drifting ocean buoys. In: Ocean science meeting, Portland. 2010.

[45] Hormann V, Centurioni LR, Rainville L, Lee CM, Braasch LJ. Response of upper ocean currents to Typhoon Fanapi. Geophys Res Lett 2014;41:3995–4003.

[46] Reverdin G, Morisset S, Boutin J, Martin N, Sena-Martins M, Gaillard F, Blouch P, Rolland J, Font J, Salvador J, Fernandez P, Stammer D. Validation of salinity data from surface drifter. J Atmos Ocean Tech 2014;31:967–83.

[47] du Penhoat Y, Reverdin G, Kartavtseff A, Langlade MJ. BODEGA surface drifter measurements in the western equatorial Pacific during TOGA COARE experiment. Notes Tech Sci Mer ORSTOM-Nouméa 1995;11. 241 pp.

[48] Tolstosheev AP, Lunev EG, Motyzhev VS. Development of means and methods of drifter technology applied to the problem of the Black Sea research. Oceanology 2007;48(1):138–46.

[49] Rousselot P. Étude des données de bouées dérivantes à chaîne de thermistances. 2017. AtlantOS project.

[50] Davis R. Drifter observations of coastal surface currents during CODE: the statistical and dynamical views. J Geophys Res 1985;90: 4756–72.

[51] Poulain P-M, Bussani A, Gerin R, Jungwirth R, Mauri E, Menna M, Notarstefano G. Mediterranean surface currents measured with drifters: from basin to subinertial scales. Oceanography 2013;26(1):38–47. https://doi.org/10.5670/oceanog.2013.03.

[52] Herbers THC, Jenssen PF, Janssen TT, Colbert DB, MacMahan JH. Observing ocean surface waves with GPS-tracked buoys. J Atmos Ocean Tech 2012;29:944–59.

[53] Merlivat L, Brault P. CARIOCA buoy: carbon dioxide monitor. Sea Technol 1995;36(10):23–30.

[54] Kuwahara VS, Strutton PG, Dickey TD, Abbott MR, Letelier RM, Lewis MR, McLean S, Chavez FP, Barnard A, Morrison JR, Subramaniam A, Manov D, Zheng X, Mueller JL. Radiometric and bio-optical measurements from moored and drifting buoys: measurement and data analysis protocols. In: Ocean optics protocols for satellite ocean color sensor validation, revision 4, vol. VI. 2003.

[55] Motyzhev SV, Lunev EG, Tolstosheev AP. The experience of barometric drifter application for investigating the world ocean Arctic region. Phys. Oceanogr. 2016;4:47–56. https://doi.org/10.22449/1573-160X-2016-4-47-56.

[56] Garrett JF. The performance of the FGGE drifting buoy system. Proc. COSPAR Symp. on FGGE System Performance and Early Results. Adv. Space Res 1980;1:87–94.

Further Reading

[1] Argos JTA. 16th meeting on argos joint tariff agreement summary report. 1996. Henley-on-Thames, 84 p.

[2] Centurioni L, Braasch L. Directional surface gravity waves properties from low-cost drifting buoys. Geneva: DBCP-31 Workshop; 2015.

[3] DBCP. Drifting buoy co-operation panel. In: 15th session, Wellington, final report, JCOMM. 1999. 56 pp.

[4] Franklin B. Sundry marine observations. Trans Am Philos Soc Ser 1785;1(2):294–329.

[5] Gentemann CL. Three way validation of MODIS and AMSR-E sea surface temperatures. J Geophys Res Oceans 2018;119:2583–98. https://doi.org/10.1002/2013JC009716.

[6] Large WG, McWilliams JC, Niiler PP. Upper ocean thermal response to strong autumnal forcing of the northeast pacific. J Phys Oceanogr 1986;16:1524–49.

CHAPTER

5.7

Innovations in Profiling Floats

Patrice Brault

nke instrumentation, Hennebont, France

5.7.1 BACKGROUND

The first use of the technique of stabilized full water immersion and monitoring equipment is attributed to John Swallow. In 1955, using acoustics on a ship, he directed the monitoring of aluminum tubes, originally intended for the building of scaffolds, adjusted to balance the desired level of immersion.

The 1970s (Rossby and Webb) saw the use of acoustic emission sources drifting at low frequency (around 260 Hz) and exploiting the characteristics of waveguides that can occur in the ocean on some dives (SOFAR channel: SOund Fixing And Ranging) [1].

In 1985, the availability of the Argos satellite communication system, together with the World Ocean Circulation Experiment program, allowed the development of new instrumentation based on the opposed mode principle (the drifting float receives while the fixed source emits), whose generic name is RAFOS (= SOFAR-1-1). The simplest instruments were floats (tubes of glass and unicycles that drift down to 4000 m and drop a ballast at the end of life [after 12–18 months usually]).

At the beginning of the 1990s, during the same period, the new concept of Autonomous Lagrangian Circulation Explorer (ALACE) emerged, which consisted in bringing the float regularly to the surface to position and possibly create a profile of temperature measurements on the way up; the movement observed between two locations at the surface is considered as corresponding to the drift in immersion.

All subsurface floats naturally balance in open water by adjustment of their density to the density of seawater at the desired immersion. The increase in density of the float with immersion is less than the increase in sea water density, which guarantees a stable state in open water that requires no energy. The profiling floats have a constant mass and variable volume. The variation in volume allows the full water balance according to the density of seawater. These floats are called isobars since the effect of pressure is dominant in the balance. The profiler works like a submarine "hot air balloon": it adjusts its depth and its emergence via control of its volume, but moves under the action of the ambient current.

The international Argo program for the measurement of changes in the ocean's heat content and salinity was launched in 1999. The initial objective of the Argo program was to operate 3200 profiling floats in ice-free water from 60°N to 60°S that measure pressure, temperature, and salinity in the upper 2000 m of the ocean.

Argo deployments began in 2000, and in November 2007 the millionth profile was collected. At the end of November 2017, 3835 floats were active worldwide and 3024 papers have been published.

The ocean floats used as part of the Argo program have enabled the following:

- Production in large quantities on recurrent and sustainable funding brings the necessary conditions in industrial focus and contribution in terms of reliability for this oceanographic vector.
- Provision to the scientific community of all data validated by the person responsible for each program has generated more than 2000 publications. Scientists use different data acquired in the same area at different times by different profilers and the result of the temporal series of Eulerian data; the contents of the various publications showed that the same sets of data could be used for understanding several phenomena over many scales of oceanography.
- To be accepted by all bordering countries, the choice has been made to obtain only physical measurements of seawater, temperature, conductivity, and pressure, and to deploy the instruments outside the exclusive economic zone (EEZ); when drifts reach EEZ areas, the acquired data collected in the economic zones are excluded from the public data set.
- Traceability of data use thanks to a systematic check of Argo program data usage in publications, which is important for the credit of float fleet operators and program sustainability.

5.7.1.1 Argo in Brief

Argo profiling floats have a core mission typically defined as:

1. Dive to a parking depth.
2. Stabilization and drift with acquisition of PTS triplet at low rate, once every 12 h, drifting depth is typically 1000 dbar.
3. After 9 days, dive to the profile depth at 2000 dbar.
4. Ascent to surface in acquisition mode. For PROVOR and ARVOR, to obtain typically 92 PTS computed by averaging.
5. Surfacing for positioning and transmission through satellite.

The cycle is repeated until energy exhaustion. Argo recommends an energy capacity of at least 150 profiles.

The processing, archiving, and availability of data is provided by an operational and effective network led by two main and redundant processing centers located in France and the United States. Financially, the goal achieved was to acquire a PTS profile at a cost of €100, which is the cost of a measurement made from a ship by an expendable probe. The Argo program is a great success for the community.

The sensors are temperature, pressure, and salinity computed from the measurement of conductivity as per "TEOS 10" (Table 5.7.1).

TABLE 5.7.1　Argo SBE41CP Parameters Range, Accuracy, Resolution and Drift

SBE41CP	Pressure	Temperature	Salinity
Range	0/2100 dbar	−2/+35°C	10/42 psu
Accuracy	±2.4 dbar	±0.002°C	±0.005 psu
Resolution	1 dbar	0.001°C	0.001 psu
Estimated drift	2 dbar/an	0.02°C/an	< −0.003 psu Observed over 2–6 years[a]

[a]Sea-Bird scientific, accuracy, and stability of Argo SBE 41 and SBE 41CP, CTD Conductivity, Temperature and Depth sensors, December 9, 2008, Version 1.

5.7.1.2 Argo Profiling Float Strengths, Weaknesses, Opportunities, and Threats Analysis

5.7.1.2.1 Profiling Float Strengths

- Simple vector that does not require continuous control by an operator.
- Long mission and extended autonomy, over 4 years.
- Capacity to incorporate various sensors in the payload.
- Silent vector.
- Powerful Argo program.

5.7.1.2.2 Profiling Float Weaknesses

- Nonselectable trajectory.
- Not possible to recover data during the dive (except via inductive modem on ice tether platform [ITP]).
- In a very large percentage of cases, the floats cannot be recovered for sensor check and recalibration, refurbishing the vectors for new deployment, or failure analysis.

5.7.1.2.3 Profiling Float Opportunities

- The critical situation of the global context of ocean and earth warming requires accurate and operational instruments, such as the Argo network.
- Profiling floats are powerful instruments and easy to operate worldwide.
- Observation satellites provide very powerful observations with high acquisition rates and high-capacity processing. However, satellites cannot acquire submarine parameters; profilers allow it by going deep under the surface: in that respect, they are complementary.
- Funding is limited but recurrent. It allows continuous improvement in the long term and ensures very reliable and well-industrialized platforms.

5.7.1.2.4 Profiling Float Threats

- Funding stopped by a major funding partner.
- Acceptability of pollution caused by expendable instruments.
- End of free data sharing throughout the community that provides the measurements.

5.7.2 DEMANDS FROM THE USER COMMUNITY

The demand for "innovations," driven by new mission features and areas, is constant and has raised new technological challenges, several of which are addressed in this chapter. These can be summarized as follows:

5.7.2.1 New Mission Features

- Acquire more variables.
- Improve accuracy.
- Operate longer to reduce profile cost.
- Simplify the deployment so it can be made by crews on vessels of opportunity.
- Improve the smart features integrated in the profiling float to transmit relevant information. This means having the ability to modify the sampling strategy automatically according to the data previously acquired.

5.7.2.2 New Areas

After a limitation to the open ocean and between 60 degrees south and north, new areas are targeted with similar profiling floats potentially outside the scope of the Argo program:

- Deeper diving, from 2000 dbar to 4000 and 6000 dbar.
- Ice-covered areas.
- Marginal seas.
- Coastal areas.

5.7.3 BIOGEOCHEMICAL-ARGO

Biogeochemical-Argo will be an extension of the Argo array that aims to equip floats across the entire network with a standard set of biogeochemical (BGC) sensors.

5.7.3.1 Array

A global Biogeochemical-Argo array would enable direct observation of the seasonal-to-decadal scale variability in net community production (the rate at which lifeforms capture and store chemical energy as biomass), the supply of essential plant nutrients transported from deep waters to the sunlit surface layer, ocean acidification, hypoxia (low oxygen levels), and ocean uptake of carbon dioxide. Biooptical sensors would supplement satellite observation of the oceans' color by providing measurement of chlorophyll, light scattering deep into the ocean interior throughout the year, in cloud- and ice-covered areas or during the dark polar winter. A global Biogeochemical-Argo system would enable a transformation in our understanding of ocean biogeochemistry, climate interactions, and marine resources. It would provide data from regions and under conditions not accessible to satellites, and the global coverage would provide an integrated vision of the systems that drive the world's ocean ecosystems [2]. Science ministers from the G7 nations endorsed the concept in May 2016 in Tsukuba, Japan. The Biogeochemical-Argo project (http://www.biogeochemical-argo.org) uses the heritage of several regional projects involving the deployment of BGC profiling floats, initiated in the early 2000s.

A set of regional and smaller-scale programs is in operation around the world. These programs have already equipped almost 10% of the Argo array with BGC sensors. Although these programs are already sponsored by various organizations and operate independently, they must make their data available for the Argo community and process them according to Argo standards. The Scientific Committee on Oceanographic Research working group 142 (http://www.scor-int.org/SCOR_WGs_WG142.htm) has laid the foundation for a global observing system by validating sensor operation and developing software tools.

Among these regional programs and initiatives we find:

- The first floats deployed with biooptical sensors: Apex in 2003 and PROVOR Carbocean project in 2004.
- The Southern Ocean Carbon and Climate Observations and Modeling project.
- The Remotely Sensed Biogeochemical Cycles in the Ocean project in the North Atlantic subpolar gyre.
- The Novel Argo Ocean Observing System in the Mediterranean Sea.
- The Integrated Physical Biogeochemical Ocean Observation Experiment in the Kuroshio region of the North Pacific.
- The Australia–India Joint Indian Ocean Bio-Argo project.

5.7.3.2 General Scientific Context

Biogeochemical-Argo is poised to address a number of grand challenges in ocean science and in the management of ocean and global resources, topics that are difficult, if not impossible, to address with our present observing assets. Further, these are topics of immediate importance to society in the face of a changing climate and the need for greater protection of ocean resources. The observing system described here will enable significant advances to be made in these areas. The key questions are stated hereafter, with links provided to specific pages of the Biogeochemical-Argo web portal for more details.

5.7.3.2.1 Ocean Science Research

- Will ocean carbon uptake continue at the same relative rate as the ocean warms? (Carbon uptake)

- How does the volume of oxygen minimum zones (OMZs) change in time? How does this affect the cycling of nitrate (NO_3)? (OMZs and NO_3 cycle)
- What is the variability and trend in ocean pH? How does the changing carbonate saturation state affect BGC processes? (Ocean acidification)
- What are the interannual variations in the biological carbon pump? Will its strength be reduced in a warmer ocean? (Biological carbon pump)
- What is the composition of phytoplankton communities? How will it affect higher trophic levels and carbon cycling? (Phytoplankton communities)

5.7.3.2.2 *Ocean Management*

- Does real-time data improve management of living marine resources? (Living marine resources)
- Does an improved ocean carbon budget lead to greater constraints on terrestrial carbon fluxes and a better understanding of global actions to reduce atmospheric CO_2? (Carbon budget verification)

5.7.3.3 Variables and Sensors

Biogeochemical-Argo focuses on a core of six essential ocean, ecosystem, or climate variables (Table 5.7.2).

The plan is to deploy 1000 BGC floats over a period of 4 years: it is expected that a float lasts 4 years and a rate of 250 units per year will enable the network to be built and maintained [2]. This has direct consequence on the required maturity level (a.k.a. Technology Readiness Level) of the sensors. Choice of current technology and products can be found in a user's guide for selected autonomous BGC sensors [3]. Table 5.7.3 provides an example of a typical sensor package embedded on a PROVOR Argo float.

5.7.3.4 Argo Data Management

In parallel with these pilot projects, the community has begun to develop shared data management procedures. The successful data management system of the existing Argo program serves as a basis for BGC parameters. The Argo Data Management (ADMT) team has developed a parallel file structure for the core temperature, salinity, and

TABLE 5.7.2   Biogeochemical-Argo Core Variables Versus Essential Variables

	Oxygen	Nitrate	pH	Chlorophyll-*a*	Backscat-erring	Irradiance
	(Doxy)	(Nitrate)	(pH)	(CHLA)	(BBP\<nnn>)	Down_IRR\<nnn>
Essential ocean variables	✓	✓	✓			
Biological ecosystem essential ocean variables				✓		
Biochemistry ecosystem essential ocean variables	✓			✓	✓	✓
Essential climate variables either oceanic or atmospheric	✓	✓	✓	✓		✓

Source: C. -Schmechtig-E-Argo-User-2017LOV [3].

TABLE 5.7.3   Biogeochemical Sensors Frequently Embedded on Board PROVOR BGC Floats

Reference (Parameter)	Manufacturer	Volume	Weight in Air	Operating Depth
4330 (dissolved oxygen)	Xylem (AADI)	87 cm³	175 g	6000 m
REM-A optical sensors set, including ECO-FLBBCD (fluorescence, backscattering, colored dissolved organic material) and OCR 504 (three wavelengths and PAR radiometer)	SBE Scientific (Wetlabs)	3627 cm³	5.50 kg	2000 m
Deep Suna (nitrate)	SBE Scientific (Satlantics)	1384 cm³	1.8 kg	2000 m

REM-A is an integration of optical sensor achieved for the REOMOCEAN ERC project [3]. Led by Herve Claustre Laboratoire océanographique de Villefranche UPMC.

pressure data and for associated BGC data. Annual meetings of the ADMT and members of the regional BGC float programs are optimizing this new data management and distribution system.

5.7.3.5 A Biogeochemical-Argo Profiling Float: PROVOR CTS4

The PROVOR CTS4 is compliant with BGC float use. This float is based on the PROVOR base already produced in large quantities. Compared to the conventional PROVOR CTS3 version for core Argo missions, a dual electronic architecture, additional batteries, additional floatability, and Iridium transmission have been added (Fig. 5.7.1 and 5.7.2).

5.7.3.5.1 Acquisition Modes

The PROVOR CTS4 float allows the balancing of power consumption and data resolution by controlling the sensor power supply, the sampling rate, and the thickness of slices within which data will be processed. These parameters can be defined independently for each sensor for five depth intervals (zone) (Fig. 5.7.3).

5.7.3.5.2 Example of Mission Possibilities

When using the PROVOR CTS4, 1 to 10 profiles can be programmed per cycle. For each profile, the following can be defined: parking depth, starting profile depth interval min.-max., time schedule for surface (over several days), sampling strategy for each sensor, and positioning and communication option. These parameters can be entered before the equipment is launched and modified during the mission via remote connection (Fig. 5.7.4).

5.7.3.6 Technological Challenges for BGC Floats

BGC floats are compliant with the definition of Biogeochemical-Argo but improvements will need to be carried out in the near future on both the sensors and the profilers.

5.7.3.6.1 Sensors: Multisourcing Issue

The range of available parameters is increasing dramatically and the so-called OoT (ocean of tomorrow) projects have produced interesting and innovative solutions. These can include access to new parameters and a dual source for existing sensors. A solution where only one sensor model is used and recognized should be avoided: several types for the same parameter must be validated. Achieving this is useful for science and leads to healthy competition between manufacturers, meaning sensors must be sufficiently validated, including during sea experiments. It is essential that scientists contribute to their assessment.

5.7.3.6.2 Mechanical Constraints for Sensors

Basically, a profiling float must resist depth pressure but must also have a lower compressibility than seawater. This implies that the sensors shall not significantly alter the compressibility feature of the profiling float. Two types of solutions are possible:

- The first one consists in integrating the sensor in the float, with only a sensitive part in contact with seawater. In this case, buoyancy can be affected by the added load and may need to be adjusted by reducing the internal weight, adding buoyancy, increasing the volume, or using syntactic foam. The mechanical constraints on the sensor are minimized, but the integration task must be completed carefully because to reduce the impact on weight, the housing of the electronic function is suppressed and the electronics may be disturbed as a result.
- The second one is to use a standalone sensor connected to the float by means of watertight plugs and a cable. This solution is more flexible and a very large combination of sensors can be quickly mounted; the sensors can easily be calibrated by the manufacturer and replaced in case of a problem during production. In this case highly reliable cables and connectors must be selected; the profiling float must be fitted with the sockets required for the plugs. In the case of PROVOR BGC, buoyancy adjustment is possible in a large range by additional floating syntactic foam and lead ballast modification.

To facilitate standalone sensors fitted externally, the following will be preferred:
- Neutral buoyancy.
- Housing with low compressibility factor.
- Compactness.

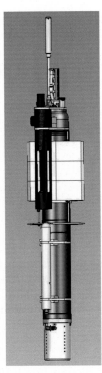

FIGURE 5.7.1   PROVOR CTS 4 with Suna and rem C pack. *Courtesy: nke.*

FIGURE 5.7.2   PROVOR CTS4 at sea. *Courtesy: G Dall'Olmo PML.*

Acquisition modes

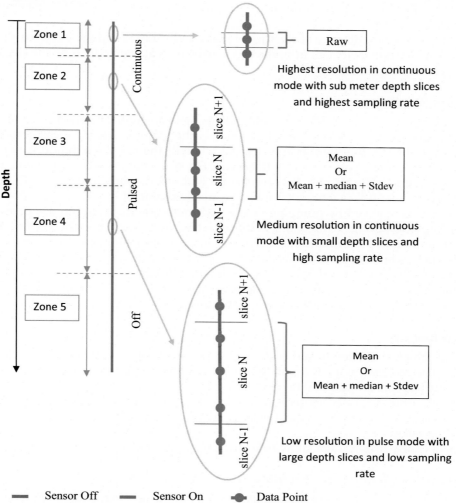

FIGURE 5.7.3   BGC sampling modes for a PROVOR CTS4. *Courtesy: © nke.*

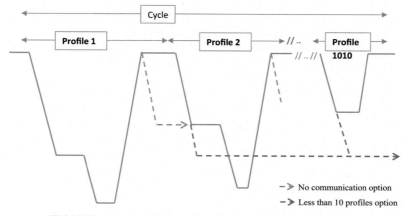

FIGURE 5.7.4   Profiles example for PROVOR. *Courtesy: © nke.*

New sensors developed during the OoT projects, such as the NeXOS project, can be mounted on the PROVOR. For example, the TriOS Matrixflu optical multifunctional sensor developed in NeXOS is a compact sensor that can be integrated in the future [4],[5].

It is recommended to ensure the absence of electrical voltage on the boxes and the sensitive parts of the measuring cells: a potential present on a metal element in contact with seawater can generate dramatic electrolytic phenomena; complete loss of the profiler may result.

5.7.3.6.3 Long-Term Stability

The deployment plan is to deploy 1000 BGC profiling floats, with an annual batch of 250 units and an expectation of 4 years' average duration of operational life. Confidence of the various sensors' measurement stability must be established. As far as possible, calibration checks are recommended after use at sea but difficult to achieve due to the position of the floats after several years of drift in the ocean. For conductivity and temperature the data acquired at 2000 dbars enable a checking of the stability by comparison with ocean climatology data. The conductivity sensor manufactured by Sea-Bird scientific is integrated into a reduced volume of water, protected by antifouling sleeves, and the volume of water is renewed by pumping before each measure: this solution has demonstrated its performance. For "oxygen" optode, several acquisitions in the air when at the surface enable control of the upper calibration point with air saturation. To perform this, dissolved oxygen sensors are now mounted on the top of a small mast.

For other sensors, such as optical, precautions must be considered. Self-control devices to monitor the calibration of the measurements must be forecast. Validations by calibration after use at sea and recovery (as far as possible) are very rich in the knowledge of the longevity of measurement stability. It is probable that cleaning procedures for the optical windows against biofouling and particles will be necessary.

5.7.3.6.4 Electrical Budget

Power is generally achieved efficiently and cost-effectively with a set of nonrechargeable lithium battery packs. The energy is mainly used to control the float, activate the hydraulic actuator, acquire measurements, and transmit data to a satellite. Electrical consumption of the sensors is critical.

For a PROVOR BGC, around 150 Ah are dedicated to the float and control, and around 100 Ah have been added to supply optional sensors. Autonomy is reasonably large but particular care must still be taken in defining missions and choosing sampling rates.

Each design team of course uses its own know-how to best optimize energy needs, including the choice of sensors.

5.7.3.6.5 Communication Channels for Profiling Floats

Profiling floats need communication for data transmission, remote sensing, and positioning when the float is at the surface. To be usable on profiling floats, a satellite communication system must allow the, preferably bidirectional, transmission of data by means of a low-power transmitter (a few watts at most) with an omnidirectional antenna. Global coverage and if possible a continuous service providing a backup of the data received, a reduced need of energy, and an optimized cost of transmission are important factors for choice.

Today, two satellite systems are mostly used: Argos and Iridium.

Argos has undeniable benefits associated with simplicity, robustness of the system, continuity of the service, capacity to locate a transmitter worldwide, its performance in data transmission (flow of the order of 0.1 bits per second), and a limit to the amount of data transmitted. The time spent on the surface, between 8 and 10 h, distorts measures of drift in immersion and induces an increased development of biofilm.

The Iridium satellite communication system is based on the use of a constellation of 66 satellites in low orbit and has led to reconsider these constraints for much less energy consumption. The Iridium system used in Short Burst Data (SBD) mode can transfer 1000 conductivity, temperature, and depth (CTD) points in 10 min; this can be compared to the 8 h of transmission needed by the Argos system to transmit 100 CTD points. In Router-Based Unrestricted Digital Interworking Connectivity Solution (RUDICS) mode the rate can be further increased.

Location is provided on board the float by a GPS receiver, while other Global Navigation Satellite Systems such as GALILEO and BEIDOU will be used soon. To be exhaustive, an analysis of each case taking into account the costs of the equipment, the transmitter, the positioner and the antennas, the cost of the transmission, and the processing of the data must be carried out.

For core Argo missions, Argos and Iridium SBD are mostly used; the volume of data to transmit is less than 10 kB. For all applications where large volumes of data such as for the BGC floats for which 100 kB must be often transferred, Iridium in RUDICS mode is the cost-efficient solution.

In respect of the constraints specific to profiling floats, the challenge of the future is technological and will relate to the increase in the transmission's data rate, which will allow an increase in the size of the transmitted files for pictures and sound.

Today, profiling floats are well connected to the two GDACS (Coriolis/Ifremer/France and US GODAE/USA) Argo databases but are still autonomous platforms with poor connection capability in comparison with modern data networks.

5.7.3.6.6 Costs

For an instrumental and expendable system such as the profiling float, the cost of one measurement profile (purchasing price + operating costs/quantity of validated measurements) is one of the most significant parameters.

5.7.4 NEXT GENERATION OF BGC FLOATS AND IMPROVEMENTS CONSIDERED

5.7.4.1 Sensors and Acquisition Unit Connection: Simplification

The large number of serial interfacing protocols used on ocean sensors generates high integration costs and is an obstacle for experimentation. During the OoT projects, solutions to these questions have been developed. If a universal solution that enables a connection between any sensor and a central acquisition unit is an answer, the fact remains that it would be preferable for the sensor manufacturer to modify its interface to comply with a universal standard. Unfortunately, this standard is not yet fully identified (Section 6 and in particular Chapter 6.1 addresses these aspects in detail. An application on a smart hydrophone is provided in Chapter 4.1).

Sensors aboard the ocean profilers are diverse and so have different and heterogeneous interface specifications. One can find simple sensors that generate few data types such as temperature or pressure sensors, but also others like hydrophones and video cameras that generate very substantial and specific data flows. It is therefore often necessary, for a new implementation, to develop a specific driver and sometimes adapt connection schematics.

"Plug-and-play" interconnection solutions have been demonstrated with success, especially during the projects that were part of the OoT call for proposals [6].

"Universal" solutions can be easy to implement and allow quick integration of new sensors but can require complex interface circuits, energy, and represent a significant cost. In the profiler's case and in a context of commercial competition, the production cost is critical and it can be difficult to bear such additional costs. This must be considered when the cost of the sensors is sometimes only a few hundred euros.

In a context where one hopes to integrate an increasing number of sensors that are subject to changes, inexpensive and fast integration is desirable. To define a solution widely established in sensors for the environment and oceanography, a choice must be made, validated, and adopted by the largest number: only then will integration be facilitated. It will be a genuine commercial asset if a sensor bears this type of interface.

The power consumption of the interface is an important point to consider. For matters of electrical consumption, no operating systems such as Linux or Windows are embedded in the PROVOR profilers. These issues are themselves partially the same as for gliders, but with a lower cost constraint because of higher onboard computing power.

5.7.4.2 Sensor Web Enablement

The data acquisition and data archival procedures usually vary significantly depending on the acquisition platform. This lack of standardization ultimately leads to information silos, preventing the data from being effectively shared across different scientific communities. Important steps have been taken to improve both standardization and interoperability, such as the Open Geospatial Consortium's Sensor Web Enablement (SWE) framework. Within this framework, standardized models and interfaces to archive, access, and visualize the data from heterogeneous sensor resources have been proposed. However, due to the wide variety of software and hardware architectures presented by marine sensors and marine observation platforms, there is still a lack of uniform procedures to integrate sensors into existing SWE-based data infrastructures. A framework aimed to enable sensor plug-and-play integration into existing SWE-based data infrastructures was achieved during the NeXOS project. An analysis of the operations required to automatically identify, configure, and operate a sensor have been analyzed. The metadata required for these operations are structured in a standard way. A modular, plug-and-play, SWE-based acquisition chain was proposed. More details on the architecture can be found in Chapters 6.1 and 6.2, while the evolution of standards for sensors and data systems are addressed in Chapters 6.3 and 6.4 [7].

5.7.4.2.1 *New Source of Energy*

Obtaining an unlimited source of electrical energy is an attractive idea and allows you to imagine a very long life at sea. An original innovation developed at the Jet Propulsion Laboratory (JPL) promoted by the Californian Company Seatrec intends to replace nonrechargeable battery packs with an embedded thermal electric generator to build a fully autonomous instrument.

5.7.4.2.2 *The Seatrec TREC Battery (TREC for Thermal REChargeable)*

"Technology: A class of substances called phase change materials (PCM) expand and contract substantially as they change between liquid and solid. The TREC Battery contains a quantity of PCM which undergoes a volume change as the host platform transits the ocean thermocline. This volume change generates high pressure which is used to drive a small electric generator. While the amount of energy derived from a single cooling-heating cycle is small, the process can be repeated indefinitely. For oceanic applications it is convenient to use a PCM that changes phase (melts) at a temperature near 10°C—typically within the main thermocline in tropical and subtropical oceans. PCMs that change phase at other temperatures are available." An extract from the Seatrec public website (Figs. 5.7.5 and 5.7.6).

The first SOLO-TREC prototype was deployed during 18 months, southwest of Honolulu, by a group of scientists and engineers from JPL and Scripps Institution of Oceanography.

The thermal engine solution from Seatrec has demonstrated its capacity to operate a profiling float in electrical autonomy and in good condition. It remains necessary to get sea experience feedback in nonoptimal situations and with an increased sensor load to propose a more easily acceptable mechanical design and to present a balanced financial budget.

5.7.4.3 PROVOR CTS5: A New Generation of BGC Float

The PROVOR CTS5 is the nke Instrumentation model for the next generation of BGC floats. Several mechanical, hydraulic, and transmission functions are imported from the PROVOR CTS4. The main difference comes from electronic board and firmware. To answer laboratories' requests, a new acquisition unit with more operator flexibility is foreseen [8]. The new electronic boards will also enable the feedback of measurements on the mission and the optimization of sampling. If an event on a parameter is detected, the float would be able to reconfigure mission depth, sampling rate, or cycle automatically. The challenge is to introduce "intelligence" inside the float to acquire more profiles and more significant data by optimizing energy consumption. This must be done while minimizing at maximum the risk of damage potentially carried by a firmware bug, for in a context where several entities can introduce firmware modifications and with a high level constraint on costs it is a real issue to validate evolutions and to determine responsibility in terms of loss.

Here are some examples of the use of intelligence:

- Aborting radiometer acquisition in the case of cloudy skies.
- Retroaction on mission configuration, e.g., aborting under ice operation for damage risk mitigation.
- Stabilization of a float in a bloom.

Progress of this new generation is ongoing. This technological base was already used to produce technological prototypes such as the PROVOR ICE used by Takuvik and Laboratoire océanographique de Villefranche UPMC (LOV) teams in the Baffin Sea for the project Green Edge and for the PROVOR NOSS intended to evaluate two sensors for "salinity" in the same water mass.

5.7.5 NEARBY THE BIOGEOCHEMICAL-ARGO PROJECT

5.7.5.1 Innovations in Embedded Measurements

On the basis of the same vectors, introduction, and evaluation of new sensors, new parameters are first tested for experimental purposes. If we refer to Argo, between the beginning of the ALACE experiments and the subsurface floats with multiple cycles, such as the MARVOR, 10 years have elapsed. Almost 15 years have passed since the first BGC floats with sensors were added to Argo CTD profiling floats and the launch of the Biogeochemical-Argo program in 2017. The future with new measurements will involve the implementation of other sensors and other measurements that should be tested in regional programs before being introduced into global programs: we may have a period of 10 years before such a step is taken.

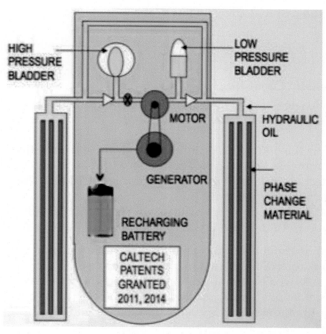

FIGURE 5.7.5   Thermal engine for a Seatrec float. reproduced from Sea Trec, © *SeaTrec*.

FIGURE 5.7.6   SOLO TREC. reproduced from Sea Trec, © *SeaTrec*.

A step-by-step description of two examples of ongoing innovations follows:

1. A passive aquatic listener (PAL): digital acoustic sensors were developed during the NeXOS OoT project, an integration including SWE on board a PROVOR [9].
2. Recently, chemical sensors based on electrochemistry techniques and "lab on a chip" chemical sensor packages have been deployed for experimental diving, also on a PROVOR as part of the SenseOcean project.

In the near future, the number of accessible parameters will increase with technological improvement and scientists' requirements. Laboratory techniques will be marinized and embedded in autonomous floats and autonomous underwater vehicles. It can be forecast that after the physical and BGC aspects, developments will focus further on chemistry and biology.

For these experiments, the issue of sensor interface remains critical and simplifications remain of strong interest.

5.7.5.1.1 PROVOR *with Acoustic Capability*

A prototype of PROVOR with a new digital hydrophone was produced during the NeXOS project Fig. 5.7.7.

The smart hydrophone, NeXOS A1, is a flexible and digital PAL designed by SMID. (see Chapter 4.1 for details).

The internal connection with the PROVOR profiling float has been modified to use the interface for Open Geospatial Consortium Programmable Underwater Connector with Knowledge and SWE. This PROVOR is fitted with an Iridium RUDICS connection and can typically transfer files of 100 kB. The various stages of the connection have been successfully tested in situ after integration.

SWE was fully demonstrated during a deployment at sea in the Canary Islands in May 2017.

There are numerous applications for onboard PALs; the following is a short list of examples [9]:

- Detection of underwater ambient sound sources generated by environmental (physical and biological) or geophysical (seismic, tsunami, rock fall, etc.) and manmade (ships, sonar, etc.) sources.
- Detection and classification followed by quantification of the foregoing sources.
- Improvement of quantitative precipitation forecast knowledge across the oceans.
- Microphysical and rainfall estimation over the oceans for satellite validation.

5.7.5.1.2 PROVOR *with Lab on a Chip and Silicate Sensors*

This innovation consists in embedding chemical sensors on board this profiling float. A prototype of a PROVOR profiling float with new chemical sensors using lab on a chip, electrochemical, and optode sensors was produced during the SENSEOCEAN project [10].

This PROVOR was fitted with several sensors connected through a Modbus interface; all these elements were developed as part of the SENSEOCEAN project. In addition to the CTD sensors suite, the following were mounted:

- An electrochemical silicate SiO_2 sensor [11].
- A lab on a chip able to acquire several parameters, including $NO_3{}^-$.
- 2 Optodes for pH and oxygen.

The PROVOR was deployed successfully off the coast of Villefranche-sur-Mer in the spring of 2017 for five dives down to 500 m. The sensor provided one point every 10 min, thus providing a depth resolution of approximately 25 m. The lab on a chip is designed to acquire macronutrients, micronutrients, trace metals, and carbonate system parameters; a phosphate sensor was also included in this electrochemical sensor base (Fig. 5.7.8 and 5.7.9).

Adding sensors to the platform is a complex process that influences float navigation with a potential risk to the float itself. The integration phase requires a study of the float's behavior since its stability and buoyancy are affected by the addition of new and sometimes bulky sensors such as the lab on a chip.

Several mission scenarios were simulated using nke's environment simulator to define the demonstration dive and the duration of the mission (Fig. 5.7.11).

5.7.5.1.3 PROVOR NOSS *Optical Density Sensor*

This is an innovative prototype of PROVOR NOSS to compare relative and absolute salinity sensors (Figs. 5.7.10 and 5.7.11). Knowing the absolute salinity in lieu of the currently used practical salinity will help improve the accuracy of calculations of the heat balance of seas. The NOSS sensor is one of the first underwater sensors for in situ refractive index measurement, opening up the scope of possibilities for direct access to density parameters. It represents a new tool for monitoring absolute salinity (S_A) and in situ density (ϱ) of seawater across TEOS-10. This sensor is able to measure the temperature (−2 to 35°C, <±0.006°C), pressure (0–2100 dbar, ±1 dbar), and refractive index (1.335300–1.345800, <10^{-6}) of seawater. It can also determine the absolute salinity (15–42 g/kg, ±0.005 g/kg) and density (1020–1030 kg/m³, ±0.003 kg/m³) of seawater. The NOSS sensor is able to achieve measurement in real-time up to 3 Hz. PROVOR NOSS floats, built on the PROVOR CTS4 BGC base, provided relevant observations on the physical functioning of the Mediterranean Sea by exploiting the potentiality of the coupled NOSS sensor and CTD observations. Tasks intended to improve accuracy of the Millard and Seaver tables and calibration are in progress. The patented NOSS sensor was designed during a partnership led by nke Instrumentation with Service Hydrographique et Océanographique de la Marine, Telecom Bretagne, and Ifremer.

FIGURE 5.7.7   PROVOR with hydrophone. *Courtesy: © nke.*

FIGURE 5.7.8   Sensor bracket for PROVOR with silicate, lab on a chip, and optodes. *Courtesy: Barus C. LEGOS.*

FIGURE 5.7.9   PROVOR SenseOcean Villefranche. *Courtesy: Barus C. LEGOS.*

FIGURE 5.7.10   Optical density sensor. *Courtesy: © nke.*

5.7.5.2 New Areas

New areas will be explored: the aim is to extend the area of Argo action "between 60°S and 60°N, and 2000 dbar." For great oceanic depths, a new generation of profiling floats named Deep Argo float has been designed. Improved knowledge of these water masses is important for monitoring the global heat budget.

Ice covered areas: surface reduction in the areas covered by sea ice in the Arctic and Antarctic will lead to deep modifications in all areas of oceanographic interest. Further studies will be needed; the period of the year where these areas will be covered with scattered ice will probably require a dedicated float design.

FIGURE 5.7.11   PROVOR with density and salinity, northwest Mediterranean. *Courtesy: © nke.*

Coastal zone: it was decided to exclude the EEZ to facilitate the program's acceptance. These areas are controlled by near-shore countries. The physical measurements are accepted most of the time but the studies on bioresources and acoustics are very often excluded. Technically, the difficulties stem from the extreme influence of maritime traffic in these areas, with potentially strong currents and frequent cases of grounding. Suitable profilers have been developed, such as the ARVOR C.

5.7.5.2.1 Deep Argo Float

New Argo profilers can reach depths of 4000 and 6000 m (Figs. 5.7.12 to 5.7.15 in Table 5.7.4), well under the 2000 m of the initial models of the core Argo mission. An operational depth of 2000 dbar gives access to 51% of the seawater volume, while 4000 and 6000 dbar represents 88% and 98%, respectively. The choice of the operational depth has implied significantly different developments and technology options: for an operational depth of 4000 dbar, similar solutions to those used for the 2000 dbar design can be used after some innovation or adaptation. The performance can be estimated and the nominal number of 150 Argo cycles (without dissolved oxygen sensor) can be reached for the "Deep ARVOR." For an operational depth of 6000 dbar, glass sphere flotation devices, such as those manufactured by Benthos (Teledyne group), are used today. These housings are different from those of the "ordinary" profiling Argo floats. Glass spheres have a greater stiffness than the tubular solution. The cycle of profiles can be one dive down to 4000 or 6000 m every "N" dives down to 2000 m: this solution allows more profiles to be performed, as required by the initial Argo recommendations.

Today, these deep-profiling floats are equipped to provide measurements of PTS and dissolved oxygen, but other measurements will soon be implemented for studies of great depths to satisfy the scientific demand. The deep-profiling floats will probably also be used in the industrial sector and deep oil and gas exploitation as autonomous stations for environmental monitoring during the exploitation phase. By the end of 2017, only a few dozen deep Argo floats were deployed and full maturity of these instruments has not yet been achieved.

5.7.5.2.2 Ice-Covered Areas

Navigation of profilers in ice-covered areas is associated with three main issues:

- Where is the float? "Unknown float location due to impossibility of reaching the surface when under the ice sheet."

TABLE 5.7.4   Deep Argo Floats Commercial Offer in 2017

FIGURE 5.7.12   Deep ARVOR. *Courtesy: ©nke.*

4000 m/TS SBE 41 CP and dissolved oxygen/150 cycles from 4000 m (CTD only)*
30 kg dry weight

Developed in industrial partnership with Ifremer/nke Instrumentation

FIGURE 5.7.13   Deep Ninja. *Courtesy: TSK.*

4000 m/TS SBE 41 CP/
75 mixed cycles 2000–4000*
Approx. 50 kg dry weight

Developed jointly by Jamstec and Tsurumi-Seiki Co. Ltd.

FIGURE 5.7.14   Apex Deep.

© Teledyne Webb Research

6000 m/dry weight is approximately 51 kg
Apex Deep will perform 150 profiles **from 6000 m**
with the SBE 61 CTD in continuous/Source Webb research

FIGURE 5.7.15   Deep Solo.

© Scripps Institution of Oceanography

6000 m/TS SBE 61./dry weight is 27 kg/
190 cycles from 6000 m**
Developed by Scripps Institute of Oceanography and manufactured by MRVSYS

*Sources: * nke Instrumentation and TSK estimated based on the energy system.*
***Teledyne Marine and Scripps Institution of Oceanography.*

FIGURE 5.7.16   Open pack ice. *Courtesy: Brault P. nke.*

- Is it possible, as in (Fig. 5.7.16), to reach the surface according the state of the surface layer (with ice or without ice), knowing the state of the surface to trigger surfacing and avoid damage?
- If emergence is not possible, the float must be able to store the data as long as it remains under water until surface transmission is made possible.

5.7.5.2.3 *Underwater Positioning*

Given that satellite positioning is only possible when the float is at the surface, positioning when under ice is achieved during the diving phase through reception of RAFOS signals transmitted by a network of underwater acoustic sources (SOFAR sources). Presently, the use of these sources is difficult due to cost, limited electric autonomy due to the power required to activate the sources, difficulty to maintain a network, and disturbances for marine animals. The ACOBAR project was designed using a cluster of drift buoys equipped with an underwater acoustic source that operates with lower frequency at depth and greater frequency near the surface (780 and 1560 Hz in lieu of the 260 Hz frequency used at first for long-range positioning). Using a higher frequency helps reduce resonant parts; sources are built with a piezo ceramic source and a quarter wavelength resonant cylinder made of aluminum.

Other ideas and solutions are being considered, such as positioning via the analysis of bathymetry and dead reckoning.

5.7.5.2.4 *Ice Layer Detection*

5.7.5.2.4.1 Upper Layer Temperature Analysis

The simplest method is the implementation of an analysis of the temperature of the water in the last 50 m. This solution is used for deployment in the Weddell Sea. A statistical analysis of temperature and salinity was achieved for the Baffin Sea by joint teams from LOV and Takuvik. The solution selected is to detect only the temperature value with a threshold selected as a compromise between the risk of impact with ice, the risk and possibility of loss, and the possibility of surfacing with transmission to the satellite. A compromise must be made to limit the risk of damage to the float and the risk to miss a transmission window. The threshold was selected at −1.3°C after a statistic study based on the analysis of 400 profiles and five PROVORs were deployed during the Green Edge 2016 campaign (Fig. 5.7.17).

5.7.5.2.4.2 Active Optical Detection

Another solution was developed at Takuvik [12], a joint Canada–France laboratory located at Laval University specialized in the study of Arctic ecosystems. It is based on an active optical sensor that uses laser polarimetry to detect the absence or presence of a layer of thin ice on the sea surface at a distance of 12–15 m. Ice is a strong depolarizer of light, and thus this ice detection system calculates the ratio of depolarization of a reflected polarized laser beam at 532 nm when pointed toward the sea surface. It is also possible to obtain a relative measurement of the ice thickness. In case of positive ice detection, the ascent phase is aborted and the profiling float resumes its cycle. The sensor was tested successfully under sea ice during the Green Edge 2016 campaign (Fig. 5.7.18).

5.7.5.2.4.3 Ice Tether Platform

Another solution for the polar region is the ITP used for the Integrated Arctic Ocean Observing System (IAOOS) network where a cable is hung under a buoy on ice or left floating, depending on ice conditions, and tethered by a lead weight. The profiling float is guided along the cable and transmits its data via an inductive modem to the

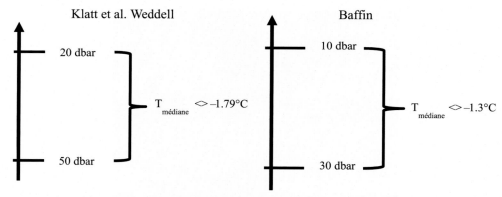

FIGURE 5.7.17   Ice-sensing algorithm adaptation. *Courtesy: Leymarie Edouard Laboratoire Oceanographique de Villefranche (France).*

FIGURE 5.7.18   PROVOR ICE with optical ice detection sensor. Note top view of the PROVOR ICE fitted with the sea ice layer detection sensor (laser source and reception). *Courtesy: Marec Claudie Takuvik Ulaval.*

surface buoy, which transmits the data via satellite. This principle is used for the PROVOR SPI in IAOOS. This solution enables regular and daily data recovery.

The IAOOS project is an Equipex project [13]. The aim of the project is to maintain 15 operational platforms designed to acquire ice, atmosphere, and Arctic Ocean parameters at the same place over a period of 7 years:

- Platforms allow measurements to be transmitted in near-real-time via Iridium satellites.
- For the ocean side, each platform is equipped with a vertically profiling float guided along a tether cable and connected to the surface buoy by an induction link. Profilers collect ocean temperature and salinity down to a depth of 800 m; additional sensors such as nitrate, chlorophyll, and pH were added at first for experimental purposes.
- Cloud and aerosol Light Detection and Ranging performs profiling of the lower atmosphere.
- A wide angle radiometer measures diffuse and direct solar flux.
- Standard meteorological parameters are measured at the surface.

5.7.6 CONCLUSIONS

Profiling floats are a family of vectors that are autonomous, easy to use, reliable, and efficient, embedding measuring instruments for in situ oceanographic observation. The success of the Argo program and the large number of profiling floats manufactured and deployed have made it possible to achieve industrial products with very good reliability and optimized cost. Argo floats acquire temperature and salinity measurements down to 2000 m deep.

The initial aim of the Argo program is to deploy 3000 profilers on the global ocean from −60°S to +60°N and outside EEZ. Today (January 2018) 3900 units are currently active.

In recent years, regional experiments and industrial works have demonstrated the opportunity to acquire several other parameters with profiling floats in the open ocean. A Biogeochemical-Argo program was officially launched in 2017 and a set of parameters capable of documenting the major oceanographic current issues have been defined.

The goal is that, with an annual deployment of 250 units, 1000 BGC profilers are active in the ocean:

- A lifespan of 4 years is intended in this plan and one of the major challenges will be to acquire sufficient confidence in the precise conservation of sensors for at least this duration.
- A second challenge will be to have several technical and industrial solutions to acquire these parameters. Technological diversity will reduce the scientific risks of a single solution and a real commercial competition will make it possible to obtain good solutions with an optimum cost.
- Finding satisfactory solutions requires a validation of sensors at sea. The use of the connection protocols resulting from works such as SWE will facilitate these implementations and thus in situ validation steps.
- In the field of profiler control, the implementation of sensors with long response times will most likely require the vectors to stabilize, with minimal energy consumption, at a fixed depth.
- It is also to improve the intelligence of the vectors to adapt the sampling and mission of the profilers to the events of the medium.
- It will be desirable to develop alternative energy solutions to lithium batteries. These solutions will have to remain suitable for air transport safety conditions and at a cost compatible with the financial context.
- If new, more efficient, satellite-based communications become available, more data and, possibly, acoustic and pictures files will be considered.

The principle of profiling floats remains relatively easy, efficient, and robust, and allows them to remain cost-efficient. A large international scientific community uses available data, builds new projects, and develops new applications in association with industry. The evolutions of the program are already maturing.

Glossary

ACOBAR project Acoustic technology for observing the interior of the Arctic Ocean
ALACE Autonomous Lagrangian Circulation Explorer
ARGOS Global satellites system for data collection and positioning
BGC Biogeochemistry
CTD instrument to measure Conductivity, Temperature and Depth
ECV Essential climate variable
EEZ Exclusive Economic Zone
EOV Essential ocean variable
eEOV Ecosystem essential ocean variable
Equipex "Equipement d'excellence," in France projects granted by "le grand emprunt" to improve research laboratories
GDAC Global data centers for Argo
GNSS Global Navigation Satellite System
IAOOS project Ice, Atmosphere, (Arctic) Ocean Observing System led by J. Pelon LATMOS and C. Provost LOCEAN
IRIDIUM Global satellite constellation of 66 cross-linked Low Earth Orbit (LEO) satellites
ITP Ice tether platform
JPL Jet Propulsion Laboratory
LATMOS Laboratoire Atmosphères, Milieux, Observations Spatiales UPMC
LOCEAN Laboratoire d'Océanographie et du Climat UPMC
LOV Laboratoire Océanographique de Villefranche UPMC
NERC Natural Environment Research Council, Southampton
nke Instrumentation Manufacturer of profiling floats PROVOR, ARVOR (http://www.nke-instrumentation.com)
NOC National Oceanography Centre, Southampton
NOSS nke Optical Salinity Sensor, seawater density
OGC Open Geospatial Consortium (http://www.opengeospatial.org/ogc)
OoT Ocean of Tomorrow, EU FP7:
    NeXOS OoT project (http://www.nexosproject.eu), led by Eric Delory, PLOCAN
    Senseocean OoT project (http://www.senseocean.eu), led by Doug Connelly, NOC
    Schema OoT project (http://www.schema-ocean.eu)
    Commonsense OoT project (http://www.commonsenseproject.eu)
PAL Passive Aquatic Listeners
PROVOR The profiling base for many profiling float applications, designed in industrial partnership between Ifremer and nke Instrumentation. Involved in production and commercialization, such as:
PROVOR CTS4 A BGC profiling float
PROVOR ICE Designed for under-ice applications
PROVOR SPI Designed for ITP, sliding profiling inductive link

PTS Pressure, Temperature, Salinity
REMOCEAN ERC REMotely sensed biogeochemical cycles in the OCEAN, European Research Council (http://remocean.eu/)
SCRIPPS Scripps Institute of Oceanography, UC San Diego (https://scripps.ucsd.edu)
SMID NeXOS Italian SME involved in submarine acoustic acquisition (http://sitepitalia.it)
SOFAR SOund Fixing And Ranging: RAFOS is SOFAR^{-1}
SOLO Sounding Oceanographic Lagrangian Observer
SOS Sensor Observation Service (http://www.opengeospatial.org/standards/sos)
SWOT Strengths, Weaknesses, Opportunities, Threats
Takuvik Laboratory comprising the University of Laval and French Centre national de la recherche scientifique
TEOS-10 Thermodynamic Equation of Sea Water 2010
TRL Technology Readiness Level
UPMC Université Pierre et Marie Curie

Acknowledgments

Thanks to the nke "profiling float" development and productions team for their contribution to this chapter. (A. David, D. Malardé, J. Sagot, D. Nogre, B. Jugeau), S. Le Reste and team (Ifremer Department of Technology Research and Development), Laboratoire Oceanographique de Villefranche sur Mer, Takuvik, and LOCEAN teams.

References

[1] Davis RE, Sherman JT, Dufour J, Scripps Institution of Oceanography. Profiling ALACEs and other advances in autonomous subsurface floats. J Atmos Ocean Technol, introduction 1, September 25, 2000. https://doi.org/10.1175/1520-0426(2001)018<0982:PAAOAI>2.0.CO;2.

[2] Johnson K, Claustre H. Biogeochemical-Argo Planning Group. The scientific rationale, design and implementation plan for a biogeochemical-argo float array. 2016. https://doi.org/10.13155/46601.

[3] Barbieux M, Uitz J, Bricaud A, Organelli E, Poteau A, Schmechtig C, Gentili B, Leymarie E, Penkerc'h C, Obolensky O, D'Ortenzio F, Claustre H. Assessing the variability in the relationship between the particulate backscattering coefficient and the chlorophyll a concentration from a global biogeochemical-argo database. J Geophys Res Oceans 2018;122. https://doi.org/10.1002/2017JC013030.

[4] Memè S, Delory E, Felgines M, Pearlman J, Pearlman F, Rio J, et al., editors. NeXOS: next generation, cost-effective, compact, multifunctional web enabled ocean sensor systems. OCEANS 2017-Anchorage; 2017 18–21 Sept. 2017. 21 September 2017. http://ieeexplore.ieee.org/document/8232104/.

[5] Ferdinand OD, Friedrichs A, Miranda ML, Voß D, Zielinski O, editors. Next generation fluorescence sensor with multiple excitation and emission wavelength: NeXOS MatrixFlu-UV. OCEANS 2017-Aberdeen; 2017 19–22 June 2017. June 22, 2017. https://doi.org/10.1109/OCEANSE.2017.8084809.

[6] Schmechtig C, et al. Status of BGC-Argo data management 27-;-E-Argo-User-2017 presentation. 2017.

[7] Martinez E, et al. Middleware for plug and play integration of heterogeneous sensor resources into the sensor web. 2017 Sensors, 17(12):2923 https://doi.org/10.3390/s17122923.

[8] Le-Traon, PY (Ifremer), et al. Equipex NAOS project: l'observation globale des océans Préparation de la nouvelle décennie d'Argo; 2000.

[9] Anagnostou MN, et al. Passive aquatic listener: a state-of-art system employed in atmospheric, oceanic and biological sciences. 2015.

[10] Senseocean successful multi sensor suite deployment on PROVOR float. www.senseocean.eu.

[11] Barus C, Chen Legrand D, Striebig N, Jugeau B, David A, Valladares M, Munoz Parra P, Ramos M, Dewitte B, Garçon V. First deployment and validation of in situ silicate electrochemical sensor in seawater; Frontiers in marine science (in revision). 2018.

[12] Lagunas-Morales J, Marec C, Leymarie E, Penkerc'h C, Babin M. Sea- ice detection for Autonomous underwater vehicles and oceanographic Lagrangian platforms by continuous-wave laser polarimetry. 2017.

[13] Dickson RR. The integrated Arctic Ocean observing system (iAOOS). 2007.

Further Reading

[1] IOCCP Report No. 2/2017. A user's guide for selected autonomous biogeochemical sensors an outcome from the 1st international IOCCP sensors summer course–instrumenting our oceans for better observations. November 2017.

CHAPTER

5.8

Innovations in Marine Robotics

Nuno A. Cruz, José C. Alves, Bruno M. Ferreira, Aníbal C. Matos
INESC TEC, FEUP–DEEC, Porto, Portugal

5.8.1 INTRODUCTION

There are many ways of getting data to understand the complex dynamic processes taking place in the world's oceans. Current technology enables large-scale observations from satellites or other remote-sensing techniques, but there is still a great demand for in situ measurements. These can be used to calibrate remote observations and compensate for their lack of vertical resolution, but their main purpose is to capture data about processes evolving on a much shorter scale. The need to capture the dynamics of many phenomena taking place in the ocean has always demanded faster, more effective and efficient methods to increase the sampling rate, both in space and time. There is a great variety of robotic systems that have been developed to facilitate these measurements, and it is no surprise that they have been playing an increasingly important role in ocean sampling.

In the last few decades, the marine robotics community has grown dramatically and has been actively seeking solutions for the most challenging issues in surface and underwater operations, such as robustness, navigation, control, and communications. Most advances have been incorporated into operational systems, and the early prototypes have steadily turned into mature solutions that are now routinely used to gather ocean data. At the surface, for example, there are many designs of Autonomous Surface Vehicles (ASVs) able to carry sensors to sample the interface layer between the ocean and the atmosphere. These operations can last for many days or weeks and data can be conveyed in real-time to a remote location, if required. In the underwater environment, Autonomous Underwater Vehicles (AUVs) are now ubiquitous versatile platforms to sample wide regions with moderate set-up and operational costs. In most applications, AUVs are routinely programmed to follow georeferenced paths while collecting relevant data, yielding efficient synoptic views of 3D fields.

For the scientific community, robotic systems are fundamental tools to characterize oceanographic dynamic processes and their performance has been steadily increasing to improve the space and time sampling density, and also the duration and range of the missions. In fact, these systems can work 24h a day in risky environments, with minor deterioration and with virtually no human intervention. Moreover, the concept of *autonomy* has been widened to consider innovative uses of the available technology. The increase of embedded computational power endows these systems with an onboard *intelligence* that permits real-time interpretation of data and new paradigms for ocean sampling. These new techniques can further benefit from the use of multiple platforms toward a global goal: they can work in cooperation, when they are assigned independent tasks, or even in collaboration, when they interact by performing synchronous activities.

This chapter describes the latest developments and future trends in marine-related robotics, with the focus on how these capabilities can contribute to a better understanding of the marine environment. One of the goals of the robotics community is to obtain more data about the oceans by extending capabilities in range and time, but this increment in quantity has also to be matched with quality for these data to be truly valuable. All operations in the marine environment have to deal with the harsh conditions faced at sea, further aggravated when the equipment is deployed in deep water, therefore all solutions have to be extremely robust to withstand operations in such unforgiving environments. Moreover, the difficulties in communicating underwater create additional challenges in terms of equipment performance and reliability, and the concept of autonomy has to include capabilities to face unanticipated events. Finally, the cost of ocean data is always a major concern, therefore robotic solutions need to be efficient, reducing operational logistics, maintenance requirements, and customization efforts.

The Center for Robotics and Autonomous System of INESC TEC in Porto, Portugal, has been involved in many R&D projects developing cutting-edge technology for the sea. Some of these achievements will be described in more detail. The maturity of technology is often described in terms of *Technology Readiness Level*, or TRL, ranging from the lowest numbers, corresponding to conceptual ideas, to the maximum TRL of 9, corresponding to fully proven, ocean ready technology. These TRL levels change as capabilities evolve over time. Although there is always some ambiguity associated with this classification, the chapter is divided into sections in a decreasing order of TRLs, so that a scientist can easily distinguish what is already available in the first section, and also anticipate what will be available in the middle to long terms in the ensuing sections.

5.8.2 ADDRESSING THE MAIN CHALLENGES

This section describes the main challenges that the robotics community has long been facing, and how they have been overcome to provide scientists with ocean data. Some of these challenges still serve as guidelines to improve performance and pursue new capabilities, but there are already many examples of solutions that have been proven in the field. Therefore the section describes high TRL devices and solutions that are ready for immediate use in ongoing programs.

5.8.2.1 Increasing Time and Space Series

The need to capture longer time series, or to extend spatial coverage, has always been a driving factor in the marine robotics community. These have been achieved through the combination of mechanical design to reduce drag, incorporation of lower power subsystems, and taking advantage of the latest developments in rechargeable batteries. For example, the Tethys AUV, developed at the Monterey Bay Aquarium Research Institute, is a 105-kg propeller-driven vehicle with an impressive range of thousands of kilometers, during months of operation. [1] reports a 3-week deployment covering 1800 km at a speed of 1 m/s, with a suite of payload sensors averaging 5 W of power consumption.

The propulsion system consumes a very significant part of the total available power of a robotic vehicle. In the case of ASVs, the range of options is larger than with AUVs, since it may also include combustion-based sources, or energy harvested from the environment. Rechargeable batteries are typically found in smaller, fully electric ASVs for operations in coastal or calm inland waters [2–4]. Fully electric ASVs are very easy to control but the limited amount of energy stored on board prevents their use in long-range missions.

The harvesting of energy from the environment has been explored, both to charge electric batteries and also to translate directly into motion, without the need to go through conversion into electric energy. Liquid Robotics, Inc. manufactures the Wave Glider, a platform that harvests energy from the waves for propulsion and from the sun for powering instruments, and has the potential for very long-term deployments. The Wave Glider uses a set of underwater wings coupled to the surface hull and transforms the vertical wave-induced motion into forward thrust, following a concept first described in [5]. Although this system is relatively recent, it already has an impressive record of thousands of miles at sea, collecting data in very harsh conditions, with some missions lasting longer than 1 year. The main drawback of this platform is the reduced velocity, with an average of 1.5 knots, which can create challenges in areas of strong surface currents.

The AutoNaut from MOST (Autonomous Vessels) Ltd. is another example of a wave-propelled ASV, but using the "wave foil technology," a combination of hull design and additional foils to produce thrust from the wave motion. The vehicle harvests solar energy to power the onboard electronics, and carries a methanol fuel cell for additional power, if required. It has a scalable mono hull, with two standard configurations, with 3.5 m (120 kg of weight) or 5 m (230 kg). The endurance of the vehicle is 3 months and the missions are preprogrammed and may be altered using a satellite link. Payload capability is 185 L/40 kg for the 3.5 m version and 500 L/130 kg for the 5 m version. The top velocities are 3 and 4 knots, respectively.

One specific technology that has evolved over the last few years is wind propulsion. Harvesting the propulsion energy from the environment is a way to overcome the limitations in energy storage and transportation. Autonomous sailboats are a particular class of ASVs that rely on wind to directly provide the propulsion energy, only needing electrical energy for the onboard electronics and rudder/sail adjustments. They are a proven technology for ocean monitoring, sampling, and surveillance and have gained particular attention for their unique ability to maintain long-term unassisted operations at the sea surface.

One key feature for enabling permanent ocean presence is the ability to withstand the harsh conditions at sea during long periods of time. Up to the present there have been some successful demonstrations of high-endurance capability with small autonomous wind-propelled crafts, with different proposals resembling conventional sailing boats or more customized and sophisticated designs. Success stories include the French Vaimos [6], the Norwegian Sailbuoy [7], and the American trimaran Saildrone [8] with an impressive record of various multimonth missions in harsh sea regions, such as the Bering Sea.

In some applications, other features are needed to complement the endurance capabilities, such as maneuverability, speed and performance upwind, as well as accuracy of course and an ability to execute more complex maneuvers such as station keeping or virtual anchoring. Examples include monitoring pollution plumes, performing scans of the seafloor, and continuously sampling the near-surface environment.

The silent nature of sailboat operation yields a great potential in marine acoustic surveillance applications, as well as detecting and tracking marine mammals across vast areas of the ocean. Examples are the missions in the Baltic Sea accomplished by the Austrian sailboat Roboat [9] for recording marine mammals [9,10], and a similar operation performed off the coast of Sesimbra in Portugal by the sailboat FASt, while towing an autonomous underwater sound recorder [11]. Fig. 5.8.2 shows a spectrogram of sounds recorded during this mission. Chapter 4.1 describes the integration of passive acoustic sensors with embedded processing on a profiling float and an underwater glider [12], and autonomous sailboats can extend the range of platforms to support this new generation of ocean sensors requiring quiet operations.

FASt is a 2.5-m long, 50-kg displacement autonomous sailing boat developed at the University of Porto and presently operated by INESC TEC (Fig. 5.8.1). The sailboat is driven by a low-power embedded computer and a minimal set of sensors required for navigation under sail (Global Positioning System [GPS], electronic compass, and wind speed and direction). Short-range communications for command and control are provided by air radio modems, and long-range ocean wide communications can be supported by satellite data services, such as Iridium. The conventional mono hull design with a deep ballasted keel and soft sail configuration provides excellent navigation performance along all points of sailing, with a minimal angle to the true wind below 50 degrees and typical speeds in the range of 2–4 knots for upwind legs [13]. Downwind is naturally the preferred course for a sailboat when precise route control is necessary. The twin rudder configuration of FASt combined with an adaptive course control algorithm permits very good maneuverability and route stability, even in adverse sea conditions. During a 30-nautical mile downwind sea trial, under 15 to 20 knots true wind, and 2 m swell, FASt has observed a cross-track error below 10 m and a top speed above 9 knots [14].

Another important feature of ASVs for ocean sampling is the ability to maintain a position in a constrained area while gathering the data of interest, usually referred as station keeping. The small size of FASt and its capacity to perform sharp turns permits a very good performance in this task, as shown in the results of a sea trial reported in [14], where the sailboat was able to maintain a navigation path within a 20-m circle during a 5-min period under 7 knots of wind.

The sailboat is highly configurable and can carry a few kilograms of payload, either in a dry compartment inside the hull, on the deck, or attached to the 1.2-m deep keel. The actual practical payload limitation is mainly dictated by the added weight distribution and the disturbances in the boat's aerodynamics and hydrodynamics that may affect its performance in terms of maneuverability, speed, and stability.

It should be noted that the intention of this section was not to provide a comprehensive state of the art in terms of marine robotic systems, but solely to review some of major achievements in terms of vehicle capabilities. Throughout the book, it will be possible to find more information on other autonomous mobile technologies, like underwater gliders in Chapters 5.1 and 8.1, and innovations in floats in Chapters 5.6 and 5.7.

5.8.2.2 Ensuring Data Quality

Data quality is probably the most important concern of every scientist collecting ocean data. It is strongly affected by the quality of the sensors used (including calibration), but also by the measurement procedures. Traditionally, oceanographers requiring in situ measurements take their instruments to the sampling site and have to pay due attention to the sampling procedure (time and location, sampling profile, etc.). With the use of robotic systems at sea, some of these concerns are overcome, while others have emerged, like contamination from the platform. In this last case, the role of the oceanographers is mostly to provide requirements for the installation, to audit the final configuration, and to identify possible constraints during operation. For example, a simple CTD installation may require a minimum AUV velocity for the minimum water flow, and limit the maximum velocity to avoid aliasing. It may also involve a detailed calibration after installation to assess a potential contamination by the vehicle's thermal signature.

As mentioned earlier in this chapter, in the last decade the prospects of operating time have increased dramatically, and there have been examples of autonomous vehicles at sea for many months [15]. Although these impressive numbers are appealing, they also convey specific concerns with data quality, mainly in terms of sensor calibration and potential biofouling. These have been addressed with joint efforts between scientists, engineers, and equipment manufacturers [16,17]. See also Chapter 4.3 for a more detailed discussion of biofouling.

With the proper installation of the payload sensors, the robotic platform can ensure a very accurate time synchronization and geolocalization of data, difficult to match with traditional techniques. The onboard computer can

FIGURE 5.8.1   The robotic sailing boat FASt developed by the University of Porto and operated by INESC TEC.

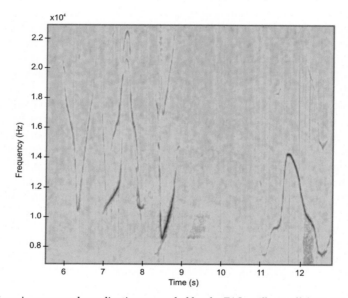

FIGURE 5.8.2   Spectrogram of marine mammal vocalizations recorded by the FASt sailboat off the coast of Sesimbra, Portugal (May 2013).

provide an absolute timestamp to all logged data, and, if needed, current off-the-shelf chip-scale atomic clocks can reach accuracies in the nanosecond range for many hours of operation.

For long-range navigation, AUVs may use a long baseline (LBL) acoustic positioning system based on two beacons installed in the operation area, such as described in [18] or [19]. This system has been extensively tested in field trials and its accuracy fully characterized, as shown in [20]. Under static conditions, positioning accuracy is better than 1 m for ranges to beacons up to 500 m. This figure deteriorates with the increase of the vehicle velocity, reaching about 3 m for a nominal velocity of 1 m/s and typical geometric dilution of precision in the operation area up to a maximum range of 3–4 km. To extend coverage, the beacons may be installed in moving platforms, forming a *moving baseline* as described in [21], and taking advantage of the availability of GPS signals at the surface to maintain accurate positioning.

One way of improving LBL navigation is to have the beacons sending pings synchronously and compute ranges just by one-way time of flight [18,22]. This provides many more measurements to the navigation filters and it only requires the AUVs' internal clock to be synchronized with the internal clocks of the beacons. With today's technology of low-power chip-scale atomic clocks, it is in fact easy to maintain all clocks globally synchronized, with neglectful drifts during the timeframe of several hours. Another advantage of this approach is power saving, as the AUV does not need to send acoustic pings for navigation, therefore this is a preferred technique for long-term operations [23].

5.8.2.3 Reducing Logistics and Mission Risks

The logistics associated with the deployment and operation of robotic systems at sea are directly related to the size and weight of the vehicles and support equipment. Naturally, this impacts operational costs, therefore there is continuous pressure to reduce the logistics footprint. In fact, the facility to operate even with modest logistics support has been a strong drive for the development of small size, portable robotic systems.

One component that has been developed to facilitate operations and reduce risks is the Launch and Recovery System (LARS). Initial LARSs have been developed and are available for larger AUVs [24], for which deployment and recovery is more complex, but recently other solutions have been proposed and demonstrated even for small AUVs [25].

To ensure self-localization of an underwater vehicle, the LBL acoustic navigation system requires the deployment of acoustic beacons in the operation area. These are typically installed in moored buoys or deployed on the seabed and require some set-up before the actual AUV mission may start. Throughout the operation, the vehicle emits coded signals and waits for the beacon replies to get ranges and estimate position. Without real-time communications, that position is only known by the vehicle. Therefore a related problem to AUV navigation is external tracking, i.e., the ability to determine, from the shore or a support vessel, the location of the vehicle in real-time. This is particularly important for safety reasons, since it allows for a mission supervisor to assess if the AUV trajectory follows the predefined plan. To track the AUV, the beacons can send specific signals and wait for the vehicle replies to get ranges, but this requires additional power wasted by the AUV and a period with no navigation data. Alternatively, with synchronous clocks, it is possible to design the navigation system of AUVs in such a way as to allow external tracking without these extra acoustic signals being emitted [26]. The idea is to use a listening-only device in the water column and determine the AUV position just by listening and interpreting the signals exchanged between the AUV and the acoustic beacons.

INESC TEC has been quite active in reducing the logistics support required for the operation of robotic vehicles. One particular solution is the concept of a *man portable buoy*, a small buoy that can be deployed by a single person and provide acoustic navigation and tracking in shallow waters [19]. Moreover, to avoid the need to moor buoys, the acoustic beacons may also be installed in station-keeping ASVs, compensating current drifts, or in moving ASVs to extend the coverage to larger areas of operations, as compared to deployments in fixed locations. Finally, with the development of an LARS that can be installed in an ASV, it is also possible to anticipate a scenario in which all equipment leaves shore to operate in a remote location without physical intervention, as proposed in [23].

5.8.2.4 Development of Modular Robotics

A great fraction of the operations of robotic systems at sea has been taking advantage of the versatility of commercially available vehicles. However, as the users become more demanding, certain scenarios require solutions for which specific configurations would be more appropriate. This encompasses a great deal of customization effort, typically at the factory, with corresponding increase in delivery time and cost. The concept of modular robotics has been proposed as a way to rapidly reconfigure a vehicle for a specific mission, therefore minimizing such inconveniences.

In terms of AUV design, a good example of modularity is the Gavia AUV, with continuous developments to accommodate newer systems [27], although other examples of modularity can also be seen in relatively large vehicles for the installation of specific high-performance sensors [28]. These examples are less often seen in small AUVs, with a few exceptions such as the Starfish from the National University of Singapore [29]. The basic configuration of Starfish is a torpedo-shaped propelled AUV, 1.7 m in length and 20 cm in diameter. Along the diameter, additional sections of 20 cm in diameter may be included to incorporate different payload sensors. Apart from the more common hardware and software modules, [30] also propose a methodology to develop modularized hydrodynamic models, so that the vehicle control system may be easily adapted to account for the installed modules.

INESC TEC, for example, has developed a long-term program for the development of small size AUVs based on modular building blocks [31]. The approach relies on modularity in terms of both hardware construction and electronics, software, and control. Probably the most obvious aspect of the modularity of these building blocks comes from the design of the hull sections. To assemble AUVs from modular components and achieve an overall smooth profile, the blocks have matching edges and constant cross-sections, with 20 cm of outer diameter. They can be divided into pressure housings, flooded extensions, and terminations. The pressure housings are dry compartments where all electronics are installed, and are the only sections requiring seals (with O-rings) and specific wall thickness to withstand the external pressure. All the others are flooded, serving mainly as mechanical support for sensors and actuators.

Using these modular building blocks, the first version of the Modular Autonomous Robot for Environment Sampling (MARES) portable AUV was built in 2007 [32]. MARES is a shallow water (100 m) hovering AUV that has been continuously updated and used in the field in many different configurations. In 2011, the versatility of the system components has been pushed further with the development of TriMARES [33], a 75-kg, three-body hybrid AUV/remotely operated vehicle (ROV) system, which was developed and delivered to a Brazilian consortium in little over 6 months. More recently, the DART (for *Deep water Autonomous Robotic Traveler*) has been assembled, with the replacement of the main pressure housing with a 4000-m rated enclosure [34]. Fig. 5.8.3 shows some pictures of these vehicles.

Given the modularity of the design, the integration of a new sensor is relatively simple. Typically, the sensor is mechanically installed in a custom-flooded section, and it is electrically connected to the electronics inside the pressure housing using one of the connection ports at the end caps. The onboard software is also configured to include the sensor data into the logging system, to have all data time stamped and georeferenced. Both MARES and TriMARES have already been tested with multiple sensors from multiple vendors, including cameras, sonars, and a variety of water quality sensors, therefore they are already at TRL 9. The DART AUV is a newer vehicle, still going through operational tests and demonstrations, so it is still at TRL 7. A summary of the main characteristics can be seen in Table 5.8.1, corresponding to the configurations of Fig. 5.8.3.

5.8.3 INNOVATIVE SOLUTIONS

This section addresses some innovative solutions in marine robotics, and their relevance to face some of the challenges in ocean sampling. Although they have already been validated in the field, at least in controlled scenarios, they are still being perfected for adoption into operational environments. Their TRLs have been steadily evolving in the midscale and will certainly move toward the high levels, therefore they can be safely considered for new projects.

5.8.3.1 Underwater Docking Stations

Underwater docking stations are generally seen as service stations for AUVs to recharge their batteries and exchange data, and they are fundamental for extended deployments at sea. Depending on the operational scenarios, they can be hanging from the surface (from a buoy or a vessel), placed along the water column, or landed on the seafloor, perhaps connected to cable observatories.

The first designs were intended for flying-type AUVs that would home in to a cone or pole, relying on ultra-short baseline acoustics for the final approximation and precision engagement with the docking station [35,36]. This approach is still being perfected for flying-type AUVs, with very good results in various operational conditions [37,38].

In the last few years, there has been a proliferation of hovering-type AUVs, with a substantial increase in maneuverability. This allows a vehicle to approach a docking station with arbitrarily low velocity, greatly reducing the mechanical impact during the docking maneuver. Typically, AUVs use a complementary navigation system for docking: a long-range navigation system for the approaching phase, followed by a short range, high accuracy, vision-based localization during the final stage [39–42].

5.8.3.1.1 A Lightweight Docking Station for the MARES AUV

The MARES AUV is a portable vehicle for shallow water operations and a lightweight docking station was developed to allow multiday operations in coastal areas [43]. Given the wide range of operational scenarios and configurations of the MARES AUV, the design also has a modular structure that can easily be reconfigured to support different vehicle configurations, deployment scenarios, and docking maneuvers. Another key aspect is the compact design, allowing for deployment and recovery using a simple rigid hull inflatable boat. The overall envelope of this configuration is 1.5 m long, 0.8 m wide, and 1 m high, with a total weight close to 40 kg. The initial structure was designed to land on the seafloor, with a low center of gravity adjusted with ballast weights, although it can later be suspended from the surface, if required. Fig. 5.8.4 shows a diagram of the main components, as well as a perspective of the docking station as viewed by the AUV.

Main components. The docking station was designed for the MARES AUV to slowly descend into the top layer cradle, where it is maintained with the aid of two electromagnets. A pressure vessel is located underneath the AUV cradle, holding all electronic boards and energy in an internal aluminum frame. It has a maximum rating of 10 bar, allowing for a maximum depth of 100 m. Inside the pressure housing, a single board computer manages all subsystems with the aid of a few Original Equipment Manufacturer interface boards. At the end caps there are spare underwater bulkheads to connect any relevant sensors. When the AUV is docked both computers can communicate using WiFi, and the batteries are charged using Wireless Power Transfer (WPT). Although the pressure housing is

FIGURE 5.8.3 Modular Autonomous Robot for Environment Sampling, TriMARES, and Deep water Autonomous Robotic Traveler (*left to right*), three vehicles assembled from the modular building blocks represented.

TABLE 5.8.1 Main Characteristics of the Modular Vehicles at INESC TEC (Fig. 5.8.4)

	MARES	**TriMARES**	**DART**
First mission	2007	2011	2017
Mode of operation	AUV	ROV/AUV	AUV
Depth rating	100 m	100 m	4000 m
Length	1.7 m	1.5 m	2.5 m
Diameter	20 cm	–	20 cm
Max. width	30 cm	80 cm	30 cm
Dry mass	35 kg	75 kg	55 kg
Default propulsion	2 horizontal 2 vertical	4 horizontal 2 vertical 1 lateral	2 horizontal 2 vertical
Horizontal velocity	0–2 m/s	0–1.5 m/s	0–2 m/s
Vertical velocity	0–0.3 m/s	0–0.3 m/s	0–0.5 m/s
Lateral velocity	Optional	0–0.3 m/s	Optional
Energy	600 Wh, Li-Ion	800 Wh, Li-Ion	800 Wh, Li-Ion
Endurance/range	10 h/50 km	8 h/40 km	30 h/110 km
Navigation	LBL acoustics IMU Visual servoing	LBL acoustics IMU	LBL acoustics IMU Visual servoing
Payload options	Cameras, sonars, CTD, turbidity, pH, DO, etc.		

AUV, autonomous underwater vehicles; *CTD*, conductivity, temperature, depth (sensor); *DART*, Deep Autonomous Robotic Traveler; *DO*, dissolved oxygen; *IMU*, inertial measurement unit; *LBL*, long baseline; *MARES*, Modular Autonomous Robot for Environment Sampling; *ROV*, remotely operated vehicle.

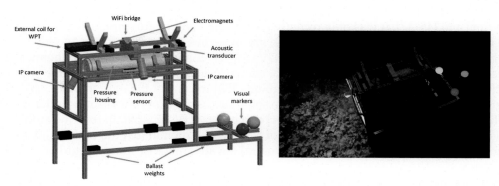

FIGURE 5.8.4 Components of the lightweight docking station and overhead view. *IP*, Internet Protocol; *WPT*, Wireless Power Transfer.

connected to the surface, it has its own Li–Po rechargeable batteries to ensure continuous autonomous operation of the whole system for a few days (naturally depending on power usage).

A visual target with the colored illuminated globes (red, yellow, green) is attached to the bottom of this structure, with sufficient offset to minimize visual occlusion during the docking maneuver. The geometry and location of the visual markers can be easily adjusted, but the most common configuration is a triangular shape with 40 cm of separation. An external LED light source was also installed to illuminate the target from above, if required. All locations and offsets can be easily changed in a few minutes by adjusting only a few set screws.

To facilitate deployment, the docking station has a downward-looking video camera and lights to confirm the flatness of the sea bottom in the deployment location. It also has pressure and attitude sensors to assess the final depth and orientation. This information is needed for the AUV navigation system to adequately perform the docking maneuver.

Surface gateway. For the real-time supervision of the docking maneuver and to provide electrical power to the docking station there is a power and Ethernet cable linking the pressure housing of the docking station with the surface. At the surface, a WiFi link establishes the connection to a control interface located on shore or on a mission support vessel.

The control interface allows the remote operation of the docking station controlling its subsystems. More specifically, it can be used to switch on and off the electromagnets that hold the AUV in place to control the intensity of the visual markers, as well as the intensity of the lights associated with the video cameras attached to the docking station. The interface also displays information about the current status of the docking station, including real-time images collected by the underwater cameras. When the AUV is docked, the interface allows the download of data stored on the AUV computational system, as well as the upload of new configurations of missions to the AUV.

Operational procedure. The deployment procedure is monitored with the control interface, ensuring that the docking station rests in a suitable orientation. The exact final position is calculated using acoustic ranging from the surface, and all information is transmitted to the AUV before launch.

Typical AUV missions are programmed as sequences of maneuvers, and the docking phase is a special case of a *GoTo* maneuver. In fact, the AUV separates this maneuver into two sequential tasks. In the first, it uses the camera to find the visual markers of the docking station. Upon positive detection, the AUV switches to a relative navigation scheme, based on visual information acquired by real-time video processing, as described in [44]. The AUV then descends toward the cradle on the docking station using visual servoing techniques. The electromagnet locking mechanism holds the vehicle in position, overcoming the small buoyancy of the vehicle, as well as any small disturbance forces. The WPT system can then be powered to transfer energy to the AUV.

When the AUV sits at the cradle, it connects to the WiFi access point at the docking station, therefore establishing a link for AUV data download and new mission upload. At the same time, a remote operator can also connect to the onboard computer via the surface gateway and proceed with a thorough checkup of the whole system before relaunch.

5.8.3.2 Real-Time Adaptive Sampling

Most autonomous robotic systems are programmed to follow a trajectory defined by a sequence of waypoints. During the mission, all data from payload sensors is time stamped and georeferenced for later analysis. Instead of waiting for the end of the mission, the objective of adaptive sampling is to process such data in real-time, and change the motion pattern of the vehicle to concentrate measurements in a region of interest. Although this concept was proposed a long time ago [45–47], it was only recently that the first successful implementations were reported using robotic marine vehicles. The first examples include the use of segmentation algorithms on video images from ROVs to track benthic boundaries [48], and also the use of AUVs for searching for the sources of chemical plumes, trying to mimic the real behavior of lobsters or bacterium in odor source localization [49–51]. In [52], the authors used adaptive sampling primitives on a Mission Oriented Operating Suite [53] to control the motion of an ASV to detect the horizontal thermal gradient. More recently, there have been some successful experiments on autonomous vertical controllers for thermocline tracking, carried out on gliders [54] and other AUVs [55–59]. In [60], the authors used the Tethys AUV to detect a coastal upwelling front off the coast of California. Later, in [59] the authors adapted the algorithms to effectively track the front by crossing it at fixed angles, taking advantage of some a priori knowledge of the typical orientation.

In another example, the Sentry AUV was used to sample the underwater plume from an oil leak in the Gulf of Mexico [61]. Initially, the vehicle carried a mass spectrometer and made transects across the plume. The trajectory was then reconfigured according to preliminary analysis of data. In one of the surveys carried out in zigzags, each segment reported elevated values of hydrocarbons from its mass spectrometer. Although this was not a case of autonomous boundary tracking, it is clearly a scenario where such implementation seems to be possible using the hydrocarbon signature to detect the boundary crossings.

5.8.3.2.1 *Adaptive Sampling of Thermoclines*

A practical successful example of adaptive sampling has been developed at INESC TEC for the case of autonomous thermocline identification and tracking in real-time using the MARES AUV, as described in [55,56]. The vehicle was instructed to follow a specific sequence of waypoints, but the depth control was continuously performing vertical *yo-yos*, for which the limits were calculated in real-time according to the temperature profiles being captured. Comparing to a typical *yo-yo* with fixed limits, this mode of operation ensures that the vehicle depth is kept close to the thermocline, taking measurements in a much denser spatial scale.

5.8.3.2.2 *Boundary Detection and Tracking*

Another successful example of adaptive sampling was the demonstration of the ability to detect and track a horizontal boundary, described in [62]. This was an expansion of the thermocline tracking maneuver to the horizontal plane, and may be useful for cases where a given scalar field has a transition zone with a higher gradient than in the rest of the region. This is a typical characteristic of many oceanographic phenomena, like fronts or plume boundaries. In this case, one ASV was used to track the boundary of a shallow water navigation channel using measurements from an altimeter (Fig. 5.8.5).

To find the maximum horizontal gradient, the vehicle needs to cross the transition zone, moving from the region where the scalar field has high values to the region with low values, and vice versa. For each crossing, a filtering process detects the point of maximum gradient, as well as the value of the gradient. The algorithm uses the location of the last crossings to estimate the direction of the boundary and keeps track of the boundary with a continuous *zig-zag* around that direction. Note that neither the evolution of the boundary line nor the amplitude of the *zig-zag* is known at the beginning of the mission. Fig. 5.8.6 shows the measured bathymetry data (in black) and the location of the maximum gradient (in red) for some of the crossings of the boundary.

5.8.3.3 Cooperative Robotics

Robotic cooperation spatially distributes resources and/or to takes advantage of robots with different and complementary capabilities. Cooperation is used to achieve coherent task execution from a team of robots performing according to the objectives of an assigned mission. Approaches are either centralized [63], in which a central entity implicitly defines the desired state of the cooperative robots, or decentralized [64,65], when no such central entity exists. In cooperative operations, robots require exchanging information on their states with regard to the operation

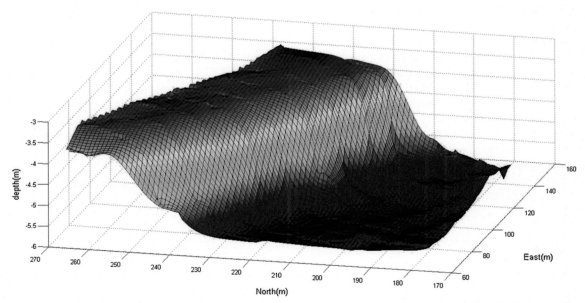

FIGURE 5.8.5 Map of a portion of the shallow water navigation channel, highlighting a clear boundary between approximately 4 and 6 m of depth (*red to blue*).

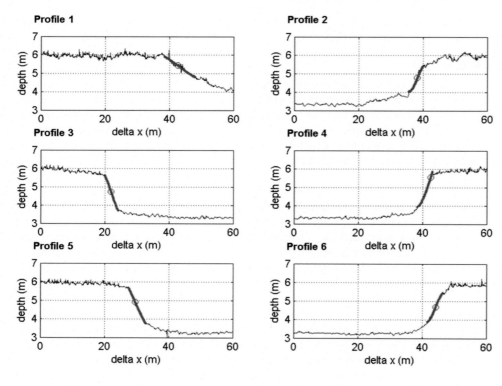

FIGURE 5.8.6 Example of the autonomous detection of maximum gradient across the slope of the navigation channel of Fig. 5.8.5. The *bold red region* corresponds to the maximum slope.

objectives. This exchange is guaranteed by means of sensing or communications, or a combination of both. The stability and robustness of the cooperative operations under constraints imposed by current technological solutions, which include delayed, low rate, and intermittent exchange of information, has attracted the attention of several researchers, encouraged by the current challenges in robotics and particularly in marine robotics [66,67].

INESC TEC has devoted effort to the cooperation of autonomous marine vehicles, and a centralized approach was developed for coordinating autonomous underwater and surface vehicles with communication capabilities. The developed approach was built so that it is robust to communication constraints, such as delays and intermittence, and differences in vehicle performances, thus enabling heterogeneous teams of robots. Additionally, the number of coordinated robots is virtually unlimited, arbitrary paths can be assigned, and variable formation geometries can also be achieved. Fig. 5.8.7 shows an example of the path of two vehicles programmed to follow parallel lines and demonstrate the accuracy of the coordination.

Several cooperative missions were accomplished by the INESC TEC team, using heterogeneous ASVs equipped with standard wireless communication devices and acoustic modems [63]. Cooperative ASVs were used to localize an underwater acoustic target, using time of arrival (TOA) only [68]. Each robot was equipped with a hydrophone, listening to acoustic signals. Localization was achieved by cooperatively moving in formation to find the position that provides the best possible estimate. The cooperative nature of this application does not rely on coordinated motion only but also on information sharing about the TOA to each ASV, therefore resulting in a cooperative estimate of an acoustic source location. One application scenario is searching for aircraft black boxes or any other devices equipped with acoustic pingers.

Multiple applications and motivating examples can be found in the literature to emphasize the need for cooperation in robotics. Some other examples, just to cite a few, include surveillance coverage, pollution containment, and deployment of navigation aids for operations in deep water [23,69].

5.8.4 FUTURE TRENDS

In the past, innovations in marine robotics have been driven by a combination of military, scientific, and industry demands. Some of these will remain pushing the technology, while more specific application scenarios are emerging as strong drivers, like seabed mining (particularly in deep water), renewable energy, response to hazard events, or the global effort to map the world's oceans.

FIGURE 5.8.7 Example of coordinated motion of Zarco and Gama Autonomous Surface Vehicles. The *circles* correspond to instants of trajectory synchronization.

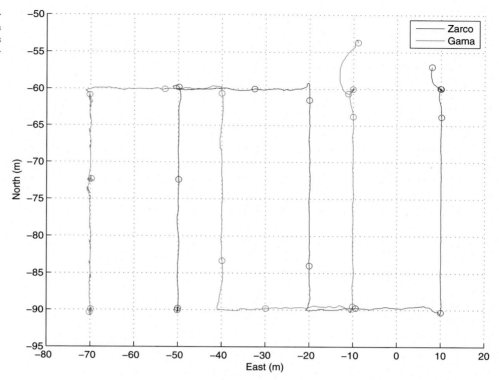

This section anticipates features for marine robotics that are expected to be fully mature in the middle to long term. They correspond to low TRL capabilities, i.e., paradigms and performance levels that have been proposed but are yet to be fully achieved with the current state of the art. Their relevance to industry and to the scientific community, and the future missions they will enable, will continue driving the commitment of the robotics community to overcome current limitations.

5.8.4.1 Autonomous Intervention

One innovation that will surely take a relevant role in future operations at sea is the concept of Intervention Autonomous Underwater Vehicles [70]. The main idea is to integrate small manipulators in AUVs so that the vehicles can perform simple intervention tasks without the need of an operator. This will have a strong impact on operational costs, particularly in deep-water operations, as the vehicles will be able to replace ROVs and avoid the considerable cost of operators and large support ships. One of the flagship designs in this respect is the Girona 500 AUV, developed by the University of Girona in Spain [71], a vehicle that has already demonstrated simple autonomous manipulation capabilities [72].

This concept requires complementary high performance in terms of perception, relative navigation, and robust control under the paradigm of floating base manipulation. In fact, there has been tremendous improvements in terms of vehicle perception of the environment. This includes, for example, short-range visual perception of distances and relative position of targets using stereo cameras with integrated processing. Such an improvement, together with increased navigation and control accuracy, will enable the implementation of light autonomous intervention capabilities in even the smallest vehicles. One application scenario can be the autonomous deployment and recovery of items at/from the sea bottom.

Some of the underlying challenges for autonomous intervention are being addressed by INESC TEC. The detection of visual targets lying on the seafloor was extensively tested during the euRathlon 2015 competition where the MARES AUV was part of a multidomain robotic team for search and rescue operations [73]. In these trials, MARES performed several sweeping maneuvers, and it was able to detect and geolocate multiple visual targets distributed in the operation area. The ability to use visual information to drive an AUV (or *visual servoing*) has also been tested for the docking maneuver, and the accuracy of the measurements was validated in indoor environments. The accuracy of the current visual positioning is 2 mm across the field of view and 15 mm along the field of view [44]. For the docking maneuver, these figures allowed the AUV to easily approach the docking station and to align itself in the cradle direction with the longitudinal position of the power transfer system and the electromagnetic holding devices. The algorithms employed for this maneuver can be perfected for light intervention tasks.

5.8.4.2 Data Muling Between Robotic Vehicles

Data muling is a technique where a mobile agent interacts with other agents to upload, download, or transport data to a different physical location. This is an important concept to overcome in the absence of long-range, high data rate communications in the underwater environment and can be used to retrieve information from seabed sensor nodes, or to establish a gateway between deep-water vehicles and the surface. There have been a few successful experiments to partially demonstrate data muling capabilities with AUVs. For example, [74] reports a system comprising an AUV and many static underwater sensor nodes with long-range acoustic and short-range optical communications. The AUV can locate the static nodes using vision and hover above the static nodes for data upload. Work [75] describes an experiment where a robot finds a sensor node, uses high-bandwidth, short-range optical communication to download data from the sensor node, and then transports these data to a base station.

The ability to provide data muling between vehicles and sensor nodes has immediate applications for retrieval of large portions of data. During long-term deployments, the vehicles can also be used to provide calibration for the sensors, and, with the advancement of intervention vehicles, even simple maintenance tasks can be performed by the vehicles. During data transfer, power can also be transferred wirelessly without docking. In the time ahead, the possibility of exchanging large amounts of data among vehicles will open up new operational scenarios for multiple vehicles. However, it will also require improvements in vehicle coordination, localization and navigation, and networking mechanisms. All of these are being addressed independently in multiple research labs around the globe, as described in previous sections.

5.8.4.3 Cooperative Adaptive Sampling

The idea of cooperative adaptive sampling extends the concept of real-time adaptive sampling, described previously, to the case where multiple assets are used for the same purpose. One obvious case is the spatial distribution of vehicles in an area of interest to accelerate the sampling process, therefore reducing the time and space aliasing of dynamic features. This is the idea proposed in [76], with the adaptation of methods for front tracking to the boundary of plumes, assuming that the problem of tracking a three-dimensional plume can be divided into multiple tracking of two-dimensional boundaries. To optimize the sampling, they propose to distribute multiple AUVs along the boundary and combine partial data to reconstruct the full boundary.

On a more advanced level, the paradigm of cooperative adaptive sampling may include sharing of information among vehicles. In this case, each vehicle can incorporate data being processed by the other team members to alter the sampling pattern in real-time. To maximize the information about a given process, different vehicles may be configured with different payloads, opening up the possibility of using networks of heterogeneous vehicles with specific roles. In the case of INESC TEC, this is the intended evolution of the algorithms developed for tracking thermoclines in the vertical plane with horizontal boundary tracking, as described earlier. The combination of both algorithms may be used by complementary vehicles resulting in a three-dimensional adaptive sampling mechanism. In fact, there are many opportunities for the operation of networked devices, and the modular vehicles described earlier can easily support this effort.

5.8.5 CONCLUSIONS

Robotic vehicles are versatile and reliable tools that are routinely used by the scientific community and industry to gather ocean data. They can sample wide regions with moderate set-up and operational costs, providing dense data with high accuracy. They are fundamental to characterize oceanographic dynamic processes and their performance has been steadily increasing with the maturity of the technology.

This chapter described the latest advances in marine robotics, starting with an overview of the current capabilities of autonomous vehicles. One of the primary guidelines for the improvement of these tools has always been to extend the duration and range of the missions, and it is now common to see data sets spanning many days of operation over hundreds of kilometers. Data quality is another major issue, and current solutions offer very accurate time and space tagging of data samples. To simplify logistics and facilitate access to the ocean, special care has been taken in the development of smaller, modular solutions.

In a following section, one chapter describes innovative solutions that have already been demonstrated in the field, but are seldom found in real operational environments. Given the state of maturity, they can already be used in application niches or included in new projects. One example is the concept of an underwater docking station, a promising solution to extend the presence at sea for very long periods. On another front, the notion of adaptive sampling intends to transfer some *intelligence* to the vehicles by changing online the sampling pattern according to the sensor data that is being collected. Finally, the simultaneous use of multiple vehicles may provide complementary, distributed data about specific ocean processes.

The final part of the chapter anticipates features for marine robotics that are still at an early stage of development. They promise to revolutionize some autonomous operations and enable many others, so it is expected that they mature in the middle to long term. Autonomous intervention is one of these capabilities, forecasting scenarios where fully autonomous vehicles perform tasks that are currently performed by ROVs, avoiding skilled operators and complex logistics. Underwater communications have been a permanent hindrance in obtaining ocean data, particularly from remote sources, and the concept of AUVs acting as data mules may alleviate these difficulties in some operational scenarios. Finally, a natural evolution of the adaptive sampling paradigm is the distribution of the underlying *intelligence* among a team of cooperative vehicles, enabling a collective perspective over a given process that may be much more than the sum of its parts.

Glossary

ASV Autonomous Surface Vehicle
AUV Autonomous Underwater Vehicle
CTD Conductivity, temperature, depth (sensor)
DART Deep Autonomous Robotic Traveler
DGPS Differential GPS
DO Dissolved oxygen
GPS Global Positioning System
IMU Inertial measurement unit
IP Internet Protocol
LARS Launch and Recovery System
LBL Long baseline
Li-Ion Lithium Ion (rechargeable batteries)
Li–Po Lithium polymer (rechargeable batteries)
MARES Modular Autonomous Robot for Environment Sampling
MOOS Mission Oriented Operating Suite
OEM Original Equipment Manufacturer
ROV Remotely Operated Vehicle
TOA Time of arrival
TRL Technology Readiness Level
WPT Wireless Power Transfer

Acknowledgments

This work was funded by the ENDURE Project (PT02_Aviso4_0015) supported by the EEA Grants Iceland, Liechtenstein, and Norway.

This work was financed by the ERDF European Regional Development Fund through the Operational Programme for Competitiveness and Internationalisation—COMPETE 2020 Programme within project "POCI-01-0145-FEDER-006961," and by National Funds through the Portuguese funding agency FCT—Fundação para a Ciência e a Tecnologia as part of project "UID/EEA/50014/2013."

Part of the research leading to these results has received funding from the European Union's Horizon 2020—The EU Framework Programme for Research and Innovation 2014–2020, under grant agreement No. 692427.

This work was funded by the STRONGMAR project, funded by the European Commission under the H2020 EU Framework Programme for Research and Innovation (H2020-TWINN-2015, 692427).

References

[1] Hobson BW, Bellingham JG, Kieft B, McEwen R, Godin M, Zhang Y. Tethys-class long range AUVs – extending the endurance of propeller-driven cruising AUVs from days to weeks. In: Proc. IEEE/OES conf. autonomous underwater vehicles AUV 2012, Southampton, UK, Sept. 2012. 2012.

[2] Vaneck TW, Rodriguez-Ortiz CD, Schmidt MC, Manley JE. Automated bathymetry using an autonomous surface craft. Navigation Winter 1996;43(4):407–19.

[3] Curcio J, Leonard J, Patrikalakis A. Scout – a low cost autonomous surface platform for research in cooperative autonomy. In: Proc. MTS/IEEE int. conf. oceans'05, Washington, D.C., USA, Sept. 2005. 2005.

[4] Cruz N, Matos A, Cunha S, Silva S. Zarco – an autonomous craft for underwater surveys. In: Proc. 7th geomatic week, Barcelona, Spain, Feb. 2007. 2007.

[5] Hine R, McGillivary PA. Wave-powered autonomous surface vessels as components of ocean observing systems. In: Proc. PACON 2007, Honolulu, HI, USA, June 2007. 2007.

[6] Ménage O, Bethencourt A, Rousseaux P, Prigent S. Vaimos: realization of an autonomous robotic sailboat. In: Robotic sailing 2013. Springer; 2013. p. 25–36.

[7] Ghani MH, Hole LR, Fer I, Kourafalou VH, Wienders N, Kang H, Drushka K, Peddie D. The Sailbuoy remotely-controlled unmanned vessel: measurements of near surface temperature, salinity and oxygen concentration in the northern gulf of Mexico. Methods Oceanogr 2014;10:104–21.

[8] Meinig C, Lawrence-Slavas N, Jenkins R, Tabisola HM. The use of saildrones to examine spring conditions in the bering sea: vehicle specification and mission performance. In: MTS/IEEE int. conf. oceans'15. 2015. p. 1–6.

[9] Stelzer R, Jafarmadar K. "The robotic sailing boat ASV Roboat as a maritime research platform. In: Proc. 22nd int. HISWA symp., Amsterdam, The Netherlands, Nov. 2012. 2012.

[10] Klinck H, Fregosi S, Matsumoto H, Turpin A, Mellinger DK, Erofeev A, Barth JA, Shearman RK, Jafarmadar K, Stelzer R. Mobile autonomous platforms for passive-acoustic monitoring of high-frequency cetaceans. In: Friebe A, Haug F, editors. Proc. 8th int. robotic sailing conf. Mariehamn, Finland: Springer International Publishing; August 2015. p. 29–37.

[11] Silva A, Matos A, Soares C, Alves J, Valente J, Zabel F, Cabral H, Abreu N, Cruz N, Almeida R, Ferreira RN, Ijaz S, Lobo V. Measuring underwater noise with high endurance surface and underwater autonomous vehicles. In: Proc. MTS/IEEE int. conf. oceans'13, San Diego, CA, USA, Sept. 2013. 2013.

[12] Golmen LG, Pearlman F, Kvalsund K, Reggiani E, Hareide NR, Østerhus S, Pearlman J, Delory E, Cyr F, Meme S. Validation and demonstration of novel oceanographic sensors on selected measurement platforms in the nexos project. In: Proc. MTS/IEEE int. conf. oceans'2017, Aberdeen, UK, June 2017. 2017. p. 1–7.

[13] Alves JC, Cruz NA. Permanent ocean presence with autonomous sailing robots – wind-propelled vessels foster long missions with precise maneuvering. Sea Tecnhol 2014;55(5).

[14] Cruz NA, Alves JC. Navigation performance of an autonomous sailing robot. In: Proc. MTS/IEEE int. conf. oceans'14, Sept 2014. 2014. p. 1–7.

[15] Meyer D. Glider technology for ocean observations: a review. Ocean Sci Discuss 2016;2016:1–26. [Online]. Available https://www.ocean-sci-discuss.net/os-2016-40/.

[16] D'Asaro EA, McNeil C. Calibration and stability of oxygen sensors on autonomous floats. J Atmos Ocean Technol 2013;30(8):1896–906.

[17] Wilson D, Koch C, Dutton A, Bennett S. Long-term water quality monitoring deployments with the Sea-Bird HydroCAT-EP. Sea-Bird Scientific; 2017. Tech. Rep.

[18] Almeida R, Cruz N, Matos A. Synchronized intelligent buoy network for underwater positioning. In: Proc. IEEE int. conf. oceans'10, Sydney, Australia, May 2010. 2010.

[19] Almeida R, Cruz N, Matos A. Man portable acoustic navigation buoys. In: Proc. MTS/IEEE int. conf. oceans'16, Shanghai, China, April 2016. 2016. p. 1–6.

[20] Almeida R, Melo J, Cruz N. Characterization of measurement errors in a LBL positioning system. In: Proc. MTS/IEEE int. conf. oceans'16, Shanghai, China, April 2016. 2016. p. 1–6.

[21] Matos A, Cruz N. AUV navigation and guidance in a moving acoustic network. In: Proc. IEEE int. conf. oceans'05, vol. 1. 2005. p. 680–5.

[22] Eustice RM, Whitcomb LL, Singh H, Grund M. Experimental results in synchronous-clock one-way-travel-time acoustic navigation for autonomous underwater vehicles. In: Proc. IEEE int. conf. robotics and automation ICRA'07, Rome, Italy, 10–14 April 2007. 2007. p. 4257–64.

[23] Cruz N, Abreu N, Almeida J, Almeida R, Alves J, Dias A, Ferreira B, Ferreira H, Gonalves C, Martins A, Melo J, Pinto A, Pinto V, Silva A, Silva H, Matos A, Silva E. Cooperative deep water seafloor mapping with heterogeneous robotic platforms. In: Proc. MTS/IEEE int. conf. oceans'17, anchorage, AK, USA, Sept. 2017. 2017.

[24] Hayashi E, Kimura H, Tam C, Ferguson J, Laframboise J-M, Miller G, Kaminski C, Johnson A. Customizing an autonomous underwater vehicle and developing a launch and recovery system. In: Proc. int. symp. underwater tech. UT'13, Tokyo, Japan, Mar. 2013. 2013. p. 1–7.

[25] Sarda EI, Dhanak MR. A usv-based automated launch and recovery system for auvs. IEEE J Ocean Eng January 2017;42(1):37–55.

[26] Cruz N, Madureira L, Matos A, Pereira FL. A versatile acoustic beacon for navigation and remote tracking of multiple underwater vehicles. In: Proc. MTS/IEEE int. conf. oceans'01, Honolulu, HI, USA, Nov. 2001. 2001. p. 1829–34.

[27] Hiller T, Steingrimsson A, Melvin R. Expanding the small AUV mission envelope; longer, deeper & more accurate. In: Proc. IEEE/OES conf. autonomous underwater vehicles AUV 2012, Southampton, UK, Sept. 2012. 2012.

[28] Taylor M, Wilby A. Design considerations and operational advantages of a modular AUV with synthetic aperture sonar. In: Proc. MTS/IEEE int. conf. oceans'11, Kona, HI, USA, Sept. 2011. 2011. p. 1–6.

[29] Sangekar M, Chitre M, Koay TB. Hardware architecture for a modular autonomous underwater vehicle Starfish. In: Proc. MTS/IEEE int. conf. oceans'08, Quebec, Canada, Sept. 2008. 2008. p. 1–8.

[30] Shuzhe C, Soon HG, Hong EY, Chitre M. Modular modeling of autonomous underwater vehicle. In: Proc. MTS/IEEE int. conf. oceans'11, Kona, HI, USA, Sept. 2011. 2011. p. 1–6.

[31] Cruz NA, Matos AC, Ferreira BM. Modular building blocks for the development of AUVs – from MARES to TriMARES. In: Proc. int. symp. underwater tech. UT'13, Tokyo, Japan, Mar. 2013. 2013.

[32] Cruz NA, Matos AC. The MARES AUV, a modular autonomous robot for environment sampling. In: Proc. MTS/IEEE int. conf. oceans'08, Quebec, Canada, Sept. 2008. 2008.

[33] Cruz NA, Matos AC, Almeida RM, Ferreira BM, Abreu N. TriMARES – a hybrid AUV/ROV for dam inspection. In: Proc. MTS/IEEE int. conf. oceans'11, Kona, HI, USA, Sept. 2011. 2011.

[34] Cruz NA, Matos AC, Almeida RM, Ferreira BM. DART – a portable deep water hovering AUV. In: Proc. MTS/IEEE int. conf. oceans'17, Anchorage, AK, USA, Sept. 2017. 2017.

[35] Singh H, Bellingham JG, Hover F, Lerner S, Moran BA, der Heydt KV, Yoerger D. Docking for an autonomous ocean sampling network. IEEE J Ocean Eng October 2001;26(4):498–514.

[36] Stokey R, Allen B, Austin T, Goldborough R, Forrester N, Purcell M, Alt CV. Enabling technologies for REMUS docking: an integral component of an autonomous ocean-sampling network. IEEE J Ocean Eng October 2001;26(4):487–97.

[37] Hobson B, McEwen R, Erickson J, Hoover T, McBride L, Shane F, Bellingham J. The development and ocean testing of an AUV docking station for a 21″ AUV. In: Proc. MTS/IEEE int. conf. oceans'07, Vancouver, BC, Canada, Oct. 2007. 2007.

[38] Hydroid, a Kongsberg Company. Underwater mobile docking of autonomous underwater vehicles. In: Proc. MTS/IEEE int. conf. oceans'12, Hampton Roads, VA, USA, Oct. 2012. 2012. p. 1–15.

[39] Kondo H, Okayama K, Choi JK, Hotta T, Kondo M, Okazaki T, Singh H, Chao Z, Nitadori K, Igarashi M, Fukuchi T. Passive acoustic and optical guidance for underwater vehicles. In: Proc. MTS/IEEE int. conf. oceans'12, Yeosu, Korea, May 2012. 2012. p. 1–6.

[40] Wirtz M, Hildebrandt M, Gaudig C. Design and test of a robust docking system for hovering AUVs. In: Proc. MTS/IEEE int. conf. oceans'12, Hampton Roads, VA, USA, Oct. 2012. 2012.

[41] Maki T, Shiroku RT, Sato Y, Matsuda T, Sakamaki T, Ura T. Docking method for hovering type AUVs by acoustic and visual positioning. In: Proc. int. symp. underwater tech. UT'13, Tokyo, Japan, Mar. 2013. 2013.

[42] Yoshida H, Ishibashi S, Yutaka O, Sugesawa M, Tanaka K. A concept design of underwater docking robot and development of its fundamental technologies. In: Proc. IEEE/OES int. conf. AUV 2016, Tokyo, Japan, Nov. 2016. 2016. p. 408–11.

[43] Cruz NA, Matos AC, Almeida RM, Ferreira BM. A lightweight docking station for a hovering AUV. In: Proc. int. symp. underwater tech. UT'17, Tokyo, Japan, Mar. 2017. 2017.

[44] Figueiredo A, Ferreira B, Matos A. Vision-based localization and positioning of an AUV. In: Proc. MTS/IEEE int. conf. oceans 2016, Shanghai, China, Apr. 2016. 2016. p. 1–6.

[45] Consi TR, Atema J, Goudey CA, Cho J, Chryssostomidis C. AUV guidance with chemical signals. In: Proc. IEEE symp. autonomous underwater vehicle tech. AUV'94, Cambridge, MA, USA, Jul. 1994. 1994. p. 450–3.

[46] Burian E, Yoerger D, Bradley A, Singh H. Gradient search with autonomous underwater vehicles using scalar measurements. In: Proc. IEEE symp. autonomous underwater vehicle tech. AUV'96, Monterey, CA, USA, Jun. 1996. 1996. p. 86–98.

[47] Willcox JS, Bellingham JG, Zhang Y, Baggeroer AB. Performance metrics for oceanographic surveys with autonomous underwater vehicles. IEEE J Ocean Eng October 2001;26(4):711–25.

[48] Barat C, Rendas MJ. Benthic boundary tracking using a profiler sonar: a mixture model approach. In: Proc. MTS/IEEE int. conf. oceans'03, San Diego, CA, USA, Sep. 2003. 2003. p. 1409–16.

[49] Farrell JA, Li W, Pang S, Arrieta R. Chemical plume tracing experimental results with a REMUS AUV. In: Proc. MTS/IEEE int. conf. oceans'3, San Diego, CA, USA, Sep. 2003. 2003. p. 962–8.

[50] Pang S, Farrell JA. Chemical plume source localization. IEEE T Syst Man Cybern Part B Cybern October 2006;36(5):1068–80.

[51] Naeem W, Sutton R, Chudley J. Chemical plume tracing and odour source localization by autonomous vehicles. J Navig May 2007;60(2):173–90.

[52] Eickstedt DP, Benjamin MR, Wang D, Curcio J, Schmidt H. Behavior based adaptive control for autonomous oceanographic sampling. In: Proc. int. conf. robot. Autom, Rome, Italy, Apr. 2007. 2007.

[53] Newman PM. MOOS – mission orientated operating suite. USA: Massachusetts Institute of Technology; 2008. Tech. Rep. 08.

[54] Woithe HC, Kremer U. A programming architecture for smart autonomous underwater vehicles. In: Proc. IEEE/RSJ int. conf. intelligent robots and systems IROS'09, St. Louis, MO, USA, Oct. 2009. 2009.

[55] Cruz NA, Matos AC. Reactive AUV motion for thermocline tracking. In: Proc. IEEE int. conf. oceans'10, Sydney, Australia, May 2010. 2010.

[56] Cruz NA, Matos AC. Adaptive sampling of thermoclines with autonomous underwater vehicles. In: Proc. MTS/IEEE int. conf. oceans'10, Seattle, WA, USA, Sept. 2010. 2010.

[57] Zhang Y, Bellingham JG, Godin M, Ryan JP, McEwen RS, Kieft B, Hobson B, Hoover T. Thermocline tracking based on peak-gradient detection by an autonomous underwater vehicle. In: Proc. MTS/IEEE int. conf. oceans'10, Seattle, WA, USA, Sept. 2010. 2010.

[58] Petillo S, Balasuriya A, Schmidt H. Autonomous adaptive environmental assessment and feature tracking via autonomous underwater vehicles. In: Proc. IEEE int. conf. oceans'10, Sydney, Australia, May 2010. 2010.

[59] Zhang Y, Bellingham JG, Ryan JP, Kieft B, Stanway MJ. Two-dimensional mapping and tracking of a coastal upwelling front by an autonomous underwater vehicle. In: Proc. MTS/IEEE int. conf. oceans'13, San Diego, CA, USA, Sept. 2013. 2013.

[60] Zhang Y, Godin MA, Bellingham JG, Ryan JP. Using an autonomous underwater vehicle to track a coastal upwelling front. IEEE J Ocean Eng July 2012;37(3):338–47.

[61] Camilli R, Reddy CM, Yoerger DR, van Mooy BAS, Jakuba MV, Kinsey JC, McIntyre CP, Sylva SP, Maloney JV. Tracking hydrocarbon plume transport and biodegradation at Deepwater Horizon. Science October 2010;330:201–4.

[62] Cruz NA, Matos AC. Autonomous tracking of a horizontal boundary. In: Proc. MTS/IEEE int. conf. oceans'14, St. John's, NFL, Canada, Sep. 2014. 2014.

[63] Ferreira BM, Matos AC, Cruz NA, Moreira AP. Coordination of marine robots under tracking errors and communication constraints. IEEE J Ocean Eng January 2016;41(1):27–39.

[64] Ghabcheloo R, Aguiar AP, Pascoal A, Silvestre C, Kaminer I, Hespanha J. Coordinated path-following in the presence of communication losses and time delays. SIAM J Control Optim 2009;48(1):234–65.

[65] Almeida J, Silvestre C, Pascoal AM. Cooperative control of multiple surface vessels with discrete-time periodic communications. Int J Robust Nonlinear Control 2012;22(4):398–419.

[66] Xiang X, Jouvencel B, Parodi O. Coordinated formation control of multiple autonomous underwater vehicles for pipeline inspection. Int J Adv Rob Syst 2010;7(1):3.

[67] Antonelli G, Arrichiello F, Casalino G, Chiaverini S, Marino A, Simetti E, Torelli S. Harbour protection strategies with multiple autonomous marine vehicles. In: Hodicky J, editor. Modelling and simulation for autonomous systems. Springer International Publishing; 2014. p. 241–61.

[68] Ferreira BM, Matos AC, Cruz NA. Optimal positioning of autonomous marine vehicles for underwater acoustic source localization using TOA measurements. In: Proc. int. symp. underwater tech. UT'13, Tokyo, Japan, Mar. 2013. 2013.

[69] Matos A, Cruz N. Positioning control of an underactuated surface vessel. In: Proc. MTS/IEEE int. conf. oceans'08, Quebec, Canada, Sept. 2008. 2008.

[70] Ridao P, Carreras M, Ribas D, Sanz PJ, Oliver G. Intervention AUVs: the next challenge. In: Proc. 19th IFAC world congress, Cape Town, South Africa, Aug. 2014. 2014. p. 12146–59.

[71] Ribas D, Ridao P, Mag L, Palomeras N, Carreras M. The Girona 500, a multipurpose autonomous underwater vehicle. In: Proc. IEEE int. conf. oceans'11, Santander, Spain, June 2011. 2011.

[72] Palomeras N, Peñalver A, Massot-Campos M, Negre PL, Fernández JJ, Ridao P, Sanz PJ, Oliver-Codina G. I-AUV docking and panel intervention at sea. Sensors (Basel) 2016;16(10).

[73] Matos A, Martins A, Dias A, Ferreira B, Almeida J, Ferreira H, Amaral G, Figueiredo A, Almeida R, Silva F. Multiple robot operations for maritime search and rescue in euRathlon 2015 competition. In: Proc. MTS/IEEE int. conf. oceans'16, Shanghai, China, Apr. 2016. 2016. p. 1–6.

[74] Dunbabin M, Corke P, Vasilescu I, Rus D. Data muling over underwater wireless sensor networks using an autonomous underwater vehicle. In: Proc. IEEE int. conf. robotics and automation ICRA'06, May 2006. 2006. p. 2091–8.

[75] Doniec M, Topor I, Chitre M, Rus D. Autonomous, localization-free underwater data muling using acoustic and optical communication. Heidelberg: Springer International Publishing; 2013. p. 841–57.

[76] Petillo SM, Schmidt H. Autonomous and adaptive underwater plume detection and tracking with AUVs: concepts, methods, and available technology. In: Proc. 9th IFAC conf. maneuvering and control of marine craft, MCMC'2012, Arenzano, Italy, Sept. 2012. 2012.

6

From Sensor to User—Interoperability of Sensors and Data Systems

6.1

Sensor Interoperability Protocol for Seamless Cross-Platform Sensor Integration

Joaquin del Rio Fernandez[1], Daniel Mihai Toma[1], Enoc Martinez[1], Simon Jirka[2], Tom O'Reilly[3]

[1]UPC, Universitat Politècnica de Catalunya, Vilanova i la Geltrú, Spain; [2]52° North Initiative for Geospatial Open Source Software GmbH, Münster, Germany; [3]Monterey Bay Aquarium Research Institute (MBARI), Moss Landing, CA, United States

OUTLINE

6.1.1 INTRODUCTION

In marine science, a high number of measurement systems are deployed that rely on different, heterogeneous communication protocols. These protocols define a set of layers addressing specific aspects of the communication flow, from the physical/hardware interface to data formats. Connecting disparate devices into a network typically requires specialized software that can translate commands and data protocols between the individual instruments and the platforms on which they are installed [1]. The platforms typically require extensive manual configuration to match the driver software and other operational details of each network port to a specific instrument. Some data protocols, such as the National Marine Electronics Association, will facilitate the configuration, but in any case, a specific configuration has to be made manually.

Nonstandardized communication links and interfaces lead to a considerable amount of custom software components, which cannot be reused. If a new sensor has to be integrated or an existing sensor needs to be moved from one platform to another, new specific components have to be developed. Generating a specific driver for each sensor is a time-consuming task that requires in-depth knowledge of both the sensor's protocol and the observation platform's architecture [2]. As the number of platforms and sensors within a collaborative environment grows, the number of custom components increases, as well as the infrastructure maintenance costs.

Oceanographic instruments are traditionally developed by small companies, and lack standardization of the protocols for instrument control and configuration, or data retrieval. RS232 and RS485 serial interfaces are the dominant physical layer protocols (although increasingly displaced by Ethernet), but on top of these common interfaces, manufacturers typically define their own distinct syntax and command sets for the instruments they produce [1–4].

This chapter introduces a Sensor Web architecture, illustrated in Fig. 6.1.1, that has been developed within the NeXOS project (www.nexosproject.eu) as an example of how the use of interoperable standards helps to facilitate the creation of an infrastructure for sharing instruments and oceanographic observation, and the integration of sensor data into applications [4].

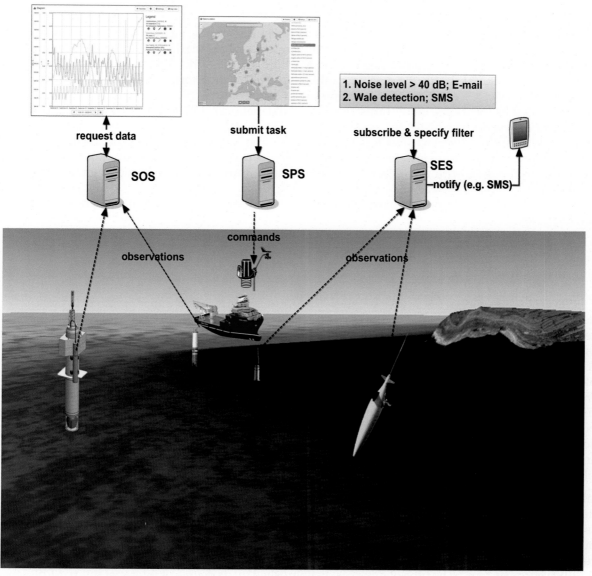

FIGURE 6.1.1   Sensor Web architecture. Data delivered by sensors is made available on the Web by enabling an interoperable data flow from sensors to the Web using standard protocols (e.g., Open Geospatial Consortium Sensor Observation Service [SOS]). Users are also able to interact with sensors and receive notifications through standardized mechanisms. *SES*, simple email service; *SMS*, short message service; *SPS*, Sensor Planning Service.

The ultimate goal should be "plug-and-play" integration, reducing human intervention to a minimum. From the operational interoperability point of view, when a sensor is deployed on an observation platform, it should be automatically detected, configured, and the data should be autonomously retrieved by the observation platform's acquisition process. For broader usage and compatibility, the data and metadata retrieved from the sensor should be encoded in a standard format and archived using uniform procedures. This would enhance not only the interoperability of the sensor, but also its data traceability, reduction of often error-prone human intervention in the process, and ease further data quality procedures, allowing to pinpoint the sensor's malfunctions [5].

Concerning the developed Sensor Web architecture, the content of this chapter is closely linked to the concepts explained in Chapter 6.2. While this chapter describes the integration of sensors into (Web-enabled) platforms, Chapter 6.2 describes those aspects of the Sensor Web architecture that cover the delivery of the collected data via the World Wide Web (hereafter "the Web") to users.

6.1.2 THE SENSOR WEB ARCHITECTURE

The vision of providing oceanographic and coastal data for Sensor Web services is not new and there are many initiatives working in that direction: Copernicus (especially the Copernicus Marine Environment Monitoring Service), the Global Earth Observation System of Systems, and the European INSPIRE Directive. Besides NeXOS, examples of recent projects and initiatives in the oceanographic and coastal monitoring area are SCHeMA (http://www.schema-ocean.eu), SenseOCEAN (http://www.senseocean.eu), Common Sense (www.commonsenseproject.eu/), EMODnet, SeaDataNet/SeaDataCloud (https://www.seadatanet.org), ODIP I/II (www.odip.eu), IMOS, IOOS, BRIDGES (http://www.bridges-h2020.eu), FixO³ (www.fixo3.eu), and others.

In contrast with building services on top of data already stored on a database or available from a server, the architecture presented herein and developed in NeXOS focuses on the optimization of information retrieval from the sensor to the user, with the objective of minimizing future efforts, thus implementing true end-to-end interoperability.

The architecture is divided into three parts, each containing different logical components, respectively to address the following tasks:

1. Propagating data from sensor to platform controller.
2. Propagating data from platform controller to the Web.
3. Making all data available through the Web to the users.

Accordingly, the following components were developed:

1. A set of software components based on the open standard Open Geospatial Consortium (OGC) Programmable Underwater Connector with Knowledge (PUCK), which facilitates the integration of oceanographic instruments on various platforms such as floats, gliders, cabled observatories, vessels of opportunity, or moorings.
2. A software component called "OGC SWE Bridge,"[1] which implements a conversion and ingestion capability to enable the flow of data from device to Sensor Web services.
3. A Sensor Web service interface for data publication, based on OGC Sensor Web Enablement (SWE) specifications. For this service interface, one element is the so-called Smart Electronic Interface for Sensors and Instruments (SEISI). As shown in Fig. 6.1.2, these interfaces comprise the OGC Sensor Observation Service (SOS) for accessing the measured data, the OGC Sensor Planning Service (SPS) for controlling SEISI sensor system parameters, and an approach for subscribing to push-based sensor data streams. (See Chapter 6.2 for more details.)

For all three tasks, the availability of a metadata file describing the device characteristics and interfaces is a prerequisite. For encoding such metadata, we used the OGC Sensor Model Language (SensorML), which is core to the SWE suite of OGC standards.

From a technical point of view, the initial activity is the deployment of a sensing device (e.g., an instrument or a set of instruments/sensors) by connecting it to a platform. In Fig. 6.1.2 such devices are shown as "SEISI sensor system" or "non-SEISI sensor system." The type of device is distinguished by its degree of interoperability. SEISI sensors have

---

[1] Sensor Web Enablement is a suite of Open Geospatial Consortium (OGC) standards that enable Web-based sharing, discovery, exchange, and processing of sensor observations, and operation of sensor systems.

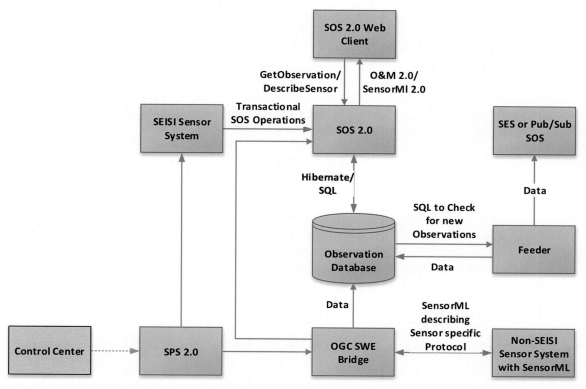

FIGURE 6.1.2 Overview of the components for the ocean and coastal Sensor Web architecture. *OGC*, Open Geospatial Consortium; *O&M*, Observations and Measurements; *SEISI*, Smart Electronic Interface for Sensors and Instruments; *SOS*, Sensor Observation Service; *SPS*, Sensor Planning Service; *SQL*, Structured Query Language; *SWE*, Sensor Web Enablement.

both greater interoperability and levels of automation. Usually, sensors found in the field are commercial off-the-shelf (COTS) with a conventional type of interfacing. We refer to such types of sensors as non-SEISI sensor systems. To operate, these sensors need a driver running on the platform to retrieve their data.

In the case of a SEISI sensor system, the system consists of a set of hardware and software components that are fully compliant with the Sensor Web architecture. This means that the implemented system is able to execute transactional operation calls on an SOS server to register itself and subsequently publish data. The way in which these transactional operations of the SOS interface are used are described in more detail in Fig. 6.1.5. Feasibility demonstrations of this architecture approach and its instantiation were performed during the NeXOS project using different types of sensors and platforms. These prototypes were developed to demonstrate the feasibility that an instrument connected directly into an Internet Protocol (IP) network is able to propagate data directly into an SOS.

For non-SEISI sensor systems, a software component called "OGC SWE Bridge" is needed as an interface to enable the data transmission to SOS servers. The SWE Bridge does the function of the conventional drivers. But in this case, the SWE Bridge is designed to be used with any kind of sensor, unlike conventional drivers designed for a specific sensor. Putting together a COTS Sensor (non-SEISI sensor systems) with the set of software components based on OGC PUCK plus the OGC SWE Bridge, the system becomes a SEISI sensor system.

Once the SOS services are up and running, users can either choose to use, or adapt, an existing "SOS Web Client" or develop their own client applications to discover, visualize, and download data that have been published on the SOS servers.

The components shown in Fig. 6.1.2 are mainly software components running on Web servers; exceptions are the SEISI and non-SEISI sensor systems that refer to the set of instruments, including cables and the host controller of the platform.

As the Sensor Web service components are described in more detail in Chapter 6.2, the following sections will focus on those components that enable the integration of sensing devices into Sensor Web infrastructures.

After this overview of the core components of the Sensor Web the following section provides some insight into how this architecture concept is implemented.

6.1.3 PROTOCOLS AND STANDARDS

The core functional requirements to achieve seamless integration of sensing devices into observation platforms are:

1. **Sensor detection:** Detect a new sensor when it is connected to an observation platform. The host platform controller should be able to detect a new sensor without human intervention.
2. **Sensor identification:** Obtain an unambiguous description of the sensor, including all the metadata to identify the sensor (unique ID, sensor model, etc.) and all information required to register the sensor (see also Section 6.1.4 of this chapter) into a service hosted on a Web server.
3. **Sensor configuration:** This requirement addresses all operations needed before the platform can start retrieving data from the sensor. This includes establishing a communication link between the platform and the sensor and applying any configuration required by the sensor (i.e., activate a specific acquisition channel, set the sampling rate, etc.).
4. **Simple measurements operations:** These operations directly relate to the retrieval of data. These operations may actively query the sensor for data or can listen for incoming sensor data (i.e., data stream). This capability also allows knowledge about the data interface provided by the sensor to be used to parse, process, and store the data.
5. **Sensor registration:** Registering a sensor to an existing Web service requires a considerable amount of metadata, organized and structured in a coherent way, including physical parameters that are measured by the sensor (observable properties), computational representation of the real-world object that is observed (feature of interest), alongside other sensor characteristics. Furthermore, the meaning of these metadata has to be made explicit and understandable by machines, thus controlled vocabularies containing formal definitions shall be used [6].
6. **Data ingestion:** Once the sensor is registered, sensor data have to be ingested by the server, where it will be stored.
7. **Resource constrained:** Any plug-and-play mechanism aimed to integrate sensors into marine acquisition platforms should be able to work in low-bandwidth, low-power, and computationally constrained scenarios.

To fulfill these requirements through the Sensor Web architecture shown in Fig 6.1.2, a set of protocols and standards is used.

6.1.3.1 Open Geospatial Consortium Programmable Underwater Connector with Knowledge Protocol

The requirement of automatic sensor detection can be achieved by using the OGC PUCK protocol within the acquisition platform controller [7]. This protocol can be implemented as an add-on in any serial or Ethernet sensor alongside any proprietary protocol, rather than replacing it. This protocol defines a set of commands that grants transparent access to an internal memory, named *PUCK payload*. This internal memory is frequently used to store sensor metadata. Another key feature of the OGC Puck protocol is its *softbreak* operation, which provides on-the-fly detection without any prior knowledge of the sensor, fulfilling requirement 1. See also Chapter 7.2.

6.1.3.2 Sensor Model Language

The SensorML encodes detailed sensor descriptions within an eXtensible Markup Language (XML)-encoded file [8]. Its main goal is to enhance interoperability by making sensor descriptions understandable by machines and shareable between intelligent nodes. Additional information related to a specific deployment can also be encoded using this standard (e.g., operational history). Thus both sensor configuration and measurement operations can be addressed with the SensorML standard. SensorML is highly flexible and modular as it can describe almost every property related to a sensor or sensor-related process. However, this flexibility and modularity can prove to be a double-edged sword, as the same information can be encoded in different ways, increasing the difficulty to generate smart processes capable of interpreting SensorML definitions. This makes it important to develop profiles to define the domain-specific application of SensorML (currently, work in progress of the informal SWE Marine Profiles Working Group). By combining the OGC PUCK protocol and SensorML so that a device can automatically deliver its description file, it is possible to automatically detect a sensor and retrieve its description encoded in a single standardized file without any a priori knowledge of the sensor. If interpreter software can interpret this file, it can identify the sensor, configure it, and retrieve its data, meeting requirements 2, 3, and 4. For this and the next two standards, see also Chapter 6.2.

6.1.3.3 Sensor Observation Service

The SOS standard provides the set of operations required to register and publish sensor data and metadata [9]. Thus SOS servers usually act as sensor data repositories/archives. This standard also provides interoperable access to real-time and archived sensor data, as well as its associated metadata using Web protocols. There are two parts to the SOS standard: the standard refers to T-SOS (Transactional SOS) action such as *InsertSensor*, *InsertResultTemplate*, and *InsertResult*, which is focused on data delivery. Other operations such as GetCapabilities or GetObservation, etc. are named SOS operations (without the T). Due to its role as a data and metadata access interface, this standard is a key piece of any Sensor Web infrastructure, fulfilling requirement 5.

6.1.3.4 Observations and Measurements

The Observations and Measurements (O&M) standard specifies an abstract model as well as an XML encoding for observations and related data, such as features involved in the sampling process [10]. It provides a uniform and unambiguous way to encode sensor measurements, fulfilling requirement 6.

6.1.3.5 Efficient XML Interchange

The Efficient XML Interchange (EXI) is a World Wide Web Consortium (W3C) format (not an OGC standard) that enables the compression of large ASCII XML files into efficient binary files, significantly reducing their size. This format has been designed to allow information sharing between devices with constrained resources, meeting requirement 7 [11].

6.1.4 IMPLEMENTATION

The following subsections describe the practical implementation of the Sensor Web architecture. While the SWE server and client components are described in Chapter 6.2, this chapter focuses on those components that enable sensor integration.

6.1.4.1 Smart Electronic Interface for Sensors and Instruments Sensor System

To ensure easy integration of sensors into SOS servers, standard services for sensor detection, identification, configuration, and execution of measuring operations have been implemented on the SEISI interface [12], consisting of a set of hardware and software components. An instrument connected to a platform controller (buoy, vehicle, etc.) is a *SEISI-Compliant Sensor System* if these platforms implement the software components shown in Fig. 6.1.3.

The SEISI sensor systems that are deployed on gliders and profiler technologies, used in global observation activities that rely on communication via costly and energy-demanding satellite links with a very low bandwidth and discontinuous link, have these standard services implemented through an OGC SWE Bridge component, located on the observatory platforms. The SEISI interface provides basic standard protocols for sensor detection, identification, configuration, and execution of measuring operations.

In the SEISI sensor system implementation shown in Fig. 6.1.3 there are four distinct functional standard services, namely, (1) SOS, (2) T-SOS, (3) SPS, and (4) PUCK, which can operate on constrained and unconstrained physical interfaces (such as serial or Ethernet). In addition to these standard services, a proprietary protocol service is implemented to cover any other sensor functionalities, which may not be fulfilled by the available standards. In the rest of this section, the functional service is presented to guarantee interoperability among the different parts of the system.

6.1.4.1.1 Data Format

As explained earlier, the OGC SWE architecture sets specific requirements in terms of data accessibility. With respect to data encoding, SWE often relies on XML-based data formats (although alternatives such as JavaScript Object Notation [JSON] are currently considered). Nevertheless, the size of XML messages is often too large for the limited bandwidth of oceanographic open-ocean platforms. Furthermore, the text nature of XML representation makes the parsing of messages by central processing unit (CPU)-limited devices more complex compared to binary formats. For these reasons, the working group of W3C [11] has proposed the EXI format, which makes it possible even for very constrained devices to natively support and generate messages using an open data format compatible with XML. Therefore, EXI schema-less encoding and processing has been implemented for all OGC SOS

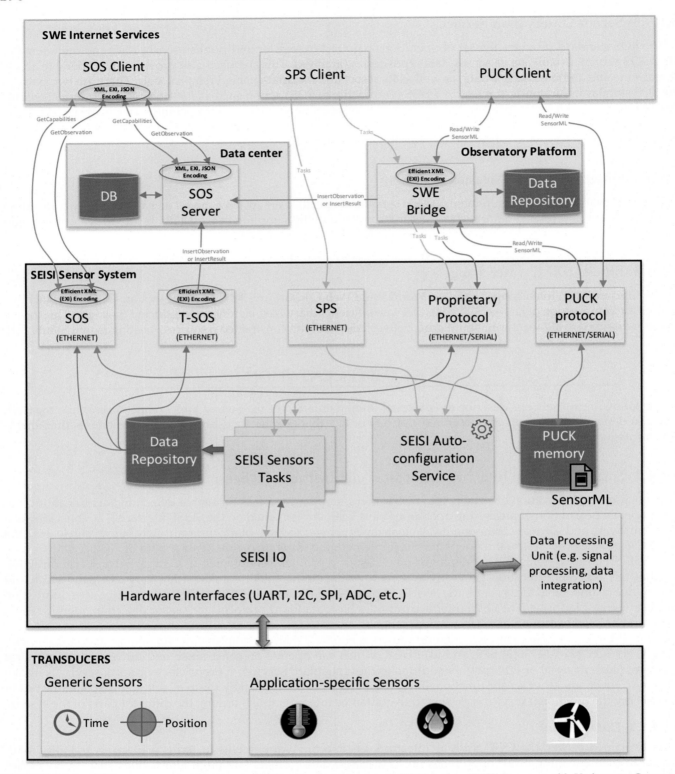

FIGURE 6.1.3 SEISI sensor system architecture and links with other components. *DB*, Data Base; *PUCK*, Programmable Underwater Connector with Knowledge; *SEISI*, Smart Electronic Interface for Sensors and Instruments; *SOS*, Sensor Observation Service; *SPS*, Sensor Planning Service; *SWE*, Sensor Web Enablement.

operations supported by the SEISI sensor system (i.e., *GetCapabilities, DescribeSensor, GetObservation, InsertSensor,* and *InsertResult,* including *OGC O&M* and *SensorML*). Moreover, EXI schema-less encoding is employed for handling SensorML-encoded metadata associated with the SEISI sensor system, provided by the PUCK protocol service [7]. Further details about EXI and schema-less processing can be found in [13].

6.1.4.1.2 Sensor Observation Service

The SOS Web service interface implemented by SEISI sensor systems offers a standard Data Access Service, as illustrated in Fig. 6.1.3.

This implementation was developed to be lightweight for deployment on platforms with limited CPU performance. Consequently, this lightweight SOS server is only implementing the core operations of OGC SOS Specification v2.0 (i.e., *GetCapabilities, DescribeSensor,* and *GetObservation*) (Fig. 6.1.4).

The *GetCapabilities* operation provides access to metadata and detailed information about the available capabilities of the service. By using HTTP GET or POST requests, the service capabilities can be retrieved from the SEISI sensor system encoded as an XML or EXI response. The capabilities response contains metadata about this service, such as information about the interface, the unique sensor identifiers, observation offerings (~data sets) of the SEISI sensor system, and a list of one or more observed properties.

The *DescribeSensor* operation provides access to the sensor metadata encoded in SensorML. The sensor description contains information about the sensor in general, but also details such as calibration data and available communication protocols, interfaces, and data formats of the SEISI sensor system. In this implementation, the SOS server does not generate the SensorML document but retrieves it from the PUCK memory of the SEISI sensor system.

The *GetObservation* operation provides access to the observations made by the sensors. On request, the SOS returns O&M/EXI-encoded observation data generated by the SEISI sensor system.

6.1.4.1.3 Transactional Sensor Observation Service

The T-SOS Web client implemented for SEISI sensor systems provides a standard Data Push Service as shown in Fig. 6.1.5. Similar to the previously introduced SOS server implementation, the SOS interface has been implemented in a lightweight manner for deployment on CPU-limited devices. The T-SOS implementation, which is focused on data acquisition, comprises the *InsertSensor, InsertResultTemplate,* and *InsertResult* operations of the OGC SOS 2.0 specification.

The *InsertSensor* operation is used to publish metadata about a new SEISI sensor system on an SOS server. The publish request can be sent by the SEISI sensor system encoded as an XML/EXI-encoded *InsertSensor* request.

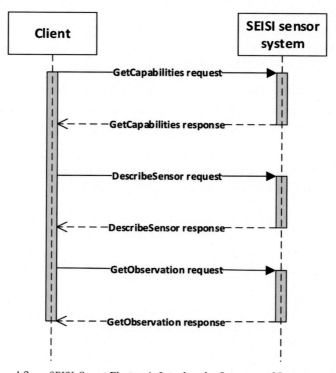

FIGURE 6.1.4   SOS data access workflow. *SEISI,* Smart Electronic Interface for Sensors and Instruments; *SOS,* Sensor Observation Service.

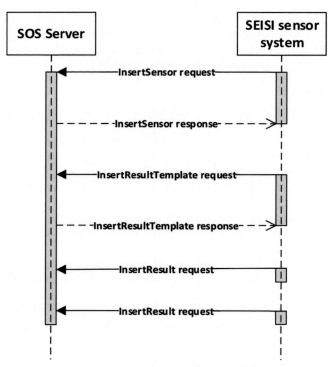

FIGURE 6.1.5   T-SOS Data Push Service procedure. *SEISI*, Smart Electronic Interface for Sensors and Instruments; *SOS*, Sensor Observation Service.

The request contains metadata about this service, such as information about the interface, the unique sensor identifiers, the observation offering of the SEISI sensor system, and a list of one or more quantities observed by this device.

The *InsertResultTemplate* operation is used by the SEISI sensor system for inserting a result template into an SOS server to describe the structure of the values delivered by the device in subsequent *InsertResult* requests. The result template can be sent by SEISI sensor systems encoded as XML/EXI requests. The *InsertResultTemplate* operation is mandatory and has to be executed before an *InsertResult* request.

The *InsertResult* operation is used by SEISI sensor systems for uploading raw values to an SOS server following the structure and encoding defined in the *InsertResultTemplate* request. The T-SOS data publication module sends the SEISI sensor system observations in an EXI/O&M encoding.

6.1.4.1.4 *Programmable Underwater Connector with Knowledge Protocol*

The OGC adopted the PUCK standard in 2012 as a new member of the SWE framework of specifications [7]. Briefly, it defines a set of standard commands that can be used between an instrument and a platform, so that the platform can identify and retrieve metadata about the instrument. OGC PUCK can be considered as a lower-level tier in a hierarchy of standards that achieve interoperability between instruments and Web-based application.

Several manufacturers of the Smart Ocean Sensor Consortium in the United States have implemented the PUCK protocol on their instruments. In Europe, NeXOS partners developed new sensors that include this protocol: TRIOS (optical sensors) [14,15], SMID (hydrophones) [16], and NIVA (carbon cycle). Platform providers such as Alseamar (SeaExplorer Glider), CMR (Sail Buoy), and NKE (Provor float) have successfully tested PUCK protocol support on the platform side.

PUCK provides a formatted electronic datasheet, which contains the information needed to identify the instrument model and manufacturer, as well as a universally unique identifier for each device. A PUCK-enabled instrument may also carry an additional payload such as digital calibration sheets and sensor history. Originally, the PUCK protocol was defined for instruments with an RS232 interface, but the current version extends the protocol to Ethernet interfaces as well. This extended "IP PUCK" protocol includes the use of Zeroconf to enable the easy installation and discovery of sensors in an IP network. The SEISI sensor system architecture uses PUCK and its payload to enable sensor detection, identification, and configuration, as illustrated in Fig. 6.1.6, and to transport a well-defined sensor protocol descriptor encoded in SensorML [8].

With PUCK, the platform host can automatically retrieve and use the SensorML-based sensor protocol descriptor when the device is installed. This "plug-and-play" implementation significantly reduces the efforts needed to integrate new sensors. Building on previous work [2], NeXOS made use of standard sensor protocol descriptions

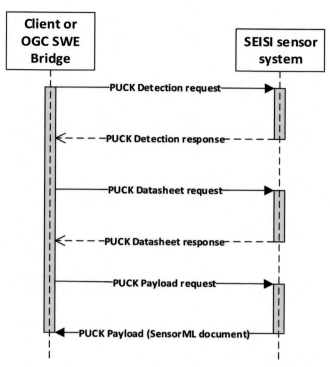

FIGURE 6.1.6 PUCK sensor detection, identification, and configuration services procedure. *OGC*, Open Geospatial Consortium; *PUCK*, Programmable Underwater Connector with Knowledge; *SEISI*, Smart Electronic Interface for Sensors and Instruments; *SWE*, Sensor Web Enablement.

and developed advanced tools to support the creation of those descriptors [17]. This approach also facilitates the integration of sensors into marine observing systems, moving toward plug-and-play-enabled oceanic sensor systems [18]. Therefore a generic platform software component such as OGC SWE Bridge can operate any instrument that is accompanied by a SensorML-based protocol description document, thus eliminating the need for instrument-specific driver software.

New sensors implementing the PUCK protocol will be able to store metadata within the instrument. However, the architecture is at the same time compatible with "conventional sensors." For such conventional sensors the SensorML document with all necessary sensor metadata can be stored directly on the platform. Thus also in this case the SWE Bridge will be configured properly to read the sensor metadata from the instrument (PUCK-enabled instrument) or from a data repository (for non-PUCK-enabled instrument).

6.1.4.1.5 *Smart Electronic Interface for Sensors and Instruments Sensor System Proprietary Protocols*

Even though there have been many incremental steps toward facilitating instrument integration into ocean and costal data management systems, there has not been a standards-based architecture that offers practical end-to-end plug-and-work capabilities. This situation is very common in marine sensor networks that do not have direct access to Web services (e.g., gliders or profiler platforms). Therefore to ensure the integration of the sensors into architectures other than the one introduced in this chapter, the SEISI sensor system concept allows the implementation of all necessary operations for sensor detection, identification, configuration, as well as the execution of measurement operations following a proprietary protocol. For that specific case, SEISI sensor systems have been equipped with a proprietary protocol based on the Standard Commands for Programmable Instrumentation (SCPI) specification [19], which provides software-level syntax and commands for operating instruments over any transport protocol, such as Ethernet and EIA232 (also known as "RS232"). Using the standardized description of the SCPI syntax and commands within SensorML documents, the OGC SWE Bridge can automatically retrieve and use this sensor protocol descriptor when the device is installed on platforms.

6.1.4.2 OGC SWE Bridge

The SWE Bridge is a software component that lets the platform and SEISI sensor system communicate based on the metadata associated with the observations and the sensor metadata encoded in SensorML 2.0. Fig. 6.1.7 shows the full

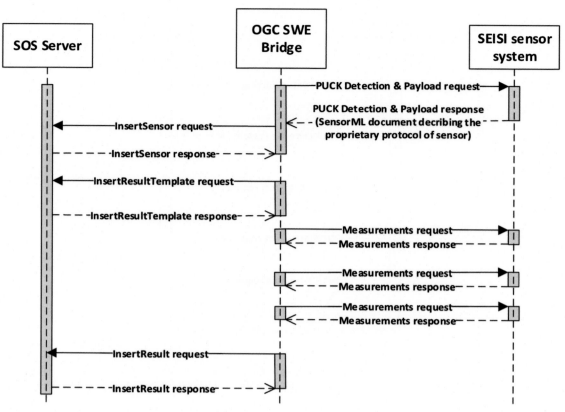

FIGURE 6.1.7   Procedures of the OGC SWE Bridge. *OGC*, Open Geospatial Consortium; *PUCK*, Programmable Underwater Connector with Knowledge; *SEISI*, Smart Electronic Interface for Sensors and Instruments; *SOS*, Sensor Observation Service; *SWE*, Sensor Web Enablement.

functionality of the SWE Bridge. The SWE Bridge is run on the platform controller, such as an autonomous underwater vehicle or a buoy. Using the OGC PUCK tools, the SWE Bridge can be configured to look for new instruments connected to the platform instrument ports. If the connected sensor is a PUCK-enabled sensor it will be automatically recognized by the SWE Bridge after reading the SensorML file stored on the sensor PUCK-dedicated memory. If the sensor is not PUCK enabled, the SensorML file can be stored locally on the platform controller's memory. Once identified, the SWE Bridge has information about which parameters are measured by the instrument, and how the platform controller can retrieve these parameters. The SWE Bridge then registers the sensor on the SOS server (*InsertSensorRequest* and *InsertResultTemplate* operations) to insert measurements (*InsertResult*) when they become available.

6.1.5 EVALUATING EFFICIENT XML INTERCHANGE

This chapter has introduced Sensor Web architecture with an example of its implementation through the NeXOS project for facilitating the integration of sensing devices into Web-based infrastructures. This architecture is able to fulfill the central requirements for establishing an interoperable exchange of oceanographic sensor data via the Web, covering the full chain: sensor–platform–communication link–Web interface. This chapter has focused on the integration of a device and platform level, while the Web-oriented aspects are further addressed in Chapter 6.2.

Underwater mobile platforms and the network they use are highly constrained. This means that the payload packet size is very important. To identify the efficiency of the standard protocols introduced previously, the request size for different specific use cases was analyzed (i.e., registering a new sensor at an SOS server and requesting/inserting sensor measurements). Two scenarios were considered:

- IP-based observatory platforms, such as cabled observatories or ferrybox systems: integration of a smart hydrophone developed in the NeXOS project [16].
- Non-IP-based observatory platforms, such as gliders and profilers, for which tests were developed to evaluate the integration of conductivity, temperature, and depth based on the SEISI concept.

6.1.5.1 Sensor Observation Service (Web)-Based Observatory Platforms

In IP-based observatory platforms, the communication between the SEISI sensor systems and upper layers such as the SOS server is achieved directly through the SOS and T-SOS interfaces. SOS servers running on the instrument provide data access functionality while T-SOS support enables the upload of sensor data. To evaluate the efficiency of the proposed implementation of the SOS and T-SOS, here in a smart hydrophone [16], the size of request/response messages was analyzed. The test data contained sound pressure level measurements and timestamp, encoded in XML, JSON, and EXI. Fig. 6.1.8 shows that EXI and compressed EXI resulted in the smallest response/request sizes since the messages are encoded as bit-packed EXI Body Stream. When EXI compression is applied, the response/request message sizes are at least 50% smaller in comparison with other encoding types.

6.1.5.2 Nonsensor Observation Service (Web)-Based Observatory Platforms

In non-IP-based observatory platforms, the communication between the SEISI sensor systems and upper layers such as SOS servers is achieved through the OGC SWE Bridge service running on the platform (for the evaluation a Slocum glider was used). To evaluate the efficiency of the proposed implementation of the OGC SWE Bridge in the Slocum glider, the size of messages containing timestamps, as well as conductivity, sea water temperature, and depth measurements, was measured for different formats. These formats comprise the Slocum binary format, XML, and EXI.

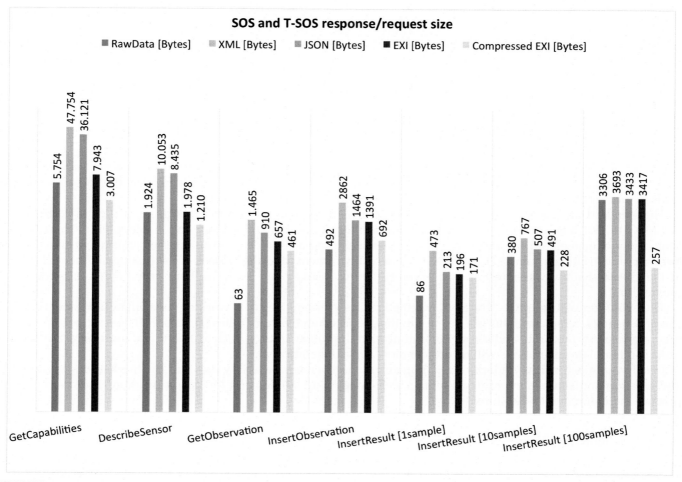

FIGURE 6.1.8 Response/request size evaluation for the SOS and T-SOS operations. *EXI*, Efficient XML Interchange; *JSON*, JavaScript Object Notation; *SOS*, Sensor Observation Service; *XML*, eXtensible Markup Language.

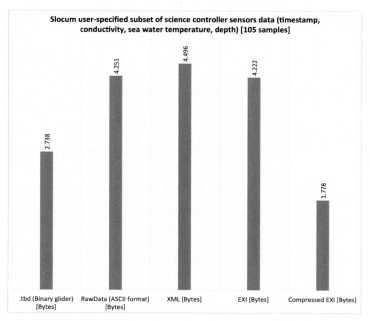

FIGURE 6.1.9 Request size evaluation for the OGC SWE Bridge operations. *OGC*, Open Geospatial Consortium; *EXI*, Efficient XML Interchange; *SWE*, Sensor Web Enablement; *XML*, eXtensible Markup Language.

As illustrated in Fig. 6.1.9, the Slocum binary format and compressed EXI generate the smallest message sizes. When EXI compression was used, the request was at least 35% smaller in comparison with the binary compression used in Slocum gliders to transmit user-specified subsets of science controller sensor data.

6.1.6 CONCLUSIONS AND FUTURE DIRECTIONS

Integration of sensors into marine platforms remains a challenging task. Will ocean sensors be as plug and play in the future as USB peripherals? To achieve such a degree of interoperability, instrument and platform manufacturers have to agree on a set of standards to be used. It is well known that in the mid and long term, the use of standards leads to more robust and cost-efficient systems.

The standard architecture described in this chapter is now being ported to other projects such as EMSODEV or EMSO-LINK, in the framework of EMSO (European Multidisciplinary Seafloor and water-column Observatory, http://www.emso-eu.org/). One of the objectives of such projects is to develop and deploy at different sites a standard set of instruments named EGIM (EMSO Generic Instrument Module). Videos of the first deployment of EGIM at the Obsea shallow water test site observatory (www.obsea.es) are available here: http://www.emsodev.eu/movies.html.

The evolution will continue. The SWE Marine Profiles Working Group is working to facilitate agreement on the use of SensorML templates. This is a key point: the proposed architecture relies on SensorML templates to describe accurately sensor metadata and sensor communication. Improvements on such standard and more accurate definitions and template generations will revert to a more rugged architecture.

Finally and most critically, adoption and implementation of interoperability standards by market actors (platform and sensor manufacturers mainly) will remain a challenge until the demand for standardization becomes common practice in procurement procedures.

Acknowledgments

This research was primarily funded by the NeXOS project. NeXOS is a collaborative project funded by the European Commission 7th Framework Programme under the title OCEAN-2013.2—The Ocean of Tomorrow 2013—Innovative multifunctional sensors for in-situ monitoring of marine environment and related maritime activities, under grant agreement No. 614102. It is composed of 21 partners, including small and medium-sized enterprises, companies, and scientific organizations from six European countries. Visit: www.nexosproject.eu.

References

[1] Waldmann C, et al. The German contribution to ESONET - integrating activities for setting up an interoperable ocean observation system in Europe. 2007.

[2] Del-Rio-Fernandez J. Standards-based plug & work for instruments in ocean observing systems. Ocean Eng 2013;39(3):430–43.

[3] European Commission Inspire, Annex III Sec. 14&15.

[4] Toma DM, et al. Smart sensors for interoperable smart ocean environment. In: OCEANS 2011 IEEE - Spain. 2011.

[5] Dawes N, Kumar KA, Michel S, Aberer K, Lehning M. Sensor metadata management and its application in collaborative environmental research. In: Proceedings - 4th IEEE International Conference on eScience. eScience; 2008. p. 143–50.

[6] Bröring A, Janowicz K, Stasch C, Kuhn W. Semantic challenges for sensor plug and play. In: Lecture notes in computer science (including subseries lecture notes in artificial intelligence and lecture notes in bioinformatics), vol. 5886. LNCS; 2009. p. 72–86.

[7] O'Reilly T, Toma DM, Del-Rio-Fernandez J, Headley K. OGC® PUCK protocol standard. 2012. [Online]. Available: http://www.open geospatial.org/standards/puck.

[8] Botts M, Robin A. OpenGIS ® sensor model language (SensorML) implementation specification. Design 2007:180.

[9] Bröring A, Stasch C, Echterhoff J. OGC sensor observation service. 2012.

[10] Cox S. Observations and measuremets-XML implementation. Measurement 2011:1–66.

[11] Schneider J, Kamiya T. Efficient XML interchange (EXI) format 1.0. W3C Recommendation; 2011.

[12] Toma DM, Del Rio J, Jirka S, Delory E, Pearlman J. Smart electronic interface for web enabled ocean sensor systems. In: 2014 IEEE sensor systems for a changing ocean. SSCO; 2014.

[13] Castellani AP, Bui N, Casari P, Rossi M, Shelby Z, Zorzi M. Architecture and protocols for the internet of things: a case study. In: 2010 8th IEEE international conference on pervasive computing and communications Workshops, PERCOM Workshops 2010. 2010. p. 678–83.

[14] Pearlman J, Zielinski O. A new generation of optical systems for ocean monitoring - matrix fluorescence for multifunctional ocean sensing. Sea Technol 2017. p. 30–3.

[15] Wollschlager J, Voß D, Zielinski O, Petersen W. In situ observations of biological and environmental parameters by means of optics-development of next-generation ocean sensors with special focus on an integrating cavity approach. IEEE J Ocean Eng 2016;41(4):753–62.

[16] Delory E, et al. Developing a new generation of passive acoustics sensors for ocean observing systems. In: 2014 IEEE sensor systems for a changing ocean. SSCO; 2014. https://doi.org/10.1109/SSCO.2014.7000383.

[17] Bröring A, Bache F, Bartoschek T, Van Elzakker CPJM. "The SID creator: a visual approach for integrating sensors with the sensor web. In: Lecture notes in geoinformation and cartography. 2011. p. 143–62.

[18] Bröring A, Maué P, Janowicz K, Nüst D, Malewski C. Semantically-enabled sensor plug & play for the sensor web. Sensors 2011;11(8):7568–605.

[19] SCPI Specification Volume 1–4. [Online]. Available: http://www.ivifoundation.org/docs/scpi-99.pdf.

Further Reading

[1] OGC. ISO 19156:2011. 2011. [Online]. Available: https://www.iso.org/standard/32574.html.

[2] Echterhoff J, Everding T. OpenGIS® sensor event service interface specification (proposed). 2008. [Online]. Available: http://portal.open geospatial.org/files/?artifact_id=29576.

[3] Simonis I, Dibner PC. "OpenGIS sensor planning service implementation specification," Implement. Specif. OGC. 2007. p. 1–21.

[4] Jirka S, Toma DM, Del-Rio-Fernandez J, Delory E. A Sensor Web architecture for sharing oceanographic sensor data. In: Sensor systems for a changing ocean (SSCO), 2014 IEEE. 2014.

C H A P T E R

6.2

From Sensors to Users: A Global Web of Ocean Sensors and Services

Simon Jirka, Matthes Rieke, Christoph Stasch

52°North Initiative for Geospatial Open Source Software GmbH, Münster, Germany

6.2.1 INTRODUCTION

A broad range of different organizations and institutions is collecting marine observation data with different objectives and technological approaches. Typical data sources are research vessels, (autonomous) gliders, and fixed-point observatories such as buoys. These platforms observe a multitude of variables relying on different types of sensor technology (e.g., electrochemical sensors, optical sensors, acoustic sensors, etc.) from a large number of manufacturers. Different communication channels (e.g., cables, mobile phone networks, satellite links, or manual transport of storage media) transfer the data to central data archives. Furthermore, these data archives follow very different technological strategies such as file-based data storage, relational databases, NoSQL databases such as MongoDB (https://www.mongodb.com), and proprietary data archiving tools. This may lead to an isolated set of data silos. If marine scientists want to use these data sets, it is rather easy for them to access the data in the infrastructure of their own institution. However, what if an analysis needs more data for the same sea region that was collected by other organizations? In this case, an easy flow of data between observatories and organizations is desirable, but usually not available yet. This chapter introduces the topic of Sensor Web technology, which uses the World Wide Web to facilitate the cross-organizational sharing of marine observation data on a global scale. First, we will provide a general introduction into the topic of Sensor Web technology followed by a more detailed presentation of the closely related Sensor Web Enablement (SWE) standards. After that, the implementation of a Sensor Web infrastructure within the European NeXOS project will be illustrated as an example. This is complemented by an overview of further related projects. Finally, this chapter concludes with a summary of practical experiences and recommendations for future work.

6.2.2 SENSOR WEB TECHNOLOGIES

6.2.2.1 What Is the Sensor Web?

The term Sensor Web was originally created in a different context. Delin defined a Sensor Web as a network of (autonomously) interacting sensors to transmit the collected data [1]. However, with the increasing use of the World Wide Web as a global communication network, the meaning of the term Sensor Web has become more inclusive [2]. Within this chapter, the term Sensor Web refers to the use of the World Wide Web to share sensors and their observations through a range of Web technologies such as eXtensible Markup Language (XML), JavaScript Object Notation (JSON), HTTP, Representational State Transfer (REST), and Web services. In this way, the World Wide Web is used to link observation data archives and sensing hardware to all consumers who need the data from these sources for their work. As a result, data available on the Sensor Web is directly available to any user interested in this data. This

---

LINKING OCEAN OBSERVING SYSTEMS TO GLOBAL INFRASTRUCTURES

There are several international efforts to build large-scale spatial data infrastructures not only on regional and national levels but also on a continental (e.g., Europewide) or even global scale. In nearly all of these efforts, the Sensor Web complements conventional geospatial data with near real-time as well as archived observation data.

Global Earth Observation System of Systems (GEOSS) is a global initiative maintained by the Intergovernmental Group on Earth Observations. Its aim is to build a global infrastructure for sharing and generating comprehensive, near-real-time environmental data and analyses. For this purpose GEOSS comprises a set of independent but coordinated systems for Earth observation, information, and processing. Besides other aspects, observation data sets are an important element of GEOSS. Thus Sensor Web technology is a relevant building block, especially for handling in situ observation data. A first implementation of a link between Sensor Web technology and GEOSS relying on a brokering approach was developed within the GEOWOW project to handle hydrological observation data [3].

Infrastructure for Spatial Information in the European Community (INSPIRE) is a European legal, organizational, and technical framework for sharing geospatial information within the European Community. It is based on a directive of the European Union [4] that is complemented by several implementing rules and technical guidance documents. It requires European member states to provide access to a defined set of geospatial information through interoperable, Web-based technologies. In this context, Sensor Web technology is the recommended approach to share observation data in an interoperable manner.

Copernicus (formerly known as Global Monitoring for Environment and Security) is a program of the European Union to build a European capacity for Earth observation. Besides Earth observation satellites, Copernicus comprises in situ data sources. On top of these data sources, Copernicus also offers processed data and a broad range of information products. Copernicus operates a marine component called the Copernicus Marine Environment Monitoring Service. Currently, the in situ component of Copernicus is not yet as advanced as the remote sensing component. However, in the future it is likely that Copernicus will also rely on Sensor Web technology for in situ observation data.

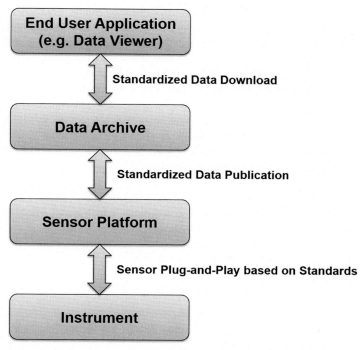

FIGURE 6.2.1 Different layers in marine applications that benefit from Sensor Web technology.

idea is closely related to the concept of spatial data infrastructures, which are actively developed in many regions and even on a global level to increase the availability of geospatial data (e.g., maps, satellite images, vector data) via the Web. Thus, the Sensor Web can be considered as an extension of spatial data infrastructures for handling sensor observation data.

6.2.2.2 The Sensor Web in an Ocean Sensing Technology Stack

Marine sensing systems and applications may benefit from Sensor Web technology on different levels and in multiple ways. This is illustrated in Fig. 6.2.1.

Chapter 6.1 introduced the application of Sensor Web standards for the integration of sensors into different platforms based on metadata standards that describe the interfaces of all kinds of sensing devices. In this chapter we focus on the transportation of observation data from such platforms to users. As Fig. 6.2.1 shows, Sensor Web technology helps to improve this flow of information on two different levels: from the sensors/sensor platforms to data archives and from the data archives to the end-users of the collected data. Before we explain how these data flows are implemented in practice, we will first have a look at the technological foundations by introducing the Open Geospatial Consortium (OGC) SWE framework, which provides specifications and standards for building interoperable Sensor Web applications.

6.2.3 SENSOR WEB ENABLEMENT

The previous section introduced the idea of Web-based infrastructures for sharing (marine) observation data. To enable the efficient creation of such infrastructures it is necessary to rely on a common language and protocol for the interaction of the different components (e.g., data sources, data consumers, catalogs). For this purpose, standardization is a central element.

6.2.3.1 Why Do We Need Standardization?

Observation systems often use very heterogeneous data access interfaces and formats. Thus the integration of new observation data into different applications would require dedicated customization of data-consuming software to the specifics of the observation systems. Considering large-scale infrastructures that rely on the World Wide Web to connect observation data sources from multiple organizations (e.g., marine research institutes from multiple countries), the use of common standards becomes an even more critical issue. For example, there are many different institutes that operate observatories in the Mediterranean Sea. If each organization offers its data through different interfaces (e.g., different Web service interfaces, FTP servers) and formats (e.g., different XML formats, Comma-Separated Values, Network Common Data Format, binary formats), the creation of data viewers or analysis tools that combine all the data sources available becomes a cumbersome task: it is necessary to develop software modules for querying each single data access interface and for decoding every data format that is used. Fig. 6.2.2 illustrates this challenge: the number of software modules that need to be developed is the result of the multiplication of the numbers of data access interfaces, data formats, and data consumers. As soon as these numbers increase it becomes obvious that this causes a significant overhead just for customizing the software that consumes the data.

Standardization helps to avoid these large overheads. Fig. 6.2.3 shows how the integration overheads are reduced if standards for data access interfaces and formats are used: in this case, each data source and customer would need to support only one single standardized data access interface and format. These standards serve as a common language to facilitate the interoperable communication between data archives/observation systems and client applications such as data viewers, analysis tools, catalogs, etc. As soon as a data consumer is able to handle the common standards, it is able to consume observation data from all sources that support these standards.

A common framework of standards for building such interoperable Sensor Web infrastructures is the OGC SWE suite of standards.

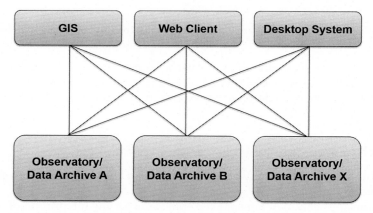

FIGURE 6.2.2   Situation without applying standards: high integration efforts needed for linking heterogeneous components. *GIS*, geographic information system.

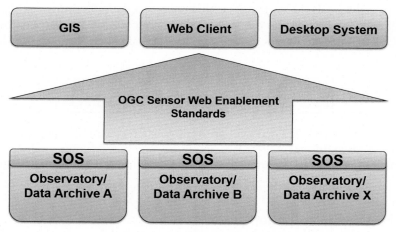

FIGURE 6.2.3   Situation if standards are used: components can be integrated as soon as the relevant standards are once implemented by the component. *GIS*, geographic information system. *OGC*, Open Geospatial Consortium; *SOS*, Sensor Observation Service.

6.2.3.2 The Open Geospatial Consortium Sensor Web Enablement Standards

The OGC SWE standards were developed by the corresponding Sensor Web Enablement Domain Working Group of the OGC with the intention to enhance spatial data infrastructures with functionality for handling sensors and observation data. Since the first version of the SWE standards, which were published about 10 years ago, several iterative, evolutionary improvements based on practical applications and experiences have led to a stable suite of standards that cover the requirements of a broad range of applications [2]. Besides marine applications, very common domains that rely on SWE standards comprise hydrology [5], environmental monitoring [6], and emergency/ disaster management [7].

The OGC SWE standards can be divided into three groups: common foundations of all standards, data model/ encoding standards, and interface standards.

Because the SWE suite comprises multiple data formats and interfaces, it is important to ensure consistency across the different standards. This is achieved on the one hand by the OGC SWE Common Data Model [8], which defines basic data building blocks (e.g., arrays, data records, and simple data types) reused by all other SWE specifications. On the other hand, the OGC SWE Service Model [9] defines common principles of the interfaces of SWE services (e.g., the DescribeSensor operation as a method to download metadata about sensors). Based on these common standards, there are two important specifications, which deal with data models and formats.

The OGC Sensor Model Language (SensorML) [10] in its most recent version 2.0 defines a model and XML encoding for describing observation processes such as sensor platforms, systems, instruments, and detectors but also nonphysical processes such as simulation models or processing steps. Thus SensorML is of high value to provide metadata about ocean-observing systems. A specific feature of SensorML is the ability to describe not only instances of sensing hardware but also provide information about a certain device type. This enables an inheritance mechanism so that the description of a device instance may inherit the properties of the corresponding device type description. As a result, SensorML enables the interpretation of observation data sets because the user is able to evaluate the technology and steps that have led to a certain measurement value. Furthermore, SensorML allows the building of discovery applications for marine sensors and platforms, e.g., catalogs and yellow pages [11]. Finally, SensorML enables the description of sensing device interfaces (e.g., commands, parameters, and outputs) so that plug-and-play solutions for connecting to these devices may be built.

Complementary to SensorML, the ISO/OGC Observations and Measurements (O&M) standard (current version is 2.0) offers a data model and encoding for the observation data delivered by sensing devices. The O&M standard comprises two specifications: on the one hand, the abstract data model (e.g., based on UML diagrams) is defined by an ISO standard [12]; on the other hand, the encoding of this data model as XML documents is specified by a corresponding OGC standard [13]. This approach leads to a higher level of flexibility, because in the future the abstract O&M data model could also be translated to other encodings such as JSON. This ensures that the O&M model can remain stable while different implementations of new O&M data encodings can be easily developed without breaking the fundamental concepts of O&M.

The SWE interface standards rely on the data model and format specifications to deliver different types of sensor-related functionality. In the core of the SWE framework, the OGC Sensor Observation Service (SOS) 2.0 standard [14] is an interface for offering access to sensor data. This interface comprises several operations that can be used to download observation data based on a broad range of thematic, temporal, and spatial filters, to retrieve the metadata about observation systems that have generated the offered observation data sets, and to publish observation data in an interoperable manner. Thus the SOS interface can be considered as a central element to enable the Web-based distribution of collected observation data. If an organization publishes its observation data through the SOS interface, other users may rely on this interface to download the offered data for any kind of further analysis, processing, or visualization.

While the SOS interface focuses on providing access to collected observation data, there is often as well a need to adjust the settings/tasks of a sensing platform to perform certain observation tasks. For example, a glider may be tasked with a route it should follow to measure certain variables along the track. Or, in a simpler scenario, the sampling frequency of a water temperature sensor should be changed. This requires the submission of these tasks to the corresponding sensors. The OGC Sensor Planning Service (SPS) interface standard 2.0 [15] covers exactly this functionality. It offers operations for submitting and managing (e.g., changing, aborting) tasks but also to check the status of task execution or the feasibility of specific tasks (e.g., to check if a glider is available and not otherwise committed so that it can follow a certain track). As the SPS interface is only intended for managing sensor tasks, it does not offer access to the results (i.e., the data) gathered during the tasks. Instead, it offers functionality to determine through which endpoint (usually SOS servers) the resulting data can be downloaded. In past years the focus of Sensor Web activities in the marine domain was especially directed to enabling data access, so that no significant marine SPS implementations have been done. There are cases in marine observations where this may be used such as in reprogramming for real-time response to unusual observations. However, for now, other application domains such as Earth observation have already given stronger consideration to the SPS specification. For the marine domain, further evaluation will be necessary.

Finally, it is necessary to mention the OGC Publish/Subscribe standard 1.0 [16]. This standard is not an element of the OGC SWE suite but as it enables push-based communication patterns for OGC Web services, it is a valuable complementary element. While the SOS follows a pull-based communication (request–response) pattern the Publish/Subscribe communication model pushes observation data to previously registered subscribers as soon as it is available. This specification can be used to deliver new observations of a certain sensor to a set of consumers without any delay. The typical workflow relying on the Publish/Subscribe standard requires that interested data consumers subscribe to certain data offered by a server. The server has then the responsibility to check new incoming data with regard to these subscriptions and to forward the matching data to the corresponding subscribers. This workflow is illustrated in Fig. 6.2.4.

Table 6.2.1 provides a summarizing overview of the SWE standards relevant for marine application scenarios.

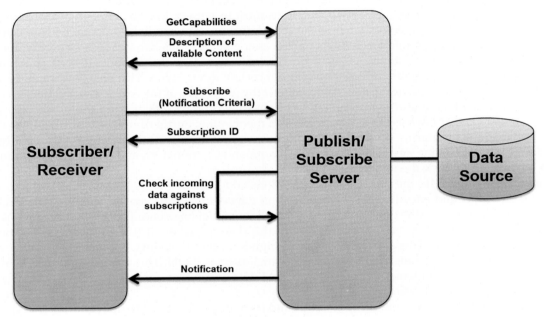

FIGURE 6.2.4 Publish/Subscribe workflow.

TABLE 6.2.1 Overview of SWE Standards Relevant for the Marine Domain

Standard	Abbreviation	Current Version	Purpose
OGC SWE Common Data Model	–	2.0	Common building blocks for all SWE standards (e.g., common data types)
OGC SWE Service Common	SWES	2.0	Common functionality and behavior of all OGC SWE interface standards
ISO/OGC Observations and Measurements	O&M	2.0	Model and (XML) encoding for observation data
OGC Sensor Model Language	SensorML	2.0	Model and (XML) encoding for metadata about observation processes (including sensors)
OGC Sensor Observation Service	SOS	2.0	(Web service) interface for pull-based access to observation data and related metadata
OGC Sensor Planning Service	SPS	2.0	(Web service) interface for controlling observation processes (e.g., configuring sensors)

OGC, Open Geospatial Consortium; *SWE*, Sensor Web Enablement.

In the context of integration and harmonization of European Spatial Data Infrastructures, it has to be noted that the ISO/OGC O&M as well as the OGC SOS standards are also part of the European INSPIRE framework. Through dedicated INSPIRE technical guidance documents, these standards are officially recommended to be used as a data model/format for observation data [17] and as the INSPIRE Download Service [18].

6.2.4 SENSOR WEB IMPLEMENTATION

This section describes the implementation of a comprehensive Sensor Web infrastructure in the European NeXOS project as an exemplar. The objective of NeXOS was to develop new, cost-effective, innovative, and compact integrated multifunctional sensor systems, which can be deployed from mobile and fixed ocean-observing platforms, as well as to develop downstream services for the Global Ocean Observing System and to support European policies.

6.2.4.1 Sensor Web in the NeXOS Project

Within the NeXOS project a range of innovative sensor and sensor platform technologies was developed [19–21]. A central objective of NeXOS was to address not only the sensor development itself but also to increase the value of the new sensor technologies by enabling a seamless plug-and-play data flow to the users via the World Wide Web. The first building block of this chain is the Smart Electronic Interface for Sensors and Instruments (SEISI) firmware described in Chapter 6.1, which is deployed on sensing hardware or platforms to enable the automatic configuration of new sensors and internal sensor–platform communications. Thus the challenge of NeXOS was to enable the flow from the SEISI firmware to Sensor Web servers and from there to users. Furthermore, a submission of tasking/configuration change requests to the sensors was needed. The following two subsections describe how these two workflows were implemented and successfully demonstrated using the OGC SWE standards.

6.2.4.1.1 Data Publication

The flow for data publication within NeXOS is illustrated in Fig. 6.2.5. The first transmission step is from the sensor platforms to a central data repository realized by an SOS server (based on the 52°North SOS 4.x development line).

The publication comprises three steps. In a first step, the SOS InsertSensor operation is called by the SEISI firmware running on the platform. With this method, the sensor platform registers itself or a sensor attached to it at the SOS server by submitting a SensorML document containing a description of the platform and/or sensor. Based on this information, the SOS server is capable of relating the data published in the future to the specific platform or sensor. For publishing the measured data, the workflow relies on the result-handling operations of the SOS specification. The reason for selecting these operations instead of the InsertObservation operation offered by the SOS standard is the more compact data representation with the result handling operations that avoids the transmission of redundant data elements. To use these handling operations, it is first necessary to inform the SOS server about the structure

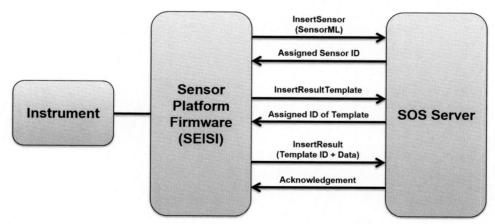

FIGURE 6.2.5 NeXOS data publication workflow. *SEISI*, Smart Electronic Interface for Sensors and Instruments; *SensorML*, Sensor Model Language; *SOS*, Sensor Observation Service.

of the observation data that will be published and about all properties that remain the same for all observations delivered by the platform or sensor (e.g., the identifiers of the variables and in case of stationary observatories the feature of interest that defines the geolocation of the sensor). The transmission of this common information happens by calling the InsertResultTemplate operation of the SOS server. After that, the SOS server is able to interpret all subsequently transmitted observation data messages. To submit new observation data, the SEISI firmware of the sensor platform will then regularly call the InsertResult operation offered by the SOS server. Updates of the sensor configuration may also be sent by the platform to the SOS server. The corresponding SensorML-based metadata documents are registered through the UpdateSensorDescription operation while new data structures are described in updated result templates submitted by calling again the InsertResultTemplate operation. It is important to note that in the

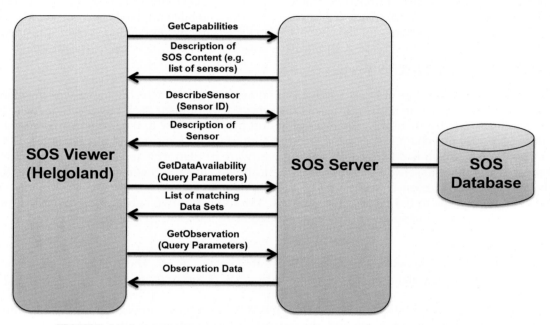

FIGURE 6.2.6   NeXOS data access workflow for users. *SOS*, Sensor Observation Service.

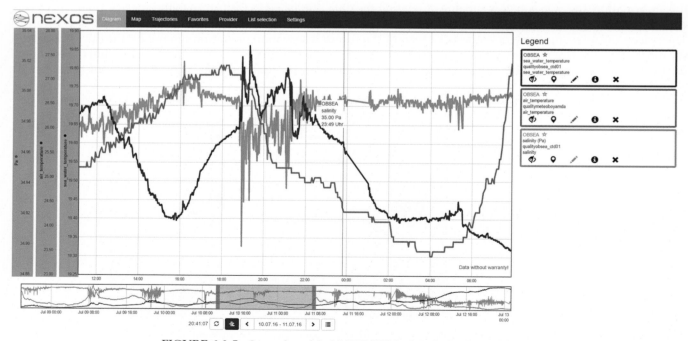

FIGURE 6.2.7   Screenshot of the NeXOS/52°North Helgoland SOS viewer.

example of NeXOS that we are using for illustration here, the project has developed a complementary way to transmit SWE messages in a more compressed manner. Many typical SWE-based infrastructures rely on XML-based messages. However, in the NeXOS project, the integration of sensors into platforms with low-bandwidth communication capabilities required a more compact data transmission. Thus the communication between SEISI and SOS servers did not rely on a plain XML format but on a compressed version of binary encoded XML messages as defined by the Efficient XML Interchange format of the World Wide Web Consortium (see Chapter 6.1 for details). This reduces the communication overhead, which is especially critical for sensor platforms with limited connectivity (e.g., gliders communicating via satellite links).

The second important flow of information is the delivery of observation data to users such as scientists. For this purpose, another part of the SOS interface is used (Fig. 6.2.6). The first operation that is called in this context is the GetCapabilities operation of the SOS. Through this operation it is possible to download an overview of contents available on a specific SOS server. A central element in this context is a listing of all so-called ObservationOfferings of the SOS server, which can be considered as metadata about the data sets. To retrieve more detailed metadata about certain data sets, clients may call the DescribeSensor operation that delivers SensorML-encoded metadata about the sensor(s) that has/have generated the data set. In addition, INSPIRE-compliant SOS servers may offer the so-called GetDataAvailability operation, which queries SOS servers for data sets that fulfill certain user-defined criteria (e.g., searching all data sets that contain salinity data for a specific time period). After these operations have been used to identify a suitable data set, the GetObservation operation of the SOS server is used, which allows downloading of O&M-encoded observation data that fulfills a set of request criteria (e.g., temporal filters, spatial filters, observed variables, etc.).

As the SOS interface is based on machine-readable formats such as XML, users do not usually interact with SOS servers directly. Instead, data viewers or analysis tools are used, which hide the complexity of the SOS interaction from the user. An example of such a tool is the 52°North Helgoland SOS viewer, which is based on HTML and JavaScript (Fig. 6.2.7). It can be executed on a broad range of devices (e.g., smartphones, tabled PCs, desktop PCs) and provides users with convenient functionality for browsing through the data offered by an SOS server, to view the data (e.g., as diagrams or on maps), and to download the data for further processing and analysis. More information on the viewer and the code are available at https://github.com/52North/helgoland.

6.2.4.1.2 Sensor Tasking

In some cases it may be necessary to change the measurement process of a sensor or a platform so that it delivers the data a user needs. This could comprise, for example, the adjustment of the sampling rate of a sensor or the activation/deactivation of a device. For controlling measurement processes, the OGC SPS is a suitable Web service interface. However, for NeXOS a process was needed that allowed sensors to actively check if there were changes of settings for them. For example, a glider can only check for parameter adjustments if it is at the sea surface so that it can use a satellite communication link. Thus the following communication workflow was designed (Fig. 6.2.8):

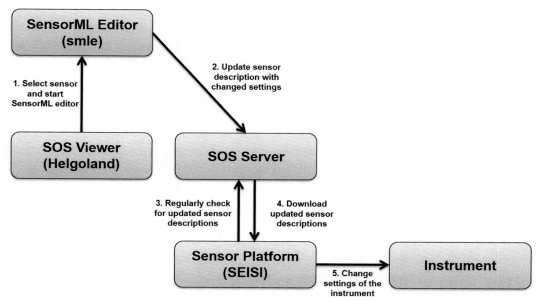

FIGURE 6.2.8 NeXOS sensor-tasking workflow. *SEISI*, Smart Electronic Interface for Sensors and Instruments; *SensorML*, Sensor Model Language; *SOS*, Sensor Observation Service.

A user may select a certain sensor via the Helgoland SOS viewer. If configuration adjustments are performed for this sensor, the user can log in and open the SensorML editor "smle" developed by 52°North (see later) to modify certain sensor parameters within a SensorML document. If the user has completed the adjustments of the settings, the updated SensorML document is submitted via the UpdateSensorDescription operation to an SOS server. This SOS server is regularly queried by the SEISI firmware of the sensor platform through the DescribeSensor operation to check if updated SensorML documents for the platform or sensors attached to the platform are available. If an updated SensorML document has been found, SEISI checks it for adjusted settings and updates the internal configuration of the platform or sensor.

6.2.4.1.3 Enabling Interoperable Access to Existing Data Archives

Besides the new Sensor Web-enabled sensors developed in the NeXOS project, there are many existing ocean-observing systems that rely on different technologies for data collection and dissemination. However, it is important also that these existing, non-SWE-compliant data sources can be integrated into the Sensor Web infrastructure implemented by NeXOS to allow common visualization and analysis. For this purpose, the 52°North SOS implementation can be deployed not only on top of these sensor systems but instead as interfaces for databases containing observation archives. In this way, data archives can also be added as a valuable source of information to a global Sensor Web. Fig. 6.2.9 illustrates this approach: the 52°North SOS offers several ways to consume data from existing databases. The most important step in this process is to achieve a mapping between the internal data model of the 52°North SOS and the structure of the databases. One typical way is to create a so-called object relational mapping, which defines which elements of the SOS data model (e.g., measurements, timestamps, sensors, features, etc.) can be found in which table and column of the database. As this mapping is not always directly feasible (e.g., complex database joins may be necessary to map elements of a data model to multiple tables in a database) another approach is to define database views in the original database that are equivalent to the default data model of the 52°North SOS implementation.

Both approaches have been successfully evaluated in practice and the decision of which approach to use needs to be taken based on the specifics of each individual deployment scenario. Within NeXOS, the approach of creating database views was applied to a database containing FerryBox data maintained by the Helmholtz-Zentrum Geesthacht. Fig. 6.2.10 shows the visualization of the data delivered by this SOS server in the Helgoland SOS viewer.

6.2.4.1.4 Provision of Metadata

The suite of Sensor Web components developed and applied in NeXOS is completed by the 52°North SensorML editor smle. This tool has been developed as a joint effort of the NeXOS and FixO3 projects. It offers functionality for

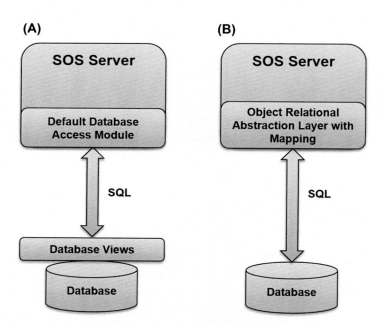

FIGURE 6.2.9   Different approaches to link existing databases as data source to an SOS server. (A) Deployment of default SOS version on top of database that exposes views resembling the default SOS data model. (B) Deployment of SOS on original database with object relational mapping that translates between the default SOS data model and the model of the database. *SOS*, Sensor Observation Service; *SQL*, Structured Query Language.

entering and managing a comprehensive set of metadata about sensor platforms and sensors. A link to the FixO³/ Esonet Yellow Pages allows the importing of sensor type descriptions so that common metadata for all sensors of the same type can be prefilled by the editor. Management of the SensorML metadata documents is achieved through an instance of the 52°North SOS server implementation. Fig. 6.2.11 shows a screenshot of the smle user interface.

6.2.4.1.5 Experiences

Experiences with Sensor Web technology in the NeXOS exemplar show that there are ways to significantly reduce the overheads for integrating new observation platforms and sensors into Web-based infrastructures. On the one

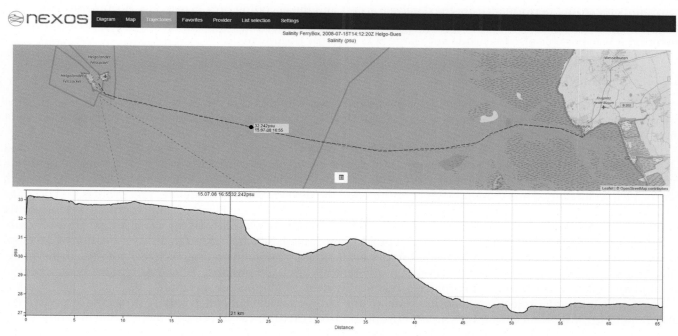

FIGURE 6.2.10   Ferrybox data displayed in the NeXOS/52°North Helgoland SOS viewer.

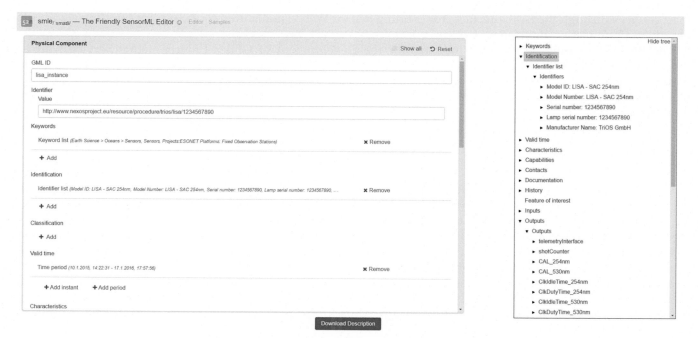

FIGURE 6.2.11   Screenshot of the 52°North smle, SensorML editor.

hand, the SEISI developments of the Universitat Politècnica de Catalunya demonstrate that it is possible to automatically register new sensors and to upload their data to SOS servers without human intervention. As soon as new sensors or data have been published, data viewers can immediately access the new information resources. For example, if a new sensor is attached to an observatory, a data viewer will automatically display it as soon as the viewer connects to the SOS server. The efficiency of data transmission (i.e., data volume and scalability) was also evaluated positively: the developed approach can run on powerful cable-based connections but works as well for resource-constrained devices such as gliders that send their data through very expensive and bandwidth-constrained satellite communication links. The scalability for delivering data to client applications was not evaluated in detail. However, the SOS servers were tested with data volumes up to 100 million observations in their internal databases. The main constraints regarding performance are the database access and the encoding of SOS response documents. However, by applying streaming-based XML encoding (i.e., without creating a tree representation of the full XML document in memory) it was possible also to encode very large SOS response documents.

6.2.4.2 Further Projects

Besides the NeXOS project, which has been the exemplar used in this chapter, there is a broad range of ongoing projects that deal with the application of Sensor Web technology in the marine domain. For conciseness, a subset of these projects is described briefly in this section.

The two European projects FixO^3 (Fixed point Open Ocean Observatory, http://www.fixo3.eu/) network and ODIP II (Ocean Data Interoperability Platform, http://www.odip.eu/) contributed to the SWE development in a cross-cutting manner. Both projects aimed at promoting the interoperable sharing of marine observation data through Sensor Web technology. Thus these projects contributed to the development of guidelines and profiles that describe how to apply the SWE standards in ocean observing systems. In addition, the FixO^3 project contributed significantly to an editor for sensor metadata based on SensorML (52°North smle), as well as tools for visualizing marine observation data (e.g., 52°North Helgoland). The ODIP II project contributed especially through the coordination of development activities that resulted in a series of prototypes evaluating and demonstrating SWE technology. Furthermore, ODIP II contributed to a harmonization between different SWE-related projects, as well as to the advancement of vocabularies needed, for example, to provide semantically interoperable sensor metadata.

The SeaDataCloud project is a continuation of the previous SeaDataNet I and II projects (http://www.seadatanet.org/). Within SeaDataNet, a pan-European infrastructure was developed by national oceanographic data centers and research institutions that connects more than 100 marine data centers. This infrastructure provides functionality such as discovery of and access to data resources for researchers. The SeaDataCloud project will also rely on Sensor Web standards for providing sensor metadata, for offering INSPIRE-compliant download services through the SOS interface, for creating Sensor Web visualization tools, and for building standards-based data publication interfaces.

The BRIDGES (Bringing together Industry for the Development of Glider Environmental Services) project (http://www.bridges-h2020.eu/) has a more specific focus. Its scope is the development of an autonomous glider platform for marine research. Within this more general scope, the BRIDGES project also deals with the development of recommendations for how SWE standards should be applied for efficient usage with the marine glider platforms.

6.2.5 FUTURE WORK

There are several projects, such as NeXOS and FixO^3, which have developed metadata editors for the SensorML-based description of marine sensing devices. Based on such editors, the provision of sensor metadata is already easier. However, further improvements of usability of SensorML editing tools will be a continuing priority. Furthermore, there should be stronger use of the resulting metadata documents. Easy-to-use search engines for sensors and their collected data, as well as the integration of metadata into visualization and analysis tools for observation data, would be another priority to further increase the value of marine data and metadata.

Many Sensor Web standards are based on rather heavyweight technologies such as SOAP and XML. New paradigms such as REST and JSON, ideally in combination with concepts such as Linked Data, may help to further reduce communication and development overheads. Thus REST/JSON bindings of the relevant OGC SWE standards are an area of future work.

As mentioned in other chapters of this book section, new standards and protocols such as Message Queue Telemetry Transport (MQTT) or the OGC SensorThings API are emerging from the field of Internet of Things (IoT) technologies. As these technologies offer new opportunities for enabling more efficient and lightweight communication patterns, it is worth investigating further their applicability to marine sensing applications. For example, MQTT-based communication flows might offer a complementary means for linking marine sensing platforms with data archive servers. Necessary extensions to these technologies might comprise modifications to handle the advanced requirements for metadata in scientific application contexts. However, with these extensions, IoT technologies are likely to be a central part of the next evolution step of marine Sensor Webs.

The new OGC Publish/Subscribe standard was released in 2017. At the moment the first prototypical implementations are available, but further evaluations and most likely advancement of the Publish/Subscribe specification are needed. This comprises, for example, a REST/JSON binding for the Publish/Subscribe standard (currently only a SOAP binding is available), as well as advancements to better accommodate the needs of event (stream) processing applications (e.g., the generation of higher-level information such as events derived from incoming data streams needs further refinements of the standard).

Finally, projects such as ODIP II, FixO³, BRIDGES, and NeXOS have undertaken steps for the creation of vocabularies to increase the semantic interoperability in marine Sensor Webs. This has led to already comprehensive vocabularies that can be used for semantically referencing, for example, from SensorML-based metadata documents. However, software tools making use of these capabilities (e.g., metadata editors, discovery tools) are not yet widely available. Thus besides further enhancing the existing vocabularies, it will be necessary to develop software packages that make use of semantic interoperability through terms defined in vocabularies.

6.2.6 CONCLUSION

This chapter introduced the Sensor Web and presented an approach for using the OGC SWE standards as a technology for implementing interoperable Sensor Webs. The NeXOS project demonstrated how SWE can be applied in practical marine sensing systems facilitating the Web-based delivery of data from the hardware via data archives to the users. During the last few years, SWE standards have proven to be a reliable and stable technology that reduces integration efforts for linking ocean sensing platforms to data servers and for building end-user applications that allow the discovery, visualization, and analysis of marine observation data. The available open source implementations can be reused and further enhanced by all interested parties. This is, for example, the case with European projects such as FixO³, ODIP II, BRIDGES, and SeaDataCloud, which also make a significant contribution to the advancement of the introduced concepts and software packages. Thus Sensor Web and SWE are highly recommended for all operators of marine sensing equipment that want to enable a reliable and fast delivery of collected data to users with a low amount of necessary integration work.

For everyone who is interested in a more detailed tutorial on how to apply Sensor Web technology, a Sensor Web tutorial has been developed by 52°North: https://52north.github.io/sensor-web-tutorial/.

Glossary

BRIDGES Bringing together Research and Industry for the Development of Glider Environmental Services
CMEMS Copernicus Marine Environment Monitoring Service
CSV Comma-Separated Values
EXI Efficient XML Interchange
FixO³ Fixed point Open Ocean Observatory network
FTP File transfer protocol
GEOSS Global Earth Observation System of Systems
GEOWOW GEOSS Interoperability for Weather, Ocean and Water
GMES Global Monitoring for Environment and Security
GOOS Global Ocean Observing System
HTTP Hypertext transfer rotocol
INSPIRE Infrastructure for Spatial Information in the European Community
IoT Internet of Things
ISO International Organization for Standardization
JSON JavaScript Object Notation
MQTT Message Queue Telemetry Transport
NetCDF Network Common Data Format

NeXOS Next generation, Cost-Effective, Compact, Multifunctional Web Enabled Ocean Sensor Systems Empowering Marine, Maritime and Fisheries Management
O&M Observations and Measurements
ODIP Ocean Data Interoperability Platform
OGC Open Geospatial Consortium
REST Representational State Transfer
SDI Spatial Data Infrastructure
SEISI Smart Electronic Interface for Sensors and Instruments
SensorML Sensor Model Language
SOS Sensor Observation Service
SPS Sensor Planning Service
SWE Sensor Web Enablement
UPC Universitat Politècnica de Catalunya
W3C World Wide Web Consortium
XML Extensible Markup Language

Acknowledgments

The work on this chapter was cofunded through the European NeXOS (Next generation, Cost-effective, Compact, Multifunctional Web Enabled Ocean Sensor Systems Empowering Marine, Maritime and Fisheries Management) project. NeXOS is funded by the Seventh Framework Programme (FP7) for Research and Innovation (FP7-OCEAN-2013) of the European Union under grant agreement number 614102.

References

[1] Delin K. Sensor webs. NASA Tech Briefs; 1999. p. 80.

[2] Bröring A, Echterhoff J, Jirka S, Simonis I, Everding T, Stasch C, et al. New generation sensor web enablement. MDPI Sens March 1, 2011;11(3):2652–99.

[3] Andres V, Bredel H, Busskamp R, Jirka S, Looser U, Schlummer M, et al. Interoperability between GRDC's data holding and the GEOSS infrastructure. In: HIC 2014-11th International Conference on Hydroinformatics; 17 August 2014-21 August 2014. New York City (NY, USA): CUNY Academic Works; 2014.

[4] Directive 2007/2/EC of the European Parliament and of the Council of 14 March 2007 establishing an Infrastructure for Spatial Information in the European Community (INSPIRE). 2007.

[5] GEOWOW Consortium. In: OGC discussion paper: OGC sensor observation service 2.0 hydrology profile (OGC 14–1004). Wayland (MA, USA): Open Geospatial Consortium Inc; 2014.

[6] Jirka S, Bröring A, Kjeld P, Maidens J, Wytzisk A. A lightweight approach for the sensor observation service to share environmental data across Europe. Trans GIS May 27, 2012;16(3):293–312.

[7] Stasch C, Walkowski AC, Jirka S. A geosensor network architecture for disaster management based on open standards. In: Digital Earth Summit on Geoinformatics 2008: tools for global change research; 12 November 2008 -14 November 2008. Potsdam (Germany): Wichmann; 2008.

[8] Robin A. OGC Implementation specification: SWE common data model 2.0.0 (08–094r1). Wayland (MA, USA): Open Geospatial Consortium Inc.; 2011.

[9] Echterhoff J. OGC implementation specification: SWE service model 2.0.0 (09–001). Wayland (MA, USA): Open Geospatial Consortium Inc.; 2011.

[10] Botts M, Robin A. OGC implementation specification: sensor model language (SensorML) 2.0.0 (12–1000). Wayland (MA, USA): Open Geospatial Consortium Inc.; 2014.

[11] Jirka S, Bröring A, Stasch C. Discovery mechanisms for the sensor web. MDPI Sens April 16, 2009;9(4):2661–81.

[12] ISO TC 211. ISO 19156:2011–Geographic information – observations and measurements - International Standard. Geneva (Switzerland): International Organization for Standardization; 2011.

[13] Cox S. OGC implementation specification: observations and measurements (O&M) - XML implementation 2.0 (10–025r1). Wayland (MA, USA): Open Geospatial Consortium Inc.; 2011.

[14] Bröring A, Stasch C, Echterhoff J. OGC implementation specification: sensor observation service (SOS) 2.0 (12–1006). Wayland (MA, USA): Open Geospatial Consortium Inc; 2012.

[15] Simonis I, Echterhoff J. OGC implementation specification: sensor planning service (SPS) 2.0.0 (09–000). Wayland (MA, USA): Open Geospatial Consortium Inc; 2011.

[16] Braeckel A, Bigagli L, Echterhoff J. OGC implementation standard: publish/subscribe interface standard 1.0-Core (13–131r1). Wayland (MA, USA): Open Geospatial Consortium Inc; 2016.

[17] INSPIRE MIG sub-group MIWP-7a. Guidelines for the use of observations & measurements and sensor web enablement-related standards in INSPIRE - version 3.0. Ispra, Italy. INSPIRE Maintenance and Implementation Group (MIG); 2016.

[18] INSPIRE MIG sub-group MIWP-7a. Technical guidance for implementing download services using the OGC sensor observation service and ISO 19143 filter encoding - version 1.0. Ispra, Italy. INSPIRE Maintenance and Implementation Group (MIG); 2016.

[19] Jirka S, Toma DM, del Rio J, Delory E. A Sensor Web architecture for sharing oceanographic sensor data. In: Sensor systems for a changing ocean (SSCO) 2014 at the sea tech week 2014; 13 October 2014-17 October 2014. Brest (France): IEEE; 2014.

[20] Pearlman J, Jirka S, del Rio J, Delory E, Frommhold L, Martinez S, et al. Oceans of tomorrow sensor interoperability for In-situ ocean monitoring. In: 2016 MTS/IEEE OCEANS''16; 19 September 2016-23 September 2016. Monterey (CA, USA): IEEE; 2016.

[21] Toma DM, del Rio J, Martínez E, Jirka S, Delory E, Pearlman J, et al. Applying ogc sensor web enablement to ocean observing systems. In: GSW 2016-geospatial sensor webs conference 2016; 29 August 2016-31 August 2016; Münster, Germany. 2016. CEUR-WS.org.

CHAPTER

6.3

Evolving Standards and Best Practices for Sensors and Systems—Sensors

Sergio Martinez[1], Jay Pearlman[2,3], Eric P. Achterberg[4], Ian Walsh[5], Eric Delory[6]

[1]LEITAT Technological Center, Barcelona, Spain; [2]FourBridges, Port Angeles, WA United States; [3]IEEE, Paris, France; [4]GEOMAR Helmholtz Centre for Ocean Research Kiel, Kiel, Germany; [5]Sea-Bird Scientific, Philomath, OR, United States; [6]Oceanic Platform of the Canary Islands (PLOCAN), Telde, Spain

6.3.1 INTRODUCTION

The use of standards is fundamental for enabling interoperability across networks and observation systems. Standards are also critical for understanding the conditions under which data were taken and how the sensors were calibrated. Without this knowledge, providing data to models and to end-users could lead to information that is neither valid nor useful for applications. However, even by following standards, a stable calibration is not guaranteed as the ocean environment is harsh and sensors can be impacted by biofouling or physical and/or thermal stress that can change even the best calibration over time. In optimum circumstances, sensors can be recalibrated at the end of a mission and then before and after mission calibrations can be compared to examine changes that have occurred. There are, however, missions such as Argo, where the floats and sensors are not retrieved. Thus for such operations, standards that support performance and calibration stability are even more important.

Another motivation for standards and best practices is supporting applications that address the societal impact of oceans and the marine environment. Over the last decade, there have been agreements on the need for ocean sustainability. In Europe, these include, for example, the Marine Strategy Framework Directive (MSFD) and the Common Fisheries Policies. These motivated the development of new sensors to meet a Good Environmental Status by 2020 and to try to fully understand the effects of climate change on the marine environment, as well as minimizing pollution and unsustainable use.

A third factor in sensor evolution and the need for standards and best practices is the increasing complexity of sensors and the ability of new ones to monitor conditions that were heretofore inaccessible. With the interest in climate change, many years ago the focus for sensor development was on sensors that could measure physical conditions: temperature, salinity, pressure (depth), and underwater noise. During the last decade, there has been increasing interest in the chemistry and biology of the oceans; the monitoring of oxygen, nutrients, and pH is needed to understand ecosystem dynamics. With this, sensors became more complex with greater demands for interoperability. Currently, sensor observations are now being added involving the monitoring of organism numbers and reproductive vitality. For example, new sensors were developed and validated by the European Oceans of Tomorrow projects [1] and the IOOS/OOI networks (http://www.whoi.edu/ooi_cgsn/sensors) in the United States and elsewhere. There is a natural evolution because the advances in lightweight electronics, memory capability, and high-speed communications with low costs have changed the measurement paradigms in many areas, as illustrated by the chapters in this book.

Looking forward, in our current digital era the information society is replete with new concepts that affect both sensors and data such as the evolving concepts of Big Data, linked data, and the Internet of Things (IoT). We can now count hundreds of millions of devices that can communicate to Web platforms or data systems through large area networks (Ethernet, WiFi, satellite, etc.). In the long term, sensors without access to communications or those that are human operated at a detailed level are becoming beautiful museum pieces.

Thus we see a technology environment with a rapid evolution potential, which must reach across domains of physical oceanography, biogeochemistry, and biology/ecosystem sciences where observation practices, data formats, quality standards, and even observation resolution vary across domains. Coming back to the issues of interoperability, there is a need for a basis describing the way sensors should perform, what methods should be used for their calibration, how measurement uncertainties are assessed, and what interfaces need to exist to have the data flow from sensor to systems to networks and ultimately to data repositories and users. Standards and best practices are evolving and will play increasingly important roles in supporting interoperability and the reliability of marine sensor observations.

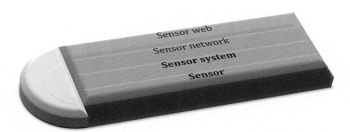

FIGURE 6.3.1   Sensors in the observations system environment.

Fig. 6.3.1 shows sensor functional interfaces going from observations to data communication to data archives. This chapter focuses on the lower layers relating to sensor standards and Chapters 6.1 and 6.2 address the communication, network, and Web interfaces.

- Sensor: An entity that reacts to some sort of stimulus, for instance physical, chemical, or biological.
- Sensor system: A number of sensors aggregated as a single unit, providing a single access interface.
- Sensor network: Collection of deployed sensor resources (sensors and/or sensor systems) connected by an internal communication link, proving to be an easy-to-use external link.
- Sensor Web: An open complex adaptive system organized as a network of open sensor resources, which pervades the internet and provides external access to sensor resources. By open sensor resources, we include any open system, including sensor networks that are a source of sensor data or sensor metadata.

6.3.2 STANDARDS AND BEST PRACTICES

For sensors, standards address the specifications and processes necessary for their outputs to be fit for purpose. Standards may be either technical or regulatory. Regulatory standards, which may be technical or policy related, are often issued by governments and compliance is required. The mandatory regulations define a benchmark with which products must comply to gain access to a market while meeting the authorities' and consumers' demand for safe and high-quality products. Sometimes regulatory standards can be an obstacle to market access for companies in developing countries, and new or changing technical regulations can create unnecessary barriers.

Technical standards are formulated through a consensus process, typically among technical experts, which may involve social as well as technical aspects of the issues at hand. Because of the complex groups of stakeholders for many applications, the process of creating a standard is slow. The end benefit, however, can be significant through improved interoperability and having a process that supports fitness for purpose. The International Organization for Standardization (ISO) defines itself as: "creating standards documents with requirements, specifications, guidelines or characteristics that can be used consistently to ensure that materials, products, processes and services are fit for their purpose." (https://www.iso.org/standards.html).

In addition to standards, there is an array of best practices that are in use for ocean observation. A definition of best practices was created by a best practices working group and validated at an international forum in 2017 [2]:

> A community best practice is a methodology that has repeatedly produced superior results relative to other methodologies with the same objective.

Best practices may come in any of a number of format types—best practices, standard operating procedures, manuals, operating instructions, etc.—with the understanding that the document content is put forth by the provider as a community best practice and it has been used by more than one organization. Like standards, the concept of fitness for purpose is seen in the definition through the use of the statement of superior results with the "same objective." This is important as measurement techniques in the cold Arctic may be different to the warm tropics for a similar measurement objective such as monitoring of temperature or nutrients.

The difference between standards and best practices is the formality of their formulation and their breadth of support across the international community, with ISO and other standards organizations having more formal processes for achieving consensus. There is also a difference in the time to create standards and best practices with the latter generally taking less time and being easier to update and evolve. That, of course, is both a benefit and a potential weakness as the stability of an observation methodology over time is important so that measurements can be compared. Best practices are not always accessible when they are created within an organization and not published openly. This is a challenge that is being addressed through the expansion of a best practice repository at the International Oceanographic Data and Information Exchange" (IODE) of the "Intergovernmental Oceanographic Commission" (IOC) of UNESCO (UNESCO/IOC-IODE [3]).

For standards, there are well-recognized organizations that create and are central repositories for standards. These include:

IEC: The International Electrotechnical Commission is a not-for-profit, quasigovernmental organization founded in 1906. IEC is a leading global organization of international standards for electrical, electronic, and related technologies. It works collaboratively with ISO and the International Telecommunication Union, as well as with regional and national organizations to coordinate global standards. IEC has experts from industry, government, academia, and consumer groups that participate in formulating and disseminating standards.

IEEE: The Institute of Electrical and Electronics Engineers is an association composed of engineers and scientists and allied professionals. For this reason the organization is known simply as IEEE. It has about 430,000 members in 160 countries. IEEE addresses standards for electronic interfaces and communications.

ISO: The International Organization for Standardization is an independent, nongovernmental organization, the members of which are the standards organizations of the 162 member countries. It is the world's largest developer of voluntary international standards and facilitates world trade by providing common standards between nations. Nearly 20,000 standards have been set covering everything from manufactured products and technology to food safety, agriculture, and healthcare.

OGC: The Open Geospatial Consortium, an international voluntary consensus standards organization, originated in 1994. In the OGC, more than 500 commercial, governmental, nonprofit, and research organizations worldwide collaborate in a consensus process encouraging development and implementation of open standards for geospatial content and services, Sensor Web and IoTs, geographic information system data processing, and data sharing.

There are standards across the value chain of sensors to user information. The foregoing standards organizations and others collaborate for selected standards so that there is improved uniformity in definitions and process descriptions. The remainder of this chapter will focus on those addressing the sensors and metrology processes for high-quality observations.

6.3.2.1 Sensor and Metrology Standards

The need for increased confidence in analytical measurements has gained prominence in recent years, as data produced by the oceanographic community and regulatory organizations have become more critical for informing the decisions of governments and international organizations on the functioning of the global climate system, such as the Intergovernmental Panel for Climate Change [4], and marine water quality.

The approaches for method validation for instrumental laboratory techniques as suggested by IEC and ISO can be used in combination or individually and include [5]: (1) interlaboratory comparisons of samples; (2) calibration using reference standards; (3) comparison of independent analytical methods; (4) systematic assessment of the influencing factors; and (5) assessment of the uncertainty of results. The transfer of these method validation approaches to sensor measurements is not necessarily straightforward. The deployment at a single location of multiple sensors from different laboratories forms an alternative sensor validation approach, and the sensors can be different makes or models, or similar. However, such comparisons are a large investment and may be used when a new type of sensor is being introduced, but are not likely to be justified for sequential calibration of production sensors.

The validation of sensor measurements is not straightforward and our approaches are still underdeveloped. Predeployment calibration of the sensor using traceable standards and subsequent analyses of certified reference material are required. The determination of repeatability of a sensor predeployment through replicate measurements on a single day of a sample or standard with a matrix that matches the seawater environment is furthermore important. Even better is performance verification through repeat measurements over longer periods of time, which provides an assessment of precision over intermediate (days to months or longer) time periods.

The repeatability (or intermediate precision) analysis and measurement of certified reference material again needs to be conducted postdeployment to allow for assessment of changes in precision and also bias of the sensor during the deployment period. The pre- and postdeployment sensor calibration and verification measurements should be stored as these data will allow for the assessment of the performance of an individual sensor unit over its lifetime, but also comparison of data between different sensor systems for the same analyte.

TABLE 6.3.1

Variable (Essential Ocean Variables)	Sensor	Pre- and Postcalibrations Needed	Intermediate Calibrations Needed
Temperature	CTD	R	R
Conductivity (salinity)	CTD	R	R
Phosphate	LOC	Y	Y
Nitrate	LOC	Y	Y
Nitrate	Hyperspectral	Y	N
Oxygen	Optode	Y	Y
pH	Optode	Y	Y
pCO$_2$	Optode	Y	Y
Sound	Hydrophone	Y	N
Marine mammal signature	Hydrophone	R	N

R (recommended) is given in the pre- and postcalibration because for the most part sensor drift is slow and steady and offsets are easily seen, so if there is no offset then the user should follow the recommended service interval. For all of the instruments in this table, efforts should be made to confirm instrument operation through analysis of contemporaneous samples. Y indicates calibration needed and N indicates not needed. *CTD*, Conductivity, temperature, and depth; *LOC*, lab-on-a-chip.

Most laboratory instruments are calibrated on a daily basis, with certified reference materials preferably analyzed at a similar frequency. For in situ ocean sensors the pre- and postcalibrations comparisons are not always possible. For a single measurement from a ship or in a coastal observation, the duration of the observation may be short, hours or a day. If the calibration and intermediate precision are important, these can be addressed for such intervals. This approach is not possible for in situ sensors with pre- and postcalibrations that could possibly be 1 year apart as is typical for fixed point observatories or long duration autonomous operations. The issues of platform availability are pervasive for ocean observing and thus for calibration. Earth-observing space-based systems, which have a similar challenge of lack of physical access, address calibration through built-in calibration techniques and use of remote sensing of known locations on Earth. Frequency of calibration and access to sensors is a subject that will be touched on briefly in this chapter and is a focus of continuing efforts for the ocean sensor and system community. For example, for Argo floats, the sensor is not retrieved and postcalibration opportunities do not exist. For some observations, such as exposure of oxygen sensors to atmospheric oxygen, the community has been addressing calibrations throughout the mission. For many of the sensors, widely accepted solutions do not yet exist.

There are some sensors that do not need intermediate calibration such as plankton imagers and so each sensor class must be addressed based on its process. Some examples are provided in Table 6.3.1.

6.3.3 CALIBRATION AND PERFORMANCE METHODOLOGIES—EXAMPLES

This chapter provides three exemplars of sensor classes that are used in ocean observations: (1) conductivity, temperature, and depth (CTD) sensors that monitor conductivity (salinity), temperature, and pressure; (2) passive acoustic sensors for noise and biology; and (3) chemical sensors for biogeochemistry.

6.3.3.1 Exemplar for Physical Conductivity, Temperature, and Depth Measurements

The CTD instrument is the backbone measurement system in oceanography as it provides the physical measurements that supply the context for all other measurements made in the ocean. Collecting the highest accuracy data from a CTD requires careful handling of the instrument and attention to calibration. High precision calibration is often critically important as gradients in both time and space can be extremely small.

To calibrate a sensor, it is placed in a precisely controlled environment (Fig. 6.3.2). The output of the sensor is collected at the same time as the environment is measured with a reference sensor. The reference sensor is carefully calibrated and has a well-known history. To gain the careful calibration and history, the reference is calibrated against physical standards such as the triple point of water and the melting point of gallium, or an agreed-upon standard such as International Association for Physical Sciences of the Ocean (IAPSO) standard seawater.

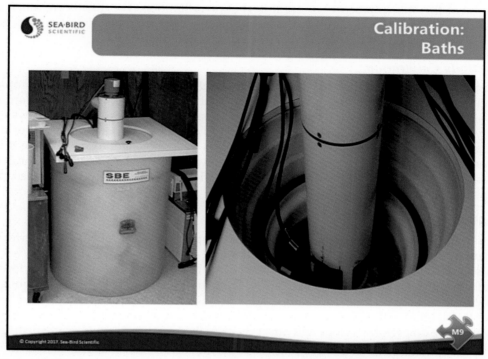

FIGURE 6.3.2 This bath design is common to all of Sea-Bird's calibration activities. The baths are highly insulated and well stirred, and they typically hold temperature to better than 0.0005°C.

Sea-Bird uses circulating baths for all of its calibration activities. The baths are highly insulated and well stirred, and they typically hold temperature to better than 0.0005°C. Sample ports in the baths allow for drawing of water samples for salinity without interfering with the stirring or creating temperature transients. In addition, the baths are equipped with the means to change the partial pressure of dissolved oxygen for oxygen sensor calibrations.

6.3.3.1.1 Temperature Calibration

For calibration, one runs into the problem of knowing exactly what the temperature of a particular object is. If we only had thermometers to rely on, how do you know which one is right? Instead, we use physical standards. The Celsius temperature scale decrees that water freezes at 0°C and boils at 100°C; however, the freezing and boiling points are subject to uncertainties such as atmospheric pressure. So, instead of the freezing and boiling points, we use two other points:

- The triple point, the temperature at which water exists as a liquid, a vapor, and a solid. The triple point of water is measured in a specially constructed cell that contains no air, only H_2O, and occurs at 0.010000°C. Because of a pressure effect, the temperature at the depth where we actually take the measurement is 0.00997°C
- The melting point of extremely pure gallium, 29.764600°C. Because of a pressure effect, the temperature at the depth where we actually take the measurement is 29.76458°C. This pins down the other end of the oceanographic scale.

We calibrate platinum reference thermometers at these points and then calibrate reference SBE 3 temperature sensors with the platinum thermometers. This allows us to trace the temperature measurement used to calibrate all other thermometers back to the physical standards.

Fixed point cells owe their name to the fact that when they are in the proper condition their temperature is fixed by the physics of the materials they are constructed of to be a single temperature. The triple point cells are maintained in a water bath very near their natural temperature. This allows them to last a long time. The gallium cells are melted slowly in an oven; the temperature where the gallium changes phase from solid to liquid is used as the calibration temperature.

As was previously mentioned, a platinum thermometer is calibrated in the fixed point cells and then used to calibrate the SBE 3 reference thermometers. The platinum thermometer is susceptible to calibration shift due to impact or vibration; because of this it is impractical to use it in routine calibration. The SBE 3s are much more robust. By careful selection of the SBE 3 and the accumulation of a drift history, very accurate calibrations can be accomplished. The reference SBE 3s are then used in the baths to calibrate production instruments.

6.3.3.1.2 Conductivity Calibration

Unlike temperature, a primary standard for the conductivity of seawater is more difficult to come by. In recognition of this, IAPSO commissions the Ocean Scientific International Corporation to provide standard seawater. Ocean Scientific sends ships out into the North Atlantic with large tanks to collect seawater. The seawater is filtered and adjusted in salinity to be 35.000. It is then sealed in vials or bottles and shipped to laboratories worldwide to be used in standardizing laboratory salinometers. Because everyone uses the same water to standardize their salinometers, we are all synchronized with Ocean Scientific. The standard seawater service has been active for decades under the auspices of various committees of scientists. It was first produced by a laboratory in Copenhagen and is sometimes referred to as Copenhagen water.

6.3.3.1.3 Pressure Calibration

For instruments that have a strain gauge pressure sensor, a complete pressure calibration is performed at Sea-Bird, using our Digiquartz pressure sensor as a secondary standard.

For instruments that have a Digiquartz pressure sensor, a true calibration of the sensor is performed by the pressure sensor manufacturer. The quality of the Digiquartz is such that an adequate calibration requires a local gravity survey and dead weight tester parts that are certified by the National Institute of Standards and Technology. These requirements, plus the stability of the Digiquartz sensor, do not make the maintenance of this capability cost effective for Sea-Bird. However, we do perform a slope and offset check of the pressure sensor in these instruments, using our Digiquartz pressure sensor as a secondary standard.

6.3.3.1.4 Postcalibration

While thermometers are very robust and low maintenance, they still require regular calibration to make sure they are on their historical drift trajectory. A sensor that has a surface that interacts with the seawater, such as conductivity or dissolved oxygen, is another matter. These require careful handling, attention to calibration, and field calibration to assure the highest quality data.

SBE 3 thermometers are essentially trouble free. They are mechanically robust and are unaffected by extremes in temperature up to 60°C. Temperature sensors tend to drift in offset—that is, the measurements drift in a uniform way over the entire range of measurement. For temperature sensors, the drift direction is dependent on the instrument electronics, and is unique to each temperature sensor. This drift typically continues in the same direction for the entire life of the instrument.

Conductivity sensors usually lose sensitivity as they drift. The drift takes the form of a slope. This is because the conductivity measured by the cell depends on the cell dimensions, which typically change due to fouling. Note that conductivity cell drift is often episodic rather than linear. This is because fouling events often cause the most significant drift. Perhaps the sensor passes through an oil film when it enters the water, or sits on deck in a warm place full of seawater, growing bacteria on the cell surface. Conductivity slope is determined by calculating calibration coefficients using data from one calibration date and applying the coefficients to data from another calibration date.

6.3.3.2 Exemplar for Passive Acoustic Measurements and Metrology

Passive acoustic measurements at sea cover a number of interests, from assessment of noise as a nuisance to population estimates of marine organisms, and seismics/geohazards. Therefore the sensors for each of these applications may have different requirements. Guidance is provided to ensure quality measurements are scattered and differ in terms of reference types. Requirements may also differ across regions, mainly because of factors such as water column stratification, salinity, local gradients, and ambient noise levels. Applications using the measurement of underwater sound as an oceanographic variable are also still emerging.

For underwater noise measurements, recommendations have been released in Europe from the Technical Study Group for Noise to support the European MSFD [6–8], providing monitoring guidance and background information for the monitoring of noise in European seas. Considerations are made on equipment, calibration, deployment, data storage, and metadata. A comprehensive resource is also available from the National Physical Laboratory in the United Kingdom [9,10] and several standards addressing a number of purposes have been and are currently being developed, again revealing the aforementioned diversity in approaches versus requirements [11].

When it comes to making measurements in the open ocean from autonomous systems, reliability concerns and the uncertainties with respect to quantifying noise (sound pressure levels) and bandwidth lead to the choice of robust multipurpose systems. Calibrations pre- and postdeployment, as for any other measuring device, are highly recommended. In situ calibration during deployment is rarely performed yet should be a consideration when a mobile system can be used that can remain steady in the vicinity of a fixed system, e.g., where one has a detector and the other a known noise source, or for intercomparison with a reference hydrophone. A recent contribution was also made on how to measure underwater sound for different purposes (noise and marine mammal monitoring) in fixed open-ocean observatories, in the framework of the FixO³ project [12].

Hydrophone calibration standards were reviewed in Refs. [9,10]. An example of a standard is IEC 60565:2007, which describes different methods for hydrophone calibration to cover different frequency ranges. The methodology consists of a combination of measurements by comparison in air for low frequencies (20 Hz to 5 kHz) and in water for higher frequencies (above 5 kHz) (Fig. 6.3.3). Several precautions need to be taken before calibration, one of them being the proper wetting of the transducing element to avoid the formation of thin bubble layers that may affect the results by several decibels. However, maximum uncertainty generally comes from the reference microphone (when in air) or hydrophone, which can often reach ±2–3 dB.

While the foregoing calibration methods are generally performed in laboratory facilities, handheld calibrators are sometimes available for a limited number of frequencies and intensities that allow for hydrophone verification before and after deployment. These are often based on microphone calibrators and hydrophone couplers. Note, however, that not all hydrophone systems have a handheld calibrator that can be acquired off the shelf, which is to be considered and checked with the manufacturers prior to the purchase of lower-cost systems.

In situ calibration or validation is fraught with several difficulties, as for most in situ sensors, which motivates pre- and postdeployment calibration or verification. For example, the foregoing mobile versus fixed platform calibration or intercomparison faces several sources of uncertainties. One is distance, as both systems are difficult to maintain within a short and fixed range, where wavefronts may differ. Another is linked to the differences in potential interferences (mechanical, electrical) from the host platforms.

6.3.3.3 Exemplar for Chemical Measurements

Metrology for chemical laboratory measurements is well developed, with a range of available technical and statistical approaches and certified reference materials. In many cases, national institutes have been instrumental in the development and dissemination of best practices for laboratory measurements, with institutes including the

FIGURE 6.3.3 Tank verification, here of the NeXOS A1 hydrophone. For medium frequencies (several kHz), pre- and postdeployment calibration can be complemented with comparison with reference hydrophones in a controlled environment, here with a RESON TC4032. Ideally, both hydrophones would be swapped to have the exact location in the tank to avoid mutual masking. *Credit Ludovic Fouille.*

National Physical Laboratory (UK), Physikalisch-Technische Bundesanstalt (Germany), and the National Institute of Standards and Technology (USA). We currently see a strong emphasis on improvements of measurement metrology for oceanographic chemical measurements, which until recently were not a major focus. The realization that the oceans play a critical role in our climate system, and provide us with a range of important ecosystem services, is leading to the development of enhanced observational capabilities, which require new metrology approaches for oceanographic measurements.

In recent years there have been significant advances in the development of analytical techniques for oceanographic chemical measurements, and the motivation to increase the number of constituents (analytes) measured in seawater via innovative analytical applications continues to grow. For example, novel lab-on-a-chip (LOC) miniaturized sensor systems have been developed in collaboration with the National Oceanographic Center's sensor group for phosphate, Fe, and pH on research cruises and time-series sites (Fig. 6.3.4). However, with an increasing number of established methods adapted for global oceanographic studies, more emphasis is now required on the determination and documentation of accuracy, repeatability (within laboratory), and reproducibility (between laboratories) and more rigorous uncertainty estimates to accompany reported data.

Chemical measurements in the oceans are challenging for many of the analytes because of their low concentrations, and the seawater matrix often interferes with the instrumental analysis. The analytical techniques therefore have to be very sensitive and selective, or some form of sample preconcentration has to be utilized prior to detection of the analyte. In some cases, the salt matrix interferes with the chosen instrumental detection techniques, as it does for observing trace metal analysis at subnanomolar concentrations using flow injection chemiluminescence techniques. Then matrix removal approaches are to be used, which in many cases also result in the preconcentration of the analyte and hence an improvement of the detection limit. Matrix removal and preconcentration techniques for marine chemical sensors has been difficult to establish and only successfully mastered for trace metal analyses using electrochemical systems [13]. Consequently, for a range of analytes, sensors are not sufficiently sensitive to allow for in situ measurements.

We can learn a great deal from the approaches developed to assess uncertainty in chemical measurements in the laboratory, and will have to assess whether they can be transferred to sensor measurements in the ocean. It is common practice for the uncertainty estimate of a chemical measurement to be based on the instrumental precision associated with the analysis of a single sample. However, this can result in an important underestimation of the uncertainties as the precision is determined over a short period of time and hence ignores intermediate and long-term errors, and also overlooks bias as no accuracy assessment using certified reference materials is included. We therefore require more sophisticated approaches for method validation for both instrumental laboratory measurements and in situ sensor measurements.

The approaches for method validation for instrumental laboratory techniques include: (1) interlaboratory comparisons; (2) calibration using reference standards or certified reference materials; (3) comparison of results

FIGURE 6.3.4   Laboratory calibration of lab-on-chip phosphate sensor. The figure shows overall set-up with in situ phosphate sensor, calibration standards, and laptop computer for control. *Produced by National Oceanography Centre Southampton, UK; Credits for the photos: Mr. Dominik Jasinski.*

achieved with independent analytical methods; (4) systematic assessment of the factors influencing the results; and (5) assessment of the uncertainty of the results based on practical experience and scientific understanding of the theoretical principles of the method. The approach equivalent to (1) can be done by deploying multiple sensors from different laboratories at a single location and the sensor can be different makes or models, or similar. This approach has been utilized successfully during comparison exercises of novel sensor systems for trace metal analyses (VIP, Idronaut) in the Gullmar fjord in Sweden [14], where a range of similar sensor units was deployed and their performance assessed and compared. In addition, in Kiel fjord [15,16] a range of sensors was simultaneously deployed, also with multiple pCO₂ and pH sensors using different measurement principles (i.e., fluorescent optode measurements [17] and LOC spectrophotometric analysis for pH [18]). These studies also compared the results obtained using the in situ sensors with results obtained from laboratory analyses (approach (3)) of discrete samples that were collected at regular intervals (c. 2 h) during sensor deployment. The laboratory analyses had been verified using certified reference materials.

The in situ sensor validation studies in Gullmar and Kiel fjords also involved the deployment of sensors for standard oceanographic variables, such as temperature, salinity (conductivity), depth (pressure), and oxygen. This approach (4) allowed the novel chemical sensor observations to be interpreted and verified in the context of oceanographic observations of, e.g., density and oxygen concentrations. The assessment of uncertainty of the sensor measurements should also be based on the long-term experience with the sensor device, with considerations for its underlying measurement principles (5). The in situ utilization of certified reference materials (approach (2)) has to our knowledge so far not been attempted.

As mentioned earlier, for chemical sensors that are deployed remotely on moorings or autonomous vehicles (gliders, Argo floats, wave gliders), the verification of the performance of the analytical units over time is challenging. Pre- and postdeployment calibration and repeatability measurements will provide information on changes in sensitivity, precision, and bias due to possible sensor deterioration and biofouling issues, but it will be unclear how these processes developed over the period of the deployment, and data quality assessment may be hampered.

A number of approaches can be utilized to improve the data produced by remotely deployed sensor systems, and assess the quality of the data produced by the sensors. A possibility is to compare the data obtained by sensors to data from discretely collected samples obtained using vessels. For sensors deployed on rosette frames of research vessels, calibration against samples that have been discretely collected using oceanographic sampling bottles (e.g., Niskin) provides a robust solution. This is used for the calibration of sensors for oxygen, conductivity, and, for example, fluorescence chlorophyll. Also, sensor data obtained using moving platforms such as gliders can be compared with station data from vessels in case the geographical locations of data collection are sufficiently close (e.g., less than 10 nm), and the data are collected at depths well below the winter mixed layer (e.g., >1 km) to avoid shorter-term variations in the observed variable. A variation of this approach is currently used in the US Southern Ocean BioArgo program, which utilizes a new type of Nernstian pH sensor [19] based on ion-sensitive field-effect transistors (SeaFET, SAtlantic) on vertically migrating Argo platforms. Here the observed pH data from the sensors are compared against pH climatologies obtained from dissolved inorganic carbon (DIC) and total alkalinity (TA) measurements from ship surveys. The scarcity of carbonate chemistry data for the Southern Ocean makes this a sensible approach. Argo floats and some other platforms cannot be retrieved following deployment and verification of sensor data obtained in regions of the ocean with very stable conditions (e.g., 3 km depth for nutrients or pH in the Southern Ocean or Pacific Ocean) may be the only option.

In situ calibrations can play an important role when looking at the sensor trends. Oxygen measurements made by optode sensors can be referenced against atmospheric oxygen by exposing the optode to air while the sensor platform comes to the ocean surface [20]. While this approach provides a reference for saturation oxygen concentrations, the opposite end of the calibration involving a zero oxygen concentration exposure is not provided (nor are intermediate oxygen concentrations).

The comparison of sensor data with available vessel-obtained data is a sensible approach. However, the assessment of data uncertainty is challenging because of potential differences due to temporal and also spatial offsets in data collection. Sensor drift during deployments is difficult to verify, and any post-deployment correction will be even more challenging to conduct. The drift can be a few percentage points of the signal due to, e.g., changes in the sensitivity of a detector, but may also result in orders of magnitude changes of the signal due to deterioration of the reagents used for the analysis or pronounced biofouling of sensor surfaces. A recent development for in situ nutrient measurement using LOC sensor approaches uses in situ calibration of nutrient measurements with nutrient standards deployed on the platforms. This approach will allow for correction of changes in sensor sensitivity and also facilitates improved assessment of sensor performance during deployment.

The carrying of chemical standards with different concentrations on platforms may be a possibility for nutrients and a range of other variables, but for carbonate chemistry variables such as pH, DIC, and TA, this is not an option as calibration is conducted using certified reference materials. A possible next step for pH, DIC, and TA but also nutrient sensors may be the in situ utilization of certified reference materials. This approach will, however, require assessment of the long-term stability of these reference materials during deployment.

The potential for improvement of measurement uncertainty of oceanic sensors through regular in situ calibration with on-board standards and bias verification using certified reference materials opens the way to the provision of uncertainty calculations as conducted for laboratory measurements. For example, the Nordtest is becoming commonly used for chemical laboratory measurements [21], and this simple but rigorous assessment of uncertainty combines a repeatability assessment through repeat measurements of a laboratory standard, with a bias assessment using a certified reference material.

6.3.4 SENSOR–PLATFORM INTERFACE STANDARDS

The foregoing exemplars address standards and processes for sensor calibration and performance assessments. There is another set of standards for sensors that cover the interface between the sensor and its host platform. In many cases, the interfaces of platforms and sensors are designed tailored to the specific platform and sensor. The reason, in part, is to reduce the demands that more flexibility would make on the system processors and memory.

Most sensor networks require careful manual installation and configuration by technicians to assure that software components are properly associated with the physical instruments that they represent. Instrument driver software, configuration files, and metadata describing the instrument and its capabilities must be manually installed and associated with a physical instrument port. Sometimes these manual procedures must be performed under physically challenging conditions, increasing the chances of human error. This opens the door for errors to be introduced in the sensor configuration file or the calibration units. Such errors make measurements useless.

There have been efforts to standardize the sensor platform interface and automate it to reduce potential for errors and to improve interoperability when sensors are used on multiple platforms. One example is PUCK.

The OGC PUCK Protocol Standard defines a standard protocol for connecting data acquisition instruments to Sensor Web networks over a serial (RS232) or Ethernet-based communication physical layer. PUCK protocol was originally developed for oceanographic applications and successfully implemented over TCP/IP together with Zeroconf and Sensor Interface Descriptor.

The set of standard PUCK protocols ensures true end-to-end "plug-and-play" capability for sensor networks and eliminates the requirement to write and install drivers for each sensor. PUCK was used successfully in a number of the Oceans of Tomorrow projects [1]. Details of the implementation are provided in Chapters 6.1 and 7.2.

6.3.5 LOOKING FORWARD

The field of sensor metrology is vital to the ability to trust and compare measurements over long periods of time. This is an evolving field whose maturity is challenged by the expanding requirements to observe the whole of ocean ecosystem dynamics, including physical, chemical, and biological elements that interact at a large range of temporal and spatial scales. In addition, transitioning established analytical techniques from the laboratory to the field is a challenging process because of the complex marine operating environment. Instruments used in the marine environment often have to be designed to work under high pressure, as well as performing while subjected to biofouling and within the power and communication limitations of most autonomous platforms. In all cases, accessing deployed instruments is economically challenging in terms of both monetary and human effort, which impacts the opportunities for calibration revalidation. In the extreme, the sensors on Argo floats are almost never retrieved and in most cases only the initial calibrations are available. Thus the need for autonomous ocean observations demands a step change in the measurement metrology of physical, chemical, and biological sensors.

There are a number of factors that impact an instrument's calibration over time, and this is often a function of the sensor methodology as well as design. For example, thermistors and hydrophones are not particularly sensitive to biofouling while for almost all other sensors, biofouling has a very high impact. Improvements in biofouling prevention will probably be the most impactful change that will improve calibration stability for most instruments. Other uncertainties exist that result from the conceptually simple problem that the ocean is undersampled. For example, frontal and mesoscale dynamics may result in significant variance between sampling events, even using rapid vertical profiling. This type of uncertainty will be reduced as data production and ingestion rates increase to allow for upstream quality control metrics to correctly evaluate the data.

Finally, oceanographic sensors and systems will be positively influenced by the general trends of improvements in data storage, satellite and remote communication bandwidth, electronics, materials, and microfluidics. While these trends will impact all oceanographic measurements, we expect that the greatest impact will occur in the remote oceans, such as in the Southern Ocean and Arctic Ocean and in the central basins where shipping routes are minimal. In those places we expect to see important advances in long-term stability of observations and reduced uncertainties done with autonomous operations, for which a more robust and widely propagated set of standards and practices is needed.

References

[1] Pearlman J, Jirka S, del Rio J, Delory E, Frommhold L, Martinez S, O'Reilly T. Oceans of Tomorrow sensor interoperability for in-situ ocean monitoring. In: OCEANS 2016 Monterey CA. 2016. p. 1–8. https://doi.org/10.1109/OCEANS.2016.7761404.

[2] Simpson P, Pearlman F, Pearlman J. Evolving and sustaining ocean best practices Workshop, 15– 17 November 2017, Intergovernmental oceanographic commission, Paris, France. In: Proceedings. AtlantOS/ODIP/OORCN Ocean Best Practices Working Group. 2017. 74 p.

[3] Pearlman J, Buttigieg PL, Simpson P, Munoz Mas C, Heslop E, Hermes J. Accessing existing and emerging best practices for ocean observation a new approach for end-to-end management of best practices. In: OCEANS 2017 – Anchorage, Anchorage, AK. 2017. p. 1–7.

[4] IPCC. Climate change 2013: the physical science basis. In: Stocker TF, Qin D, Plattner G-K, Tignor M, Allen SK, Boschung J, Nauels A, Xia Y, Bex V, Midgley PM, editors. Contribution of working group I to the fifth assessment report of the Intergovernmental Panel on climate change. Cambridge University Press; 2013. p. 1535.

[5] ISO/IEC. General requirements for the competence of testing and calibration laboratories. International Organization for Standardization/International Electrotechnical Commission; 2005.

[6] Dekeling RPA, Tasker ML, Van der Graaf AJ, Ainslie MA, Andersson MH, André M, et al. Monitoring guidance for underwater noise in European seas, Part I. Luxembourg: Executive Summary; 2014. EUR 26557 EN.

[7] Dekeling RPA, Tasker ML, Van der Graaf AJ, Ainslie MA, Andersson MH, André M, et al. Monitoring guidance for underwater noise in European seas, Part II. Luxembourg: Monitoring Guidance Specifications; 2014. EUR 26555 EN.

[8] Dekeling RPA, Tasker ML, Van der Graaf AJ, Ainslie MA, Andersson MH, André M, et al. Monitoring guidance for underwater noise in European seas, Part III. Luxembourg: Background Information and Annexes; 2014. EUR 265556EN.

[9] Robinson SP, Lepper PA, Hazelwood RA. Good practice guide for underwater noise measurement - 133. 2014.

[10] Calibration of hydrophones and electroacoustic transducers: a contribution to the OES standards initiative. In: Robinson SP, Theobald PD, Foote KG, editors. OCEANS 2014-TAIPEI. April 7–10, 2014.

[11] Standards Committee S12N. Quantities and procedures for description and measurement of underwater sound from ships - Part 1: general requirements. Acoustical Society of America; 2009.

[12] van der Schaar M, André M, Delory E, Gillespie D, Rolin JF. Passive acoustic monitoring from fixed platform observatories. 2017. FixO3 project report - European Union Seventh Framework Programme (FP7/2007-2013) Grant Agreement no [312463] http://www.fixo3.eu/download/Deliverables/D12.6_FINAL_20170710.docx.

[13] Howell KA, Achterberg EP, Tappin AD, Braungardt CB, Worsfold PJ, Turner DR. Voltammetric in situ measurements of trace metals in coastal waters. Trends Anal Chem 2003;22:828–35.

[14] Braungardt CB, Achterberg EP, Axelsson B, Buffle J, Graziottin F, Howell KA, Illuminati S, Scarponi G, Tappin AD, Tercier-Waeber M-L, Turner DR. Analysis of dissolved metal fractions in coastal waters: an inter-comparison of five voltammetric in situ profiling (VIP) systems. Mar Chem 2009;114(1–2):47–55.

[15] Geißler F, Achterberg EP, Beaton AD, Hopwood MJ, Clarke JS, Mutzberg A, Mowlem MC, Connelly DP. Evaluation of a ferrozine based autonomous in situ lab-on-chip analyzer for dissolved iron species in coastal waters. Front Mar Sci 2017;4:322. https://doi.org/10.3389/fmars.2017.00322.

[16] Staudinger C, Strobl M, Fischer JP, Thar R, Mayr T, Aigner D, Müller BJ, Müller B, Lehner P, Mistlberger G, Fritzsche E, Ehgartner J, Zach PW, Clarke J, Geißler F, Mutzberg A, Müller JD, Achterberg E, Borisov SM, Klimant I. A versatile optode system for oxygen, carbon dioxide, and pH measurements in seawater 1 with integrated battery and logger. Limnol Oceanogr Methods 2018. [Submitted for publication].

[17] Clarke JS, Achterberg EP, Rerolle VMC, Kaed Bey SA, Floquet CFA, Mowlem MC. Characterisation and deployment of an immobilised pH sensor spot towards surface ocean pH measurements. Anal Chim Acta 2015;897:69–80.

[18] Rerolle VMC, Floquet CFA, Harris AJK, Mowlem M, Bellerby RRGJ, Achterberg EP. Development of a colorimetric microfluidic pH sensor for autonomous seawater measurements. Anal Chim Acta 2013;786:124–31.

[19] Martz TR, Connery JR, Johnson KS. Limnol Oceanogr Methods 2010;8:172.

[20] Bittig HC, Körtzinger A. Technical note: update on response times, in-air measurements, and in situ drift for oxygen optodes on profiling platforms. Ocean Sci 2017;13(1):1–11. https://doi.org/10.5194/os-13-1-2017.

[21] Magnusson B, Näykki T, Hovind H, Krysell M. Handbook for calculation of measurement uncertainty in environmental laboratories. 3rd ed. 2012.

CHAPTER

6.4

Evolving Standards for Sensors and Systems—Data Systems

Adam Leadbetter

Marine Institute, Oranmore, Ireland

6.4.1 INTRODUCTION

Standardized data management principles in the ocean observation domain have been in place since the formation of the International Oceanographic Data and Information Exchange of the Intergovernmental Oceanographic Commission (IODE-IOC) in the early 1960s. Throughout its history, IODE-IOC has existed to facilitate and promote the discovery, exchange of, and access to marine data and information. This includes metadata, products, and information in real-time, and near real-time and delayed mode, through the use of international standards and in compliance with the IOC Data Exchange Policy for the ocean research and observation community and other stakeholders.[1] However, there has been in recent years a great deal of progress in computer science activities with which the international ocean data management community is only now beginning to catch up. The data standards available have to evolve rapidly to take into account the emergence of new data analysis paradigms such as Big Data analytics and the related Semantic Web [1] and Linked Data [2] movements emerging from the World Wide Web Consortium's activity of the data Web. As these data standards evolve, so must the ways in which data are captured from sensors by data logging platforms and how data are transferred to and integrated into data systems.

The emergence of these new data analysis and data integration paradigms has coincided with an exponential growth in the number of ocean-observing sensors being deployed. This growth has been facilitated by the availability of new instrumentation platforms that may be deployed at a lower cost than the traditional oceanographic observation platform of the research vessel or the mooring. Platforms of this type include: Argo floats [3]; autonomous underwater vehicles such as Autosub [4] and subsea gliders [5]; and devices operating in the coastal area accessing low-powered Wide Area Networks[2] for communications.

This chapter will provide the reader with an overview of the technical developments from the Big Data and Semantic Web paradigms; demonstrate advances in data delivery formats and data integration platforms; and detail areas that are still being researched and highlight areas of future research as the emerging data platforms settle down into operational models.

6.4.2 THE BIG DATA PARADIGM

The common definition of Big Data incorporates the characteristics of [6]:

- **Volume:** The quantity of generated and stored data.
- **Variety:** The type and nature of the data.
- **Velocity:** The speed at which the data are generated and processed to meet the demands and challenges that lie in the path of growth and development.
- **Variability:** Inconsistency of the data set, which can hamper processes to handle and manage it.
- **Veracity:** The quality of captured data, which can vary greatly, affecting accurate analysis.

Within the Earth sciences domain, the problem of *volume* has been addressed by projects such as eReefs [7], which provides both a Semantic Web ontology [8] to describe the data sets available to the system and a brokering layer to bring those data sets into common processing and display tools. Ocean forecast and hindcast, modeling projects that

---

[1] http://iode.org/index.php?option=com_content&view=article&id=385&Itemid=34.

[2] For example the LoRa system https://www.lora-alliance.org/.

are run on large-scale high-performance computing platforms, are also challenging in terms of the volume of data they output.

Many projects have looked to provide a variety of oceanographic information through single points, such as the SeaDataNet [9] project and its follow-on, SeaDataCloud[3] (2016–20), in Europe and the Biological and Chemical Oceanography-Data Management Office [10] project in the United States. A common theme of the approach to addressing the issue of *variety* in these projects, as with eReefs, is to use Semantic Web techniques (in particular well-managed controlled vocabularies published online) to ensure the data descriptions are interoperable.

SeaDataNet has also included controlled vocabularies to describe the *veracity* of data that have been quality controlled in post hoc procedures by national oceanographic data centers. The contents of these vocabularies have been incorporated by the IOC's IODE and the Joint IOC-World Meteorological Organization Technical Commission for Oceanography and Marine Meteorology (JCOMM) into the Ocean Data Standards process [11].

However, the problem of the *velocity* of ocean data has been historically restricted to, at best, near-real-time applications due to the constraints placed on systems by the harshness of the operating environment. The deployment of a number of subsea cabled observatories across the globe has allowed for the development of real-time data systems and to explore emerging methods of delivering data through new standards evolving from the Internet of Things approach. Other sections in Chapters 6.1 and 6.3 of this volume also address recent projects developing innovations in this regard.

By combining the approaches from these projects and their solutions to the issues that the Big Data paradigm exposes, an evolution in data standards for marine sensing can be realized. This chapter will now expand on the issues of variety, and how to make the output of sensors interoperable; how to deliver data in a structured, modern, lightweight format; and how these data may be connected in a single data access platform.

6.4.3 A WEB OF DATA

The World Wide Web is now a well-established platform for publishing sensor data, as shown by the large effort put into the delivery of the Sensor Web by organizations such as the Open Geospatial Consortium (OGC) [12]. One of the challenges of Big Data is to address the variety inherent within data sets, a common problem in the marine domain where an ex situ water sample may yield many tens of measurements. The publication of structured data on the Web is one mechanism for addressing this problem. Berners-Lee proposed that the World Wide Web could be used to publish structured data in a manner similar to the publication of documents on the Web, leading to the related concepts of the Semantic Web and Linked Data [1,2]. This section will concentrate on the Linked Data paradigm. The four fundamental principles of Linked Data are (Fig. 6.4.1):

1. Use Uniform Resource Identifiers (URIs) as names for things.
2. Use Hyper-Text Transfer Protocol (HTTP) URIs so that interested parties can look up those things. This is analogous to the use of HTTP Uniform Resource Locators as the addresses by which Web pages may be looked up in a browser.
3. A standards-compliant document is returned when a user accesses a URI. In the World Wide Web, documents are returned in Hypertext Markup Language (HTML), in Linked Data, and the Semantic Web, and the documents are returned in a serialization of the Resource Description Framework (RDF). RDF describes a model in which all information is encoded in a graph of triples of the form subject-predicate-object, for example, ocean-contains-saltwater. RDF also supports its own query language, known as SPARQL, which can be used over standard Web connections.
4. Finally, Linked Data publishers should provide links from their named data resources to other data resources. This is an analog of providing links off-page in an HTML Web page. The state of the art in Linked Data for the marine science domain is highlighted next.

Ref. [13] provides four use cases for Linked Data and the Semantic Web in the observation of the marine environment (Fig. 6.4.2). These are: standardization of terminology; enhanced browsing of data within portals; discoverability of data within search engines; and the overarching interoperability of data. For sensor-based applications, it is mainly interoperability that is the focus, but the discoverability of data remains important because once it is collected it needs to be used, interpreted, re-presented, and amalgamated to enhance its position

[3] https://www.seadatanet.org/About-us/SeaDataCloud.

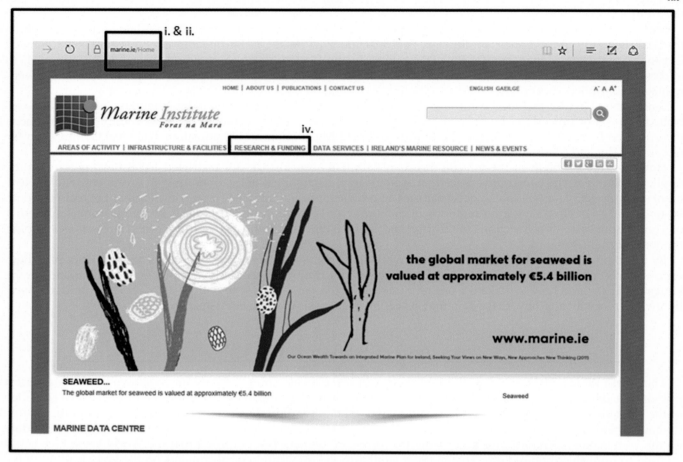

FIGURE 6.4.1   The four fundamental principles of Linked Data viewed in the context of a standard web page.

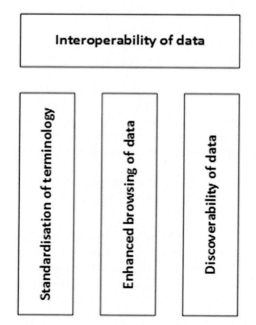

FIGURE 6.4.2   The four use cases for Linked Data and the Semantic Web in the observation of the marine environment. Note that three of the use cases support the overarching use case of the interoperability of data.

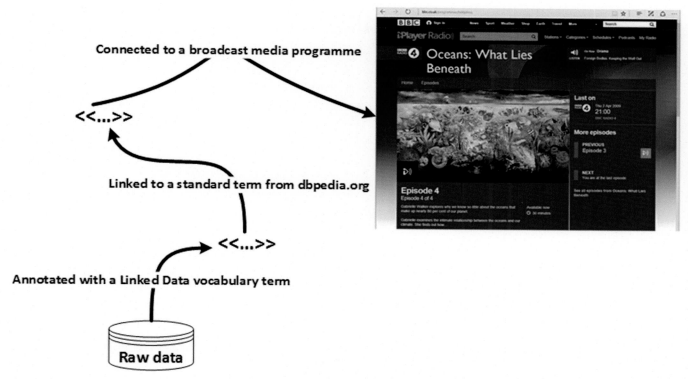

FIGURE 6.4.3   An overview of the connections which Linked Data can provide from raw observation data to popular consumption in the broadcast media.

on the data value chain. Interoperability is also important in republishing collected data for multiple applications, for instance for a legislative sampling framework and for a regional spatial data infrastructure where the data standards may differ.

The Linked Ocean Data concept was introduced by Ref. [14], was refined by Ref. [15], and further expanded by Ref. [16]. The initial Linked Ocean Data cloud consisted of 18 nodes, and by the final publication numbered over 20. The Linked Ocean Data cloud centers around a number of vocabulary servers used by international projects to address the use case of standardization of terminology (including the NERC Vocabulary Server and Chemical Analytical Services Thesaurus; the Marine Metadata Interoperability Ontology Register and Repository (MMI-ORR); and CSIRO's SISSVoc server). Higher-level, or more general, vocabularies are mapped into from these domain-specific vocabularies, including: NASA's Global Change Master Directory and Quantities, Units, Dimensions and Data Types ontology; European Environment Agency vocabularies; Life Sciences ID; ChemDPlus; Open Biomedical Ontologies; DBPedia; GeoNames; Heritage Data; and the British Ordnance Survey. Data providers such as SeaDataNet in Europe and the United States projects Biological and Chemical Oceanography Data Management Office and Rolling Deck to Repository use these vocabulary terms to standardize their metadata and data definitions. Finally, media outlets such as the British Broadcasting Corporation have begun to use the Linked Data techniques in publishing their websites and a small number of links have been created between the vocabularies and the resources describing broadcast media programs. The long-term goal of this final linkage is to have data available to support the broadcast media programs and to increase the visibility of data products to the general public. This final goal is illustrated in Fig. 6.4.3.

6.4.4 BEYOND PLAIN OLD XML

The interoperability data transport standard of choice has traditionally been XML as witnessed by the standards issued by the OGC. There are many benefits to the use of XML, including the ability to define the structure and required elements of a document using XML Schema Definitions. However, the size of an XML document is often quite large because it requires both opening and closing tags to define the beginning and end of every element.

```json
1  {
2    "$schema": "http://json-schema.org/draft-04/schema#",
3    "name": "Instrument example",
4    "type": "object",
5    "properties": {
6      "name": {
7        "type": "string"
8      },
9      "serialNumber": {
10       "type": "string"
11     },
12     "instrumentType": {
13       "type": "string",
14       "format": "uri"
15     }
16   },
17   "required": ["name", "serialNumber"]
18 }
```

```json
1  {
2    "name": "Idronaut CTD",
3    "serialNumber": "1234-5678",
4    "instrumentType": "http://vocab.nerc.ac.uk/collection/L05/current/130/"
5  }
```

FIGURE 6.4.4   A simple JSON Schema example for a document describing an instrument. Detailed are the instrument name, serial number and a connection to a controlled vocabulary term describing the make and model of the instrument.

As such, Web developers have begun to look for and use alternative data transport formats for everyday tasks. Efforts have been made to make XML more suitable for dynamic data exchange, such as the Efficient XML Interchange format[4] as used in the European Commission NeXOS project with OGC Sensor Web Enablement standards. The modern, de facto standard for data transport on the Web is JavaScript Object Notation (JSON) [17]. Until very recently, the publication of data from formal standards using JSON was not as desirable as in XML due to the lack of Schema Definitions for JSON. However, in 2010 JSON Schema was proposed, and it was refined in 2013, which has increased its usage.[5] An example of the usage of JSON Schema is shown in Fig. 6.4.4.

During 2015, the OGC's Sensor Web Enablement Domain Working group discussed a JSON encoding of the ISO/OGC Observations and Measurements standard (OM-JSON). The O&M standard is the delivery mechanism of sensed data in the Sensor Web Enablement suite.[6] JSON encoding is complete with a JSON Schema definition to ensure that fields are used correctly and consistently. There has been an effort by the IOC's Ocean Acidification community to validate the OM-JSON encoding and also to provide feedback, in particular in the realm of the analytic process that is key to that user group.

In addition to the JSON Schema approach, a JSON publication pattern for Linked Data, known as JSON-LD, is also available. JSON-LD defines a header to the JSON document, termed the "context." The JSON-LD context defines the Linked Data URIs of each attribute within the JSON document. Leadbetter et al. [16] show a scheme that can take a OM-JSON document, add a context, and produce a full Linked Data resource for that observation. However, there is work to be done in completing the JSON-LD mapping for OM-JSON and in validating the Linked Data records against the published ISO and OGC O&M standard. This approach allows structured data to be created for every observation a sensor makes in a relatively lightweight, transportable format. Here, Linked Data is being used to address the fourth use case, that of data interoperability, presented by Lassoued and Leadbetter [13]. However, this begs the question: how do we convert sensor output into these structured data? Some insights to answering this question are presented later, and in Chapter 6.1 of this volume.

---

[4] https://www.w3.org/TR/exi/.

[5] http://json-schema.org/.

[6] http://www.opengeospatial.org/ogc/markets-technologies/swe.

6.4.5 BORN CONNECTED

The concept of creating structured data at the point of capture has been discussed by Fredericks [18], Leadbetter [15], and Leadbetter et al. [19]. The concept may be considered analogous to the transition of analog records (for example, film-based photography) to digital through the application of new technologies. Initially, the concept was termed "Born Semantic" as the idea had been to create a Semantic Web representation of data outputs from sensors. However, as the emphasis has shifted toward a Linked Data representation of the data, the naming of the concept has evolved to "Born Connected." In general, the idea is to produce a structured, standards-based Linked Data record as close to the point of acquisition as possible. This comes with a set of challenges, namely, what standards to adhere to and what architecture can be employed to practically achieve this when operating under the constraints of the marine environment.

We have already considered the data standards from the OGC, and in particular the Sensor Web Enablement suite. Therefore as we are considering the output from sensors, the Sensor Observation Services and in particular O&M are the logical choices of structured data format. However, O&M is not the only standard within Sensor Observation Services, as Sensor Model Language (SensorML) is used to describe the devices and their set-up, including calibration details; and the ubiquitous OGC GetCapabilities document[7] is used to describe the devices and observations available from the Sensor Observation Service. There is also a fork of the Sensor Observation Service that makes use of the de facto Internet of Things communication standard, Message Queue Telemetry Transport (MQTT), to publish O&M records (George Percivall, personal communication, December 5, 2016) further reinforcing this choice. The remainder of this section will consider some of the architectural and social challenges that are faced in the implementation of Born Connected O&M and SensorML data.

The architecture for Born Connected sensor data systems must take into account the varied bandwidth available to sensors for communicating their results. In oceanographic applications, there is a full spectrum from the broadband connections of fiber-optic-cabled observatories through to narrowband satellite communications from autonomous underwater vehicles. This spectrum is expanded if we consider future use cases for Internet of Things sensors on drifting buoys operating close to the shore, and communicating their positions via very narrowband communications protocols such as the LoRA or SigFox networks. This is compounded by the internal communications on, for example, autonomous underwater vehicles where, due to (among other considerations) battery power limitations, the bandwidth between devices, data loggers, and communications devices is limited. The economic and power costs of sending more data than necessary also has to be considered.

There have therefore been three proposed approaches to Born Connected. The first, as presented by Fredericks [18], is to embed SensorML directly into the devices through collaboration with instrument manufacturers and incorporating terms published on the MMI-ORR. This approach has the advantage that SensorML comes as part of the Original Equipment Manufacturer's offering. However, there are a number of technical and social difficulties to this approach. First, SensorML is either expensive to communicate as it is a large XML document, or only available at deployment or retrieval time of an instrument. There is work ongoing (Mike Botts, personal communication, October 29, 2016) to convert the SensorML model to JSON in line with the earlier OM-JSON work, which may go some way to mitigating this, as does the OGC PUCK protocol (see Chapter 6.1; [20,21]). Second, there is an issue with giving the instrument manufacturers enough incentive to join in with such a proposal. The effort to provide SensorML directly from instrument firmware is a "nice-to-have" and should be pursued (see other chapters in this section for more details on rationale and solutions), yet it is not crucial to the scientific validity of the instrument outputs.

A second option is to use Big Data techniques, particularly those developed by large-scale Web companies reliant on fast message passing. For example, the LinkedIn platform operates on tools that form queues of messages and act upon these messages. This processing pipeline has been made available as open source software through the Apache Software Foundation as Apache Kafka (for the messaging queues) and Apache Storm (for the processing portion). The Apache Software Foundation also supports the Apache Spark data processing tool. These packages are designed to be massively scalable, working across many processing cores [22]. The Galway Bay Subsea Observatory data system uses the approach of adding each reading from its various instruments to message queues on the Apache Kafka software, and these messages are enriched through stream processing software, which feeds the data along a series of message queues until its final archival, or exposure over HTTP or MQTT protocols. The enrichment of the data streams consists of ensuring the timestamps are standardized and mapping the names of data channels to terms from Linked Data vocabularies. One of these stream processing routes can take a pattern written in the Grok arbitrary text to a structured data filter, formatting the data to a JSON structure, which can easily be mutated to OM-JSON, with

---

[7]https://wiki.52north.org/SensorWeb/GetCapabilities.

the OM-JSON document containing terms published on the NERC Vocabulary Server. The NERC Vocabulary Server terms are chosen in this example in preference to the MMI-ORR terms due to the use of the former in the European Commission's SeaDataNet and SeaDataCloud projects, the European Marine Observation and Data Network, and the European INSPIRE Spatial Data Infrastructure. In other examples, the choice of vocabulary service may differ, and there are stored mappings on both services between the NERC Vocabulary Server and the MMI-ORR. Filter patterns for a number of instrument feeds are registered on GitHub.[8] This approach has the benefit of low-latency translation of raw instrument outputs into O&M. However, it has not been proven against binary data outputs such as from an Acoustic Doppler Current Profiler, and requires a well-maintained register of the filter patterns to correctly translate the raw instrument output. However, writing a filter is a fairly straightforward process and can be undertaken by domain data managers, rather than instrument manufacturers.

The European Commission's SenseOcean project has developed a hybrid approach, taking portions of both the foregoing architectures [23]. Working in a consortium of data managers and instrument developers and manufacturers, the project has adopted an approach where SensorML is transmitted from a remote platform only when it changes and at other times is referenced by a globally unique identifier to connect the metadata to the data in the data center and to minimize the bandwidth used for communications. On receipt of the observation data from the remote platforms, the SenseOcean data system makes it available as O&M records through a Sensor Observation Service. This may, in the long term, prove to be a good and sustainable approach to building Born Connected applications. Other projects (e.g., NeXOS) with a similar approach for reducing costs while maintaining a standard Sensor Web-enabled end-to-end information chain have been discussed in Chapters 6.1 and 6.2.

6.4.6 FROM DATA CATALOGS TO DATA ACCESS

We have seen how the use cases of Linked Data in marine science applications for standardization of terminology and for interoperability of data have been fulfilled through the use of controlled vocabulary terms and Born Connected data systems. The remaining use case of discoverability of data, either through browsing interfaces or search patterns, must now be addressed.

There have been many attempts to catalog marine science data in many different projects. Some of these include the previously mentioned SeaDataNet project in Europe; the IOC's Ocean Data Portal; and the Biological and Chemical Oceanography Data Management Office. In some cases these catalogs feed off each other, and can create a dense data ecosystem. However, they are generally not optimized for discovery within popular internet search engines, meaning that in many cases users need to be pre-aware of the existence of the data catalog before they can find any data within it. Another general observation of these catalogs is that they only provide users with metadata records, and the data files must be downloaded before they can be explored and checked if they are truly what the user wants. In the case of large data sets this can be a time-costly mistake to make. Therefore the use of applications that can make data available to users in preview form is highly desirable.

The Open Data movement, typified by such portals as data.gov, data.gov.uk, and data.gov.ie, and the Big Data paradigm both advocate the availability of data and the ability to do analyses of data sets remotely, albeit on differing scales. For example, the data.gov.ie portal allows users to preview data sets that are offered in given standard formats, including Comma-Separated Values (CSV) files.[9] Where data are made available through Born Connected systems, or even well-managed delayed mode systems, using well-defined controlled vocabularies in common data formats, this data preview is simplified. Data access brokers, such as the Erddap software developed by the US National Oceanic and Atmospheric Administration [24] or the GEO Data Access Broker [25], can make this task easier by mapping from relational databases, nonrelational data stores, and flat files to a given number of highly used data files, including JSON, CSV, and Network Common Data Format.

In the same way as there are standards for publishing the observational data, there are standards for publishing these catalog records. Many of the Open Data systems use the Data Catalog Vocabulary terminology as published by the World Wide Web Consortium [26], but this is not well recognized by internet search engines. The major search engines are more aware of the schema.org vocabulary [27]. In September 2016, the schema.org vocabulary for data sets was expanded for Google Science Datasets,[10] specifically designed to help search engines to understand the

---

[8] https://github.com/IrishMarineInstitute/grok-raw-inst.

[9] An example may be seen at https://data.gov.ie/dataset/marine-institute-buoy-wave-forecast.

[10] https://developers.google.com/search/docs/data-types/datasets.

metadata describing a data set, and allowing discovery and potentially preview of those data sets. However, previewing the data is only the first step in what such a system may offer.

6.4.7 TOWARD AN INTEGRATED DIGITAL OCEAN

The oceans are global and thus there is a need to create a digital ocean on an international basis. There are many national and regional developments that provide building blocks. An example (and case study) is the work ongoing in Ireland. Building on top of the ideas presented previously, the Marine Institute in Ireland has developed a Digital Ocean Platform that is designed to make it as simple as possible for data providers to contribute their observation results to a portal. As such, there are no minimum data requirements for a provider to join the network beyond providing a web-based interface for access to the observations or information.

The portal connects to various Web services (including ERDDAP from the Marine Institute; custom Application Programming Interfaces from the Commissioners of Irish Lights and Dublin City University; OGC Web Map and Web Feature Services from University College Cork; and the Environmental Protection Agency) via a series of ingestion scripts. These scripts allow the display of data from the multiple originators in a user interface built on Open Source components (as illustrated in Fig. 6.4.5): jQuery for the general UI components; Leaflet for the mapping interface; HighCharts for graphing; and three.js for WebGL display of 3D modeling of bathymetric features such as bays and shipwrecks and dynamic representation of oceanographic model results.

The data are presented to the end-user in a manner beyond the simple catalog entries (Fig. 6.4.6). A combination of maps, charts, and card widgets (see the waves, tides, and weather data to the right-hand side of Fig. 6.4.6) are available to summarize data. These cards combine in situ, model, and archived data in a range of formats. They are reusable and can be quickly integrated into other applications, and lead through to a data download area for offline analysis. Map layers come from a range of sources such as the INFOMAR seabed survey, Failte Ireland (the Irish tourist board), and Dublin City Council's Dublin Bay Biosphere project. All inputs to the Digital Ocean Platform

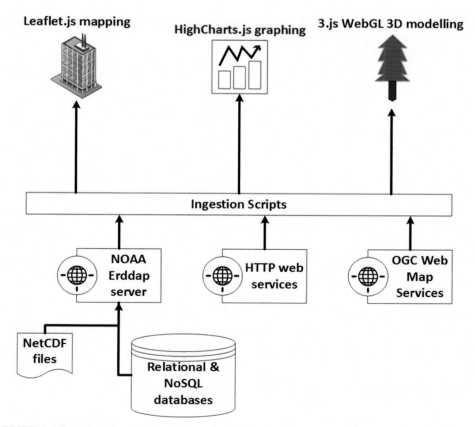

FIGURE 6.4.5 A high-level architectural representation of Ireland's Integrated Digital Ocean platform.

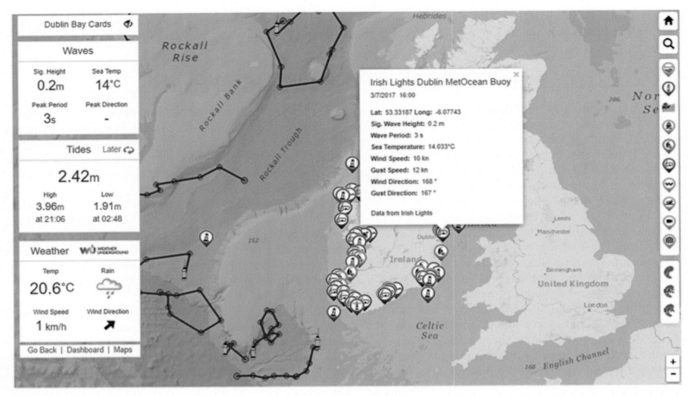

FIGURE 6.4.6   An example of marine data from multiple organisations displayed in Ireland's Integrated Digital Ocean Portal (http://www.digitalocean.ie).

are also monitored for data outages in a traffic-light dashboard system. It may be possible to replicate this approach through a "black box" application such as Data Turbine.[11]

6.4.8 CONCLUSION

In this chapter, it has been shown that the real issue for in situ sensor-based observations of the marine environment in terms of adopting the Big Data paradigm is in the variety axis. There are limitations on the Big Data volume axis due to bandwidth issues, for which we have demonstrated some strategies for mitigation; and this chapter has shown strategies for high-velocity data from in situ sensors. There remains some work to be done in the field of Born Connected systems that truly addresses the issue of variety in data from as close to the point of capture as possible. This work includes deciding the best architectural approach to reach this new ideal situation, but the approaches shown by the SenseOcean or the NeXOS projects may prove to be fruitful paths to follow in the future. It has also been shown that the use of Linked Data publishing techniques to address the issues of the Big Data variety axis standardizes the terminology used to describe data files; facilitates interoperability between data files; and improves the discoverability of data files. Advances toward a truly integrated view of ocean data have also been shown to be in progress.

References

[1] Berners-Lee T, Hendler J, Lassila O. The semantic web. Sci Am 2001;284(5):28–37.

[2] Bizer C, Heath T, Berners-Lee T. Linked Data-the story so far. Semantic services, interoperability and web applications: emerging concepts. 2009. p. 205–27.

[3] Roemmich D, Gilson J. The 2004–2008 mean and annual cycle of temperature, salinity, and steric height in the global ocean from the Argo Program. Prog Oceanogr 2009;82(2):81–100.

[4] Griffiths G, Millard NW, McPhail SD, Stevenson P, Challenor PG. On the reliability of the Autosub autonomous underwater vehicle. Underw Technol 2003;25(4):175–84.

[11] http://dataturbine.org/.

[5] The Economist. 20,000 colleagues under the sea. June 9, 2012. p. 72–3.

[6] Hitzler P, Janowicz K. Linked Data, Big Data, and the 4th paradigm. Semant Web 2013;4(3):233–5.

[7] Yu J, Leighton B, Car N, Seaton S, Hodge J. The eReefs data brokering layer for hydrological and environmental data. J Hydroinform 2016;18(2):152–67.

[8] Car NJ, Fitch PG, Lemon D. Scoping study: eReefs work package 2–interoperable data and information systems Technical report. CSIRO; 2012. ISBN: 978-1-922173-07-2.

[9] Schaap DM, Lowry RK. SeaDataNet–Pan-European infrastructure for marine and ocean data management: unified access to distributed data sets. Int J Digit Earth 2010;3(S1):50–69.

[10] Allison MD, Chandler CL, Groman RC, Wiebe PH, Glover DM, Gegg SR. The biological and chemical oceanography data management Office. In: AGU fall meeting abstracts, vol. 1. December 2011. p. 1602.

[11] Schaap D, Lowry R, Buck J, Fichaut M, Glaves H. SeaDataNet controlled vocabularies for describing marine and oceanographic datasets: a joint proposal by SeaDataNet and ODIP projects. 2015. Available from: http://www.oceandatastandards.org/images/stories/ods_docs/prop201501.doc.

[12] Botts M, Percivall G, Reed C, Davidson J. OGC® Sensor Web Enablement: overview and high level architecture. In: International conference on GeoSensor networks. Berlin Heidelberg: Springer; 2006. p. 175–90.

[13] Lassoued Y, Leadbetter A. Ontologies and their contribution to marine and coastal geoinformatics interoperability. In: Bartlett D, Celliers L, editors. Geoinformatics for marine and coastal management. CRC Press; 2016.

[14] Leadbetter A, Arko B, Chandler C, Shepherd A, Lowry R. Linked Data: an oceanographic perspective. J Ocean Technol 2013;8(3):7–12.

[15] Leadbetter A. Linked Ocean Data. In: Narock T, Fox P, editors. The semantic web in earth and space science: current status and future directions. Amsterdam: IOS Press; 2015.

[16] Leadbetter A, Shepherd A, Smyth D, Fuller R, O'Grady E. Where Big Data meets Linked Data: applying standard data models to environmental data streams. In: IEEE international conference on Big data, Washington DC, 5th-8th December. 2016.

[17] Firtman M. Programming the mobile web. Sebastapol (CA): O'Reilly Media; 2010.

[18] Fredericks J. Persistence of knowledge across layered architectures. In: Diviacco P, Fox P, Pshenichny C, Leadbetter A, editors. Collaborative Knowledge in Scientific Research Networks. 2015. p. 262–82.

[19] Leadbetter A, Cheatham M, Shepherd A, Thomas R. Linked Ocean Data 2.0. In: Diviacco P, Leadbetter A, Glaves H, editors. Oceanographic and marine cross-domain data management for sustainable development. IGI Press; 2017.

[20] Pearlman J, Jirka S, del Rio J, Delory E, Frommhold L, Martinez S, O'Reilly T. Oceans of Tomorrow sensor interoperability for in-situ ocean monitoring. In: OCEANS 2016 MTS/IEEE monterey. IEEE; September 2016. p. 1–8.

[21] Jirka S, del Rio J, Toma D, Martinez E, Delory E, Pearlman J, Rieke M, Stasch C. SWE-based observation data delivery from the instrument to the user - sensor web technology in the NeXOS project. Geophys Res Abstr 2017;19:EGU2017–7769.

[22] Gittens A, Devarakonda A, Racah E, Ringenburg MF, Gerhardt L, Kottaalam J, Liu J, Maschhoff KJ, Canon S, Chhugani J, Sharma P, Yang J, Demmel J, Harrell J, Krishnamurthy V, Mahoney MW, Prabhat. Matrix factorization at scale: a comparison of scientific data analytics in Spark and C+MPI using three case studies. In: IEEE international conference on Big data, Washington DC, 5th-8th December. 2016.

[23] Kokkinaki A, Darroch L, Buck J, Jirka S. Semantically enhancing SensorML with controlled vocabularies in the marine domain. In: The geospatial sensor webs conference, August 29–31, Muenster, Germany. 2016.

[24] Simons RA. ERDDAP. Monterey (CA): NOAA/NMFS/SWFSC/ERD; 2016. https://coastwatch.pfeg.noaa.gov/erddap.

[25] Nativi S, Craglia M, Pearlman J. Earth science infrastructures interoperability: the brokering approach. IEEE J Sel Topics Appl Earth Obs Remote Sens 2013;6(3):1118–29.

[26] Maali F, Erickson J, Archer P. Data Catalog Vocabulary (DCAT). W3C Recommendation; 2014.

[27] Ronallo J. HTML5 microdata and schema.org. Code4 Lib J 2012;16.

7

Challenges and Approaches to System Integration

CHAPTER

7.1

Understanding the Requirements of the Mission and Platform Capabilities

Timothy Cowles

College of Earth, Ocean, and Atmospheric Sciences, Oregon State University, Corvallis, OR, United States

OUTLINE

Challenges and Innovations in Ocean In Situ Sensors
https://doi.org/10.1016/B978-0-12-809886-8.00007-7

7.1.1 INTRODUCTION

Given the critical role that the ocean plays in global weather and climate, there are compelling scientific and social needs to document multidecadal trends and variability in ocean conditions, particularly the oceanic response to the atmospheric forcing created by anthropogenic greenhouse gases. To meet those scientific and societal imperatives, there are numerous essential ocean properties and processes the values and amplitudes for which must be measured continuously, on timescales ranging from seconds to hours, across broad horizontal scales, and from the air–sea interface to the seafloor. As has been described in earlier sections of this book, over the past several years a new generation of measurement tools (sensors) and instrumented platforms has been developed that greatly enhances our ability to document physical and biogeochemical ocean properties and processes. However, to move from a one-time measurement with a new sensor to an extended deployment of multiple sensors on one or more ocean platforms requires careful planning, design, construction, and operation, usually under budgetary constraints and project completion deadlines. This chapter will address this transition to extended deployments of multiple sensors and platforms, from the development of the scientific or societal justification or motivation to the successful integration of platform hardware and sensor data streams into a system that meets the project objectives. As the size, complexity, deployment duration, and cost of the proposed observing system expands beyond the scope of a short-term oceanographic experiment or expedition, this detailed development process will require both system engineering and project management expertise. Chapter sections will focus on the application of engineering and management processes in the development of requirements from the overarching science questions to the application of those requirements to the specification and procurement of the appropriate sensor technologies and platform components that meet technical-readiness criteria. These considerations directly impact the capability of the respective instrumented platforms to deliver data to the users as originally prescribed in the science objectives of the project (meeting the performance metrics within the predefined testing and evaluation protocols). In addition, chapter sections will address the identification of risks, the challenges of system integration, the need (and process) for upgrades to a deployed system as sensor technology evolves, and, finally, the continued maintenance of deployed systems in fluctuating budgetary environments.

There is a long record of successful mooring arrays of limited to moderate duration, providing a sound technical basis for decadal-scale observing systems, particularly those measuring physical properties of the ocean with technically mature sensors (high technical readiness level, "TRL"). The past two decades have seen considerable growth in the development of biogeochemical sensors which have the potential to provide extended time-series observations as the respective sensors reach the necessary technical maturity. Although there is great interest in incorporating new technologies in ocean-observing systems, there are risks to be considered when balancing potentially positive outcomes against the potential failures to meet requirements due to adoption of immature sensors or types of platforms. These risks will be discussed in this chapter.

7.1.2 FROM SCIENCE OBJECTIVES TO SAMPLING REQUIREMENTS

An observing system is conceived as a solution, or set of solutions, to a set of scientific or societal problems that have several constituencies or sets of stakeholders. The process to build and operate a system to address those problems usually occurs within a finite budget established by the sponsoring entity (usually a government agency or ministry). The overall size and complexity of the observing system may range from a few moorings at one site (less than $1M [US] to build and install, probably within one calendar year) to multiple moored, cabled, and autonomous systems at multiple sites (more than $200M [US] to build and install over a period of 2–4 years). Within this range of small to large is a concomitant range of annual operational costs. Regardless of size and complexity, however, each installed system will have finite budget limits and accountability to the sponsoring entity.

To fulfill as many of the stakeholder objectives as possible within the budgetary constraints, it is most cost-effective for the project team to use well-proven tools from systems engineering and project management. Such tools facilitate establishing traceable and verifiable paths from the project objectives, through the detailed specifications for sensors and platforms, to the final delivery of meaningful ocean observations to the stakeholders.

As one possible example of this trajectory from objectives through requirements, a group of stakeholders (perhaps a combination of scientists, marine-policy makers, leaders of regional governments) have justified the need to document the time series of fluctuations in ocean properties and processes offshore of a major river delta to understand:

- the health impacts of the ocean dispersal of river-borne pollutants;
- the marine-safety impacts of changes in coastal submarine topography;
- the marine-fisheries impact of regional hypoxic zones;
- the response of coastal ecological dynamics to seasonal, annual, and decadal variations in the linkage between coastal and offshore circulation and exchanges across a shelf–slope boundary.

The set of stakeholders also includes the potential users of the information that will be obtained from the observing system. Each group within the stakeholders must be involved in establishing the overarching themes of the proposed system, with a breakdown of those themes into sets of objectives. It is vital that all stakeholders participate in this process and have ownership in the scientific outcomes, once the components of the observatory are operating.

In the simple previous example regarding the temporal variations in a river delta system, the themes include basic ocean science, human health, public safety, and fisheries resource management. The observational objectives and specific questions that flow from those broad themes form the basis for the scientific and technical requirements of the components of the observing system. As an example, let us examine the fourth point in the earlier list—the response of coastal ecological dynamics to temporal variability in circulation and cross-shelf exchange. This broad topic area contains a multitude of possible scientific objectives.

7.1.2.1 Specificity in Science Objectives

> What are the responses of coastal ecological dynamics to seasonal, annual, and decadal variations in the linkage between coastal and offshore circulation and exchanges across a shelf–slope boundary?

This multidimensional question needs to be made more specific so that science objectives can be stated clearly. Which measures of coastal ecological dynamics will be included in the observed properties (nutrients, plankton, fish, benthos)? What physical variables and processes, on what horizontal (1–10 km) and vertical scales (cm to 10s of meters), must be measured to resolve the variability in coastal and offshore circulation? For example, the cross-shelf penetration of subsurface nitrate will vary seasonally and annually in response to local stratification and the fluctuations in onshore advection. Therefore, it may be essential to measure the vertical distribution of nitrate at a few cross-shelf locations frequently enough to characterize the range of nitrate variability, while also measuring the onshore/offshore flow within several vertical depth bins. Similarly, which aspects of atmospheric forcing must be measured across the domain of the system and over what spatial scales? What is the hydrology in the nearby land and the sea–land interface? These few questions illustrate only some of the specific details to be extracted from this large topic area. To further illustrate this point, a small subset of examples is listed below.

A Subset of Possible Science Objectives:

- Document, on timescales of hours, the vertical structure of coastal circulation across the continental shelf and beyond the shelf–slope boundary.
- Document, on timescales of hours, the vertical structure of offshore circulation (beyond the shelf–slope boundary).
- Document, on timescales of hours, wind speed and direction above the sea surface, across the continental shelf, and beyond the shelf–slope boundary.
- Document, on timescales of minutes, the vertical structure of temperature and salinity across the continental shelf and beyond the shelf–slope boundary.
- Document, on timescales of minutes, the vertical structure of phytoplankton biomass across the continental shelf and beyond the shelf–slope boundary.
- Document, on timescales of hours, the vertical structure of water column oxygen content across the continental shelf and beyond the shelf–slope boundary.
- Document, on timescales of hours, the vertical structure of water column nitrate content across the continental shelf and beyond the shelf–slope boundary.

As one can see from this partial list of objectives derived from an overarching topic or theme, the specific statement of each objective raises many technical questions that must be addressed in the translation of that objective into a set of requirements. What horizontal spacing of observing platforms is required to resolve cross-shelf circulation? What depth resolution is required to obtain current velocity, estimates of phytoplankton biomass, concentration of oxygen, concentration of nitrate? What horizontal spacing of meteorological data is required to resolve surface-wind forcing across the domain? The answers to these types of questions are required, usually through a scientific analysis or product trade study, to provide the justification for decision-making about horizontal and vertical spacing, as well as for specific technical requirements that lead to acquisition of sensors with the appropriate measurement response time, accuracy, resolution, drift, etc. Answers to these questions also establish the initial estimates of the scope of the project (# of moorings, # of autonomous vehicles, # of sensors, data transmission needs, etc.), with the associated cost estimates.

From the overarching science objective to document the spatial and temporal variability of water temperature, the following requirements statements could emerge:

- For fixed measurements on a moored platform, temperature shall be capable of being sampled at a frequency of 0.1 Hz (6 times per minute).
- For profilers and autonomous vehicles, temperature shall be capable of being sampled at a frequency of 1 Hz.
- Subsurface temperature measurements shall have a resolution of 0.0001°C.
- Surface water temperature measurements shall have a resolution of 0.001°C.
- Temperature measurement shall have an annual drift of no more than 0.01°C.
- Surface-water temperature measurements shall have an accuracy of ±0.005°C, inclusive.
- Subsurface temperature measurements shall have an accuracy of ±0.002°C, inclusive.

This partial list of requirements would need to be expanded to develop the specifications for a solicitation for bids from vendors, with additional requirements to address power limits, depth range (pressure housings), and the extent of the need for water flow past the sensor. Other requirements could be the types of underwater electrical or optical connectors, data output formats, mounting configuration (the difference between horizontal and vertical mounting may influence sensor performance), etc. Comparable lists of requirements would be developed, in advance of sensor procurement, for each parameter to be measured (current direction and velocity, oxygen, chlorophyll fluorescence, pH, etc.). The description of a measurement requirement for a parameter (accuracy, precision, drift, time response, etc.), along with the other requirements noted earlier, allows the development of a set of specifications to be used in the solicitation for the appropriate sensor.

There are multiple benefits to a well-documented "objective to requirements" process, which permits clear traceability from the original objectives to the deployed sensor, its sampling configuration, and the acceptability of the delivered data. The stakeholders (including the financial sponsor of the system) expect to have confirmation that the system was built to meet the requirements, as verified by testing, and demonstrated through the usability of the delivered data.

The establishment (and stakeholder approval) of requirements for platforms, sensors, data delivery, etc., must, of course, occur within the cost and schedule constraints of the project. The engineering and management team may need to iterate on scope, cost, schedule, along with the requirements to fulfill project objectives.

7.1.3 THE ROLE OF SYSTEM ENGINEERING

Large, complex projects, such as ocean observatories, usually represent a significant financial investment by a sponsoring organization or government. The managers of the project, therefore, are responsible and accountable for the effective use of those funds and the delivery of specified performance. Fortunately, for the stakeholders and managers of the observatory, the mature processes and protocols within system engineering [1] and project management [2] provide the necessary guidelines to meet the sponsor's standards for accountability and alignment to project objectives.

Although not commonly used in ocean science outside the domain of "large projects," there are well-proven sets of processes from systems engineering and project management that provide technical and management guidance to the leaders of a project. In many cases, the primary sponsor of the observatory construction and operation may require adoption of, and compliance to, established system engineering and project management protocols in advance of the design phase. These protocols are embedded within a timeline of project development, often represented by a "V" diagram that summarizes the discrete steps, with approval "gates," through which a project must be managed (Fig. 7.1.1). The V-diagram also provides a graphical representation of the verification and validation links between the definition phase of the project (left branch of the "V") and the integration phase (right branch of the "V"). These essential verification and validation tests emerge from the rigor of the systems engineering protocols that provide explicit assessment processes (test plans, test execution) for each technical management step within the project (Fig. 7.1.2).

Even in the absence of a contractual requirement with the sponsor to adopt engineering and management protocols, project leaders for a new ocean-observing system are strongly advised to hire skilled systems engineers and project managers early in the design phase, and use their training and experience to establish proven processes and procedures that can guide successful implementation (Fig. 7.1.1). These management and evaluation steps (Fig. 7.1.2) provide structure and accountability within the project, and greatly increase the likelihood of successful completion of the project within the cost and schedule constraints dictated by the sponsor and/or the stakeholders [1]. Note that each step from initial acquisition request to final product has both technical management and technical assessment processes, each with traceable documentation of plans and tests.

The elaboration of science and technical requirements from the overarching science themes must be conducted with a clear understanding of the specific cost constraints and scheduled completion date for the observing system. For example, a detailed breakdown of requirements, and the subsequent technical specifications, for a well-instrumented array of moorings may reveal higher than expected costs for the integration of all the different sensors into the platforms. Alternatively, an engineering evaluation of the requirements and specifications may reveal unacceptable levels of risk in the design/implementation, including certain fabrication or construction steps, as well as introducing potential delays to project completion, each of which may result in increased project costs. It is not uncommon to encounter many dozens of such cost and schedule issues during the planning and design phases of the project. It is often necessary to revisit the overall project design and alter the system configuration or overall scope (number of moorings, number of sensors, etc.) to meet the cost and schedule constraints of the project. As illustrated in Fig. 7.1.2, technical evaluation steps are essential for assuring that the requirements and system structure can be verified (Fig. 7.1.1), and that valid products (meaningful ocean data) are provided to all the users of the system. Ultimately, the verification test procedures confirm that requirements have been met, showing that the platform or system was built properly (according to specifications). The validation-test procedures confirm that the proper platform or system was built (actually delivers the correct data product to the user).

Although "requirements to data delivery" is the primary focus of this chapter, the successful management and engineering of the path to data delivery relies upon a sound foundation of project documentation, accessible in a secure digital document repository. Each of the steps and processes shown in Figs. 7.1.1 and 7.1.2 should be documented under a project-wide protocol for document version control. The foundational documents include the overall project design and concept of operations, specific plans for engineering management, data management, project execution,

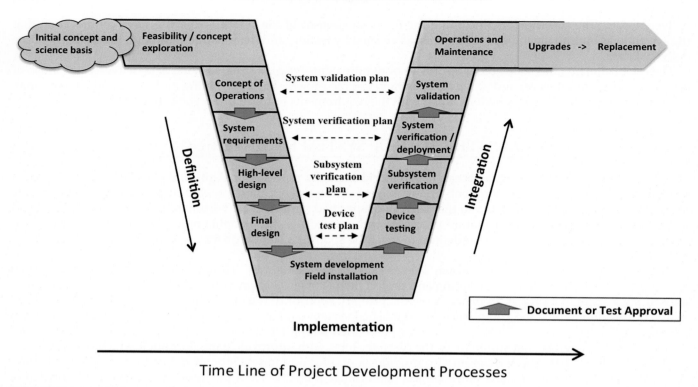

Time Line of Project Development Processes

FIGURE 7.1.1 A V-diagram of project development, with sequential steps of initial concept through final project completion. Time moves along the x-axis of the figure, and the planning and design steps of project development on the left side of the "V" are then verified through testing after implementation, as shown on the right side of the "V." Testing and evaluation to specific predetermined metrics is required to obtain approval to move from step to step along the time line.

and control of system configuration [3]. Other important records in the document repository include all requirements, specifications, test protocols, test results, maintenance protocols, bills of materials, financial reports, performance, validation records, etc. The guiding principle for documentation is based on the expectation that portions of the system may need to be rebuilt in the future by a different technical team than originally designed and built the system.

7.1.4 FROM REQUIREMENTS TO PLATFORM (SYSTEM) CAPABILITY

The foundational objective of any observing system is the delivery of meaningful data to the community of users represented by the stakeholders of the system. Data acquisition and delivery by an observing system is dependent upon the ability of managers and engineers to sustain fidelity to system requirements, while overcoming obstacles that emerge as an observatory moves from planning to operations.

7.1.4.1 Issues Related to Technical Readiness of Sensors and Platforms

Unlike sensors and machinery built for terrestrial or freshwater applications, marine sensors and equipment must function in a caustic habitat, while subject to oscillating and turbulent fluid motion, often under many atmospheres of pressure. Marine scientists and ocean instrumentation vendors have repeatedly overcome these challenges and continue to explore new opportunities for in situ sensing of essential ocean properties. There is, however, a relatively small market for newly developed in situ ocean sensors, thus limiting the opportunities for investment in the extended testing and redesign so essential for maturation of new technologies. As we transition from short-term deployments using a few well-tested sensors to extended deployments with well-tested as well as relatively new sensors, we face the risk of asking more of the new sensors than their technical readiness levels (TRLs, [4–6]) will permit. It is important for the project engineering team to assess TRLs in advance of acquisition, and identify performance risks when appropriate.

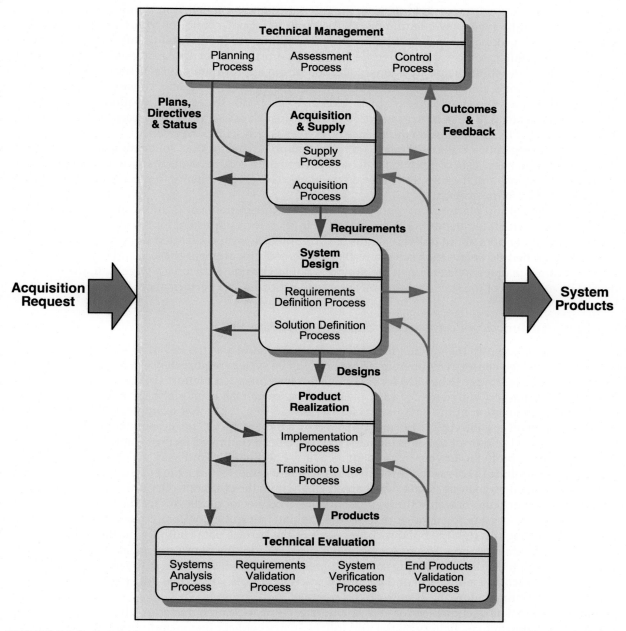

FIGURE 7.1.2 Steps to the system-engineering process. *Source: ANSI/EIA-632-1998, Processes for Engineering A System. Reprinted with permission Copyright © 2003 SAE International. Further distribution of this figure is not permitted without prior permission from SAE.*

The project team and stakeholders may agree that the project should work with a vendor in the maturation of a critical sensor technology, particularly in field-testing of sensor performance and reliability. There are shared risks and costs to be negotiated between the sponsor and the vendor in such a situation, and there may be unacceptable consequences of such risk and cost sharing on the completion schedule for the observing system. However, a project–vendor partnership may facilitate the inclusion of a key ocean variable in the range of data products provided by the observatory.

7.1.4.2 Acquiring the Sensors and Platforms

Once the requirements have been developed and verified, the component parts of the observing system can be built or purchased that can meet those requirements. The procurement step (e.g., for 10 temperature sensors) should consist of a product solicitation that contains explicit specifications for those temperature sensors that are directly derived from the measurement requirements for temperature. In addition, the solicitation should specify the other critical factors, including depth range, power requirements, data-output format, connector type, timeline for product delivery, production history, production quality data for that sensor, and product warranty information. Must the instrument orientation be limited to horizontal or vertical on the platform to meet specifications? Does the sensor require a minimum water flow to meet specifications? Selection of the appropriate product from those submitted should be based on the vendor's ability to meet the specifications and delivery schedule, consistent with a matrix containing all selection criteria. When it is necessary to use existing sensors due to cost or acquisition lead time, some adaptation of the requirements may be necessary in specifications. In such cases, the project team and sponsor will need to review the impacts of these changes on the science objectives and system performance.

For system components, e.g., buoy hulls, without a commercial supplier, the product solicitation may be for fabrication, rather than direct purchase. That component will have performance requirements that also must be specified in the solicitation for fabrication, and the selection process would also rely on compliance to specifications (including cost) within of a matrix of selection criteria.

On-time delivery of sensors and platform components is crucial for integration, verification, and validation of the system within the cost and schedule constraints of the project. It is, therefore, important for the project management team to be in contact with vendors about delivery status and compliance with contracted delivery schedules. The project team will reduce schedule delays by having a well-documented procurement process and experienced procurement staff and contract managers. Even when the project is not large enough to have a large staff, the functions need to be covered by members of the staff to assure meeting both schedule and performance.

7.1.4.3 Identifying All Junction Points in the Signal and/or Data Path

The system requirements identify the specific measurements to be made and the technical specifications of the platforms on which the selected sensors will operate. There also will be system requirements that specify the type of data transmission and/or data storage to be used for each different instrumented platform in the system. It is important to develop detailed data-flow diagrams for each measured parameter on each platform to assure that signal (data) input/output at each junction in the data path is fully specified (Fig. 7.1.3). For example, a chlorophyll fluorescence sensor is mounted on a moored platform and powered by a battery pack. The measurement of fluorescence intensity by the sensor is a set of output voltages (signals) that have been calibrated against fluorescence standards before deployment and recalibrated following instrument recovery. The signals (S in Fig. 7.1.3) must travel from the sensor, through the sensor-bulkhead connector, through the instrument cable, through the bulkhead connector of the underwater data controller, undergo none/some signal processing within the data controller, and travel wirelessly or via cable from the underwater data controller through the bulkhead connector of the surface data controller. There, they undergo none/some processing for data storage within the data controller, or compressed into a data transmission packet, before passing through the bulkhead connector of the surface data controller to the transmitter for satellite data transfer to the shore station. Upon receipt by the shore station, the fluorescence data packet is queued for further processing through the data processing software pipeline, then made available to the users of the system. As described in Fig. 7.1.3, each of the physical and communication interfaces within this data path must be clearly specified and subjected to testing before the system can be accepted for operations.

This multistep process must be documented for every parameter being measured, including the background engineering data from each platform (e.g., battery status, tilt, roll, etc.), to avoid confusion and potential oversight of a sensor–platform interface or sensor–controller interface during system integration. Different technical groups within a project team may have responsibility for different segments of the data path. Care must be taken to assure that the signal input/output (I/O) and physical aspects (connectors, cables) of every junction along the data path, for every data product, has clear specifications and ownership within a technical group of the project team. Cost overruns and schedule delays may occur during final system integration if any of the junction points along a data path are not addressed during the planning and design phase.

7.1.4.4 Data Transmission, Data Integrity, Data Quality

Data transmission from the deployed platform to a receiver at a shore station, with shore-based validation of data packet integrity and data quality, is the final step in verifying that the capabilities of the platform meet the system requirements. In some cases, all data streams are recorded on the platform (hard drive, or solid-state memory) and downloaded and tested at the shore station following recovery of the platform. The project engineering team should evaluate data-transmission options between the deployed platform and a shore station to find the most cost-effective solution given the hourly or daily volume of data and the transmission cost per kilobyte. For ocean observatories

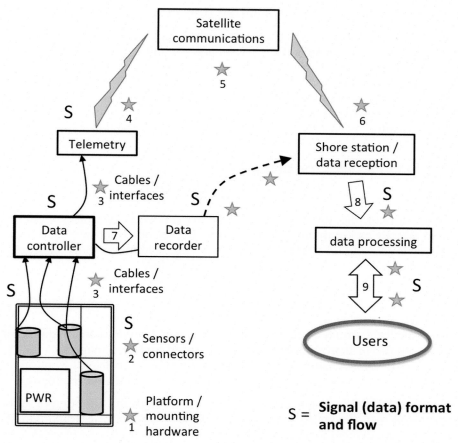

FIGURE 7.1.3   Schematic diagram of hypothetical signal flow (S) from a deployed sensor to the final user. Each numbered star (⭐) represents a physical or communication interface (cable/connector) that must be specified, must have a technical "owner," and must have an explicit completion date within the project schedule. Signal format also must be specified (and owned) through each interface to assure clear acquisition by the user.

with seafloor cabling (power distribution and Internet Protocol [IP] data communications), the data transmission protocols from cabled sensors and platforms differ from those used on moored or mobile platforms, but have the same verification testing requirement for data integrity and quality. Regardless of type of platform or location, predeployment and postdeployment verification must include tests for possible influence of other systems on the platform, as interference from adjacent instruments (e.g., power supplies) could compromise data integrity. Deployment verification will also impact the error and uncertainty analyses of the data.

A discussion of postacquisition data processing for improved data quality is outside the scope of this chapter, but the topic of data quality and the extent of data quality processing should be addressed during the planning and design phases of the project. The challenges of data handling, data processing, and data management often are underestimated, with costly consequences later in the project. It is extremely important to engage all stakeholders, especially the science users of the observatory data, in the formation of data processing and data quality requirements for each of the measured variables. Sensors for the measurement of a number of important ocean properties (or their proxies) have only recently been developed, and there are not long data records for these variables, making it difficult to develop a cost-effective data processing protocol. In addition, highly resolved time series for multiple variables create data handling challenges for both the project team analysts and the science users of the system. It is, however, essential to develop the requirement and the cost estimate for a justifiable level of data processing. The summation of these requirements and the associated cost estimates will provide the stakeholders and management team with critical information for a realistic assessment of observatory scope within the sponsor's cost constraints.

System requirements should specify that effective data-monitoring tools be working at each shore station that receives telemetered data or executes the data download from recently recovered platforms. System requirements should specify the performance metrics to be tracked for each sensor's data stream, including, e.g., number of data packets per hour, number and length of data gaps, and out-of-range engineering and scientific data. These metrics assist the technical team in evaluating the reliability of sensors and system elements in the data path.

7.1.4.5 Defining, Understanding, and Implementing Best Practices and Common Standards

Several ocean-observing systems have now been built (including, Ocean Networks Canada, Dense Oceanfloor Network System for Earthquakes and Tsunamis [DONET; Japan], Ocean Observatories Initiative [OOI; US], European Multidisciplinary Seafloor Observatory [EMSO; EU], Fixed-point Open Ocean Observatory network [FixO3; EU]; Refs. [7–10]), creating a sound foundation of experience in overcoming challenges in ocean-observatory construction and operation. The technical websites of these programs should be consulted for best practices. For example, a compendium of best practices is being developed under the Atlantic Ocean Observing Systems (AtlantOS) project (https://www.atlantos-h2020.eu).

7.1.5 CHALLENGES IN FULFILLING REQUIREMENTS WITHIN COST AND SCHEDULE

It is not unusual for creators of the initial project idea to have an overly optimistic view of the steps (and cost) required to establish and operate an ocean observatory. In the absence of a rigorous requirement-development process, this optimism can create cost and schedule challenges not usually found in short-term ocean experiments. Cost and schedule challenges emerge from several areas, including this subset:

- System owners oversell the capabilities of the proposed observing system,
- System creators may have unrealistic expectations for recently developed sensors to sustain performance during extended deployments,
- Stakeholders change some (or many) requirements after the system is under construction or in operation,
- Observatory team lacks skilled system engineers and experienced project managers,
- Observatory team lacks a strong acquisition process for equipment (contracts with vendors),
- Stakeholders redefining the final configuration or "baseline" of the system after finalizing cost and schedule.

Each of the challenges created by these gaps can be overcome through the adoption of system engineering and project management protocols (e.g., Ocean Networks Canada, Ocean Observatories Initiative). It is the responsibility of the stakeholders and the management team to staff the project with the right mix of expertise to assure that adoption. Clear development and communication of system requirements by the project team and the stakeholders can minimize the offset between user expectations for system performance and the capabilities of the installed system.

Given the multidisciplinary range of ocean properties and processes that could be selected to meet the overarching objectives of the observatory, and the desire to incorporate new sensing approaches in the observatory, it is essential that the team conduct a careful assessment of the TRL for the sensor type for each proposed parameter [11]. TRLs are relied upon by many large entities (National Aeronautics and Space Administration [NASA]: [4–6]) to reduce procurement risk and performance risk when evaluating sensors and technical solutions. The timeline to "maturity" for new sensor technology is usually longer than expected, and procurement of an immature sensor could lead to schedule delays or an unusable data stream. This is risk/benefit trade as sensors that are at TRL 8 may be ready for operations, but have not been demonstrated in a continuously operational environment. If the sensor provides a unique measurement that is critical to the mission, there are questions of "willingness to wait" or options for replacement with degraded performance. There are formal methodologies that have been evolved over time that can help with making these decisions [2].

Stability of the project baseline can be maintained through compliance to policies and procedures established during the design phase of the project. These procedures usually conform to standard "large project" protocols (e.g., Refs. [1,2]), with the goal of having consistent execution of tasks in construction, integration, and operation across the project team.

7.1.6 ADDRESSING AND MITIGATING RISK

Risk (or opportunity) refers to anything that may negatively (or positively) affect the cost, schedule, or deliverable products/performance of system [12]. For example, there is always the risk of stormy weather during an oceanographic cruise. There are risk-assessment protocols to assist managers and engineers in estimating the degree of uncertainty and level of impact that a sudden storm might have on a deployment cruise, and to provide an estimate of the cost to recover if the risk "event" occurs. Every component and process within the observatory is subject to some degree of risk or opportunity, with an ever-evolving risk profile as the project moves through construction and into operations. There are risks associated with hiring the right staff members, conducting sensor procurements on schedule, estimating software completion dates, contracting the most cost-effective vendor for sensor recalibration, etc. For example, there is a risk that an essential buoy component will not be delivered on schedule. The project manager must assign a likelihood to this risk, based on prior performance of the vendor and status reports from the vendor. The project manager also must assign an estimate of the impact to the project budget or schedule if this risk is realized, that is, the vendor fails to deliver on time. Finally, the project manager must have a plan for mitigating that risk, either through schedule adjustments or reallocation of funds to keep the project on track. Compilation of project risks, with assignment of likelihood and impact (or consequence in terms of cost and/or schedule), forms the project risk register. Project managers should revisit the risk register on a regular basis to reassess older risks, eliminate risks that no longer pose a threat, add emerging risks, and determine possible mitigation steps for the risks that still pose a threat to cost and schedule. A thorough knowledge of the risk register and the potential mitigation steps can inform decision-making during the execution of the project.

7.1.7 OBSERVATORY LIFE CYCLE CONSIDERATIONS FOR SUSTAINING COMPLIANCE TO REQUIREMENTS

The initial objectives of most ocean observatories likely include an expectation for multiyear or decadal-scale observations of ocean properties and processes. Extended deployment durations require a number of important considerations in the planning and design phases of the project, particularly with respect to the timing and expense of the required maintenance cycles. Compliance to system requirements, particularly sustained delivery of good data to system users, remains the central focus of the observatory and guides management decisions during operations. The continuous alignment and realignment of observatory performance to operational requirements during maintenance cycles requires assessment relative to predefined and approved performance metrics for sensors, platforms, data transmission and storage, data processing, etc.

The primary cost driver in the construction and operation of a marine observatory is the minimum expected (or acceptable, affordable) duration of deployment, as that deployment duration sets the schedule for ships to recover and redeploy the observatory platforms. Oceanographic research vessels have daily operational costs that range from $10K–40K (US), depending on the size and capability of the vessel, so it is possible for an annual 30-day maintenance cruise to have over $1M (US) in vessel costs alone. A moderate to large observatory may require multiple maintenance cruises to service deployed systems every year. Therefore, longer intervals between maintenance cruises have a major impact on annual observatory costs. The hardware and labor costs of equipment refurbishment and replacement remain relatively fixed for any given observing-system size and complexity, but it can be expected that a moderate to large system will have annual maintenance expenses in excess of $10M–20M (US).

The frequency and extent of maintenance cycles are influenced by several factors, including the timeline for degradation in performance of some sensors due to biofouling, power limitations of batteries, reagent use by chemical sensors, mechanical stresses on platforms and cabling, and the cost of ship time for platform recovery, servicing, and redeployment. All such factors (and related costs) should be assessed and incorporated during system planning and design, so that the deployed system can be managed within the anticipated maintenance budget.

7.1.7.1 Sensor Performance Lifetimes

Project engineers and scientists must build upon the knowledge and expertise of the user community, as well as the requirements and subsequent specifications for each sensor deployed within the system, in their projections of necessary service intervals. The team should establish target performance criteria for acceptable data delivery, based on sensor specifications and user experience, that provide realistic deployment durations. For example, the data quality from certain optical sensors deteriorates during extended deployments due to biofilm formation on the sensor windows, although antifouling approaches [13] have improved deployment durations in recent years (http://www.nexosproject.eu/the-project/developments/sensor_anti-fouling). The surfaces of moored or anchored platforms (and their sensors) and autonomous vehicles are subject to colonization by marine invertebrates (e.g., barnacles, sponges, hydroids), with the potential to compromise performance. For example, the velocity and trim of an autonomous vehicle (glider) can be impacted by the colonization of invertebrates during an extended deployment (weeks to months) [14]. The degree of biofouling or invertebrate colonization also will differ across the spectrum of water depth and ecosystem type (shallow and coastal to deep and oligotrophic). The resulting range of time estimates of service intervals, therefore, is a major consideration in the estimation of annual maintenance costs for different parts of the observing system. Once the observing system is operating, the management team can evaluate the actual performance of deployed sensors and platforms, and propose alterations to the maintenance plans for the overall system that fit within the constraints of available funding.

7.1.7.2 Structural Integrity of Platform Components

The components of the observing system (platforms, interfaces, etc.) are subject to numerous stresses during extended deployments, some of which may then require a more robust design than would be used in a short-term deployment. For example, the shackle (or universal joint) between the mooring cable and the surface float is likely to experience higher mechanical stress during a strong winter storm season than during the other seasons of deployment. Estimates of mean time between failures (MTBF) for observatory components, therefore, must be included in the assessment of realistic maintenance intervals. For example, how many months of motion or vibration can a sensor cable and its connectors withstand before replacement is prudent? This estimate may vary for different sensors and platforms, but the shortest MTBF will likely guide the evaluation of service frequency.

As outlined by Galván et al. [15], there are systematic approaches for addressing the many aspects of survivability of deployed systems. The primary elements are reliability, availability, maintainability, and safety (RAMS), although this order of terms does not imply that safety is anything but first in any project's construction or operational activities.

The terms can be further specified as follows:

- Reliability is a product's or system's ability to perform a specific function and may be given as design reliability or operational reliability. Predicted reliability will be able to be tested against system performance following installation and operations. Periodic assessment of reliability is an essential step in evaluating the need for replacement of a sensor or system component.
- Availability is the ability of a system or component to remain in a functioning state. Declines in availability will have a negative impact on data access by users, and will compromise the perceived value of the overall system to the stakeholders.
- Maintainability is determined by the ease with which the product or system can be repaired or maintained. Initial designs for the system should have this capability in mind.
- Safety is the requirement not to harm people, the environment, or any other assets during a system's life cycle. Each step in the design process as well as the implementation process must have explicit guidelines for compliance with the project's safety plan.

Thoughtful investment by project managers and system engineers in RAMS will improve project performance within the constraints of cost and schedule.

7.1.7.3 Observatory Servicing Costs

For most ocean-observing systems, servicing of deployed platforms, including recovery and redeployment, will require a ship, technical personnel, and a number of days at sea. There may also be the need and additional expense of the use of a remotely operated vehicle to retrieve and replace instrumentation on the seafloor. Several questions may guide the overall budget for servicing: what size and capabilities must the vessel possess? Can some elements of the observing system be serviced with a small vessel? Can autonomous vehicles such as gliders be directed to navigate close to shore for recovery? Can cost-sharing approaches with other observing systems be used to minimize long-term costs? (http://www.ofeg.org/np4/home.html, http://www.eurofleets.eu/np4/home.html)

7.1.7.4 Sensor Refurbishment and Calibration

For observatories with extended time-series objectives, there will be repeated cycles of instrumented-platform deployment, recovery, and redeployment through the lifetime of the observatory. The sensors on these platforms will have been procured with known product life cycles that permit a number of recalibrations and redeployments, while sustaining the specified measurement accuracy and precision. The operations team for the observatory will require specific maintenance contracts with vendors, or well-controlled maintenance procedures within the observatory laboratories, to assure that each sensor is appropriately cleaned, calibrated, and prepared for redeployment. Each refurbished and recalibrated sensor must be subjected to the same predeployment test protocols used in prior deployments. It is recommended to conduct simultaneous cross-calibration of multiple instances of the same sensor type to identify any unique performance features of a particular unit. The overall annual budget for operations must include the cost (including labor) for sensor refurbishment and calibration across the observatory. Compliance with identified "best practices" is highly recommended [16].

7.1.7.5 Platform Refurbishment, Reconstruction, and Redeployment

Surface floats (buoys), subsurface instrument frames, brackets and mounting hardware, mooring cables, etc., must all be refurbished or replaced, then reconstructed and tested according to that unit's specific configuration requirements during each interval between recovery and redeployment. The engineering team can estimate, based on experience and specifications, which platform components can survive more than one deployment cycle to minimize, to the degree possible, the "rebuild" costs of each observatory unit.

As was mentioned with sensor refurbishment and calibration, it may be beneficial to include refurbishment within the initial procurement contract for platform components. Alternatively, it may be more cost-effective to negotiate an external refurbishment contract for particular items that are too difficult or expensive to maintain by the operating institution.

7.1.7.6 Sustaining Scientific Relevance

An ocean-observing system with multiyear or decadal-scale scientific objectives must stay relevant within the evolving needs and requirements of observatory stakeholders. In particular, the scientific questions that initially justified the observatory may change to accommodate new sets of questions about the changing planet. Additionally, advances in sensor technology may provide opportunities for greater resolution of one or more essential ocean properties. It is, therefore, important for stakeholders to reexamine the scientific bases for the observatory on a periodic basis (3–5 years), and adjust, if necessary, the operational requirements of the observatory. The stakeholders should also put a periodic "sensor reassessment" process into the maintenance schedule. This process would reevaluate the deployed sensor and its performance against the potential provided by emerging sensor developments. Promising new sensors could then be evaluated by engineering staff to assess gaps between the new opportunities and the requirements and specifications for sensor performance and data delivery. Those gaps can then be evaluated and assessed by the stakeholders (or an expert panel) for consideration of next steps, including recommendations for adoption, or for further testing after some interval of sensor maturation, as has been done through the addition of biogeochemical sensors on Argo Floats (Bio–Argo) program (e.g., Ref. [17]). Stakeholders must remain aware, however, of the lengthy time required for maturation of new technology, particularly for extended marine deployments. Even for very promising new sensors, there are risks that include changes in hardware mounting, power, and/or data interfacing, and timelines for production, integration, and refurbishment.

7.1.7.7 Budgetary Challenges

It is inevitable that sponsoring stakeholder(s) (usually agencies or governments) will, at some point in the construction or operation phase, reduce the expected allocation of funds for the observing system. The project management team, along with other stakeholders, should anticipate this inevitability by prioritizing the possible responses to budget reductions. In the construction phase of a project, permanent budget reductions will force the stakeholders to agree on changes to the system requirements, based on the changes to what can be built under the new budget constraints, absent an alternative source of funds. Such changes could include, but are not limited to, reduction in the number of platforms, the number of deployed sensors, the number of autonomous vehicles, the timeline for initial installation, and the number of personnel involved in the construction of the observatory.

It is more likely that budgetary constraints will arise after the observatory components have been installed. As with budget changes during construction, the project management team should have a prioritized list of actions based on stakeholder and external expert feedback. Those priorities may be based upon a ranking of scientific outcomes (particular suites of data products) or upon the performance statistics of different observatory elements during Annual Review of the system. It may be prudent for the management team, with stakeholder concurrence, to create budget reduction scenarios that address, for example, 5%, 10%, or 20% reductions in the next year's operating budget. Such reduction scenarios might include some combination of scope reduction, extension of service intervals, less ship time, and reductions in personnel. Each scenario should include clear statements of the anticipated adverse effects of a budget reduction on compliance with the project's operational requirements, especially with respect to data delivery to users.

7.1.8 SUMMARY

The successful installation and ongoing operation of an ocean observatory requires alignment of stakeholder objectives within a strong science/societal justification for the expenditure of funds (usually many millions of dollars, euros, or equivalent). The technical and management team of the observatory must convert the scientific/technical objectives into requirements, which then lead to clear specifications for sensors and platforms. Because delivery of data products is the single most important activity of the observatory, identification and selection of appropriate sensors (and sensor platforms) must have a well-defined procedure for assessing the compliance of proposed sensors with specifications. Acquisition of sensors and observatory components requires a robust procurement process, with personnel experienced in contract development and maintenance, and with postacquisition test procedures to check for vendor compliance to contract terms. Stakeholders can be assured of verification and validation of requirements through specific testing protocols (pre- and postdeployment) with clear metrics for compliance. Close cooperation of all parties in developing clear system requirements from science objectives is critical to optimizing data delivery to the users. This path from objectives to requirements to product (data to users) can be successfully traversed by a project team that adopts well-proven management and engineering protocols and recognizes that all changes to initial concept and design are likely to have cost and schedule impacts.

References

[1] BKCASE Editorial Board. In: Adcock (EIC) RD, editor. The guide to the systems engineering body of knowledge (SEBoK), vol. 1.8. Hoboken (NJ): The Trustees of the Stevens Institute of Technology; 2017. [BKCASE is managed and maintained by the Stevens Institute of Technology Systems Engineering Research Center, the International Council on Systems Engineering, and the Institute of Electrical and Electronics Engineers Computer Society] www.sebokwiki.org.

[2] PMBoK: the guide to the project management body of knowledge. 5th ed. Newtown (PA): Project Management Institute; 2013. https://www.pmi.org/pmbok-guide-standards/foundational/pmbok.

[3] Ocean observatories initiative technical data package. 2016. http://oceanobservatories.org/technical-data-package/.

[4] European Commission. Technology readiness levels (TRL) (PDF)., G. Technology readiness levels (TRL). In: Horizon 2020 – work programme 2014-2015 general annexes, extract from Part 19-Commission decision C(2014)4995. 2014. http://ec.europa.eu/research/participants/data/ref/h2020/wp/2014_2015/annexes/h2020-wp1415-annex-g-trl_en.pdf.

[5] Mankins JC. "Technology readiness levels: a white Paper" (PDF). NASA, Office of Space Access and Technology, Advanced Concepts Office; 1995. https://www.nasa.gov/pdf/458490main_TRL_Definitions.pdf.

[6] United States Department of Defense. Technology readiness assessment (TRA) guidance (PDF). 2011. http://www.acq.osd.mil/chieftechnologist/publications/docs/TRA2011.pdf.

[7] Ocean Networks Canada. http://www.oceannetworks.ca/.

[8] DONET (Dense Oceanfloor Network System for Earthquakes and Tsunamis). https://www.jamstec.go.jp/donet/e/.

[9] EMSO European Multidisciplinary Seafloor and water column Observatories. http://www.emso-eu.org.

[10] FixO3 Fixed-point Open Ocean Observatories. http://www.fixo3.eu.

[11] Brasseur L, Tamburri M, Plueddemann A. Sensor needs and readiness levels for ocean observing.. In: Hall J, Harrison DE, Stammer D, editors. Proceedings of OceanObs'09: WPP-306. Venice (Italy): ESA Publication; 2010.

[12] Hulett D. Practical schedule risk analysis. NY: Routledge; 2009.

[13] Lobe H. Recent advances in biofouling protection for oceanographic instrumentation. In: OCEANS'15 MTS/IEEE Washington. 2015. https://doi.org/10.23919/OCEANS.2015.7401854.

[14] Moline M. Evaluation of glider coatings against biofouling for improved flight performance ONR Report 2011http://www.dtic.mil/docs/citations/ADA547644.

[15] Galván B, Marco AS, Rolin J-F. NeXOS contribution to the adaptation of system analysis engineering tools for mature and reliable ocean sensors. In: Sensor systems for a changing ocean (SSCO), 2014. IEEE; 2014. https://doi.org/10.1109/SSCO.2014.7000370.

[16] FixO3. Fixed-point open ocean observatories. Handbook of best practices. 2016. http://www.fixo3.eu/download/Handbook%20of%20best%20practices.pdf.

[17] Xing X, Claustre H, Uitz J, Mignot A, Poteau A, Wang H. Seasonal variations of bio-optical properties and their interrelationships observed by Bio-Argo floats in the subpolar North Atlantic. J Geophys Res Oceans 2014;119:7372–88. https://doi.org/10.1002/2014JC010189.

Further Reading

[1] Brito M, Smeed D, Griffiths G. Underwater glider reliability and implications for survey design. J Atmos Ocean Tech 2014;31:2858–70. https://doi.org/10.1175/JTECH-D-13-00138.1.

[2] Cowles T, Delaney J, Orcutt J, Weller R. The ocean observatories initiative: sustained ocean observations across a range of spatial scales. Mar Technol Soc J 2010;44:54–64.

[3] Gawarkiewicz GG, Todd RR, Plueddemann AJ, Andres M, Manning JP. Direct interaction between the gulf stream and the shelfbreak south of New England. Nat Sci Rep 2012;2:553. https://doi.org/10.1038/srep00553.

CHAPTER

7.2

Challenges and Constraints of Sensor Integration Into Various Platforms

Douglas Au, Tom O'Reilly

Monterey Bay Aquarium Research Institute (MBARI), Moss Landing, CA, United States

7.2.1 INTRODUCTION

Marine-sensor applications may be oriented to scientific research, public health, fisheries management, and security, to name a few. In this chapter, we describe some challenges of designing the sensor and host platform systems needed to implement these applications, and some technical approaches based on experience at the Monterey Bay Research Institute and elsewhere.

A given marine application has mission objectives and goals that must be met. The data acquired by the application sensors must meet certain requirements to meet the goals. Data requirements may specify time and space sampling scales and resolution, revisit intervals to a particular region, Eulerian versus Lagrangian sampling, accuracy and precision, allowed data telemetry latency, and other parameters. Application designers (scientists and engineers) must select sensors and platforms that meet these requirements. In some cases the requirements may be relaxed if the cost or technical difficulty of meeting them is prohibitive.

We define a "sensor" to be a physical device that consists of one or more sensing elements and a communications interface to a "host" platform such as buoy, vehicle, or observatory cable to which the sensor is physically attached and electronically interfaced. The sensing element (a k a transducer or detector) converts a physical phenomenon into a measurable signal. The signals are transferred to the host platform through the communications interface. Optional elements of the sensor include a power interface that receives electrical power from the host platform, and computing elements that process the sensed signal before transfer to the host platform through the communications interface. The term "instrument" can be used interchangeably with our definition of "sensor." The vast majority of today's oceanographic sensors communicate with the host computer using the RS232 serial interface standard, which is well suited to common oceanographic applications involving limited power and long cable lengths (up to 15 m). RS485 or in some cases RS422 may be used for cable lengths exceeding 15 m. Ethernet sensor interfaces are becoming more common for applications in which power is not severely limited (e.g., cable-to-shore observatories), yet Ethernet is still fairly rare in oceanographic sensors.

A "host platform" physically carries one or more sensors. Host platforms can be stationary (e.g., moorings and seafloor instrument nodes) or mobile (ships, autonomous underwater vehicles [AUVs], autonomous surface vehicles [ASVs], aerial drones, etc). A platform has a power subsystem that provides electricity to operate attached sensors, as well as other platform subsystems such as telemetry and propulsion. The platform has one or more *sensor ports* through which a sensor is attached; the port includes communication (often RS232) and power channels for the sensor (except in the case of self-powered sensors).

A *systems approach* should be taken when determining the platform and sensor characteristics best suited for a given application (also see Chapter 7.1). The choice of which approach to use depends on the scale of the project and implementing organization. For large-scale missions with many end-users undertaken by large organizations, a very formal approach could be considered, such as that described by standard International Organization for Standardization (ISO)/ International Electrotechnical Committee (IEC)/ Institute of Electrical and Electronics Engineers (IEEE) 29148: Systems and software engineering—life-cycle processes—requirements engineering [1]. Sensor deployments aboard autonomous marine platforms are analogous in several respects to sensor deployment aboard spacecraft, with similar mission design and trade-off considerations, so similar design approaches can be used [2]. But these approaches may be too expensive and cumbersome for some projects and organizations, in which case processes that are more lightweight can be considered. Note that although not all marine-sensor applications require extensive mechanical and electrical engineering development, virtually all of them do require software development. So-called agile methods should be considered for software development projects. Agile approaches that include Extreme Programming (XP), Scrum, and Kanban were originally developed for commercial software projects, and emphasize continuous engagement with users, changing user requirements, the need for frequent software releases, focus on essential software components, and deemphasize formal requirements documentation and avoidance of costly "future proofing." [3]

Virtually all of these methods include the notion of a *concept of operations document* or a CONOPS; the CONOPS is a narrative description of system functionality and characteristics from a user point of view. The CONOPS provides a shared understanding of the system for engineers and users, and takes various forms depending on development approach; it is a formal document in ISO/IEC/IEEE 29148, but a collection of informal "user stories" in XP. In any case, the CONOPS is essential to deriving user requirements for a system.

For example, a CONOPS for an AUV carrying genomic sensors could describe the vehicle survey area and depth, how frequently it is deployed, how long it remains at sea, the specific kinds of data collected, and how the data is returned to shore and distributed to users, among other characteristics.

When a suitable CONOPS has been created (and the CONOPS may undergo revisions), the science users and engineers can begin to consider candidate sensors and host platforms.

During the course of specifying mission goals and evaluating platforms and sensors, some candidate systems may be available "off the shelf," whereas in other cases new platforms or sensors may be developed. Considerations during a "buy versus build decision" can include whether the off-the-shelf system adequately meets the mission requirements, as well as the availability and cost of technical expertise for new system development and operation.

7.2.2 SENSOR CHARACTERISTICS

Sensor candidates, whether existing or yet to be designed, will be identified based on the application's required measurements and compatibility with available platforms. Sensor characteristics can influence selection of host platform; conversely, available host platforms may influence sensor selection.

Many of the characteristics in Table 7.2.1 are self-explanatory, and we provide some more detail later.

TABLE 7.2.1   Sensor Characteristics

Characteristic	Relevance
Size, mass, buoyancy	Host compatibility
Power requirements; battery versus external power source	Host compatibility, deployment duration
Pressure tolerance	Maximum depth
Thermal range (transport and operational)	Ocean environment; deep/polar water T~−2°C, tropical water T~38°C
Shock, vibration resistance, orientation (transport and operational)	Transport, deployment, recovery
Power and data interfaces	Host platform compatibility
Internal clock	If no internal clock, host driver must retrieve, time-stamp, and log data
Internal data log	If no internal log, host driver must retrieve and log data
Command protocol, output data format, data size	Driver software design, host telemetry bandwidth
Data accuracy, precision	Must meet mission requirements
Deployment duration	Limited by data stability, calibration requirements, onboard consumables (e.g., reagents, fluids), biofouling tolerance, log capacity, battery life
Sensor response/dwell time	May influence host sample-intake design, mobile-host hover/drift requirements
Interference sensitivity	Compatibility with host, other host sensors
Replacement cost	Risk of host platform loss

7.2.2.1 Sensor Deployment Duration

Most applications specify sensor-data accuracy and precision, which may change during deployment due to *sensor drift* or biofouling. Drift can be caused by physical changes within the sensor itself and can be corrected by periodic calibration. Some calibration procedures can be performed aboard a deployed platform, whereas other calibrations must be performed in a laboratory or by the manufacturer; the latter case requires sensor recovery when accuracy or precision drifts out of specification. Biofouling will affect data quality from many sensors and may necessitate recovery after several weeks. Finally, some sensors may carry consumables such as reagents for in situ chemical processing, and can no longer make measurements when the consumables run out. Batteries for internally powered sensors can also be considered to be consumables.

7.2.2.2 Sensor Response Time

Sensor response time specifies the amount of time needed to make a measurement; this may range from microseconds to hours, depending on the sensor. Special considerations may be needed for sensors with long response time, as interference caused by other devices may be more likely during a long measurement. If the host is moving, the sensor's data may be spatially "smeared" over the distance traveled during the measurement. If smearing is not acceptable then the host may need to stop or drift during acquisition. Alternatively, a mechanical fluidic device that quickly takes a water sample that is subsequently analyzed might be required; this allows a mobile platform to keep moving during the measurement but adds mechanical complexity.

7.2.2.3 Internal Clock and Data Log

Many sensors provide an internal clock, which time-stamps data records as they are acquired. It is often necessary to characterize the sensor clock's accuracy and drift, and clock corrections may be necessary in postprocessing the data, especially when comparing data between multiple sensors. The sensor may also provide internal storage for the time-stamped data. If not, the data must be output to the host, which must provide "driver" software to capture and log the data. If the sensor does not provide a clock, the driver must also time-stamp the data with sufficient accuracy, which will depend on the application. See Section 7.2.6.3 for more-detailed discussion of these issues.

7.2.2.4 Sensor Replacement Cost

The sensor-replacement cost may determine whether the instrument is selected for deployment on a particular platform in a particular location. For example, it may be too risky to deploy an expensive sensor on a buoy in regions with a high incidence of vandalism or thievery, or the risk of vehicle loss may be deemed too great, given known system reliability or environmental factors.

7.2.3 NEW SENSOR DESIGN

In some cases a new kind of sensor must be designed and built for an application, and application engineers may have considerable control over the sensor's mechanical, electrical, and software interfaces. It is often beneficial to select standard interfaces when practical, because standards can result in interoperability with a greater variety of host platforms and applications (Chapters 6.1 and 6.3). The new sensor command protocol should be kept as simple as possible to facilitate automated control; a stateless protocol that does not prompt the host to answer questions is usually straightforward to automate. In particular the Standard Commands for Programmable Instruments protocol should be considered for adoption [4].

7.2.4 HOST PLATFORM CHARACTERISTICS

The host platform for an application will be selected based on the mission requirements as well as characteristics of associated sensors. Some platforms may carry only one kind of sensor, whereas other platforms carry multisensor payloads. In the latter case, compatibility between payloads must be considered (e.g., potential electrical, optical and acoustic noise, power budget, etc). Host platforms have a wide variety of characteristics that must be considered when integrating sensors.

TABLE 7.2.2   Host Platform Characteristics

Characteristic	Relevance
Size, mass, buoyancy	Sensor compatibility, Experimental requirements
Power requirements; battery versus external power source	Sensor compatibility, platform processing, grounding, and connectorization
Pressure tolerance	Maximum depth
Thermal range	Transport and operation Deep/polar water T~−2°C Tropical water T~38°C
Shock, vibration resistance, orientation	Transport, deployment, recovery
Power and data interfaces	Instrument compatibility
Platform time reference	Time synchronization of samples, time-stamp synchronization for multiple sensors
Internal data log	If host must retrieve and log data
Command protocol, output data format, data size	Driver software design, host telemetry bandwidth
Data accuracy, precision	Must meet mission requirements
Deployment duration	Limited by data stability, calibration requirements, onboard consumables (e.g., fuel, battery capacity), biofouling tolerance, log capacity
Host platform speed/dwell time	May influence selection of sensor frequency or sampling duration
Interference sensitivity	Compatibility with instruments, other host sensors
Replacement cost	Risk of host platform loss

Many of the characteristics in Table 7.2.2 are self-explanatory, and we provide some more detail later.

Sensor-package volume and mass are constrained by the platform's available-payload volume and buoyancy. If a platform is submersible, then the sensor package needs to be appropriately depth-rated as well.

Length of platform deployment influences requirements with respect to biofouling control, sensor stability over time, and sensor consumables. The power budget of the host platform will determine the amount of available power for sensor payloads. This may determine the options for sensors, particularly when power is limited. Positional accuracy of the host platform should be consistent with the mission concept of operations and experimental requirements. Hydrodynamic characteristics can determine the placement of the sensors on the platform to ensure the sample/measurement is from the intended water mass.

Host platform can be either fixed in position or tethered, as in the case of moorings, mobile in the horizontal plane such as ASV or ships, or mobile in the horizontal and vertical planes such as AUVs.

The speed at which mobile-platform candidates operate must be compatible with the desired frequency of sensor sampling.

7.2.5 FRAME OF MEASUREMENT, EARTH FRAME, WATER-MASS FRAME

The host platform must be selected based on the required observation frame of reference. For example, if the desire is for a Lagrangian measurement, the platform must be able to travel at the same speed and direction as the water mass being measured. If the experiment is intended to be Eulerian, then the platform must be able to maintain position with respect to an Earth frame of reference independent of the surrounding environmental conditions. In both cases, the sensor's sampling frequency must be understood with respect to the speed of the water mass and the speed at which the particular variable being measured can change.

Additional consideration should be paid to the platform telemetry system's bandwidth and cost, as some sensor data require significant processing. If extensive sensor data must be processed on board, then additional processors will impact power requirements for the host platform.

As communications speeds continue to improve, real-time networking of sensors and platforms will become a reality. As this capability develops, the need for managing multiple platforms and sensors will become a necessity. The requirement to synchronize sampling across multiple distributed platforms will also be an important capability.

Energy density and storage also continue to improve, allowing for longer duration deployments. The need for increased platform and sensor reliability will be an important consideration. Methods for platforms and sensors to self-diagnose and recover from nonterminal-fault conditions will become an important feature as will the ability for sensors to self-calibrate and provide performance information for the duration of deployments.

7.2.5.1 Mobile Platform Characteristics

Mobile platforms have unique characteristics that will come into play when integrating sensor packages with them. Their maximum and minimum horizontal and vertical speed will impact the response-time requirement for a sensor package. The depth range will determine the maximum-depth rating of the sensor package. The lateral range could create various geolocation and position requirements as well as the ability to remain on station. The ability to move with the water mass being observed and/or station-keep provides the opportunity for both Lagrangian and Eulerian observation.

7.2.5.2 Physical, Electrical, and Mechanical Interfaces

The physical interface between platform and sensors will consist of appropriate brackets and fastening devices such as cables and connectors. Choice of materials will be important depending on corrosion resistance, strength, and deployment duration. Titanium and stainless steel can be selected when strength and corrosion resistance are required. Plastics offer less strength but obvious corrosion resistance and lower cost. Plastics can also electrically isolate the instrument package if required. The positioning of the instrument package may be critical to avoid interference from the platform itself as well as other instrumentation. This positioning may be important for acoustic sensors depending on the frequency and intensity of the acoustic signals as well as the width and the depth of the volume of water being interrogated. Optical instruments could be impacted by interference and shading from poor positioning as well. Bubble entrainment is a good example of a negative effect of poor instrument placement.

The electrical interface will consist of a cable and connector that interfaces the instrument package to the platform. This may not be required if the instrument package is standalone or battery powered. In both cases attention must be paid to the type of grounding required for the platform. Mixing grounding schemes could result in the instrument corroding or otherwise being damaged due to stray electrical currents. Attention must be paid to whether the instrument has a ground connection to the instrument housing or seawater. This may not be compatible with cabled high-power platforms such as cable observatories or remotely operated vehicles.

Connectors tend to be the most common single point of failure in marine-observing systems. Careful selection based on deployment requirements is imperative. Connectors are also a challenge with respect to integration and maintenance. Standardization of connectors could be a possible solution. Common connector types and even common connector pinout selections could be of benefit in large-scale systems.

The method that the host platform employs to detect ground-fault conditions should be understood when integrating instruments. Care must be taken to ensure that the instrument is compatible with these methods and does not cause unwanted interactions with the host's ground-fault detection system.

7.2.6 ISSUES COMMON TO PLATFORMS AND SENSORS

7.2.6.1 Biofouling

Marine systems are subject to biofouling, which can result in decreased performance on mobile platforms (due to increased drag and fouling of control surfaces) and degraded sensor performance (due to clogging of water intake ports and blocking of optical or other detectors). Depending on deployment duration requirements, steps may be taken to limit biofouling by use of toxic paints, mechanical wipers, ultraviolet (UV) light, or electrolytic techniques (see Chapter 4.3).

7.2.6.2 Interference

Electrical, acoustic, or optical "noise" generated by platforms and sensors can have a significant effect on the signals received by sensors. Electrical noise can impact sensitive electronics and may cause the need for additional shielding or signal conditioning to filter out this noise. Acoustic noise can impact sensors such as acoustic doppler current profilers (ADCPs) that rely on acoustics as their main sensing method. Signal and data postprocessing may be required to filter out acoustic interference.

LEDs, sunlight, or other light sources can interfere with optical instrumentation, and careful placement or shielding may be required. If the noise source occurs only when certain devices are powered on, it may be possible to coordinate device operation to avoid the interference. For example, if a conductivity, temperature, and depth instrument (CTD) pump is known to cause electrical interference with the measured signal from an ADCP, then the two instrument schedules can be specified such that they do not operate simultaneously. Alternatively, ADCP data acquired during known CTD operation periods can be rejected in postprocessing.

7.2.6.3 Clock Accuracy, Synchronization, and Drift

Each sensor measurement or "record" will have an associated time-stamp. In most cases, users want to compare data from the host platform's sensors using a common time base. A time-stamp is generated by a digital clock, either on the host platform or internal to the sensor. Digital clocks are not absolutely accurate. The clock accuracy will drift with time, and the drift rate may depend on clock quality, ambient temperature, power-supply stability, oscillator age, and other factors. The drift may be linear or nonlinear, depending on how much these factors change. The required clock accuracy and precision ("jitter") varies greatly with application—e.g., seismologic measurements may demand accuracy of 1 ms or better. If the clock accuracy does not meet this requirement, then either it must be periodically resynchronized with a reliable time base (e.g., National Institute of Standards & Technology [NIST]) during deployment, or the drift over the deployment duration must be measured and accounted for in postprocessing. Host platforms on the surface may synchronize their clock using Global Positioning Systems (GPS) signals; GPS provides a very stable source and synchronization results in accuracies of microseconds to milliseconds, depending on the protocol used. A host platform that is connected to the Internet (e.g., on a cable-to-shore observatory) can use standard protocols such as Network Time Protocol (NTP) or IEEE 1588 to achieve accuracy at the millisecond-to-microsecond level. But it may not be trivial to synchronize a sensor clock if the sensor is connected to the host with RS232.

7.2.6.3.1 Host Clock Time Base

Sensor driver-software time-stamps each sensor record with the system-clock time as it is received through the sensor port. But the host time-stamp represents the time at which the record was received through the sensor port, not the time at which the sensor firmware actually acquired the data; this latency may be trivial or significant, depending on the nature of the measurement.

7.2.6.3.2 Individual Sensor Clock Time Base

If the sensor has an internal clock, the sensor firmware time stamps each record with that clock's current time. The record may be logged to the sensor's internal storage (if available), or output to the host, in which driver software retrieves and logs it. The host driver software may also add a host clock time-stamp. The accuracy and drift rate will likely vary from one sensor clock to another, especially if the clocks come from different manufacturers. Thus, each sensor driver should periodically synchronize the sensor clock to the host-clock time base by issuing the appropriate sensor command. This synchronization is not "standard" and its accuracy should be verified, e.g., the sensor firmware may introduce latency during the clock-setting operation. We should expect synchronization accuracy on the order of 1 s for this approach.

Alternatively, the sensor clocks can be allowed to "run free" during the deployment, the clock drift measured immediately after recovery, and the data time-stamps corrected in postprocessing, as described in the following discussion.

In case clock synchronization is not feasible during deployment (e.g., the platform is always underwater), the following procedure is recommended for each clock (host and sensors) that will be used in data analysis:

- Before deployment, set clock time to NIST time. If the host can be temporarily connected to the Internet, its clock can easily synchronized with NIST by using NTP protocol. Many sensor clocks have only an RS232 serial connection, and the sensor protocol usually includes a method to manually set the clock or to synchronize with the attached host's clock. The sensor driver could implement this time-synchronization command. Note the time t_1 at which synchronization was performed.
- After recovery, note the offset between each clock and current NIST time t_2. The average drift rate is for the clock is then (offset)/$(t_2 - t_1)$. If drift is assumed to be linear, clock offset as a function of time can be plotted as a line between t_1 and t_2. The time-stamp on each record can be adjusted accordingly. If drift rate was constant throughout the deployment, then timing error will be completely removed by this procedure.

Assuming that Linux or Unix is the host operating system, the host-clock drift can be easily monitored. A Linux system actually contains two clocks—a battery-powered "real-time clock" (RTC) and a "system clock." The RTC is not referenced by Linux while the system is running, but keeps time continuously, even when the system is powered off. The system clock (also known as "software clock") is a software counter based on the timer interrupt; the system clock only exists while Linux is running, and is initialized to RTC time at boot time. The system clock is accessed by standard Linux utilities like "date" and the sensor driver to time-stamp data. The two clocks drift at different rates, and the RTC is usually more accurate than the system clock. Linux provides utilities to periodically resynchronize the system clock with the RTC, but note that some applications may not behave well if the system clock experiences large jumps or moves backward. As a less obtrusive alternative, a script on the host can periodically log the offset between RTC and system clocks (e.g., by invoking the Linux "hwclock" utility once per hour)— these offsets can later be used to correct the host-clock time-stamps in postprocessing. If the RTC drift rate is well known, then the RTC can be at least approximately corrected for drift, leading to better time-stamp accuracy.

7.2.7 SENSOR DRIVER SOFTWARE

In some cases, a sensor's data may not be needed during deployment, but only retrieved after recovery; if the sensor also has internal data storage, it can internally log its data without transferring it to the host. But in many cases the sensor does not have internal storage, or data will be used in some way during the deployment, e.g., transferred to shore through the host's telemetry system or used by other components aboard the host for "adaptive sampling." In these cases, the host must execute sensor "driver" software that communicates with the sensor through the sensor-port interface, commonly RS 232. The driver must implement command protocols that are recognized by the sensor, including the command set to configure the device and acquire its data and metadata. The command protocol also specifies other details such as command–response timeouts and Acknowledge/Negative Acknowledge (ack/nak) indicators.

In some cases, the sensor manufacturer may provide driver software with the device, but that software is usually applicable only to specific commercial desktop operating systems. This driver may be used if the sensor is being deployed from a ship or on a cable-to-shore observatory that provides users with a serial port interface to the sensor—in this case the user's desktop machine may host the manufacturer-supplied driver software. Nevertheless, of course, many sensors are deployed on low-power hosts without a direct Internet connection such as AUVs, gliders, and buoys; therefore, sensor integrators must usually develop driver software targeted for a specific "embedded" host operating system such as Linux, and for specialized system resource interfaces such as power, logging, and telemetry. Fig. 7.2.1 schematically shows the driver in relation to interfaces on the host platform and sensor device. Unfortunately, for driver-software developers, each sensor manufacturer generally defines their own unique protocol, with few standards in the industry; thus, a different driver must be developed for each kind of sensor. A given manufacturer may base a line of sensor models on similar protocols, but even then there are likely to be differences between the different models.

Sometimes a given sensor must be integrated on more than one kind of host, e.g., on a buoyancy-driven glider, AUV, and coastal buoy. Each host may have a different operating system, resource interfaces, and supported programming languages. Some hosts may have a preexisting sensor driver, and new sensor-driver types can be implemented by following the approach used by the existing drivers. But in the following, we assume that new driver-software architecture is being designed from scratch.

7.2.7.1 Driver Requirements

Initialize sensor and interfaces: The driver may first connect host power to the sensor (unless the sensor is battery powered), and initialize the communication port, e.g., set baud, parity, and stop bits for RS 232. The driver may issue configuration commands to the sensor itself, e.g., to set gains, operation mode, or other internal-sensor parameters. If the sensor includes an internal clock, the driver might also synchronize the clock to that of the host platform.

Acquire and log sensor data and metadata: The driver must acquire sensor data and log it to host storage. Some sensors "stream" data continuously through their communication interface after an initial trigger sent by the driver, whereas other sensors are "polled," i.e., they provide a single sample in response to the appropriate command issued by the driver. The driver may time-tag each acquired sample using the host clock before logging, especially if the sensor lacks an internal clock. In many cases, the "raw" data format should be preserved even if the data will be

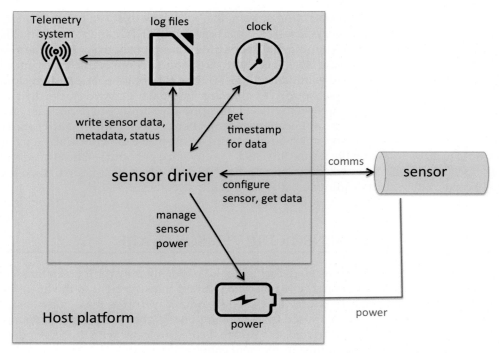

FIGURE 7.2.1   Some interfaces between a sensor driver, the host platform, and the sensor device.

processed further by the driver or other host-software components, so that the raw data can be reprocessed later if necessary. The driver might "push" data to the host's telemetry system for relay to shore (or the telemetry system might "pull" the logged data).

Metadata capture: The driver should also capture and log the sensor *metadata*, which is information about the sensor and its data; the metadata puts the data into context and so can be crucially important to data interpretation. *Static metadata* like device serial number and calibration coefficients do not change during the life of the deployment. *Dynamic metadata* describes changeable sensor settings that can be changed by the driver. Most sensor manufacturers define a protocol to retrieve the sensor's static and dynamic metadata, and the driver should retrieve and log those at the appropriate times; static metadata should be retrieved during sensor initialization, and dynamic metadata whenever the driver changes the sensor configuration.

Sensor health and status: The driver should have some means to log the sensor's health and status. Health and status information are usually provided by commands in the sensor's protocol, in which case the information format and content is manufacturer specific. Some host platforms are designed to autonomously detect and respond to faults and failures during a deployment; in such cases, the driver could include a "self test" function that is invoked by the host system. This function would likely be more complex than simply retrieving status through a single sensor command, but might require the driver to interpret data quality or device responsiveness.

Variety of sensors: As noted previously, each kind of sensor responds to a specific command protocol, and thus usually requires implementation of a specific driver (but see the section later on the "universal driver" approach, also described in Chapter 6.1).

Host environments: Some driver software is required to run in only one host environment, such as a particular AUV, which provides specific interfaces to resources such as power, timekeeping, serial ports, etc. In other cases, the drivers may be required to run on a variety of hosts, e.g., AUVs, ASVs, and buoys.

7.2.7.2 Driver Development Approaches

7.2.7.2.1 *Computer Languages and Design; Procedural Versus Object Oriented*

Possible software architectures depend in part on the implementation languages that are available for the target host environment. For example, if only the C language is supported, a "procedural" architecture is the natural choice. In this architecture, the code for a specific sensor driver calls generic library functions as needed. Developers

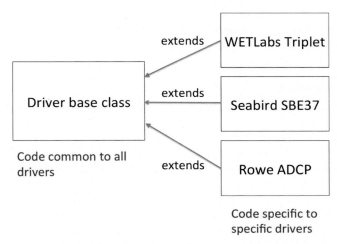

FIGURE 7.2.2   Sensor-driver application framework class hierarchy.

commonly develop a new driver by first copying the code of an existing working driver, and then modifying the copy as appropriate. Although such a scheme may work when only a few simple drivers are involved, the approach has some drawbacks:

- This approach unnecessarily duplicates code, resulting in more lines of software to maintain and larger executable code size.
- After a new driver has been created by copying code of the "original" driver, a bug may be discovered in the original. Simply fixing the bug in the original code does not fix the same bug in the copies. Likewise, later improvements might be made in the original code, but these improvements must be manually replicated in the copies.

Languages such as Java, C++, and Python support *object-oriented* design, which provides several important benefits over procedural designs, including increased code reuse through class inheritance, modularity, and encapsulation that lead to software that is more easily modified, and others [5]. Debate between adherents of procedural versus object-oriented design can be almost religious in nature, but the benefits of the latter outweigh those of the former in our experience and opinion.

7.2.7.2.2 Application Frameworks

Object-oriented languages can be used to develop a driver *application framework* [6]. The framework can be viewed as a "partial" driver consisting of generic classes and functions needed by all drivers. The framework includes a base class that captures procedures common to all sensor types, e.g., the sequence of steps to provide power to a sensor and initialize its communication interface. Generic base-class functions can call "abstract" or "virtual functions," which initialize and configure the device, acquire and log its data and metadata, and can be implemented in different ways in a subclass for each kind of sensor. A complete sensor driver is created by extending the framework base class and "filling in" the abstract and virtual functions with code appropriate to the command protocol of the particular sensor (Fig. 7.2.2). Thus, specific sensor code is called by the generic-framework code, inverting the procedural library approach. There are many benefits to the application-framework approach. Code in the base class is executed by *all* sensor drivers by default, ensuring consistent procedures, e.g., for sensor initialization, logging, etc. Because the base class implements most of the functionality needed for any sensor, relatively minor amounts of additional subclass code are required to completely implement a specific sensor driver. In our experience with driver frameworks, the generic-base class typically implements about 75% of a complete driver. Developers can focus on making the base-class robust and efficient, and driver subclasses inherit these virtues. Likewise, later improvements to the base class are inherited as well.

Some sensors may be deployed on multiple-host platforms, in which each host provides different interfaces to power, timekeeping, data logging, etc., and maybe even a different operating system. For example, a CTD may need to be integrated with an AUV and at other times with a buoy. One approach is to write completely different driver software for the same sensor on different hosts, duplicating the sensor-protocol logic for each host type. However, it is possible to minimize code duplication by developing a single-driver framework for multiple hosts, and defining abstract ways of accessing host resources (e.g., using a Java Interface, or C++ multiple-inheritance). For example, a sensor driver can switch power to its sensor through functions in an abstract "system power" interface. The system-power interface can be implemented appropriately for different hosts, but the sensor-driver code is isolated from those details. Similar interfaces can be defined and implemented for other host-specific resources. This approach avoids unnecessary code duplication of sensor-protocol logic for different hosts.

7.2.7.2.3 Universal Driver Approach

Driver software is typically developed for each kind of sensor using the programming language of choice to implement the sensor protocol. For example, the necessary command-and-response logic may be expressed in Java, C, or Python. An alternative approach has been developed by engineers at the Polytechnic University of Catalonya's Technological Development Center for Remote Acquisition and Data Processing System (SARTI) lab, who make extensive use of the Open Geospatial Consortium's Sensor Web Enablement (OGC SWE) standards [9]. The SARTI engineers have developed a "universal driver," also known as the "SWE bridge driver" [7]. This driver can operate any sensor the command protocol for which can be expressed with an Open Geospatial Consortium (OGC) Sensor Model Language (SensorML) document. The universal driver loads the SensorML file associated with the sensor, interprets the SensorML, then performs sensor initialization and data acquisition using the sensor-specific command strings and parameter values defined by the SensorML file (Fig. 7.2.3). This approach eliminates the need to write sensor-specific code in a language such as Java or C++, but it does require that SensorML be written to describe each kind of sensor. The SensorML is generally not dependent on host operating system or other environmental details—the only requirement is that the universal driver can parse and interpret the SensorML file.

The host runs an instance of the universal driver for each installed sensor. If a sensor implements the Open Geospatial Consortium Plug-and-Work (PUCK) standard [8], the universal driver can retrieve the SensorML from the device itself through the sensor port. Thus, the driver can automatically operate any PUCK-enabled sensor plugged into the port without a priori configuration on the host, resulting in a "plug and work" system. If the sensor is not PUCK-enabled, the driver retrieves the SensorML from a designated location in the host's file system that corresponds to the port on which the sensor is installed.

The universal driver must be compatible with the platform's operating system, resource interfaces, and programming language. SARTI's universal driver is written in "C" and runs on a specific embedded operating system, and it may be necessary to implement the driver for other environments. However, once the universal driver—a single program—is available on the host platform, significant savings in software development can be realized. This approach has been demonstrated to simplify end-to-end sensor configuration when used within the OGC SWE approach, as described in Chapter 6.1.

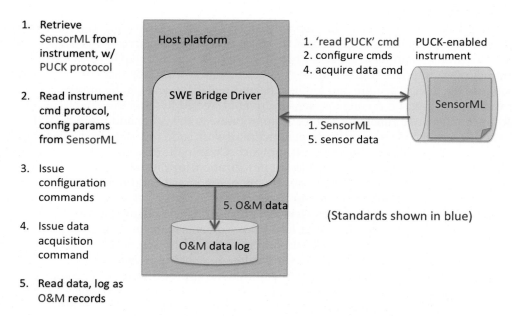

SWE Bridge Driver procedure:

1. Retrieve SensorML from instrument, w/ PUCK protocol

2. Read instrument cmd protocol, config params from SensorML

3. Issue configuration commands

4. Issue data acquisition command

5. Read data, log as O&M records

Host platform

SWE Bridge Driver

5. O&M data

O&M data log

1. 'read PUCK' cmd
2. configure cmds
4. acquire data cmd

1. SensorML
5. sensor data

PUCK-enabled instrument

SensorML

(Standards shown in blue)

FIGURE 7.2.3   SWE bridge driver ("universal driver") approach.

7.2.8 OTHER HOST SOFTWARE

In addition to sensor drivers, other host software may be needed for such functions as data summarization and/or compression (in case of limited telemetry bandwidth—see Chapter 4.1 for more details on passive acoustics pre-processing). Other software includes adaptive sampling applications, in which data produced by one or more sensors can trigger other actions such as vehicle-navigation changes or triggering of other sensors, and automatic fault detection and response. We expect the need for these applications will increase as mission durations increase due to improvements in platform autonomy, energy sources, and component endurance.

7.2.9 INTEGRATION AND TEST

Sensors and their drivers may first be "unit tested" on the bench, perhaps using just a lab power supply and laptop computer as substitutes for the actual host platform. The host computer on the bench might be identical to the one deployed, or might be a desktop Linux™, MacOS™, or Windows™ computer that runs the driver. Unit testing can expose driver bugs and other sensor-specific issues.

Once all sensors have been unit-tested, they are ready for integration with the deployable host platform. When physically plugging sensor power-communication cables into the platform's sensor ports, the following steps must be carried out for each port:

- Note which sensor is connected to the port, using a unique sensor identifier, e.g., the serial number. This information will typically be entered into a configuration file that resides on the host, so that the system can execute the appropriate sensor driver on each port, and associate acquired sensor data with the correct device. Operators typically maintain a database that contains sensor descriptions, using the unique sensor identifier as the database key. If OGC PUCK protocol is used, then this step and some of the others may be performed automatically [8].
- Configure the port for communication with its attached sensor (e.g., set baud, data bits, parity, stop bits, and flow control); if the sensor driver is responsible for port configuration, ensure that the driver is setting the proper parameter values. Note that although some sensors communicate at a fixed baud, others can be configured for different baud rates.
- Configure the port to provide power to the attached sensor. In some cases, sensors will have their own internal batteries and this step may be skipped.

These steps must typically be carried out manually by a human operator, can be time-consuming and tedious, and are subject to human error. We have commonly experienced these problems:

- Sensor's unique identifier is not visibly printed on the external sensor housing; in this case, it might be necessary to retrieve the serial number through the sensor's command protocol. It is important to clearly label the sensor housing with its identifier.
- Operator enters wrong sensor identifier for the port, even though sensor type is correctly identified. For example, although Seabird™ CTD "A" and "B" both exist, "A" is actually attached to the port but operator mistakenly enters the "B" identifier. Thus, the driver can successfully operate the sensor, but incorrect metadata (serial number, calibration coefficients, etc.) are associated with the data log. This error is subtle and not easily exposed by testing, and can lead to inaccurate data products, e.g., through use of incorrect calibration coefficients.
- Operator misreads the host sensor-port number, resulting in a driver–sensor mismatch, i.e., the driver is unable to communicate with the sensor. It is important to clearly label the host serial ports and cables.

Although most of these problems can be rectified during integration and test, they can result in a lengthier, more expensive process. In some cases, integration occurs at sea aboard ship, and the physiological challenge of rough sea state can increase the likelihood of errors. Automated schemes such as SARTI's approach avoid these problems (Chapter 6.1).

Once the sensors have been physically integrated with the host platform and the drivers and other software are running, the system should be rigorously tested before deployment. Based on our experience, we recommend that formalized, reproducible test procedures be used. One set of unit tests should verify proper functionality of individual sensors. An *acceptance test procedure* (ATP) is a formalized script to verify proper functionality of a sensor before and after it has been integrated with the platform. The ATP typically issues each command in the sensor's protocol and verifies proper response; verification may be done by software or by human inspection. In the case of data-acquisition commands, the data values should be checked for validity. In some cases, the sensor may be immersed to validate in-water data values. The ATP can be repeated anytime, especially when changes—intentional or otherwise—have been made to the system. The ATP may be in the form of a script read by a human operator, who then actually issues commands to the sensor, or can be a "shell script" or "command file" that is executed by a computer. The latter is easier and more foolproof to repeatedly execute, but must be developed and tested before use.

Other tests exercise full system-level functionality and end-to-end operation, whereas system drivers, telemetry, and other subsystems are operated as they would be during an actual deployment. Onboard logs as well as telemetry streams are monitored and examined to verify correct functionality. Ideally, these tests should be performed in water (e.g., in a test tank) to verify water tightness and buoyancy, as well as functionality. Ideally, the ambient temperature during the system test should match the minimum temperature expected during deployment. But in some cases, "dry testing" is also useful or more feasible. For example, in a dry "parking lot test" the host platform and its sensors are set out in the open air where a cell or satellite telemetry link is available and left to operate for extended periods. Similarly, it is sometimes not feasible to test the entire system on shore at expected deployment temperatures. Thermal testing is usually needed at temperatures colder than room temperature, but a sufficiently large refrigerated volume with power and telemetry connections may not be available. In these cases, a "cold soak" test is advised, in which system components are stored at low temperature for at least 24 h before the full-system test. The cold soak can expose issues with electronic tolerance to cold, e.g., solder problems.

For deep systems, pressure-tolerance tests are performed in a pressure chamber, but these chambers are often too small to accommodate the fully integrated system. Hence, individual components and housings must be tested individually.

A *readiness review* should be conducted after final system integration and prior to deployment. Postintegration ATP results and other relevant information are presented at the review to verify that a platform and its sensors are functioning properly, that the sensors are acquiring valid data, and to identify any remaining issues. The review should end in a "go-no-go" recommendation from the reviewers.

7.2.10 DIRECTIONS FORWARD

Commercial sectors such as mobile computing, electric self-driving cars, medical technology, and other areas are producing rapid improvements in battery power density, small energy-efficient sensors, wireless communication bandwidth, autonomy software, system reliability, and other areas. As these technologies improve and new ones emerge, we expect to see a new class of ocean-observing systems consisting of both stationary and mobile platforms that carry a wide variety of sensors, with a high degree of autonomy, reliability, and endurance. These systems will form networks with one another, in which platforms coordinate their sensor measurements at various temporal and spatial scales. Mobile platforms will be capable of both Eulerian and Lagrangian operations. These networks will provide views of natural marine systems at much higher accuracy and resolution than today's methods do. Realization of these large-scale systems will likely require increased software, mechanical, and software standards, which often reduce implementation time, simplify automation and operation, and improve reliability. In the software domain, use of technologies such as application frameworks, object-oriented design, and third-party libraries—e.g., for machine learning and image recognition—will help to reduce software development effort and time.

Glossary

ADCP Acoustic doppler current profiler
ASV Autonomous surface vehicle
AUV Autonomous underwater vehicle
CTD Conductivity, temperature and depth instrument

References

[1] Technical report. ISO, IEC. IEEE. 29148: 2011-Systems and software engineering-Requirements engineering. 2011.
[2] Larson WJ, Wertz JR. Space mission analysis and design. Torrance (CA, US): Microcosm, Inc.; December 31, 1992.

[3] Beck K. Extreme programming explained: embrace change. Addison-Wesley Professional; 2000.

[4] Standard Commands for Programmable Instrumentation (SCPI) Consortium. http://www.ivifoundation.org/scpi/default.aspx.

[5] Booch, Grady, Maksimchuk RA, et al. Object-oriented analysis and design with applications. 3rd ed. Addison-Wesley; 2007.

[6] Fayad ME, Schmidt DC, Johnson RE, editors. Implementing application frameworks. John Wiley and Sons; 1999.

[7] del Río J, Toma DM, Martínez E, O'Reilly TC, Delory E, Pearlman JS, Waldmann C, Jirka S. A sensor Web architecture for integrating smart oceanographic sensors into the semantic sensor Web. IEEE J Ocean Eng 2017.

[8] OGC PUCK standard protocol. http://www.opengeospatial.org/standards/puck.

[9] OGC Sensor Web Enablement standards. http://www.opengeospatial.org/ogc/markets-technologies/swe.

Further Reading

[1] O'Reilly TC, Headley K, Graybeal J, Gomes KJ, Edgington DR, Salamy KA, Davis D, Chase A. MBARI technology for self-configuring interoperable ocean observatories. In: OCEANS 2006. IEEE; 2006. p. 1–6. [Application frameworks, driver design, metadata, PUCK, etc].

[2] Gardner AT, Collins JA. Advancements in high-performance timing for long term underwater experiments: a comparison of chip scale atomic clocks to traditional microprocessor-compensated crystal oscillators. In: Oceans, 2012. IEEE; 2012. p. 1–8. [Timing precision and accuracy, chip-scale atomic clocks, power costs].

[3] Pallares O, Bouvet P-J, del Rio J. TS-MUWSN: time synchronization for mobile underwater sensor networks. IEEE J Ocean Eng 2016;41(4): 763–75. [Time-synchronization with IEEE 1588 via acoustic link].

CHAPTER

8

Glider Technology Enabling a Diversity of Opportunities With Autonomous Ocean Sampling

Xu Yi[1,2], Scott Glenn[1], Filipa Carvalho[4], Clayton Jones[3], Josh Kohut[1], Janice McDonnell[4], Travis Miles[4], Greg Seroka[4], Oscar Schofield[1]

[1]Center of Ocean Observing Leadership, Department of Marine and Coastal Sciences, School of Environmental and Biological Sciences, Rutgers University, New Brunswick, NJ, United States; [2]IMBER Regional Project Office, State Key Laboratory of Estuarine and Coastal Research, East China Normal University, Shanghai, China; [3]Teledyne Webb Research, Falmouth, MA, United States; [4]Center of Ocean Observing Leadership, Department of Marine and Coastal Sciences, School of Environmental and Biological Sciences, Rutgers University, New Brunswick, NJ, Canada

Gliders have been demonstrated to be robust and effective tools for addressing fundamental and applied science needs (see Chapter 5.1). As the platforms have matured over the decade they offer the opportunity to assess how these systems are changing how and what these platforms can fulfill. A major advantage of gliders is their modularity, which allows them to evolve rapidly as new sensors are incorporated into the platforms, as an autonomous underwater vehicle is only as useful as the sensors it carries. Historically, the types of sensor available to gliders were limited by the size and energy consumption of the sensor. Fortunately, there is a revolution occurring in instrument miniaturization, which is now opening the door for potentially a new suite of sensors for gliders. Another opportunity is that as the platforms have matured they are now robust enough to be deployed during periods when traditional shipboard sampling is not feasible. This includes extreme weather events when dangerous conditions limit how and when ships can conduct science operations. This is of special value as these storms play a significant role

Challenges and Innovations in Ocean In Situ Sensors
https://doi.org/10.1016/B978-0-12-809886-8.00008-9

in structuring the physics, chemistry, and biology of the ocean. Finally, as marine science enters the 21st century it is critical to entrain the general public in the adventure of exploring this planet's oceans. Ocean robots offer a unique tool to engage the imagination of the public and we believe they can provide an effective means for entraining the next generation of oceanographers. Based on our experience of close to 15 years of Slocum glider operations (see Chapter 5.1), we focus this chapter on providing examples of how they will address these issues. We specifically discuss the process of incorporating a new sensor in a glider and the insights gained by deploying a glider in a hurricane to fill critical observation gaps. Finally, we provide a range of education/outreach topics that directly leverage off glider observations in the sea.

8.1 INTEGRATION OF A NEW SENSOR INTO A SLOCUM GLIDER

Chlorophyll fluorometers are now a standard sensor used by the oceanographic community for mapping phytoplankton biomass in the oceans. Despite their utility, standard chlorophyll measurements have limitations, which include being sensitive to numerous physiological processes in the phytoplankton [1], making a universal correlation with phytoplankton biomass difficult to impossible. Additionally, while a proxy for biomass, it does not, as conventionally measured, allow for estimates of photosynthetic rate processes, which is a critical piece of knowledge given the extremely high turnover rates in phytoplankton. In the mid-1980s, benchtop instruments were built to measure the kinetics of fluorescence, and this allowed for estimates of the optical cross-section of photosynthesis, photosynthetic quantum yields, and photosynthetic electron transfer [2,3]. These measurements are sensitive to a cell's physiological state and since their advent have allowed oceanographers to study how changes in the environment drive changes in phytoplankton physiology and corresponding cellular rate processes [2,4]. Since its introduction has become an increasingly important shipboard measurement, however, the ability to provide sustained measurements over time has not been possible and to date could not measure photosynthetic performance under in situ conditions.

In a partnership between Teledyne Webb, Satlantic, and Rutgers University, an existing commercially available desktop Fluorescence Induction and Relaxation (FIRe) instrument was substantially modified so it could be carried and operated onboard a glider [5] (Fig. 8.1). This required a complete redesign of the optics and electronics from the large benchtop model to fit within the internal compartment dedicated for science sensors. The standard glider science bay for carrying sensors has a diameter that is approximately 8.25 inches and a length of 22 inches. Beyond the redesign of the sensor it was necessary to integrate the instrument directly to the glider's computers to allow the sensor to be operated remotely during a glider mission. These considerations are similar to other sensor integrations that have been done, and it points out the importance of having a team that includes the academic partner, as well as the sensor and glider manufacturers.

A major consideration for determining amenable sensors for gliders, beyond their physical size, is their overall power requirements. Missions are most valuable if they are of sufficient length to allow them to map a geographic or time domain area of scientific value given the slow travel speeds of the glider. For the FIRe, the sensor is based on fluorescence signals being induced by flashes from blue (450 nm) light-emitting diodes (LEDs). The computer-controlled LED driver delivers pulses with varied duration from 0.5 μs to 50 ms, which ensures fast saturation of the photosystem II complex within the phytoplankton that is within the single photosynthetic turnover (<100 μs). The fluorescence signal, isolated by the red (680 nm) interference filter, is detected by a sensitive avalanche photodiode module [6]. This measurement is power intensive requiring a peak of ~6 W of power, roughly 9 times more energy compared to standard fluorometers, which can considerably shorten the glider mission lifetime from months to days

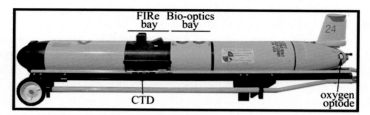

FIGURE 8.1 The FIRe Slocum glider. In this glider there are two science bays. One bay contains the FIRe system and photosynthetically active radiation. The second bay contains WetLabs Ecopucks providing measurements of optical backscatter and standard chlorophyll and colored dissolved organic fluorescence. Oxygen is measured in the rear compartment of the glider. *CTD*, conductivity, temperature, and depth; *FIRe*, fluorescence induction and relaxation.

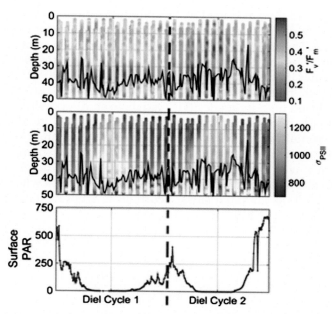

FIGURE 8.2   The daily changes in phytoplankton physiology as measured by a glider-mounted FIRe sensor. *Top panel* shows the maximum quantum yield (Fv'/Fm') as a function of depth and time. *Middle panel* shows the optical cross-section of photosystem II as a function of time and depth (σPSII). *Bottom panel* shows the photosynthetically available radiation at the sea surface as a function of time. *FIRe*, fluorescence induction and relaxation; *PAR*, photosynthetically active radiation.

if the sensor is run continuously. This requires the operator to optimize the duty cycle of the sensor to maximize the lifetime of the mission. For the FIRe, one strategy is to operate the glider only at night, which minimizes the impacts of light-induced nonphotochemical quenching of chlorophyll fluorescence and doubles the length of the mission, while also collecting data that allow for the calculation of maximum quantum yield for photosystem II activity (Fig. 8.2). To operate in this mode, the operator turns the FIRe on and off via the glider, which is possible due to integration of the sensor into the glider hardware/software. Fig. 8.2 shows 2 days of FIRe data collected in the coastal waters of the West Antarctic peninsula. The decreases in the maximum quantum yield (Fv'/Fm', top panel) and in the optical cross-section of photosystem II (σ_{PSII}, middle panel) during daylight hours (bottom panel) reflect physiological adjustments to the ambient light conditions experienced by the in situ phytoplankton. Taken together these results show that we can use gliders to estimate fundamental biological rate processes.

8.2 USING GLIDERS TO STUDY PROCESSES DURING EXTREME EVENTS

Few studies have investigated the variability of the physics in coastal waters during a hurricane due to the severe operational conditions. This is problematic as it has been shown that the prediction of hurricane intensity at landfall is strongly dependent on resolving realistic conditions in the water column that currently cannot be resolved by operational ocean-atmosphere models [6,7]. This gap is to a large extent an indictment of the lack of in situ observations. The few available in situ observations in the deep ocean, combined with satellite imagery, have shown that an intense, slowly moving hurricane may cool the sea surface by 2–6°C due to hurricane-induced mixing [8]. This cooling has been shown to influence hurricane intensity.

One of the major factors that influence tropical cyclone intensity is the interaction between the atmosphere and the ocean. Air/sea interactions are dominated by atmospheric forcing through the air/sea interfacial fluxes of heat, momentum, and buoyancy. Vertical mixing in response to wind forcing can feed back on the storm given the potentially large temperature change if deeper waters are mixed with the surface. So estimating the heat budget of surface water under a hurricane is important in understanding the heat transport, as well as for predicting hurricane intensity and path. Slocum gliders are effective tools for sampling storm conditions, especially in the coastal ocean [9,10,13] but it is only recently that they have been demonstrated to fill observational gaps during a hurricane [6,7,14]. Beyond the standard measurements of temperature, salinity, oxygen, and optical backscatter, gliders can also provide measurements of depth- and time-averaged water column currents (see later). A single hour-long sampling segment during a glider mission might collect 5–10 profiles depending on water column depth (see Chapter 5.1 for a description of glider operation). The difference between the calculated horizontal displacement from the final predive location and the actual surfacing location divided by the time underwater provides an estimate of depth- and time-averaged velocity of the water that the glider experienced.

In a recent study, glider measurements of depth-averaged water-column currents were combined with surface currents collected by shore-deployed high-frequency (HF) radars to study nearshore circulation patterns during the approach of Hurricane Irene to the northeast of the United States [6]. The Mid-Atlantic Bight (MAB) is unique in that it is characterized by an extremely strong vertical stratification, with summer surface water temperatures ranging from 25 to 28°C and summer bottom waters being 8–10°C [11]. With an extremely tight thermocline (meters), it is effectively a two-layer system. Given this, [6] combined full water column depth-averaged currents from gliders and HF radar surface current to estimate bottom currents along the glider track when the hurricane approached and passed overhead. This assumes that the HF radar surface currents are representative of the surface layer above the thermocline (defined as the maximum vertical temperature gradient along each profile) and require that the depth-weighted average surface and bottom layer currents must equal the total depth-averaged currents experienced by the glider:

$$U_b = \frac{U_g (H_s + H_b)}{H_b} - \frac{U_s H_s}{H_b}$$

(8.1)

$$V_b = \frac{V_g (H_s + H_b)}{H_b} - \frac{V_s H_s}{H_b}$$

(8.2)

where H_s and H_b are the layer thicknesses above and below the thermocline, respectively, U_g and V_g are along- and cross-shelf depth-averaged currents, respectively, from glider dead-reckoning, U_s and V_s are surface layer-averaged currents from HF radar, and U_b and V_b are the calculated bottom layer-averaged currents. Glider data can also be used to provide model initial conditions, especially when clouds associated with the encroaching storm do not allow for satellite imagery to be collected.

Hurricane Irene formed east of the Caribbean's Windward Islands on August 22, 2011 and made initial US landfall in North Carolina as a Category 1 hurricane on August 27. It reemerged over the ocean in the MAB before a second landfall in New Jersey as a tropical storm on August 28. Irene accelerated and lost intensity as it crossed the MAB, moving parallel to the coast with the eye over inner continental shelf waters. Irene reveals the regional pattern of MAB sea surface cooling. Irene passed over the underwater glider RU16 deployed on the New Jersey continental shelf (Fig. 8.3). Glider-observed subsurface temperatures (Fig. 8.4B) indicate that initially typical MAB summer stratification was present, with a seasonally warmed surface layer and cold pool water below the thermocline.

Integrated ocean observations and calculations during Hurricane Irene (2011) reveal that the wind-forced two-layer circulation of the stratified coastal ocean, and resultant shear-induced mixing, led to significant and rapid ahead-of-eye-center cooling (at least 6°C and up to 11°C) over a wide swath of the continental shelf. Significant cooling of the surface layer (5.1°C) and deepening of the thermocline (>15m) were observed under the leading edge of the storm. Little change in thermocline depth and much less cooling (1.6°C) of the upper layer were observed after eye passage. The glider observations suggest that much of the satellite-observed sea surface temperature cooling (over ~100,000km² of continental shelf) occurred ahead of eye center.

Time series of atmospheric conditions (Fig. 8.4A) were recorded just inshore of the glider measuring subsurface ocean conditions along the track (Fig. 8.3, yellow line). Ocean surface currents measured by a CODAR HF radar network illustrated the rapid response of the surface layer to the changing wind direction. The cross-shelf components of the currents from CODAR data (Fig. 8.4C, red line) at the glider location indicate that the onshore surface currents (positive values) began building before the eye entered the MAB, increasing to a peak value >50cm/s toward the coast before eye passage. After the eye, the winds changed direction and within a few hours the cross-shelf surface currents switched to offshore (negative values). Despite the strong observed surface currents, the depth-averaged current (Fig. 8.4C, green line) reported by the glider remained small during the storm's duration, with peaks barely exceeding 5cm/s. These bottom layer currents (Fig. 8.4C, blue line) were estimated based on Eqs. (8.1) and (8.2) and suggest offshore transport as the eye approached and onshore after eye passage. This resulted in significant shear across the thermocline and affected the storm surge. These results show the value of glider data collected during an extreme event that is not possible using traditional ship sampling techniques.

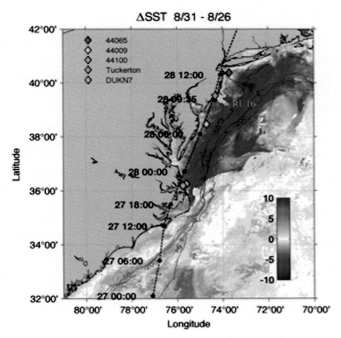

FIGURE 8.3   SST difference map post-Irene (8/31) minus pre-Irene (8/26) with National Hurricane Center best track (*black dots* connected by *dashed line* labeled with August date and UTC time), weather buoys/stations (*colored diamonds*), underwater glider RU16 track (*yellow line with green dot* signifying location at eye passage), and bathymetry (*fine black lines*). *SST*, sea surface temperature; *UTC*, coordinated Universal Time.

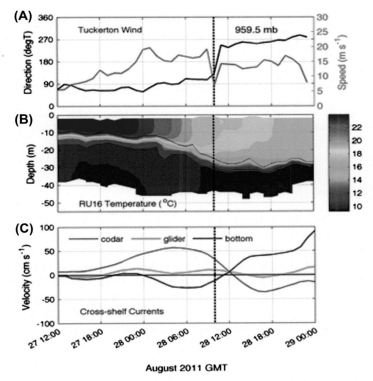

FIGURE 8.4   (A) Tuckerton WeatherFlow, Inc. station 10 m wind speed (*green*) and direction from (*black*) with vertical *black dashed line*/label indicating the time/value of the minimum air pressure associated with the passage of Hurricane Irene's eye. (B) Glider temperature cross-section during storm conditions with lines indicating top (*black*) and bottom (*magenta*) of thermocline. (C) Cross-shore currents for the surface layer (*red*) from CODAR HF radar, depth averaged (*green*) from the glider, and bottom layer (*blue*) inferred by requiring surface and bottom layer transports to equal the depth-averaged transport.

8.3 ADVANTAGES AND EXAMPLES OF USING GLIDER TECHNOLOGY FOR EDUCATION AND OUTREACH

Robots are extremely effective tools for education/outreach (EO), and therefore we have been working with diverse collaborators to develop tools/curricula that use glider technology. We treat this effort with equal importance to the science and technology development in our team. Since 1996, we have focused on creating meaningful science experiences for middle and high school classrooms through data-rich activities that highlight how science research practices are conducted. These kinds of contribution are critical to current education reform efforts needed for compliance with the Next Generation Science Standards (NGSS). The goal of the NGSS is to move science instruction away from disconnected facts and toward interrelated ideas, which learners can use to explain scientific concepts and solve problems [12]. Observing systems, especially glider technology, represent an exciting new paradigm for these internet-based ocean explorations.

Our efforts early on were enabled through the National Science Foundation's Centers for Ocean Science Education Excellence Networked Ocean World (COSEE NOW) program that focused on building an online network of scientists and educators focused on using ocean data from emerging Ocean Observing Systems technologies for public education. The focus was on engaging learners in real-time data across a broad range of audiences, including community colleges, the K-12 formal education community, and informal learning institutions. COSEE NOW focused on surveying, summarizing, and distributing knowledge from educators and scientists on their use of ocean data with the overarching goal to build the ocean science community's ability to use real-time data in education and public outreach. This has resulted in myriad EO efforts, which we highlight in a range of programs that rely on glider data next.

8.3.1 Podcasts: Ocean Gazing

COSEE NOW and National Public Radio contributor Ari D. Shapiro worked together to develop a podcast series focused on ocean-observing technology called Ocean Gazing. The audio series consisted of 52 episodes interviewing prominent scientists and educators involved in ocean-observing science and technology development. Five of the episodes focused specifically on glider technology. Companion lesson plans were developed to help educators bring this cutting-edge science to their classrooms and inspire a generation of scientists and technologists. The Website, despite being a decade old (coseenow.net/podcast/),continues to receive visits from educators working in formal and informal learning contexts and who are interested in integrating research and data in their teaching.

8.3.2 Ocean Science Extended Laboratory Education Programs at Liberty Science Center, New Jersey

We have partnered with the museum at the Liberty Science Center in Jersey City, New Jersey. Our partnership has focused on creating programs that are then taught by the museum during 2-day programs focused on specific themes or topics for up to 25 students. These programs continue to be offered with one of the more popular classes being "seasonality in the ocean," which is anchored by glider data and technology.

8.3.3 Floor Activities With Ocean Robots

In addition, a simple glider activity called *Exploring Ocean with Robots* has been used as a floor activity at Liberty Science Center. This program uses hands-on activities to explain how buoyancy is used to propel the Slocum gliders, and introduces guests to the types of research data gliders collect. Center volunteers present the program 3–4 days a week to school groups and family groups of mixed-age children and adults. At the end of the program the Center distributes sheets with a picture of a glider that children can color, and with links to the Rutgers University glider blogs and the Center of Ocean Observing Leadership room. Finally, as an alternative approach for connecting guests with glider research once they leave the museum, Liberty Science Center has created an audio clip guests can access via their cell phone. The clip utilizes audio collected as part of the Ocean Gazing podcast series of scientists discussing their research with Slocum gliders. It has been available to guests since August 2011 via the National Science Foundation (NSF)-supported Science Now Science Everywhere capability in the museum.

8.3.4 Polar Interdisciplinary Coordinated Education

More recently, our team has developed additional educational initiatives through the NSF-funded Polar Interdisciplinary Coordinated Education (Polar-ICE) program. This project aims to build the capacity of polar scientists in communicating and engaging with diverse audiences while creating scalable, in-person, and virtual opportunities for educators and students to engage with polar scientists and their research through data visualizations, data activities, educator workshops, webinars, and student research symposia.

Polar-ICE is working to help educators gain access to polar data, data activities, lesson plans, and media, explaining polar science and technology. Polar-ICE supports educators in developing skills in how to use real scientific data in their classrooms, as well as support them in utilizing online data software tools to help students learn how to orient to data, as well as interpret and synthesize data observations. Students (grades 6–16) participating in Polar-ICE programs conduct polar-related science investigations to enhance their comfort with using and analyzing data, as well as presenting their results to broad audiences (Fig. 8.5). The objective is to engage students in authentic experiences in the process of science to develop positive identities in science, technology, engineering, and math, and ultimately contribute to the lifelong trajectory of identity development as scientists. Next we discuss some key initiatives of Polar-ICE.

8.3.5 Science Investigator Program

The Science Investigator (Sci-I) program seeks to increase the understanding of teachers and students in grades 6–9 regarding the authentic process of science by supporting them through developing, conducting, and presenting a polar-related open-ended science investigation. By mirroring the process of science through such investigations and by developing personal relationships with scientists throughout the project, we hope to increase students' identification with and engagement in science. The Sci-I program starts with a 4-day professional development workshop for middle school teachers. The intention of the workshop is to unpack the nuances and realities of the process of science while enabling the teachers to experience first-hand participating in an open-ended polar science investigation. The workshop investigates various aspects of the Palmer Long Term Ecological Research project as the data are easily available online, allow for interesting time series analyses, and are interdisciplinary in nature. A real-time glider is often the most popular. Through hands-on activities, group discussions, scientist panels, and field trips the teachers explore the following daily themes: developing truly testable questions, finding and diving into data, making sense of data, and communicating initial results. Twenty-one educators from New Jersey and California participated from 10 different schools, with approximately 1500 students in 2015. Each teacher has at least one other teacher in their school participating to increase successful implementation rates. An effort is made to increase the diversity of the students. A selected group of students attends the annual Student Polar Research Symposium to present their research to their peers, teachers, students, and polar scientists from other schools in the project.

8.3.6 Looking Ahead: Workshops on Teaching With Data With College Professors

The Ocean Observatory Initiative (OOI) Teaching with Data project leverages data and links it to user-friendly, interactive, online activities. Use of real data enables students to deepen their understanding of a concept while

FIGURE 8.5 A grade school data team focused on analyzing real data collected from polar systems. Gliders are consistently a popular focus for students.

enhancing their scientific and data skills. The project merges available OOI data, data visualization theory, user interface best practices, and current learning research to create OOI Data Explorations. The explorations are short (15–20 min) and use near-real-time, professionally collected data to illustrate how scientists use data. Through professional development workshops and online resources, the project develops and then supports use of ocean data in undergraduate classrooms. Teachers can thus easily and effectively integrate OOI data into their teaching. OOI data are directly relevant to the content of 14 of the 16 chapters in *Essentials of Oceanography*, the market-leading textbook used by tens of thousands of students enrolled in 2-year community colleges and 4-year universities. This project focuses on mapping OOI Data Exploration to the 14 relevant chapters providing a direct link to the tens of thousands of community college and university undergraduates across the United States. Glider data are one of key data types.

In conclusion, gliders have matured to become central tools for the oceanographer. An added benefit of these systems beyond their ability to collect data in extreme and remote locations is that they can simultaneously provide a powerful tool for education and public outreach. The ability for shore-side scientists–teachers–citizen scientists to access data/imagery during an experiment as it is happening provides a powerful and compelling tool. Ultimately, this will hopefully help a wider cross-section of society to become ocean literate and better understand the process of conducting science. This is critical as humanity will increasingly have to confront many challenges associated with accelerating observed changes in the ocean.

Acknowledgments

This work was not possible without the dedicated members of COOL. The FIRe integration work was supported by National Science Foundation Palmer LTER Program (grant 0823101), National Oceanographic Partnership Program (grant NA05OAR4601089), and NASA Ocean Biology and Biogeochemistry Program (grant NNX16AT54G). Filipa Carvalho was funded by a Portuguese doctoral fellowship from Fundação para a Ciência e Tecnologia (grant DFR-SFRH/BD/72705/2010) and a Teledyne Graduate Fellowship. The hurricane work was supported in part by the NOAA IOOS award to MARACOOS (NA11NOS0120038), Environmental Protection Agency (EP-11-C-000085), the New Jersey Department of Environmental Protection (WM13-019-2013). The education/outreach work is being supported by the National Science Foundation (PLR-1440435).

References

[1] Falkowski PG, Raven J. Aquatic photosynthesis. Oxford: Blackwell Scientific; 1997 (375p.).

[2] Falkowski PG, Koblfzek M, Gorbunov M, Kolber Z. Development and application of variable chlorophyll fluorescence techniques in marine ecosystems. In: Chlorophyll a Fluorescence in aquatic sciences: methods and applications. Springer; 2004. p. 757–78.

[3] Kolber ZS, Prášil O, Falkowski PG. Measurements of variable chlorophyll fluorescence using fast repetition rate techniques: defining methodology and experimental protocols. Biochim Biophys Acta Bioenerg 1998;1367:88–106.

[4] Suggett DJ, Moore CM, Geider RJ. Estimating aquatic productivity from active fluorescence measurements. In: Chlorophyll a fluorescence in aquatic sciences: methods and applications. Springer; 2004. p. 103–27.

[5] Gorbunov MY, Falkowski PG. Fluorescence induction and relaxation (FIRe) technique and instrumentation for monitoring photosynthetic processes and primary production in aquatic ecosystems. In: Photosynthesis: fundamental aspects to global perspectives. Proceedings of 13th International Congress of photosynthesis. 2004. p. 1029–31.

[6] Glenn SM, Miles TN, Seroka GN, Xu Y, Yu F, Forney RK, Roarty H, Schofield O, Kohut J. Stratified coastal ocean interactions with tropical cyclones. Nat Commun 2016. https://doi.org/10.1038/ncomms10887.

[7] Seroka G, Miles T, Xu Y, Kohut J, Schofield O, Glenn S. Hurricane Irene (2011) Hurricane Irene sensitivity to stratified coastal ocean cooling. 2016. Monthly Weather Review https://doi.org/10.1175/MWR-D-15-0452.1.

[8] Price JF. Upper ocean response to a hurricane. J Phys Oceanogr 1981;11:153–75.

[9] Glenn SM, Jones C, Twardowski M, Bowers L, Kerfoot J, Webb D, Schofield O. Studying resuspension processes in the Mid-Atlantic Bight using Webb Slocum gliders. Limnol Oceanogr 2008;53(6):2180–96.

[10] Ruiz S, Renault L, Garau B, Tintoré J. Underwater glider observations and modeling of an abrupt mixing event in the upper ocean. Geophys Res Lett 2012;39:L01603.

[11] Schofield O, Chant R, Cahill B, Castelao R, Gong D, Kahl A, Kohut J, Montes-Hugo M, Ramadurai R, Ramey P, Xu Y, Glenn SM. Seasonal forcing of primary productivity on broad continental shelves. Oceanography 2008;21(4):104–17.

[12] Krajcik J, Codere S, Dahsah C, Bayer R, Mun K. Planning instruction to meet the intent of the next generation science standards. J Sci Teach Educ 2014;25(2):157–75.

[13] Miles T, Glenn SM, Schofield O. Spatial variability in fall storm induced sediment resuspension on the Mid-Atlantic Bight. Continental Shelf Research 2012. https://doi.org/10.1016/j.csr.2012.08.006.

[14] Miles T, Seroka G, Glenn S. Coastal ocean circulation during Hurricane Sandy. Journal of Geophysical Research 2017;122(9):7095–114. https://doi.org/10.1002/2017JC013031.

Further Reading

[1] Schofield O, Jones C, Kohut J, Miles T, Saba G, Webb D, Glenn S. Building a coordinated community fleet of autonomous gliders for sampling coastal systems. Marine Technol. Soc. 2015;49(3):9–16.

Index

Note: 'Page numbers followed by "f" indicate figures, "t" indicate tables and "b" indicate boxes.'